MATHEMATICS IN INDUSTRY 16

THE EUROPEAN CONSORTIUM
FOR MATHEMATICS IN INDUSTRY

E C M I

SUBSERIES

For further volumes:
http://www.springer.com/series/4650

Bastiaan Michielsen
Jean-René Poirier
Editors

Scientific Computing in Electrical Engineering SCEE 2010

With 176 Figures, 115 in color and 36 Tables

Springer

Editors

Bastiaan Michielsen
ONERA
2 avenue Édouard Belin
31055 Toulouse
France
Bastiaan.Michielsen@onera.fr

Jean-René Poirier
LAPLACE-ENSEEIHT
2 rue Charles Camichel
31071 Toulouse
France
poirier@laplace.univ-tlse.fr

ISBN 978-3-642-22452-2 e-ISBN 978-3-642-22453-9
DOI 10.1007/978-3-642-22453-9
Springer Heidelberg Dordrecht London New York

Library of Congress Control Number: 2011942998

Printed on acid-free paper

Springer is part of Springer Science+Business Media (www.springer.com)

Preface

This book presents an account of the Scientific Computing in Electrical Engineering conference, SCEE 2010, which took place in Toulouse, September 2010. The SCEE series of conferences covers many aspects of mathematics applied to electrical engineering, including electronics, electrical networks and electromagnetics. It started as a national meeting in 1997 in Germany and the first two meetings were organised under the auspices of the Deutscher Mathematiker Vereinigung. The title SCEE appeared for the first time in 2000 and since then the conference has been held every other year and in different European countries.

The organisation of the 8th conference was provided by the Toulouse branch of Onera, the French aerospace laboratory, and the ENSEEIHT, situated in the heart of Toulouse, was kind enough to make one of its lecture halls available. This 8th edition of the SCEE conference was further sponsored by

ABB,	Switzerland	http://www.abb.com
AWR,	Finland	http://web.awr.com
MunEDA,	Germany	http://www.muneda.com
CST,	Germany	http://www.cst.com

Their financial and material support is gratefully acknowledged.

The scientific programme of the conference was organised by the programme committee, which consisted of:

Dr. Andreas Blaszczyk (ABB Corporate Research, Switzerland)
Prof. Gabriela Ciuprina (Polytehnica University of Bucharest, Romania)
Dr. Georg Denk (Infineon, Germany)
Prof. Michael Günther (University of Wuppertal, Germany)
Dr. Jan ter Maten (NXP Semiconductors, The Netherlands)
Ir. Bastiaan Michielsen (Onera, France)
Prof. Ursula van Rienen (University of Rostock, Germany)
Prof. Vittorio Romano (University of Catania, Italy)
Dr. Janne Roos (Helsinki University of Technology, Finland)
Prof. Wil Schilders (TU/e & NXP Semiconductors, The Netherlands)
Prof. Thomas Weiland (TU Darmstadt & CST, Germany)

The programme committee attended to the reviewing of the proposed contributions. This resulted in 41 plenary talks and 34 poster presentations. The programme committee was also responsible for inviting specialist speakers to introduce the sessions. We were happy to have the following invited talks (in order of their appearance at the conference):

- Guillaume Sylvand (EADS IW, France),
  "From quasi-static to high frequencies: An overview of numerical simulation at EADS"
- Tim Davis (University of Florida, USA),
  "Sparse matrix methods for circuit simulation problems"
- Heidi Thornquist (Sandia National Laboratories, USA),
  "Advances in Parallel Transistor-Level Circuit Simulation"
- Maurizio Repetto (Politecnico Torino, Italy),
  "Tonti diagrams and algebraic methods for the solution of coupled problems"
- Patrick Dular (Université de Liège, Belgium),
  "Magnetic model refinement via coupling of finite element subproblems"
- Naoufel Ben Abdallah (Université Paul Sabatier), who was our invited speaker for the device modelling session, passed away in the summer of 2010, just two months before the conference. His time slot in the conference programme was left unfilled
- Jörg Ostrowski (ABB, Switzerland),
  "Transient Full Maxwell Computation of Slow Processes"
- Helmut Gräb (TU München, Germany),
  "From Sizing over Design Centering and Pareto Optimization to Tolerance Pareto Optimization of Electronic Circuits"
- Joost Rommes (NXP, The Netherlands),
  "Challenges in model order reduction for industrial problems"

All authors of papers accepted for presentation at the conference were also invited, in a second round, to propose a contribution for publication in this post-conference book. The programme committee organised the final reviewing and you will find the 47 selected papers in the main part of this book.

As editors of this book and members of the local organising committee, we would like to thank all the authors who contributed to the conference and, later, to this book for their good work. We thank the reviewers for having gone twice through all the proposed contributions and for all their constructive remarks. We also thank Onera and the ENSEEIHT for having made this work possible and the staff of Springer Verlag for their patience in getting this book to the press.

Toulouse, France

Bastiaan Michielsen, Onera
Jean-René Poirier, ENSEEIHT

Contents

Introduction

Electromagnetic interactions are the only fundamental interactions which can be easily manipulated on the macroscopic scale. As such, electromagnetic interactions play a major role in modern life. Technological advances in information processing machines, from mobile telephones to car information systems and personal computers to cash machines, are perhaps the ones that most immediately come to mind. The advances of electromagnetic technology show no sign of stopping and will continue to influence not only our daily life but also the way we do research.

The role of applied mathematics in this process is important in two complementary ways. Firstly, mathematical analysis of the fundamental models from theoretical physics, as well as the engineering models derived from them, is essential for the proper understanding of the nature of the phenomena themselves. Understanding physical phenomena means knowing some abstract, concise, organising model for them. Secondly, the mathematics of numerical computations with the said models is essential for the reliability of the conclusions we draw concerning these phenomena. In a certain way, we always seek to master the physical phenomena we encounter, first by trying to predict them and then by trying to influence the course of events.

Following this line of reasoning, we can be a bit more precise on the role of mathematics in industry. Existence and uniqueness results are important to exactly identify what can be considered as a "consistent model" and what not. Convergence results are essential for being able to decide whether an obtained conclusion is reliable or not. In the end, the purpose is to replace time-consuming and costly real experiments with more time-efficient and cheaper simulated experiments. This ideal situation has not yet been reached in every domain but scientific computation is already an indispensable part of industrial design cycles.

As for the vast domain of electromagnetics-related technological development, one can distinguish different application domains of mathematics. The most obvious one is the computation of the electromagnetic field itself. The possibility to construct numerical representations of electromagnetic fields corresponding to given space-time distributions of electric charge has grown from the construction of elementary "analytical" solutions of canonical problems in the nineteenth century into an abundance of discretisation-based algorithms, where, roughly speaking, the concept of a point-wise converging approximation using global expansions (the analytic functions) has been replaced by globally converging approximations using local expansions. This has created a completely new and extremely vast domain of research where there seems to be no limit to the geometrical complexity for which solutions can be obtained using modern computers.

Although one may question the applicability of the system of Maxwell equations as a correct model for physics on the atomic and subatomic scale, there is no doubt that it is successful as a generic model for macroscopic electromagnetic phenomena. This however does not mean that all modelling necessarily involves solving the Maxwell equations. From a certain point of view there is a hierarchy of derived models. The most often used model derivation method is asymptotic in nature. The fundamental Maxwellian model is a space-time model, but, more often than not, in engineering applications one is interested in single frequency states. These have only an asymptotical meaning but, in fact, the asymptote is rather easily attained and it pays off to work with the so-called frequency domain Maxwell equations, which determine the complex valued vectorial amplitudes (phasors) of the time harmonic fields.

Another extremely useful asymptotic is the "quasi-static" one, which describes the field behaviour in configurations where propagation delay is negligible. This asymptotic provides the link between macroscopic electromagnetic field theory and the models for electronic circuits. Starting from the microscopic point of view, where one should replace the Maxwell equations by quantum physics models of matter and interaction, one can also climb up to the macroscopic level by using different sorts of asymptotics. The constitutive coefficients one uses in the macroscopic Maxwell equations can be obtained through statistical quantum physics, though it is also possible to make do with a phenomenological model in accordance with irreversible thermodynamics.

It should be said that all of these different models, fundamental or derived, have their own specificities. Even though certain models are derived from for example the time-domain Maxwell equations, they require a separate mathematical study. A satisfying unique-existence analysis of a time domain does not necessarily clear up the situation of the corresponding problem in the time harmonic case, a reduced two-dimensional version of the problem or a quasi-static problem. All of these problem classes present their own set of difficulties and solution methods. The same thing can be said of problems of (statistical) quantum physics and their derived models. This implies that the different communities, each focussing on one specific type of model, show a tendency to isolate themselves from the others. This is unfortunate, because it neglects the fact that we need all the models together to better understand the way things work and in order to make more reliable predictions about how new products will behave, once the designs are realised.

The SCEE series of conferences was created to counteract this isolation. The idea was and is to organise the communication between researchers working on all the above-mentioned aspects of electromagnetic and electronic phenomena, from the detailed functioning of electronic devices, where the dynamics of electrically charged particles in solid state or plasmas are studied, to network theory and circuit simulation, to macroscopic electromagnetic field, for which the modelling of machines requires considering the coupling of mechanical and thermophysical phenomena with electromagnetic ones, as well as the radiation, propagation and reception of electromagnetic waves.

Outline of the Book

In this book the readers will find papers on many of the aspects discussed above. The book is divided into five parts. The first part and the last part are both more or less generic. This first part is called *Mathematical Methods* and presents contributions which, although linked to some application domain, are closer to general applied mathematics than those in the other parts. The last part is called *Model Order Reduction*. Techniques of model order reduction (MOR) can often find use in several application domains and can be based on an application-independent analysis of a system of equations. However, some MOR methods depend on a very special property only appearing in one given class of applications.

The remaining three parts present contributions from applied mathematics, which are more closely related to their respective application domains. The second part, *Computational Electromagnetics*, examines computational methods in macroscopic electromagnetic field theory. The contributions in the third part, *Coupled Problems*, are concerned with multi-physics modelling. The part *Circuit and Device Modelling and Simulation*, deals with mathematics applied to circuit simulation, i.e. electromagnetic interactions on the scale of electronic systems, as well as with modelling on the scale of the interior of the electronic devices themselves.

Each part has its own introduction, which serves to situate the various contributions in the overall context sketched above.

Part I
Mathematical Methods

Introduction

This part is concerned with general mathematical methods of interest for numerical modelling in electrical engineering. As, in the end, any numerical modelling is based on finite dimensional models, numerical linear algebra is a common subject of interest to the whole community. As such, the first four papers show some aspects of this vast domain of research.

The first contribution in this part was written by T. Davis (an invited speaker at the conference) and E. Palamadai Natarajan. It is concerned with sparse matrices, especially those arising with the differential algebraic equations (DAE) used in circuit simulation problems. Sparse methods based on operations on dense submatrices, such as multi-frontal methods, are not effective in these cases. A software package named KLU, which was specifically written to exploit the properties of sparse circuit matrices, is presented as are results comparing it with other packages for circuit simulation.

In the next contribution, P. Benner and A. Schneider discuss a priori error estimation for singular value decomposition-based model order reduction methods. Proven error estimates are a necessary first step before a fully automatic application of such approaches can be relied on. This work presents steps towards a global a priori error estimation for this class of algorithms.

The contribution by M. Bollhöfer and S. Lanteri discusses block pre-conditioning for the solution of large linear systems resulting from the discretisation of the time-harmonic Maxwell equations. The proposed strategies combine principles from incomplete factorisation methods with complex shift of the diagonal entries of the underlying system matrices. Numerical results are presented for electromagnetic wave propagation problems in homogeneous and heterogeneous media.

The following three contributions present mathematical methods related to circuit design and analysis. The contribution by Gräb (an invited speaker at the conference) provides an overview of multi-objective sizing tasks in electronic circuit design. It is shown how statistically distributed and range-valued parameters can be included in yield optimisation and design centering. In addition, accounting for parameter tolerances by multi-objective Pareto optimisation is presented.

The next contribution by E.J.W. ter Maten et al. studies importance sampling as a means of achieving efficient Monte Carlo sampling that also properly covers tails of distributions. An optimal upper bound is derived for the number of samples needed to efficiently obtain an accurate fail probability. The contribution by L. Jansen and C. Tischendorf is concerned with the analysis of parameter-dependent differential algebraic equations. The authors show how to take benefit from the smoothness of the solution as a function of the parameters, in order to efficiently find the solutions for a range of parameter values.

The last two contributions in this part present a mathematical analysis of two specific modelling problems, one from circuit theory and one from electromagnetic field theory. R. Riaza's contribution is concerned with a new lumped circuit element, called a memristor, which is characterised by a nonlinear charge-flux relation. Some analytical properties of semi-state models of the corresponding memristive circuits are studied in terms of differential algebraic equations. More specifically, the geometric index of the DAEs arising in so-called branch-oriented analysis methods is considered. In the last paper of this part, T. Pollok et al. discuss scattering problems for the Helmholtz equation in periodic configurations. The authors develop a general mathematical analysis, valid in any dimension, and algorithms for the handling of periodic structures with local defects.

Sparse Matrix Methods for Circuit Simulation Problems

Timothy A. Davis and E. Palamadai Natarajan

Abstract Differential algebraic equations used for circuit simulation give rise to sequences of sparse linear systems. The matrices have very peculiar characteristics as compared to sparse matrices arising in other scientific applications. The matrices are extremely sparse and remain so when factorized. They are permutable to block triangular form, which breaks the sparse LU factorization problem into many smaller subproblems. Sparse methods based on operations on dense submatrices (supernodal and multifrontal methods) are not effective because of the extreme sparsity. KLU is a software package specifically written to exploit the properties of sparse circuit matrices. It relies on a permutation to block triangular form (BTF), several methods for finding a fill-reducing ordering (variants of approximate minimum degree and nested dissection), and Gilbert/Peierls' sparse left-looking LU factorization algorithm to factorize each block. The package is written in C and includes a MATLAB interface. Performance results comparing KLU with SuperLU, Sparse 1.3, and UMFPACK on circuit simulation matrices are presented. KLU is the default sparse direct solver in the Xyce^{TM}circuit simulation package developed by Sandia National Laboratories.

1 Overview

The KLU software package is specifically designed for solving sequences of unsymmetric sparse linear systems that arise from the differential-algebraic equations used to simulate electronic circuits. Two aspects of KLU are essential for these

T.A. Davis (✉)
Department of Computer and Information Science and Engineering. University of Florida, FL, USA
e-mail: davis@cise.ufl.edu

E. Palamadai Natarajan
Ansys. Inc., USA
e-mail: ekanathan@gmail.com

B. Michielsen and J.-R. Poirier (eds.), *Scientific Computing in Electrical Engineering SCEE 2010*, Mathematics in Industry 16, DOI 10.1007/978-3-642-22453-9_1,

problems: (1) a permutation to block upper triangular form [15, 17], and (2) an asymptotically efficient left looking LU factorization algorithm with partial pivoting [18]. KLU does not exploit supernodes, since the factors of circuit simulation matrices are far too sparse as compared to matrices arising in other applications (such as finite-element methods).

Circuit simulation involves many different tasks for which KLU is useful:

1. DC operating point analysis, where BTF ordering is often helpful. Convergence in DC analysis is critical in that it is typically the first step of a higher level analysis such as transient analysis.
2. Transient analysis, which requires a fast and accurate sparse LU factorization. The sparse linear factorization/solve stages typically dominate the run-time of transient analyses of post-layout circuits with a large number of parasitic devices.
3. Harmonic balance analysis, which is typically solved using Krylov based iterative methods, since the Jacobian representing all the harmonics is huge and cannot be solved with a direct method. KLU is useful in factor/solve stages involving the pre-conditioner.

Section 2 describes the characteristics of circuit matrices, which motivate the design of the KLU algorithm. Section 3 gives a brief description of the algorithm. A more detailed discussion may be found in [24]. Performance results of KLU in comparison with SuperLU [12], Sparse 1.3 [21, 22], and UMFPACK [4, 6, 7] are presented in Sect. 4. An extended version of this paper appears in [11].

In this paper, $|A|$ denotes the number of nonzeros in the matrix A.

2 Characteristics of Circuit Matrices

Circuit matrices arise from Newton's method applied to the differential-algebraic equations representing the underlying circuit [23]. A modified nodal analysis is typically used, resulting in a sequence of linear systems with unsymmetric sparse coefficient matrices with identical nonzero pattern (ignoring numerical cancellation). Circuit matrices exhibit certain unique characteristics for which KLU is designed, which are not generally true of matrices from other applications:

1. Circuit matrices are extremely sparse and remain so when factorized. The ratio of floating-point operation (flop) count over $|L + U|$ is much smaller than matrices from other applications (even for comparable values of $|L + U|$). A set of columns in L with identical or similar nonzero pattern is called a *supernode* [12]. Supernodal and multifrontal methods obtain high performance by exploiting supernodes via dense matrix kernels (the BLAS, [13]). Because their nodal interconnection is highly dissimilar and their fill-in is so low, the supernodes in circuit matrices typically have very few columns. Dense matrix kernels are not effective when used on very small matrices, and thus supernodal/multifrontal methods are not suitable for circuit matrices.

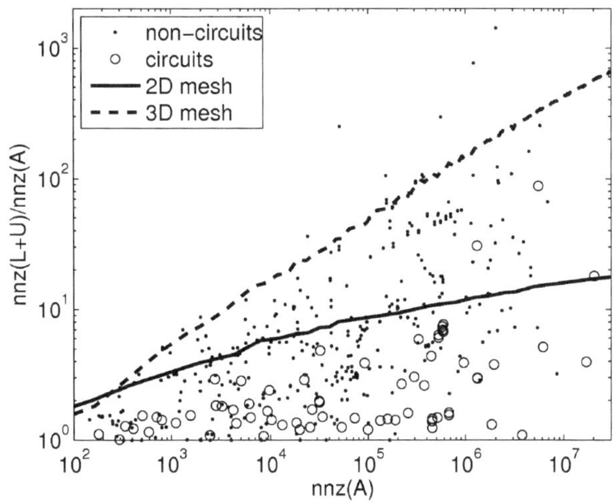

Fig. 1 Fill-in factor versus the number of nonzeros in the largest irreducible block

2. Nearly all circuit matrices are permutable to a block triangular form. In DC operating point analysis, capacitors are open and hence node connectivity is broken in the circuit. This helps in creating many small strongly connected components in the corresponding graph, and the resulting permuted matrix is block triangular with many small blocks. However in transient simulation, capacitors are not open and hence the nodes of the circuit are mostly reachable from each other. This often leads to one large diagonal block when permuted to BTF form, but still a large number of small blocks due to the presence of independent and controlled sources.

The following experiment illustrates the low fill-in properties of circuit matrices. As of March 2010, the University of Florida Sparse Matrix Collection [10] contains 491 matrices that are real, square, unsymmetric, and have full structural rank[1] (excluding matrices tagged as subsequent matrices in sequences of matrices with the same size and pattern). Of these 491 matrices, 81 are from circuit or power network simulation. Figure 1 plots the fill-in factor ($|L + U|/|A|$ versus $|A|$) for each matrix, using `lu` in MATLAB (R2010a). If the matrix is reducible to block triangular form, only the largest block is factorized for this experiment (found via `dmperm` [5]). For comparison, the two lines in Fig. 1 are 2D and 3D square meshes as ordered by METIS [20], which obtains the asymptotically optimal ordering for regular meshes.

The fill-in factor for circuit matrices stays remarkably low as compared to matrices from other applications. Very few circuit matrices experience as much fill-in as 2D or 3D meshes.

[1]A matrix has full structural rank if a permutation exists so that the diagonal is zero-free.

The properties of circuit matrices demonstrated here indicate that they should be factorized via an asymptotically efficient non-supernodal sparse LU method, which motivates the KLU algorithm discussed in the next Section.

3 KLU Algorithm

KLU performs the following steps when solving the first linear system in a sequence.

1. The matrix is permuted into block triangular form (BTF). This consists of two steps: an unsymmetric permutation to ensure a zero free diagonal using maximum transversal [14, 15], followed by a symmetric permutation to block triangular form by finding the strongly connected components of the graph [16, 17, 26]. A matrix with full rank permuted to block triangular form looks as follows:

$$PAQ = \begin{bmatrix} A_{11} & A_{12} & \cdots & A_{1n} \\ & A_{22} & & \vdots \\ & & \ddots & \vdots \\ & & & A_{nn} \end{bmatrix}$$

2. Each block A_{kk} is ordered to reduce fill. The Approximate Minimum Degree (AMD) ordering [1,2] on $A_{kk}+A_{kk}^T$ is used by default. The user can alternatively choose COLAMD [8, 9], an ordering provided by CHOLMOD (such as nested dissection based on METIS [20]), or any user-defined ordering algorithm that can be passed as a function pointer to KLU. Alternatively, the user can provide a permutation to order each block.
3. Each diagonal block is scaled and factorized using our implementation of Gilbert/ Peierls' left looking algorithm with partial pivoting [18]. A simpler version of the same algorithm is used in the LU factorization method in the CSparse package, `cs_lu` [5] (but without the pre-scaling and without a BTF permutation). Pivoting is constrained to within each diagonal block, since the factorization method factors each block as an independent problem. No pivots can ever be selected from the off-diagonal blocks.
4. The system is solved using block back substitution.

For subsequent factorizations for matrices with the same nonzero pattern, the first two steps above are skipped. The third step is replaced with a simpler left-looking method that does not perform partial pivoting (a *refactorization*). This allows the depth-first-search used in Gilbert/Peierls' method to be skipped, since the nonzero patterns of L and U are already known.

When the BTF form is exploited, entries outside the diagonal blocks do not need to be factorized, requiring no work and causing no fill-in. Only the diagonal blocks need to be factorized.

The final system of equations to be solved after ordering and factorization with partial pivoting can be represented as

$$(PRAQ)Q^T x = PRb \tag{1}$$

where P represents the row permutation due to the BTF and fill-reducing ordering and partial pivoting, and Q represents the column permutation due to just the BTF and fill-reducing ordering. The matrix R is a diagonal row scaling matrix (discussed below). Let $(PRAQ) = LU + F$ where LU represents the factors of all the blocks collectively and F represents the entire off diagonal region. Equation (1) can now be written as

$$x = Q(LU + F)^{-1}(PRb). \tag{2}$$

The block back substitution in (2) can be better visualized as follows. Consider a simple 3-by-3 block system

$$\begin{bmatrix} L_{11}U_{11} & F_{12} & F_{13} \\ 0 & L_{22}U_{22} & F_{23} \\ 0 & 0 & L_{33}U_{33} \end{bmatrix} \begin{bmatrix} x_1 \\ x_2 \\ x_3 \end{bmatrix} = \begin{bmatrix} b_1 \\ b_2 \\ b_3 \end{bmatrix}. \tag{3}$$

The equations corresponding to the above system are

$$L_{11}U_{11}x_1 + F_{12}x_2 + F_{13}x_3 = b_1 \tag{4}$$

$$L_{22}U_{22}x_2 + F_{23}x_3 = b_2 \tag{5}$$

$$L_{33}U_{33}x_3 = b_3 \tag{6}$$

In block back substitution, we first solve (6) for x_3, and then eliminate x_3 from (5) and (4) using the off-diagonal entries. Next, we solve (5) for x_2 and eliminate x_2 from (4). Finally we solve (4) for x_1.

The core of the Gilbert/Peierls factorization algorithm used in KLU is solving a lower triangular system $Lx = b$ with partial pivoting where L, x and b are all sparse. It consists of a symbolic step to determine the non-zero pattern of x and a numerical step to compute the values of x. This lower triangular solution is repeated n times during the entire factorization (where n is the size of the matrix) and each solution step computes a column of the L and U factors. The importance of this factorization algorithm is that the time spent in factorization is proportional to the number of floating point operations performed. The entire left looking algorithm is described in the algorithm below.

The lower triangular solve is the most expensive step and includes a symbolic and a numeric factorization step. Let $b = A(:, k)$, the kth column of A. Let G_L be the directed graph of L with n nodes. The graph G_L has an edge $j \to i$ iff $l_{ij} \neq 0$. Let $\mathscr{B} = \{i \,|\, b_i \neq 0\}$ and $\mathscr{X} = \{i \,|\, x_i \neq 0\}$ represent the set of nonzero indices in b and x respectively. Now the nonzero pattern $\mathscr{X}$ is given by

Algorithm 1 Left-looking LU factorization

```
L = I
for k = 1 to n do
    solve Lx = A(:, k) for x
    do partial pivoting on x
    U(1 : k, k) = x(1 : k)
    L(k : n, k) = x(k : n)/U(k, k)
end for
```

$$\mathscr{X} = Reach_{G_L}(\mathscr{B}) \tag{7}$$

$Reach_G(i)$ denotes all nodes in a graph G reachable via paths starting at node i. $Reach(S)$ applied to a set S is the union of $Reach(i)$ for all nodes $i \in S$. Equation (7) states that the nonzero pattern $\mathscr{X}$ is computed by the determining the vertices in G_L that are reachable from the vertices of the set $\mathscr{B}$.

The reachability problem is solved using a depth-first search. During the depth-first search, the Gilbert/ Peierls algorithm computes the *topological* order of $\mathscr{X}$. If the nodes of a directed acyclic graph are written out in topological order from left to right, then all edges in the graph would point to the right. If $Lx = b$ is solved in topological order, all numerical dependencies are satisfied. The natural order 1, 2, ..., n is one such ordering (since the matrix L is lower triangular), but any topological ordering will suffice. That is, x_j must be computed before x_i if there is a path from j to i in G_L. Since the depth-first graph traversal produces $\mathscr{X}$ in topological order as an intrinsic by-product, the solution of $Lx = b$ can be computed using the algorithm below. Sorting the nodes in $\mathscr{X}$ to obtain the natural ordering could take more time than the number of floating-point operations, so this is skipped. The computation of $\mathscr{X}$ and x both take time proportional to the floating-point operation count.

Algorithm 2 Solve $Lx = b$ where L, x and b are sparse

```
𝒳 = Reach_{G_L}(ℬ)
x = b
for j ∈ 𝒳 in any topological order do
    x(j + 1 : n) = x(j + 1 : n) − L(j + 1 : n, j)x(j)
end for
```

4 Performance Comparisons with Other Solvers

Five different sparse LU factorization techniques are compared:

1. KLU with default parameter settings: BTF enabled, the AMD fill-reducing ordering applied to $A + A^T$, and a strong preference for pivots selected from the diagonal.

Table 1 The thirteen test matrices with the highest run times

Matrix	Entire matrix		Largest block		Rows in	Singletons
	Rows	Nonzeros	Rows	Nonzeros	2nd largest	
	$\times 10^3$	$\times 10^3$	$\times 10^3$	$\times 10^3$	block	$\times 10^3$
Raj1	263.7	1,300.3	263.6	1,299.6	5	0.2
ASIC_680k	682.9	2,639.0	98.8	526.3	2	583.8
rajat24	358.2	1,947.0	354.3	1,923.9	172	3.4
TSOPF_RS_b2383_c1	38.1	16,171.2	4.8	31.8	654	0.0
TSOPF_RS_b2383	38.1	16,171.2	4.8	31.8	654	0.0
rajat25	87.2	606.5	83.5	589.8	57	3.4
rajat28	87.2	606.5	83.5	589.8	57	3.4
rajat20	86.9	604.3	83.0	587.5	57	3.6
ASIC_320k	321.8	1,931.8	320.9	1,314.3	6	0.3
ASIC_320ks	321.7	1,316.1	320.9	1,314.3	6	0.1
rajat30	644.0	6,175.2	632.2	6,148.3	7	11.7
Freescale1	3,428.8	17,052.6	3,408.8	16,976.1	19	0.0

2. KLU with default parameters, except that BTF is disabled. For most matrices, using BTF is preferred, but in a few cases the BTF pre-ordering can dramatically increase the fill-in in the LU factors.
3. SuperLU 3.1 [12], using non-default diagonal pivoting preference and ordering options identical to KLU (but without BTF).[2] These options typically give the best results for circuit matrices. SuperLU is a supernodal variant of the Gilbert/Peierls' left-looking algorithm used in KLU.
4. UMFPACK [4,6,7] with default parameters. In this mode, UMFPACK evaluates the symmetry of the nonzero pattern and selects either the AMD ordering on $A + A^T$ and a strong diagonal preference, or it uses the COLAMD ordering with no preference for the diagonal. For most circuit simulation matrices, the AMD ordering is used. UMFPACK is a right-looking multifrontal algorithm that makes extensive use of BLAS kernels.
5. Sparse 1.3 [21, 22], the sparse solver used in SPICE3f5, the latest version of SPICE.[3]

The University of Florida Sparse Matrix Collection [10] includes 81 real square unsymmetric matrices or matrix sequences (only the first matrix in each sequence is considered here) arising from the differential algebraic equations used in SPICE-like circuit simulation problems, or from power network simulation. All five methods were tested an all 81 matrices, except for two matrices too large for any method on the computer used for these tests (a single-core 3.2 GHz Pentium 4 with 4 GB of RAM). The thirteen matrices requiring the most amount of time to analyze, factorize, and solve (as determined by the fastest method for each matrix) are shown in Table 1. All of the matrices come from a transient analysis, since the run time

[2] Threshold partial pivoting tolerance of 0.001 to give preference to the diagonal, the SuperLU symmetric mode, and the AMD ordering on $A + A^T$.

[3] http://bwrc.eecs.berkeley.edu/Classes/icbook/SPICE/

for KLU is very low for matrices arising from a DC analysis. The table lists the matrix name followed by the size of the whole matrix and the largest block in the BTF form (the dimension and the number of nonzeros). The last two columns list the dimension of the second-largest block, and the number of 1-by-1 blocks, respectively.

A performance profile compares the relative run times of multiple methods on a set of test problems. Let the relative run time of a method on a particular problem be equal to its run time for that problem divided by the fastest run time of any method for that problem. A relative run time of 1.0 means that the method is the fastest for that problem among the methods being compared; 2.0 means that it took twice the time as the fastest method. The x axis of a performance profile is this relative run time. The y axis of a performance profile is the number of problems. A point (x, y) is plotted if a method has a relative run time of x (or less) for y problems in the test set.

The performance profiles of the four methods are shown in Fig. 2. It excludes the symbolic ordering and analysis, since this step is done just once for a whole sequence of matrices. Note that the x axis of Fig. 2 is a log scale. For most matrices, KLU (with BTF) is the fastest method. In the worst case (the Raj1 matrix) it is 26 times slower than SuperLU, but this is because the permutation to BTF used by KLU causes fill-in to dramatically increase.

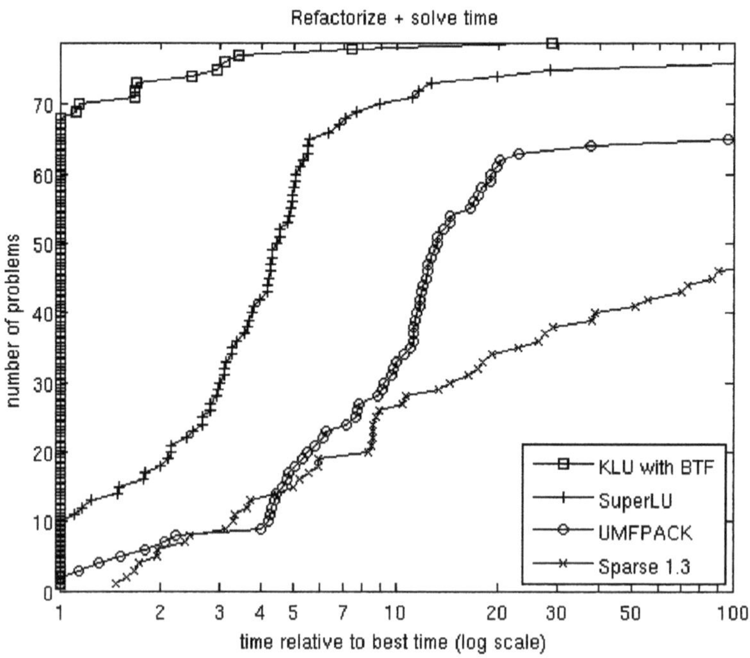

Fig. 2 Performance profile of refactorize+solve time

Table 2 Analyze+factorize+solve time in seconds, and relative fill-in ($|L+U|/|A|$) for KLU. Run times within 25% of the fastest are shown in bold. A dash is shown if the method ran out of memory

Matrix	KLU+BTF		KLU no BTF		SuperLU	Sparse 1.3
	Fill	Time	Fill	Time	Time	Time
Raj1	40.3	111.0	5.5	**4.6**	**4.2**	3,038.9
ASIC_680ks	2.6	**5.0**	2.7	7.2	**4.6**	818.1
ASIC_680k	2.1	**5.8**	2.1	7.4	**5.8**	8,835.1
rajat24	28.7	119.0	3.3	**6.0**	13.9	–
TSOPF_RS_b2383_c1	1.3	**6.5**	2.1	71.8	34.9	–
TSOPF_RS_b2383	1.3	**6.5**	2.1	72.0	34.2	–
rajat25	6.7	**8.5**	35.2	31.7	37.2	2,675.4
rajat28	6.9	**9.1**	28.4	25.4	50.0	3,503.0
rajat20	7.0	**9.1**	35.2	31.3	40.5	4,314.1
ASIC_320k	2.5	30.4	42.9	447.5	**18.1**	7,908.2
ASIC_320ks	3.2	36.6	3.2	36.4	**21.5**	684.9
rajat30	5.1	73.0	3.2	**23.8**	**22.5**	–
Freescale1	3.9	**86.8**	3.9	**85.6**	–	–

Table 3 Refactorize+solve time in seconds

Matrix	KLU+BTF	KLU no BTF	SuperLU	Sparse 1.3
	Time	Time	Time	Time
Raj1	94.4	**3.0**	**3.3**	127.4
ASIC_680ks	**3.9**	5.4	**3.5**	256.7
ASIC_680k	**4.6**	**5.1**	**4.6**	835.8
rajat24	91.2	**3.7**	12.4	–
TSOPF_RS_b2383_c1	**5.2**	40.8	10.9	–
TSOPF_RS_b2383	**5.1**	41.0	10.9	–
rajat25	**6.7**	27.0	36.8	374.4
rajat28	**7.3**	21.8	49.6	512.7
rajat20	**7.3**	26.8	40.2	657.1
ASIC_320k	28.7	429.0	**17.1**	870.1
ASIC_320ks	35.0	35.0	**20.7**	182.0
rajat30	60.5	**18.6**	**19.6**	–
Freescale1	**70.5**	**70.6**	–	–

The time for the thirteen largest matrices is shown in Tables 2 and 3. The fastest run times and run times within 25% of the fastest are shown in bold. A dash is shown if the method ran out of memory.

For sparse Cholesky factorization, the flops per $|L|$ ratio is an accurate predictor of the relative performance of a BLAS-based supernodal method versus a non-supernodal method. If this ratio is 40 or higher, `chol` in MATLAB (and `x=A\b` for sparse symmetric positive definite matrices) automatically selects a supernodal

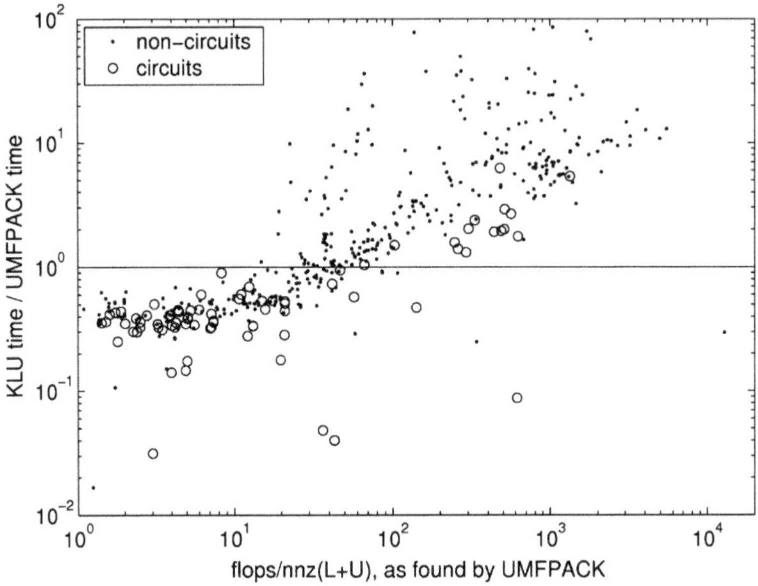

Fig. 3 Relative performance of KLU versus UMFPACK as a function of flops/$|L + U|$

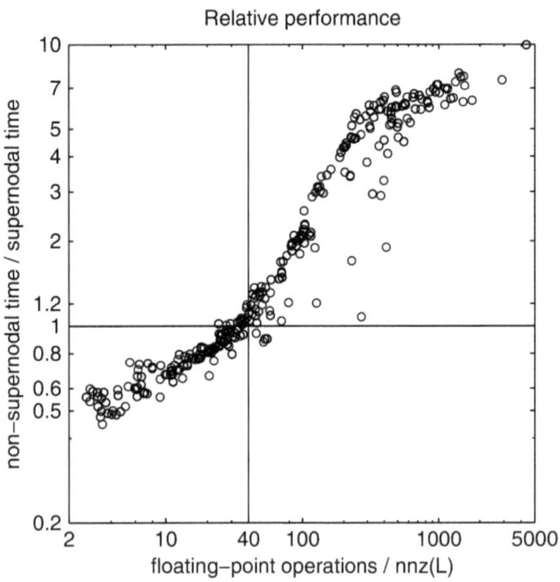

Fig. 4 Relative supernodal (BLAS-based) and non-supernodal (not BLAS-based) performance for sparse Cholesky

solver. Otherwise, a non-supernodal solver is used [3]. A similar comparison is shown in Fig. 3 between KLU and UMFPACK. If the matrix is reducible, only the largest block is factorized. Figure 4 shows the results for sparse Cholesky factorization from [3].

These results are remarkable for three reasons:

1. Circuit matrices tend to have a low flop/$|L+U|$ ratio as compared to other matrices.
2. Even when the flop/$|L+U|$ ratio is high enough (200 or more) to justify using the BLAS, the relative performance of a BLAS-based method (UMFPACK) versus KLU is much less than what would be expected if only non-circuit matrices were considered. Thus, circuits not only remain sparse when factorized, even large circuit matrices with higher flops/$|L+U|$ ratios hardly justify the use of the BLAS.
3. The flops/$|L+U|$ ratio for LU factorization (Fig. 3) is not a very accurate predictor of the relative performance of BLAS-based sparse methods as compared to non-BLAS-based methods, as it is for sparse Cholesky factorization (Fig. 4).

5 Summary

KLU has been shown to be an effective solver for the sequences of sparse matrices that arise when solving differential algebraic equations for circuit simulation problems. It is the default sparse solver in Xyce, a circuit simulation package developed by Sandia National Laboratories [19], for which it has been proven to be a robust and reliable solver [25].

Acknowledgements We would like to thank Mike Heroux for coining the name "KLU" and suggesting that we tackle this project in support of the Xyce circuit simulation package developed at Sandia National Laboratories [19, 25]. Portions of this work were supported by the Department of Energy, and by National Science Foundation grants 0203270, 0620286, and 0619080.

References

1. Amestoy, P.R., Davis, T.A., Duff, I.S.: An approximate minimum degree ordering algorithm. SIAM J. Matrix Anal. Appl. **17**(4), 886–905 (1996)
2. Amestoy, P.R., Davis, T.A., Duff, I.S.: Algorithm 837: AMD, an approximate minimum degree ordering algorithm. ACM Trans. Math. Softw. **30**(3), 381–388 (2004)
3. Chen, Y., Davis, T.A., Hager, W.W., Rajamanickam, S.: Algorithm 887: CHOLMOD, supernodal sparse Cholesky factorization and update/downdate. ACM Trans. Math. Softw. **35**(3), 1–14 (2008). DOI http://doi.acm.org/10.1145/1391989.1391995
4. Davis, T.A.: Algorithm 832: UMFPACK V4.3, an unsymmetric-pattern multifrontal method. ACM Trans. Math. Softw. **30**(2), 196–199 (2002)
5. Davis, T.A.: Direct Methods for Sparse Linear Systems. SIAM, Philadelphia, PA (2006)
6. Davis, T.A., Duff, I.S.: An unsymmetric-pattern multifrontal method for sparse LU factorization. SIAM J. Matrix Anal. Appl. **18**(1), 140–158 (1997)
7. Davis, T.A., Duff, I.S.: A combined unifrontal/multifrontal method for unsymmetric sparse matrices. ACM Trans. Math. Softw. **25**(1), 1–19 (1999)

8. Davis, T.A., Gilbert, J.R., Larimore, S.I., Ng, E.G.: Algorithm 836: COLAMD, a column approximate minimum degree ordering algorithm. ACM Trans. Math. Softw. **30**(3), 377–380 (2004)
9. Davis, T.A., Gilbert, J.R., Larimore, S.I., Ng, E.G.: A column approximate minimum degree ordering algorithm. ACM Trans. Math. Softw. **30**(3), 353–376 (2004)
10. Davis, T.A., Hu, Y.: University of Florida sparse matrix collection. ACM Transactions on Mathematical Software, **38**(1), (2011). URL http://www.cise.ufl.edu/sparse/matrices
11. Davis, T.A., Palamadai Natarajan, E.: Algorithm 907: KLU, a direct sparse solver for circuit simulation problems. ACM Trans. Math. Softw. **37**(3), 36:1–36:17 (2010). DOI http://doi.acm.org/10.1145/1824801.1824814. URL http://doi.acm.org/10.1145/1824801.1824814
12. Demmel, J.W., Eisenstat, S.C., Gilbert, J.R., Li, X.S., Liu, J.W.H.: A supernodal approach to sparse partial pivoting. SIAM J. Matrix Anal. Appl. **20**(3), 720–755 (1999)
13. Dongarra, J.J., Du Croz, J., Duff, I.S., Hammarling, S.: A set of level-3 basic linear algebra subprograms. ACM Trans. Math. Softw. **16**(1), 1–17 (1990)
14. Duff, I.S.: Algorithm 575: Permutations for a zero-free diagonal. ACM Trans. Math. Softw. **7**(1), 387–390 (1981)
15. Duff, I.S.: On algorithms for obtaining a maximum transversal. ACM Trans. Math. Softw. **7**(1), 315–330 (1981)
16. Duff, I.S., Reid, J.K.: Algorithm 529: Permutations to block triangular form. ACM Trans. Math. Softw. **4**(2), 189–192 (1978)
17. Duff, I.S., Reid, J.K.: An implementation of Tarjan's algorithm for the block triangularization of a matrix. ACM Trans. Math. Softw. **4**(2), 137–147 (1978)
18. Gilbert, J.R., Peierls, T.: Sparse partial pivoting in time proportional to arithmetic operations. SIAM J. Sci. Statist. Comput. **9**, 862–874 (1988)
19. Hutchinson, S.A., Keiter, E.R., Hoekstra, R.J., Waters, L.J., Russo, T., Rankin, E., Wix, S.D., Bogdan, C.: The Xyce^{TM}parallel electronic simulator – an overview. In: Joubert, G.R., Murli, A., Peters, F.J., Vanneschi, M. (eds.) Parallel Computing: Advances and Current Issues, Proc. ParCo 2001, pp. 165–172. Imperial College Press, London (2002)
20. Karypis, G., Kumar, V.: A fast and high quality multilevel scheme for partitioning irregular graphs. SIAM J. Sci. Comput. **20**, 359–392 (1998)
21. Kundert, K.S.: Sparse matrix techniques and their applications to circuit simulation. In: Ruehli, A.E. (ed.) Circuit Analysis, Simulation and Design. North-Holland, New York (1986)
22. Kundert, K.S., Sangiovanni-Vincentelli, A.: User's guide: Sparse 1.3. Tech. rep., Dept. of EE and CS, UC Berkeley (1988)
23. Nichols, K., Kazmierski, T., Zwolinski, M., Brown, A.: Overview of SPICE-like circuit simulation algorithms. IEE Proc. Circuits, Devices & Sys. **141**(4), 242–250 (1994)
24. Palamadai Natarajan, E.: KLU - a high performance sparse linear system solver for circuit simulation problems. M.S. Thesis, CISE Department, Univ. of Florida (2005)
25. Sipics, M.: Sparse matrix algorithm drives SPICE performance gains. SIAM News **40**(4) (2007)
26. Tarjan, R.E.: Depth first search and linear graph algorithms. SIAM J. Comput. **1**, 146–160 (1972)

Some Remarks on A Priori Error Estimation for ESVDMOR

Peter Benner and André Schneider

Abstract In previous work it is shown how to numerically improve the ESVDMOR method of Feldmann and Liu to be really applicable to linear, sparse, very large scale, and continuous-time descriptor systems. Stability and passivity preservation of this algorithm is also already proven. This work presents some steps towards a global a priori error estimation for this algorithm, which is necessary for a fully automatic application of this approach.

1 Motivation

Although model order reduction (MOR) for linear time invariant (LTI) systems is a well investigated area of research [1], most of the established approaches, e.g., Krylov subspace methods or balanced truncation methods [7], are not able to work on systems with a lot of input and output terminals. They are not easily reducible, especially really large scale ones. ESVDMOR is, besides other approaches [5], a MOR approach to reduce linear systems with a large number of terminals [2–4, 6]. Within the algorithm, approximation errors are caused at different steps. The magnitude of these errors can be influenced with the help of different decisions. Some of the correlations between these decisions and the influence on the results are well known, but a closed error analysis for the ESVDMOR approach does

P. Benner (✉)
Max Planck Institute for Dynamics of Complex Technical Systems, Sandtorstr. 1, D-39106 Magdeburg, Germany
e-mail: benner@mpi-magdeburg.mpg.de

A. Schneider
Technische Universität Chemnitz, Fakultät für Mathematik, Mathematik in Industrie und Technik, D-09107 Chemnitz, Germany
e-mail: andre.schneider@mathematik.tu-chemnitz.de

B. Michielsen and J.-R. Poirier (eds.), *Scientific Computing in Electrical Engineering SCEE 2010*, Mathematics in Industry 16, DOI 10.1007/978-3-642-22453-9_2,

not yet exist. For efficient reduction which meets the requirements placed on the reduced order model, the knowledge about this correlations is of essential relevance. The goal is to get a reduced model which is as small as possible and at the same time as good as necessary. This knowledge is essential for the industrial usage of MOR algorithms. In Sect. 2 we briefly repeat required basic knowledge including the steps of SVDMOR and ESVDMOR. We emphasize those steps which cause an approximation error in some way. The following section deals with the single errors and the known theory. We combine all influences, firstly with a lot of assumptions and for the easy cases and later for more complicated models, to get ideas about a global error bound for the ESVDMOR approach.

2 (E)SVDMOR Basics Including Error Sources

Starting point is a given (mostly by modeling in circuit simulation but also in mechanical, biological, and chemical applications) linear time-invariant continuous-time descriptor system

$$\begin{aligned} C\dot{x}(t) &= -Gx(t) + Bu(t),\ x(0) = x_0, \\ y(t) &= Lx(t), \end{aligned} \tag{1}$$

where $C, G \in \mathbb{R}^{n \times n}$, $B \in \mathbb{R}^{n \times m_{in}}$, $L \in \mathbb{R}^{m_{out} \times n}$. Vector $x(t) \in \mathbb{R}^n$ contains the descriptor variables, $u(t) \in \mathbb{R}^{m_{in}}$ is the vector of inputs, $y(t) \in \mathbb{R}^{m_{out}}$ is the output vector, and $x_0 \in \mathbb{R}^n$ is the initial value. The value n is called order of (1) defined by the number of descriptor variables and m_{in} and m_{out} denote the number of I/O terminals, respectively. System (1) has the following transfer function in frequency domain:

$$H(s) = L(sC + G)^{-1}B, \tag{2}$$

which we get from (1) for $x_0 = 0$ by applying the Laplace transform. Like mentioned in Sect. 1 we want to investigate systems with

$$m_{in/out} \sim n.$$

Further on, we define the *i-th block moment of (2)* as $\mathbf{m_i} = L(-G^{-1}C)^i G^{-1}B$, $i = 0, 1, \ldots$, in terms of $\mathbf{m_i}$ as an $m_{out} \times m_{in}$ matrix. These moments are equal to the coefficients of the Taylor series expansion of (2) about $s_0 = 0$, $H(s) = \sum_{i=0}^{\infty} m_i s^i$. For $s_0 \neq 0$ this leads to *frequency shifted moments* defined as

$$\mathbf{m_i}(s_0) = L(-(s_0C + G)^{-1}C)^i (s_0C + G)^{-1}B, \qquad i = 0, 1, \ldots$$

Thus, the Taylor series expansion including these moments is

$$H(s) = \sum_{i=0}^{\infty} m_i (s - s_0)^i.$$

To allow terminal reduction for inputs and outputs separately, w.l.o.g. we use r different (frequency shifted) block moments forming two moment or ansatz matrices, the input response matrix M_I and the output response matrix M_O, as follows:

$$M_I = \begin{bmatrix} \mathbf{m_0} \\ \mathbf{m_1} \\ \vdots \\ \mathbf{m_{r-1}} \end{bmatrix}, \qquad M_O = \begin{bmatrix} \mathbf{m_0}^T \\ \mathbf{m_1}^T \\ \vdots \\ \mathbf{m_{r-1}}^T \end{bmatrix}. \tag{3}$$

It is also possible to use different numbers of block moments to create M_I and M_O. The number r is the first possibility to influence the accuracy of the reduced model. For simplicity, we assume the number of rows in M_I and M_O of (3) to be larger than the number of columns, i.e., $r \cdot m_{out} \geq m_{in}$ and $r \cdot m_{in} \geq m_{out}$. If not, r has to be increased. Applying the SVD to these matrices, we obtain a low rank approximation

$$M_I \approx U_{I_{r_i}} \Sigma_{I_{r_i}} V_{I_{r_i}}^T \text{ and } M_O \approx U_{O_{r_o}} \Sigma_{O_{r_o}} V_{O_{r_o}}^T, \tag{4}$$

which causes an approximation error. The matrices $\Sigma_{I_{r_i}}$ and $\Sigma_{O_{r_o}}$ are $r_i \times r_i$ and $r_o \times r_o$ diagonal matrices, $V_{I_{r_i}}$ and $V_{O_{r_o}}$ are $m_{in} \times r_i$ and $m_{out} \times r_o$ isometric matrices that contain the dominant column subspaces of M_I and M_O, and $U_{I_{r_i}}$ and $U_{O_{r_o}}$ are $r \cdot m_{out} \times r_i$ and $r \cdot m_{in} \times r_o$ isometric matrices that are not used any further. The values $r_i \leq m_{in}$ and $r_o \leq m_{out}$ denote the numbers of significant singular values (SV) as well as the numbers of the virtual input and output terminals of the terminal reduced order model. Due to the fact that the important information about the dependencies of the I/O-ports is hidden in the matrices $V_{I_{r_i}}^T$ and $V_{O_{r_o}}^T$, we use these matrices to find the searched approximate factorization of B and L. Hence, $B = BI \approx B(V_{I_{r_i}} V_{I_{r_i}}^+)$, where I denotes the identity matrix and $()^+$ denotes the Moore-Penrose pseudoinverse. Using the properties of this pseudoinverse and $(V_{I_{r_i}}^T V_{I_{r_i}})^{-1} = I$ leads to $B \approx BV_{I_{r_i}} (V_{I_{r_i}}^T V_{I_{r_i}})^{-1} V_{I_{r_i}}^T = BV_{I_{r_i}} V_{I_{r_i}}^T$. Defining a matrix B_r as $B_r := BV_{I_{r_i}}$ we finally get the approximation $B \approx B_r V_{I_{r_i}}^T$. Equivalent arguments lead to $L \approx V_{O_{r_o}} L_r$ with $L_r = V_{O_{r_o}}^T L$. The approximation errors which appear in these equations are very important, see Sect. 3. Plugging in these approximations in (2), we consequently get a new internal transfer function $H_r(s)$ by using the approximation

$$H(s) \approx \widehat{H}(s) = V_{O_{r_o}} \underbrace{L_r (G + sC)^{-1} B_r}_{:= H_r(s)} V_{I_{r_i}}^T.$$

This terminal reduced transfer function $H_r(s)$ can be further reduced to

$$\tilde{H}_r(s) = \tilde{L}_r(\tilde{G} + s\tilde{C})^{-1}\tilde{B}_r \approx H_r(s) \tag{5}$$

by any established MOR method. Balanced truncation approaches are advantageous as there exists a well known error theory, see Sect. 3. We end up with a very compact terminal reduced and reduced-order model $\tilde{H}_r(s)$, i. e.

$$H(s) \approx \widehat{H}(s) = V_{O_{r_o}} H_r(s) V_{I_{r_i}}^T \approx \widehat{H}_r(s) = V_{O_{r_o}} \tilde{H}_r(s) V_{I_{r_i}}^T. \tag{6}$$

3 Bounds for Particular Approximation Errors and Global ESVDMOR Error Bound

In this section we recall known facts about the errors mentioned in Sect. 2. We give ideas how to connect these errors to a global error bound for ESVDMOR. To get an appropriate entrance in the subject matter we recall two needed matrix norms.

Definition 1 (Spectral norm). The *spectral norm* of the transfer function (2) is induced by the Euclidean vector norm and defined as

$$||H(s)||_2 = \sqrt{\lambda_{\max}(H(s)^H H(s))},$$

where H^H denotes the conjugate transpose of H and $\lambda_{\max}$ denotes its largest eigenvalue.

Another very useful and important norm is based on the Hardy Space theory.

Definition 2 ($\mathscr{H}_\infty$-norm). Let $\mathbb{C}_+$ be the open right half plane. The *$\mathscr{H}_\infty$-norm* of the transfer function (2) is defined as

$$||H||_{\mathscr{H}_\infty} = \sup_{s\in\mathbb{C}_+} \sigma_{\max}(H(s)) = \sup_{s\in\mathbb{C}_+} ||H(s)||_2, \tag{7}$$

where σ_{max} denotes the largest singular value. Because of the maximum modulus theorem we can express (7) as $||H||_{\mathscr{H}_\infty} = \sup_{\omega\in\mathbb{R}} \sigma_{\max}(H(i\omega))$.

3.1 Particular Error Bounds

Equation (4) describes a truncated singular value decomposition (SVD). We know the error caused by a SVD of M_I is

$$e_{M_I} = \left\| M_I(r) - U_{I_{r_i}} \Sigma_{r_i}^I V_{I_{r_i}}^T \right\|_2 = \sigma_{r_i+1}^I,$$

where

$$\Sigma^I = \operatorname{diag}(\sigma_i^I) \approx \Sigma_{r_i}^I = \operatorname{diag}(\sigma_j^I),$$

with $i = 1, \dots, m_{in}$ and $j = 1, \dots, r_i$, and $\sigma_1^I \geq \dots \geq \sigma_{r_i}^I \geq \sigma_{r_i+1}^I \geq \dots \geq \sigma_{m_{in}}^I \geq 0$. The same applies to M_O. Here, the notation $M_I(r)$ expresses the dependency on the number r of used block moments m_i.

Another well known error can be found in (5) if we use a suitable method which gives the information, e. g., balanced truncation (BT) methods. The use of these methods leads to a reduction based on the truncation of the so called Hankel SVs. Provided that G is invertible, we get these values by balancing the controllability and the observability Gramian of H_r in the following form:

$$P = Q =: \begin{bmatrix} \Sigma_1 & \\ & \Sigma_2 \end{bmatrix} = \operatorname{diag}(\hat{\sigma}_1, \dots, \hat{\sigma}_n).$$

Due to storage, efficiency and accuracy reasons usually one computes approximate low rank factors $P \approx P_C P_C^T$ and $Q \approx Q_C Q_C^T$. Using these factors, we compute a singular value decomposition of the form

$$Q_C^T C P_C = \begin{bmatrix} U_1 \ U_2 \end{bmatrix} \begin{bmatrix} \Sigma_1 & 0 \\ 0 & \Sigma_2 \end{bmatrix} \begin{bmatrix} V_1^T \\ V_2^T \end{bmatrix}.$$

Now we define the balancing transformations

$$T_l = Q_C U_1 \Sigma_1^{-1/2} \quad \text{and} \quad T_r = P_C V_1 \Sigma_1^{-1/2},$$

where $\Sigma_1^{-1/2} = \operatorname{diag}(\frac{1}{\sqrt{\hat{\sigma}_1}}, \dots, \frac{1}{\sqrt{\hat{\sigma}_l}})$, such that we are able to compute the reduced system as

$$(\tilde{C}, \tilde{G}, \tilde{B}_r, \tilde{L}_r) := (T_l^T C T_r, T_l^T G T_r, T_l^T B_r, L_r T_r).$$

The error for this square root variant of balanced truncation is bounded by

$$\left\| H_r - \tilde{H}_r \right\|_{\mathcal{H}_\infty} \leq 2 \sum_{k=l+1}^{n} \hat{\sigma}_k = \delta, \tag{8}$$

in case we keep the l largest $\hat{\sigma}_i$. A proof can be found, e.g., in [1]. Figure 1 shows a system with $n = 500$ states, $m_{in} = 5$ inputs, $m_{out} = 10$ outputs and it is reduced to order $l = 60$. The computed error bound is $\delta = 9.796 \cdot 10^{-3}$. The error does not even reach the bound.

3.2 Total Error Bound

Due to (6) and the triangle inequality the total ESVDMOR error in spectral norm on the imaginary axis can be expressed locally as

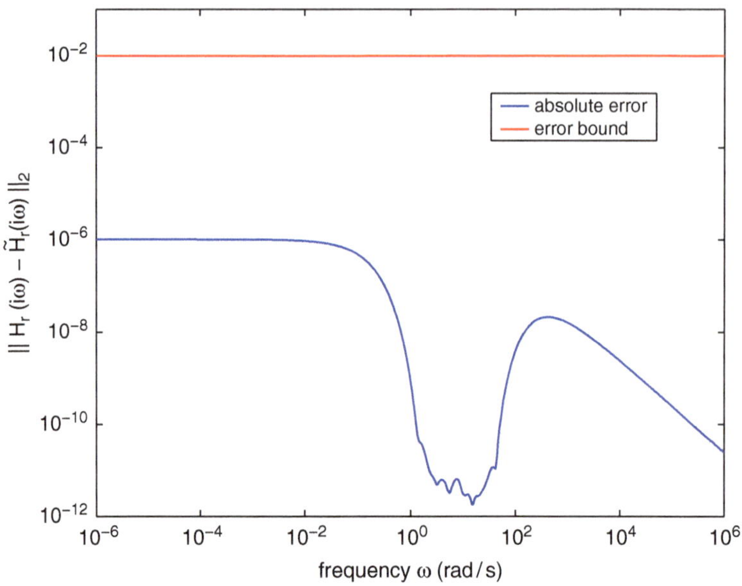

Fig. 1 Absolute error of a BT reduced system

$$e_{tot} = \left\| H(i\omega) - \hat{H}_r(i\omega) \right\|_2 \leq \underbrace{\left\| H(i\omega) - \hat{H}(i\omega) \right\|_2}_{=e_{out}} + \underbrace{\left\| \hat{H}(i\omega) - \hat{H}_r(i\omega) \right\|_2}_{e_{in}}. \tag{9}$$

The BT part (the error caused by the inner reduction e_{in}) follows from (6), (8), (9)

$$e_{in} = \left\| V_{O_{r_o}} H_r(s) V_{I_{r_i}}^T - V_{O_{r_o}} \tilde{H}_r(s) V_{I_{r_i}}^T \right\|_2 = \left\| H_r(s) - \tilde{H}_r(s) \right\|_2 \leq \delta,$$

due to the fact the spectral norm is invariant under orthogonal transformations. The terminal reduction part, also called outer reduction error e_{out}, turns out to be more complicated. To keep things simple we assume dealing with RLC circuits only, i. e., $m_{in} = m_{out} = m$, $L = B^T$, and, if $s_0C + G \geq 0$, consequently $H(s) = H(s)^T$. Due to symmetry, $M_I = M_O = U\Sigma V^T$, and also $V_I = V_O = V$. Moreover $U = V$ holds in the SVDMOR case, which means that there is only one m_i in the ansatz matrices ($r = 1$), e.g. m_0 and $s = s_0 \in \mathbb{R}$ such that

$$M_I = M_O^T = m_0 = B^T(s_0C + G)^{-1}B = U\Sigma V^T = U\Sigma U^T \approx U_r\Sigma_r U_r^T.$$

The local terminal reduction error e_{out} then is

$$e_{out} = \left\| H - \hat{H} \right\|_2 = \left\| B^T(s_0C + G)^{-1}B - U_r B_r^T(s_0C + G)^{-1} B_r V_r^T \right\|_2$$

$$
\begin{aligned}
&\overset{(U=V)}{=} \left\| B^T(s_0C+G)^{-1}B - U_rU_r^TB^T(s_0C+G)^{-1}BU_rU_r^T\right\|_2 \\
&= \left\|U\Sigma U^T - U_rU_r^TU\Sigma U^TU_rU_r^T\right\|_2 = \left\|U\Sigma U^T - U_r\Sigma_rU_r^T\right\|_2 \\
&\overset{(SVD)}{=} \sigma_{k+1}^{I/O},
\end{aligned}
$$

if we keep k singular values or terminals. The total error in the SVDMOR case in spectral norm then is

$$
e_{tot} \le \sigma_{k+1}^{I/O} + 2\sum_{j=l+1}^{n} \hat{\sigma}_j. \tag{10}
$$

In the ESVDMOR case we allow $r \ge 1$ (r times m_i within the ansatz matrices), for simplicity let us assume $r = 3$ and m_0, m_1, and m_2. Thus,

$$
\begin{aligned}
M_I = \begin{pmatrix} m_0 \\ m_1 \\ m_2 \end{pmatrix} = \begin{pmatrix} U^{(1)} \\ U^{(2)} \\ U^{(3)} \end{pmatrix}\Sigma V &= \begin{pmatrix} U_1^{(1)} & U_2^{(1)} \\ U_1^{(2)} & U_2^{(2)} \\ U_1^{(3)} & U_2^{(3)} \end{pmatrix}\begin{pmatrix} \Sigma_1 & 0 \\ 0 & \Sigma_2 \end{pmatrix}\begin{pmatrix} V_1^T \\ V_2^T \end{pmatrix} \\
&=: \begin{pmatrix} U_1 & U_2 \end{pmatrix}\begin{pmatrix} \Sigma_1 & 0 \\ 0 & \Sigma_2 \end{pmatrix}\begin{pmatrix} V_1^T \\ V_2^T \end{pmatrix},
\end{aligned}
$$

where the row partitioning in U is as in M_I, M_O and the column partitioning refers to the number of kept singular values, call this number k. We get $m_j = U^{(j)}\Sigma V^T$, $j = 1,2,3$, (which is not an SVD as $U^{(j)}$ is not orthogonal, but $\|U^{(j)}\|_2 \le 1$ holds.) Thus we can write

$$
\begin{aligned}
H(s) - \hat{H}(s) &= \sum_{j=0}^{\infty}(m_j - \hat{m}_j)(s-s_0)^j \\
&= (m_0 - \hat{m}_0) + (m_1 - \hat{m}_1)(s-s_0) + (m_2 - \hat{m}_2)(s-s_0)^2 + \mathcal{O}(s-s_0)^3.
\end{aligned}
$$

We are now able to bound the first expressions. We write $P_1 = V_1V_1^T$, hence $I - P_1 = V_2V_2^T$, thus,

$$
\begin{aligned}
m_j - \hat{m}_j &= m_j - P_1m_jP_1 = U^{(j)}\Sigma V^T - P_1U^{(j)}\Sigma V^TV_1V_1^T \\
&= U^{(j)}\begin{pmatrix} \Sigma_1 & 0 \\ 0 & \Sigma_2 \end{pmatrix}\begin{pmatrix} V_1^T \\ V_2^T \end{pmatrix} - P_1U^{(j)}\begin{pmatrix} \Sigma_1 & 0 \\ 0 & \Sigma_2 \end{pmatrix}\begin{pmatrix} I_k \\ 0 \end{pmatrix}V_1^T \\
&= U^{(j)}\begin{pmatrix} \Sigma_1 & 0 \\ 0 & 0 \end{pmatrix}\begin{pmatrix} V_1^T \\ V_2^T \end{pmatrix} + U^{(j)}\begin{pmatrix} 0 & 0 \\ 0 & \Sigma_2 \end{pmatrix}\begin{pmatrix} V_1^T \\ V_2^T \end{pmatrix} - P_1U^{(j)}\underbrace{\begin{pmatrix} \Sigma_1 \\ 0 \end{pmatrix}V_1^T}_{=\begin{pmatrix} \Sigma_1 & 0 \\ 0 & 0 \end{pmatrix}\begin{pmatrix} V_1^T \\ V_2^T \end{pmatrix}}
\end{aligned}
$$

$$= U^{(j)} \begin{pmatrix} 0 & 0 \\ 0 & \Sigma_2 \end{pmatrix} V^T + (I - P_1) U^{(j)} \begin{pmatrix} \Sigma_1 \\ 0 \end{pmatrix} V_1^T$$

$$= U^{(j)} \begin{pmatrix} 0 & 0 \\ 0 & \Sigma_2 \end{pmatrix} V^T + V_2 V_2^T U^{(j)} \begin{pmatrix} \Sigma_1 \\ 0 \end{pmatrix} V_1^T =: e_{j,1} + e_{j,2}.$$

We can now express the error as follows:

$$H(s) - \hat{H}(s) = e_{0,1} + e_{1,1}(s - s_0) + e_{2,1}(s - s_0)^2$$

$$+ e_{0,2} + e_{1,2}(s - s_0) + e_{2,2}(s - s_0)^2 + \mathscr{O}(s - s_o)^3,$$

where, when taking norms, and using $\|U^{(j)}\|_2 \leq 1$, $\|V^T\|_2 = 1$,

$$\|e_{j,1}\|_2 \leq \sigma_{k+1}.$$

Unfortunately, the terms $\|e_{j,2}\|_2$ can not be bounded in a meaningful way. But if σ_{k+1} were zero, then $V_2 V_2^T$ projects onto the nullspace of M_I, so that if σ_{k+1} is small enough, $V_2 V_2^T$ is still an orthoprojector onto the joint approximate nullspace of the first r moments. That is, the error, up to order $r - 1$, is essentially contained in the nullspace of the first r moments. Future investigations will focus on exploiting this fact to get a general a priori error bound.

4 Conclusions

In this rather theoretical work we explain and reveal all important matters to get an error bound for the ESVDMOR approach. Although, we are not able to find a universal total error bound in all cases, in (10) we find an expression for the total error in spectral norm. With the help of the results in [8], which states that for some linear RLC circuits $||H||_{\mathscr{H}_\infty} = ||H(0)||_2$, our results are interesting and provide a total a priori SVDMOR error bound in $\mathscr{H}_\infty$-norm, as

$$\left\|H - \widehat{H}_r\right\|_{\mathscr{H}_\infty} = \sup_{\omega \in \mathbb{R}} \left\|H(i\omega) - \widehat{H}_r(i\omega)\right\|_2 = \left\|H(0) - \widehat{H}_r(0)\right\|_2 \overset{(10)}{\leq} \sigma_{k+1}^{I/O} + 2 \sum_{j=l+1}^{n} \hat{\sigma}_j,$$

for these circuits.

Acknowledgements The work reported in this paper was supported by the German Federal Ministry of Education and Research (BMBF), grant no. 03BEPAE1. Responsibility for the contents of this publication rests with the authors.

References

1. Antoulas, A.: Approx. of Large-Scale Dynamical Systems. SIAMPub, Philadelphia, PA (2005)
2. Benner, P., Schneider, A.: Model order and terminal reduction approaches via matrix decomposition and low rank approximation. In: Roos, J., Costa, L.R.J. (eds.) Scientific Computing in Electrical Engineering SCEE 2008, Mathematics in Industry, vol. 14, pp. 523–530. Springer-Verlag, Berlin/Heidelberg, Germany (2010)
3. Benner, P., Schneider, A.: On stability, passivity, and reciprocity preservation of ESVDMOR. In: Benner, P., Hinze, M., ter Maten, J. (eds.) Model Reduction for Circuit Simulation, Lecture Notes in Electrical Engineering, vol. 74, pp. 267–278. Springer, Berlin (2010)
4. Feldmann, P., Liu, F.: Sparse and efficient reduced order modeling of linear subcircuits with large number of terminals. In: ICCAD '04: Proceedings of the 2004 IEEE/ACM Intl. Conf. Computer-aided design, pp. 88–92. IEEE Computer Society, Washington, DC, USA (2004)
5. Ionutiu, R., Rommes, J., Schilders, W.: Model reduction for multi-terminal RC circuits. In: Scientific Computing in Electrical Engineering SCEE (2010)
6. Liu, P., Tan, S.X.D., Yan, B., McGaughy, B.: An efficient terminal and model order reduction algorithm. Integr. VLSI J. **41**(2), 210–218 (2008)
7. Mehrmann, V., Stykel, T.: Balanced truncation model reduction for large-scale systems in descriptor form. In: Benner, P., Mehrmann, V., Sorensen, D. (eds.) Dimension Reduction of Large-Scale Systems. Lecture Notes in Computational Science and Engineering, vol. 45, pp. 83–115. Springer, Berlin (2005)
8. Stykel, T., Reis, T.: Lyapunov balancing for passivity-preserving model reduction of RC circuits. SIAM J. Appl. Dyn. Syst. **10**(1), 1–34 (2011)

Block Preconditioning Strategies for High Order Finite Element Discretization of the Time-Harmonic Maxwell Equations

Matthias Bollhöfer and Stéphane Lanteri

Abstract We study block preconditioning strategies for the solution of large sparse complex coefficients linear systems resulting from the discretization of the time-harmonic Maxwell equations by a high order discontinuous finite element method formulated on unstructured simplicial meshes. The proposed strategies are based on principles from incomplete factorization methods. Moreover, a complex shift is applied to the diagonal entries of the underlying matrices, a technique that has recently been exploited successfully in similar contexts and in particular for the multigrid solution of the scalar Helmholtz equation. Numerical results are presented for 2D and 3D electromagnetic wave propagation problems in homogeneous and heterogeneous media.

1 Introduction

The present study is concerned with the development of a high-performance numerical methodology for the computer simulation of time-harmonic electromagnetic wave propagation problems in irregularly shaped domains and heterogeneous media. In this context, we are naturally led to consider volume discretization methods (i.e. finite difference, finite volume or finite element methods) as opposed to surface discretization methods (i.e. boundary element method). Most of the related existing works deal with the second-order form of the time-harmonic Maxwell equations discretized by a conforming finite element method [14]. More

M. Bollhöfer (✉)
Institute of Computational Mathematics, TU Braunschweig, D-38106 Braunschweig, Germany
e-mail: m.bollhoefer@tu-bs.de

S. Lanteri
NACHOS project-team, INRIA Sophia Antipolis - Méditerranée research center 2004 Route des Lucioles, BP 93, F-06902 Sophia Antipolis Cedex, France
e-mail: Stephane.Lanteri@inria.fr

B. Michielsen and J.-R. Poirier (eds.), *Scientific Computing in Electrical Engineering SCEE 2010*, Mathematics in Industry 16, DOI 10.1007/978-3-642-22453-9_3,

recently, discontinuous Galerkin (DG) methods have also been considered for this purpose (see [4–6]). Here, we concentrate on the first-order form of the time-harmonic Maxwell equations discretized by a high order DG method formulated on unstructured simplicial meshes. While it keeps almost all the advantages of the finite element method (large spectrum of applications, complex geometries, etc.), the DG method has other nice properties among which, an easy extension to higher order interpolation, no global mass matrix to invert (when solving time-domain problems using an explicit time scheme), easy handling of unstructured meshes, natural treatment of discontinuous solutions and coefficient heterogeneities, nice parallelization properties (the compact nature of a DG scheme is in favor of high computation to communication ratio especially for high order interpolation methods).

The DG discretization of the first order form of the time-harmonic Maxwell equations leads to a large sparse complex system of equations that exhibits a block structure which is linked to the use of a polynomial interpolation method for the approximation of the electromagnetic field within a mesh element. For moderately large 2D problems, this system can be efficiently solved by an optimized sparse solver such as MUMPS [1]. However, for large 2D problems or for 3D problems, such a solution strategy is simply not feasible. In [8], a hybrid iterative-direct solver is proposed for the solution of the linear system resulting from the DG discretization of the 3D time-harmonic Maxwell equations. At the discrete level, this domain decomposition solver combines an iterative solver acting on a reduced linear system of equations involving interface unknowns, with a sparse direct solver within each subdomain. For moderately large 3D problems and for the lowest interpolation degrees (i.e. 0-th and 1-st order) in the DG method, the resulting hybrid iterative-direct solver is a viable solution strategy. However, for very large problems and for high interpolation degrees, the size of the subdomain problems prohibits the use of a sparse direct solver. Besides, increasing the number of subdomains to reduce the size of the local problems is generally not a proper approach since this incurs numerical scalability issues which have not been investigated so far for optimized Schwarz methods.

In this paper we will discuss an alternative way of solving the discretized time-harmonic Maxwell equations. Our approach is mainly based on the relations between the second order Maxwell equations and Helmholtz equations. For Helmholtz equations, recently numerical methods have been presented that are based on the shifted Laplacian [2, 9, 12, 13]. I.e., first an artificial damping is introduced into the equations which results in an additional imaginary shift. Then the numerical approximation is computed for the shifted system instead of the original system. Finally, the approximation is applied to the original equations. For the first order time-harmonic Maxwell equations an analogous perturbation is performed that implicitly shifts the second order systems. The numerical approximation we apply to the shifted system is based on a multilevel block incomplete factorization that uses a pivoting strategy to deal with small pivots. Furthermore, our block factorization approach is designed to deal with large blocks in order to preserve the natural block structure which is obtained from the DG discretization. Numerical experiments confirm that this approach is able to efficiently solve the time-harmonic Maxwell equations.

2 The Continuous Problem

We consider solving the normalized time-harmonic Maxwell equations in the first order form:

$$\mathrm{i}\omega\varepsilon_r \mathbf{E} - \operatorname{curl}\mathbf{H} = -\mathbf{J}_{\mathrm{E}}\ ,\quad \mathrm{i}\omega\mu_r\mathbf{H} + \operatorname{curl}\mathbf{E} = 0, \tag{1}$$

where $\mathbf{E}$ and $\mathbf{H}$ are the unknown electric and magnetic fields and $\mathbf{J}_{\mathrm{E}}$ is a known current source; ε_r and μ_r respectively denote the relative electric permittivity and the relative magnetic permeability and we assume here the case of a linear isotropic non-magnetic (i.e. $\mu_r = 1$) media. The relative electric permittivity is linked to its absolute value through $\varepsilon = \varepsilon_r\varepsilon_0$ where ε_0 is the permittivity of the vacuum. The angular frequency of the problem is given by ω. In the normalization of the equations, the electric field is unchanged, the magnetic field is given by $\mathbf{H} = z_0\overline{\mathbf{H}}$ where $z_0 = \sqrt{\mu_0/\varepsilon_0}$. With this choice, the electric and magnetic fields have the same unit i.e. V/m. Besides, $\omega = \overline{\omega}/c_0$ where $c_0 = 1/\sqrt{\mu_0\varepsilon_0}$. Equations (1) are solved in a bounded domain Ω. On the boundary $\partial\Omega = \Gamma_a \cup \Gamma_m$, the following boundary conditions are imposed:

- a perfect electric conductor (PEC) condition on $\Gamma_m : \mathbf{n}\times\mathbf{E} = 0$,

(2)

- a Silver-Müller absorbing condition on $\Gamma_a : \mathscr{L}(\mathbf{E},\mathbf{H}) = \mathscr{L}(\mathbf{E}^{\mathrm{inc}},\mathbf{H}^{\mathrm{inc}})$,

where $\mathscr{L}(\mathbf{E},\mathbf{H}) = \mathbf{n}\times\mathbf{E} - Z\mathbf{n}\times(\mathbf{H}\times\mathbf{n})$ with $Z = \sqrt{\mu_r/\varepsilon_r}$. The vectors $\mathbf{E}^{\mathrm{inc}}$ and $\mathbf{H}^{\mathrm{inc}}$ represent the components of an incident electromagnetic wave and $\mathbf{n}$ denotes the unit outward normal. Equations (1) and (2) can be further rewritten in the form:

$$\begin{cases} \mathrm{i}\omega G_0\mathbf{W} + G_x\partial_x\mathbf{W} + G_y\partial_y\mathbf{W} + G_z\partial_z\mathbf{W} = -\mathbf{J} \text{ in } \Omega, \\ (M_{\Gamma_m} - G_{\mathbf{n}})\mathbf{W} = 0 \text{ on } \Gamma_m, \\ (M_{\Gamma_a} - G_{\mathbf{n}})(\mathbf{W} - \mathbf{W}^{\mathrm{inc}}) = 0 \text{ on } \Gamma_a, \end{cases} \tag{3}$$

where $\mathbf{W} = (\mathbf{E},\mathbf{H})^T$ is the new unknown vector, $\mathbf{J} = (\mathbf{J}_{\mathrm{E}}, 0)^T$ and:

$$G_0 = \begin{pmatrix} \varepsilon_r \mathrm{I}_3 & 0_3 \\ 0_3 & \mu_r \mathrm{I}_3 \end{pmatrix},\quad G_l = \begin{pmatrix} 0_3 & N_{\mathbf{e}^l} \\ N_{\mathbf{e}^l}^T & 0_3 \end{pmatrix},\quad N_{\mathbf{v}} = \begin{pmatrix} 0 & v_z & -v_y \\ -v_z & 0 & v_x \\ v_y & -v_x & 0 \end{pmatrix},$$

with $l \in \{x,y,z\}$ while $(\mathbf{e^x},\mathbf{e^y},\mathbf{e^z})$ is the canonical basis of $\mathbb{R}^3$, and $\mathbf{v} = (v_x,v_y,v_z)^T$. I_3 is the identity matrix, and 0_3 the null matrix, both of dimension

3×3. The real part of G_0 is symmetric positive definite and its imaginary part, which appears in the case of conductive materials, is symmetric negative. In the following we denote by $G_{\mathbf{n}}$ the sum $G_x n_x + G_y n_y + G_z n_z$ and by $G_{\mathbf{n}}^+$ and $G_{\mathbf{n}}^-$ its positive and negative parts.[1] We also define $|G_{\mathbf{n}}| = G_{\mathbf{n}}^+ - G_{\mathbf{n}}^-$. In order to take into account the boundary conditions, the matrices M_{Γ_m} and M_{Γ_a} are given by $M_{\Gamma_m} = \begin{pmatrix} 0_3 & N_{\mathbf{n}} \\ -N_{\mathbf{n}}^T & 0_3 \end{pmatrix}$ and $M_{\Gamma_a} = |G_{\mathbf{n}}|$.

3 Discretization by a Discontinuous Galerkin Method

Let Ω_h denote a discretization of the domain Ω into a union of conforming simplicial elements K. We look for the approximate solution $\mathbf{W}_h$ of (3) in $V_h \times V_h$ where the functional space V_h is defined by $V_h = \{\mathbf{U} \in [L^2(\Omega)]^3 \;/\; \forall K \in \Omega_h,\ \ \mathbf{U}_{|K} \in \mathbb{P}_p(K)\}$, where $\mathbb{P}_p(K)$ denotes a space of vectors with polynomial components of degree at most p over the element K. The DG discretization of system (3) yields the formulation of the discrete problem which aims at finding $\mathbf{W}_h$ in $V_h \times V_h$ such that:

$$
\left\{
\begin{aligned}
&\int_{\Omega_h} (\mathrm{i}\omega G_0 \mathbf{W}_h)^T \overline{\mathbf{V}} dv + \sum_{K \in \Omega_h} \int_K \left(\sum_{l \in \{x,y,z\}} G_l \partial_l(\mathbf{W}_h) \right)^T \overline{\mathbf{V}} dv \\
&+ \sum_{F \in \Gamma^m \cup \Gamma^a} \int_F \left(\frac{1}{2} (M_{F,K} - I_{FK} G_{\mathbf{n}_F}) \mathbf{W}_h \right)^T \overline{\mathbf{V}} ds \\
&- \sum_{F \in \Gamma^0} \int_F (G_{\mathbf{n}_F} [\![\mathbf{W}_h]\!])^T \{\overline{\mathbf{V}}\} ds + \sum_{F \in \Gamma^0} \int_F (S_F [\![\mathbf{W}_h]\!])^T [\![\overline{\mathbf{V}}]\!] ds \\
&= \sum_{F \in \Gamma^a} \int_F \left(\frac{1}{2} (M_{F,K} - I_{FK} G_{\mathbf{n}_F}) \mathbf{W}^{\mathrm{inc}} \right)^T \overline{\mathbf{V}} ds, \quad \forall \mathbf{V} \in V_h \times V_h,
\end{aligned}
\right.
\tag{4}
$$

where Γ^0, Γ^a and Γ^m respectively denote the set of interior (triangular) faces, the set of faces on Γ_a and the set of faces on Γ_m. The unitary normal associated with the oriented face F is $\mathbf{n}_F$ and I_{FK} stands for the incidence matrix between oriented faces and elements whose entries are equal to 0 if the face F does not belong to element K, 1 if $F \in K$ and their orientations match, and -1 if $F \in K$ and their orientations do not match. For $F = \partial K \cap \partial \tilde{K}$, we also define $[\![\mathbf{V}]\!] = I_{FK} \mathbf{V}_{|K} +$

[1]If $T\Lambda T^{-1}$ is the eigendecomposition of $G_{\mathbf{n}}$, then $G_{\mathbf{n}}^{\pm} = T\Lambda^{\pm}T^{-1}$ where Λ^+ (respectively Λ^-) only gathers the positive (respectively negative) eigenvalues.

$I_{F\tilde{K}}\mathbf{V}_{|\tilde{K}}$ and $\{\mathbf{V}\} = \frac{1}{2}\left(\mathbf{V}_{|K} + \mathbf{V}_{|\tilde{K}}\right)$. Finally, the matrix S_F, which is hermitian positive semi-definite, permits to penalize the jump of a field or of some components of this field on the face F, and the matrix $M_{F,K}$ insures the asymptotic consistency with the boundary conditions of the continuous problem. Problem (4) is often interpreted in terms of local problems in each element K of Ω_h coupled by the introduction of an element boundary term called numerical flux (see also [11]). We refer to [7] for all the details on the various terms involved in this DG formulation. Within each mesh element K the electromagnetic field $(\mathbf{E},\mathbf{H})^T$ is approximated as:

$$(\mathbf{E}_h)_{|K} = \sum_{i=1}^{d_K} \mathbf{E}_i^K \varphi_i^K \quad \text{and} \quad (\mathbf{H}_h)_{|K} = \sum_{i=1}^{d_K} \mathbf{H}_i^K \varphi_i^K \tag{5}$$

where $\mathbf{E}_i^K$ and $\mathbf{H}_i^K$ are the vectors of local degrees of freedom corresponding to the basis expansion $\{\varphi_i^K\}_{i=1,\cdots,d_K}$ of $\mathbb{P}_p(K)$. In the present study, we adopt the classical Lagrange nodal basis functions defined on a simplex and we assume that the interpolation degree is uniform (i.e. the same for all the elements of the mesh). Then the resulting method is denoted as DG-$\mathbb{P}_p$.

4 Block Preconditioning

The DG discretization of the system of time-harmonic Maxwell equations (3) leads to a large sparse complex linear system of equations of the form $\mathscr{A}\mathbf{W}_h \equiv (\mathrm{i}\omega\mathscr{M} + \mathscr{C})\,\mathbf{W}_h = b$, where $\omega\mathscr{M}$ refers to the discretization of the term:

$$\int_{\Omega_h} (\omega G_0 \mathbf{W}_h)^T\, \overline{\mathbf{V}} dv$$

in (4), while $\mathscr{C}$ represents the discretization of the curl operators and the boundary conditions for the remaining integrals on the left hand side of (4). For the numerical treatment we assume that the sign of the first equation of the time-hamonic Maxwell equations is flipped to $-\mathrm{i}\omega\varepsilon_r\mathbf{E} + \operatorname{curl}\mathbf{H} = +\beta_{\mathbf{E}}\omega\varepsilon_r\mathbf{E} + \mathbf{J}$ and consistently changed in $G_0, G_\beta, G_x, G_y, G_z$. Then the matrices $\mathscr{M}$ and $\mathscr{C}$ become symmetric, thus $\mathscr{A}$ is complex symmetric. The matrix of this system exhibits a block structure which is linked to the polynomial approximation of the electromagnetic field within a mesh element (5). Up to a permutation which is induced by first taking the contributions with respect to $\mathbf{E}$ and then the $\mathbf{H}$ part we find that:

$$\mathscr{M} = \begin{pmatrix} -M_{\epsilon_r} & 0 \\ 0 & M_{\mu_r} \end{pmatrix}, \quad \mathscr{C} = \begin{pmatrix} -C_{\mathbf{EE}} & C_{\mathbf{HE}}^T \\ C_{\mathbf{HE}} & C_{\mathbf{HH}} \end{pmatrix},$$

where M_{ϵ_r} and M_{μ_r} are real symmetric positive definite block diagonal matrices whose block elements are the local mass matrices computed in each element K.

Computing a preconditioner based on an incomplete factorization of $\mathscr{A}$ happens to be prohibitively expensive. Therefore we shift the initial system by:

$$\omega \begin{pmatrix} -\beta_{\mathrm{E}} M_{\epsilon_r} & 0 \\ 0 & \beta_{\mathrm{H}} M_{\mu_r} \end{pmatrix},$$

where $\beta_{\mathrm{E}}, \beta_{\mathrm{H}}$ are chosen appropriately. This precisely refers to adding artificially $-\beta_{\mathrm{E}}\omega\varepsilon_r\mathbf{E}$ and $-\beta_{\mathrm{H}}\omega\mu_r\mathbf{H}$ to the right-hand side of (1). With respect to $\mathbf{E}$ this can be interpreted as artificial conductivity. We propose three different variants of block preconditioning. The first version consists of choosing $\beta_{\mathrm{E}} = \beta_{\mathrm{H}} = \beta$ and applying our preconditioner to the shifted system:

$$\mathscr{P}_1 = \beta\omega\mathscr{M} + \mathscr{A}.$$

The second and third variant are best understood as a discrete analogy of eliminating the magnetic field $\mathbf{H}$ from the second equation of the perturbed form of (1) and inserting it into the first equation of (1). The resulting equation thus reduces to:

$$\frac{1}{\omega(\mathrm{i}+\beta_{\mathrm{H}})}\left(-(1-\beta_{\mathrm{E}}\mathrm{i})(1-\beta_{\mathrm{H}}\mathrm{i})\omega^2\epsilon_r\mathbf{E} + \mathrm{curl}(\frac{1}{\mu_r}\,\mathrm{curl}\,\mathbf{E})\right) = -J.$$

This is essentially a vector-valued Helmholtz equation, where the operator is shifted by a multiple of the mass matrix. The discrete analogy can be described by eliminating the $\mathbf{H}$ part from $\beta\omega\mathscr{M} + \mathscr{A}$ by one block elimination step:

$$\begin{pmatrix} -\omega(\mathrm{i}+\beta_{\mathrm{E}})M_{\epsilon_r} - C_{\mathrm{EE}} & C_{\mathrm{HE}}^T \\ C_{\mathrm{HE}} & \omega(\mathrm{i}+\beta_{\mathrm{H}})M_{\mu_r} + C_{\mathrm{HH}} \end{pmatrix} \rightarrow$$

$$\mathscr{S} = -\omega(\mathrm{i}+\beta_{\mathrm{E}})M_{\epsilon_r} - C_{\mathrm{EE}} - C_{\mathrm{HE}}^T\left(\omega(\mathrm{i}+\beta_{\mathrm{H}})M_{\mu_r} + C_{\mathrm{HH}}\right)^{-1} C_{\mathrm{HE}}.$$

For the second variant block preconditioning we use $\beta = \beta_{\mathrm{E}} = \beta_{\mathrm{H}}$ to obtain the reduced system $\mathscr{P}_2$. This can be read as first shifting and then eliminating. Finally for the third variant we proceed analogously to the second one except that we first eliminate $\mathbf{H}$ from the unshifted system $\mathscr{A}$ and then shift the reduced system by $-\beta\omega M_{\epsilon_r}$, i.e., we choose $\beta = \beta_{\mathrm{E}}$ and $\beta_{\mathrm{H}} = 0$ in order to obtain the reduced system $\mathscr{P}_3$. According to the work by Magolu [13], Erlangga et al [10], shifting the operator with a real-valued β significantly improves incomplete LU preconditioning and multilevel preconditioning. For preconditioning we apply the inverse-based multilevel block ILU [3], as implemented in ILUPACK.[2] Its hallmark is the strategy of keeping the inverse triangular factors below a given bound κ. In order to deal with indefinite systems, a block factorization approach is used based on a symmetrized maximum weight matching (see [2] for details).

[2]http://ilupack.tu-bs.de.

Table 1 Direct solver PARDISO applied to A

Computation time	$\frac{nz(LU)}{nz(A)}$
5.2×10^3	90.4

Table 2 Multilevel block ILU applied to $\mathscr{P}_1 = A + \beta\omega M$

β	ILU[sec]	$\frac{nz(ILU)}{nz(A)}$	Levels	SQMR[sec]	Steps
1.5	8.8×10^2	11.7	5	3.2×10^3	620
3.0	1.7×10^2	5.4	4	1.7×10^3	387
5.0	1.0×10^2	6.2	2	2.4×10^3	574
10.0	4.6×10^1	3.3	1	1.9×10^3	1,035

Table 3 Multilevel block ILU for the reduced system $\mathscr{P}_2$ of $A + \beta\omega M$ after eliminating the **E** part first

β	ILU[sec]	$\frac{nz(ILU)}{nz(A)}$	Levels	SQMR[sec]	Steps
1.5	3.6×10^2	9.9	6	1.7×10^3	398
3.0	1.4×10^2	5.2	2	9.8×10^2	302
5.0	8.5×10^1	4.3	2	1.9×10^3	613
10.0	3.9×10^1	1.9	1	1.0×10^3	842

Table 4 Multilevel block ILU for the reduced system $\mathscr{P}_3$ of A after eliminating the **E** part first and then shifting by $\beta\omega M_{\mu_r}$

β	ILU[sec]	$\frac{nz(ILU)}{nz(A)}$	Levels	SQMR[sec]	Steps
1.5	4.9×10^2	15.2	8	9.7×10^3	1,773
3.0	3.5×10^2	9.2	5	1.9×10^3	452
5.0	2.6×10^2	6.7	4	1.3×10^3	337
10.0	1.6×10^2	6.0	3	1.2×10^3	325

Table 5 Multilevel block ILU applied to $\mathscr{P}_1 = A + \beta\omega M$ with $\omega = 9.41$, $\omega = 37.64$

	$\omega = 9.41$					$\omega = 37.64$				
β	ILU[sec]	$\frac{nz(ILU)}{nz(A)}$	Lev.	SQMR[sec]	Steps	ILU[sec]	$\frac{nz(ILU)}{nz(A)}$	Lev.	SQMR[sec]	Steps
0.75			–			8.6×10^2	11.2	5	4.0×10^3	813
1.5			–			1.4×10^2	5.4	4	2.4×10^3	607
3.0	9.1×10^2	11.9	5	3.1×10^3	613	7.2×10^1	4.8	2	4.3×10^3	1,174
5.0	6.5×10^2	6.5	4	1.6×10^3	383	4.7×10^1	3.2	1	2.8×10^3	1,762
10.0	1.0×10^2	6.3	2	2.1×10^3	489			–		
20.0	4.6×10^1	3.2	2	2.5×10^3	893			–		

5 Numerical Results

We now present the impact of shifting the initial system by a multiple of the mass matrix for a 3D problem discretized by a DG-$\mathbb{P}_1$ method. The problem under consideration is the scattering of a plane wave by a perfectly conducting unit sphere. The frequency of the incident plane wave of frequency f = 900 MHz and thus, we have $\omega = 18.84$ (after renormalization of the Maxwell equations). The computational domain is defined as the free space between the perfectly conducting sphere and an outer sphere on which the Silver-Müller absorbing condition is applied. We have used an unstructured tetrahedral mesh consisting of 46,704 tetrahedral elements. This yields a complex symmetric system of size $n = 1{,}120{,}896$. The computations were performed on a workstation equipped with an Intel Xeon E7440 CPU with frequency 2.4 GHz and 64 GB of memory. For the ILU we use a drop tolerance of 10^{-2} but limit the maximum amount of fill per row by $10\times$ the number of nonzeros per row in $\mathscr{A}$. We use an inverse bound of $\kappa = 5$ for inverse-based pivoting. As iterative solver we use the simplified QMR method which allows for the use of (complex) symmetric systems and preconditioners. The iteration is stopped, whenever the backward error satisfies $\|Ax - b\| \leq 10^{-6}(\|A\|\ \|x\| + \|b\|)$. As comparison we also add numerical results of the direct solver PARDISO[3] (see Table 1). The numerical results in Tables 2–4 confirm the efficiency of our shifted multilevel block ILU approach. They illustrate that shifting the initial system is essential for the ILU. If the shift is too small then the fill would increase drastically if there were no limit imposed. On the other hand, shifting the system too much turns the preconditioned system away from the original system. A similar observation is made in Table 5 when we halve (resp. double) ω but reverse the shifts.

References

1. Amestoy, P., Duff, I., L'Excellent, J.-Y.: Multifrontal parallel distributed symmetric and unsymmetric solvers. Comput. Meth. App. Mech. Engng. **184**, 501–520 (2000)
2. Bollhöfer, M., Grote, M., Schenk, O.: Algebraic multilevel preconditioner for the Helmholtz equation in heterogeneous media. SIAM J. Sci. Comput. **31**(5), 3781–3805 (2009)
3. Bollhöfer, M., Saad, Y.: Multilevel preconditioners constructed from inverse–based ILUs. SIAM J. Sci. Comput. **27**(5), 1627–1650 (2006)
4. Cockburn, B., Karniadakis, G., Shu, C. (eds.): Discontinuous Galerkin methods. Theory, computation and applications. In: Lecture Notes in Computational Science and Engineering, vol. 11. Springer, Berlin (2000)
5. Cockburn, B., Shu, C. (eds.): Special issue on discontinuous Galerkin methods. of J. Sci. Comput. **22-23** (2005)
6. Dawson, C. (ed.): Special issue on discontinuous Galerkin methods. Comput. Meth. App. Mech. Engng. **195** (2006)
7. Dolean, V., Fol, H., Lanteri, S., Perrussel, R.: Solution of the time-harmonic Maxwell equations using discontinuous Galerkin methods. J. Comp. Appl. Math. **218**(2), 435–445 (2008)
8. Dolean, V., Lanteri, S., Perrussel, R.: A domain decomposition method for solving the three-dimensional time-harmonic Maxwell equations discretized by discontinuous Galerkin methods. J. Comput. Phys. **227**(3), 2044–2072 (2007)
9. Erlangga, Y., Oosterlee, C., Vuik, C.: A novel multigrid based preconditioner for heterogeneous Helmholtz problems. SIAM J. Sci. Comput. **27**, 1471–1492 (2006)
10. Erlangga, Y., Vuik, C., Oosterlee, C.: Comparison of multigrid and incomplete LU shifted-Laplace preconditioners for the inhomogeneous Helmholtz equation. Appl. Numer. Math. **56**, 648–666 (2006)

[3]http://www.pardiso-project.org.

11. Ern, A., Guermond, J.-L.: Discontinuous Galerkin methods for Friedrichs systems I. General theory. SIAM J. Numer. Anal. **44**(2), 753–778 (2006)
12. van Gijzen, M., Erlangga, Y., Vuik, C.: Spectral analysis of the discrete Helmholtz operator preconditioned with a shifted Laplacian. SIAM J. Sci. Comput. **29**, 1942–1958 (2007)
13. Magolu monga Made, M.: Incomplete factorization based preconditionings for solving the Helmholtz equation. Int J. Numer. Meth. Eng. **50**, 1077–1101 (2001)
14. Monk, P.: Finite element methods for Maxwell's equations. Numerical Mathematics and Scientific Computation. Oxford University Press, New York (2003)

From Sizing over Design Centering and Pareto Optimization to Tolerance Pareto Optimization of Electronic Circuits

Helmut Gräb

Abstract This paper presents an overview of sizing tasks in electronic circuit design and their corresponding formulations as optimization problems. We will start with the general multi-objective sizing problem. Then, the inclusion of statistically distributed parameters and of range-valued parameters into the scalar problems of yield optimization and design centering will be described. Finally, a problem formulation for considering these parameter tolerances by multi-objective Pareto optimization will be presented.

1 Parameters, Performances, Simulation

This paper deals with optimization of electronic circuits which are modeled with continuous signals in time and value, and which are usually nonlinear. Circuits of this type are usually described with nonlinear differential algebraic equations and often called analog circuits. Analog circuits are analyzed based on numerical integration with one of the many successors of the SPICE simulator [4]. Modern simulators are capable of handling mixed-signal circuits with digital parts, and of handling circuits which are described not only with transistor netlists, but with hardware description languages like VHDL-AMS or Verilog-AMS. It is worth noticing that not only analog and mixed-signal circuits and systems, but digital components as well may be described in this way. Hence, simulation-based design not only refers to analog design but to a general analog design view on any type of system. It is also worth noticing that numerical simulation provides the most significant way to abstract the analog design view from the physical level to the

H. Gräb (✉)
Institute for Electronic Design Automation, Technische Universität München, Arcisstr. 21, 80333 Munich, Germany
e-mail: graeb@tum.de

B. Michielsen and J.-R. Poirier (eds.), *Scientific Computing in Electrical Engineering SCEE 2010*, Mathematics in Industry 16, DOI 10.1007/978-3-642-22453-9_4,

formal level of a performance function that maps the modeled circuit parameters $\mathbf{x} \in \mathbb{R}^{n_x}$ (simulator input) on the modeled circuit performance features $\mathbf{f} \in \mathbb{R}^{n_f}$ (simulator output, e.g., Gain bandwidth, delay):

$$\mathbf{x} \mapsto \mathbf{f} \tag{1}$$

Simulation of electronic circuits may take CPU times from seconds to minutes or hours. The cost for simulation is therefore by far dominating all other computational steps of an optimization process. This requires specific customized optimization approaches for electronic design.

We distinguish the following three types of parameters:

- Design parameters (e.g., transistor widths) $\mathbf{x}_d \in \mathbb{R}^{n_{xd}}$
- Statistical parameters (e.g., threshold voltage, oxide thickness) $\mathbf{x}_s \in \mathbb{R}^{n_{xs}}$
- Range parameters (e.g., supply voltage, temperature) $\mathbf{x}_r \in \mathbb{R}^{n_{xr}}$

2 Parameter Tolerances, Performance Specifications

Statistical parameters reflect the manufacturing variations which are transformed into a Gaussian distribution.

Range parameters reflect the circuit operating conditions. They are interval-bounded by upper bounds:

$$x_{r,i} \leq x_{r,U,i}\,,\; i = 1,\ldots,2n_{xr} \tag{2}$$

Lower bounds are transformed into upper bounds, $x \geq x_L \rightarrow -x \leq -x_L$, no bound refers to $x_{r,U} \rightarrow \infty$.

The explicit performance specifications are given as bounds in the same way:

$$f_i \leq f_{U,i}\,,\; i = 1,\ldots,2n_f \tag{3}$$

On the other hand, there are implicit specifications, which refer to conditions on the transistor channel geometries and transistor operating voltages. These implicit specifications define the constraint region of design parameters X:

$$X = \{\mathbf{x}_d \,|\mathbf{c}(\mathbf{x}_d) \geq \mathbf{0}\} \tag{4}$$

They can be computed for each circuit, e.g., by [2].

3 Sizing Tasks

Figure 1 shows the basic sizing tasks. While nominal design aims at finding design parameter values for optimum performance without considering parameter tolerances, tolerance design does include the tolerance ranges of operational parameters and the distribution of statistical parameters [1].

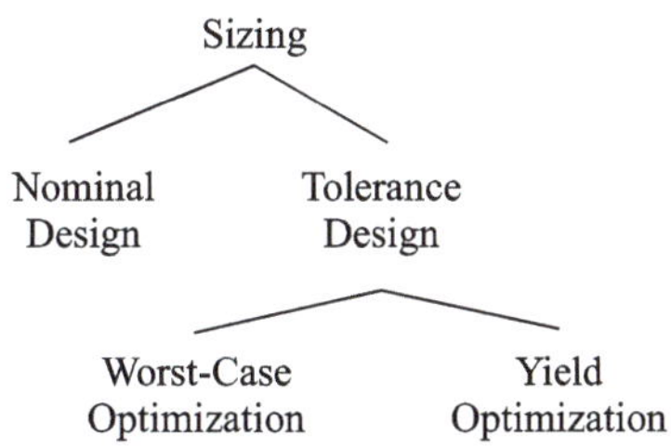

Fig. 1 Sizing tasks

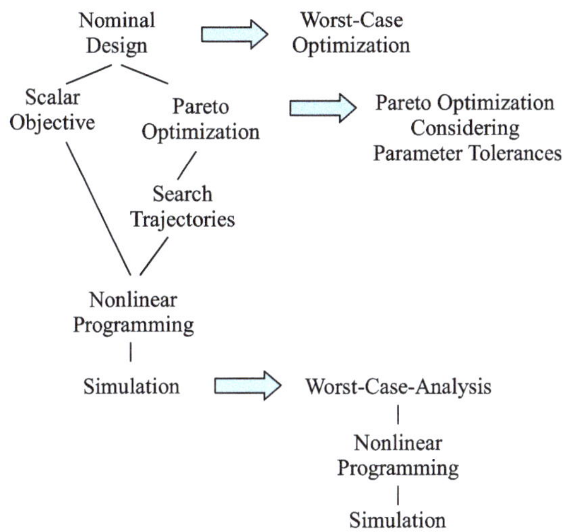

Fig. 2 Nominal design and worst-case optimization

4 Nominal Design

Circuit design inherently is a multiobjective optimization problem:

$$\min_{\mathbf{x}_d \in X} \begin{bmatrix} \vdots \\ f_i(\mathbf{x}_d) \\ \vdots \end{bmatrix} \rightarrow \mathbf{x}_d^*\,,\; f_i^* = f_i(\mathbf{x}_d^*)\,,\; i = 1, \ldots, n_f \tag{5}$$

An optimal design will always represent a certain trade-off between the competing design objectives. Nominal design therefore is either approached by a scalar objective function or by Pareto optimization, as shown in Fig. 2. If Pareto optimization is solved using deterministic methods, then the basic task consists in defining a set of search trajectories for scalar optimization problems, which in turn are solved with nonlinear programming methods like Sequential Quadratic Programming. At the bottom of this task chain, simulation is called frequently, which makes optimization cost between minutes, hours or even days.

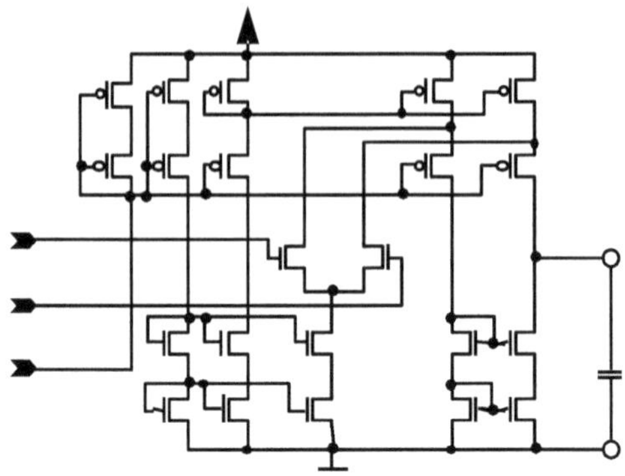

Fig. 3 Example circuit: operational amplifier

Table 1 Nominal design of operational amplifier

Performance	Specification	Initial	Design 1	Design 2
Gain	$\geq$80 dB	67 dB	100 dB	100 dB
Transit frequ.	$\geq$10 MHz	5 MHz	20 MHz	18 MHz
Phase margin	$\geq 60^\circ$	75°	68°	72°
Slew rate	$\geq$ 10 V/μs	4.1 V/μs	12 V/μs	12 V/μs
DC power	$\leq$50 μW	122 μW	38 μW	39 μW

4.1 Example

For a simple operational amplifier depicted in Fig. 3, Table 1 shows typical results of a nominal design, in this case obtained with a commercial tool [3]. The circuit has 14 design parameters and five performances given in the first column of Table 1. The CPU time for one simulation is in the range of seconds, the CPU time for the optimization is in the range of minutes.

Column two gives the considered performance features and specifications, column three typical initial values of nominal design. The last two columns give the results of two different optimization runs with different weights among the performance features. We can see that in both cases the specs are fulfilled. While design 1 has a larger safety margin with respect to the transit frequency, design 2 has a larger safety margin with respect to the phase margin. The final decision on the design depends on the application and other aspects like manufacturing variability.

References

1. Graeb, H.: Analog Design Centering and Sizing. Springer, Berlin (2007)
2. Massier, T., Graeb, H., Schlichtmann, U.: The Sizing Rules Method for CMOS and Bipolar Analog Integrated Circuit Synthesis. IEEE Trans. Computer-Aided Des. Integr. Circ. Sys. **27**(12), 2209–2222 (2008)
3. MunEDA: WiCkeD (2009). Www.muneda.com
4. Nagel, L.: SPICE2: A computer program to simulate semiconductor circuits. Ph.D. Dissertation, University of California, Berkeley (1975)

Importance Sampling for Determining SRAM Yield and Optimization with Statistical Constraint

E.J.W. ter Maten, O. Wittich, A. Di Bucchianico, T.S. Doorn, and T.G.J. Beelen

Abstract Importance Sampling allows for efficient Monte Carlo sampling that also properly covers tails of distributions. From Large Deviation Theory we derive an optimal upper bound for the number of samples to efficiently sample for an accurate fail probability $P_{\text{fail}} \leq 10^{-10}$. We apply this to accurately and efficiently minimize the access time of Static Random Access Memory (SRAM), while guaranteeing a statistical constraint on the yield target.

1 Introduction

As transistor dimensions of Static Random Access Memory (SRAM) become smaller with each new technology generation, they become increasingly susceptible to statistical variations in their parameters. These statistical variations may result

E.J.W. ter Maten (✉)
Eindhoven University of Technology, Department of Mathematics and Computer Science, CASA/LIME, P.O. Box 513, 5600 MB Eindhoven, The Netherlands
NXP Semiconductors, High Tech Campus 32 and 46, 5656 AE Eindhoven, The Netherlands
Bergische Universität Wuppertal, Fachbereich C, Wicküler Park Rm 503, Bendahler Str. 29, D-42285, Wuppertal, Germany
e-mail: E.J.W.ter.Maten@tue.nl; Jan.ter.Maten@nxp.com; Jan.ter.Maten@math.uni-wuppertal.de

A. Di Bucchianico
Eindhoven University of Technology, Department of Mathematics and Computer Science, CASA/LIME, P.O. Box 513, 5600 MB Eindhoven, The Netherlands
e-mail: A.D.Bucchianico@tue.nl

O. Wittich
RWTH Aachen, Lehrstuhl A für Mathematik, Analysis und Zahlentheorie, Schinkelstr. 4, D-52056 Aachen, Germany
e-mail: Olaf.Wittich@mathA.rwth-aachen.de

T.S. Doorn · T.G.J. Beelen
NXP Semiconductors, High Tech Campus 32 and 46, 5656 AE Eindhoven, The Netherlands
e-mail: Toby.Doorn@nxp.com; Theo.G.J.Beelen@nxp.com

B. Michielsen and J.-R. Poirier (eds.), *Scientific Computing in Electrical Engineering SCEE 2010*, Mathematics in Industry 16, DOI 10.1007/978-3-642-22453-9_5,

in failing memory. An SRAM is used as a building block for the construction of large Integrated Circuits (ICs). To ensure that a digital bit cell in SRAM does not degrade the yield (fraction of functional devices) of ICs with Megabits of memory, very small failure probabilities $P_{\text{fail}} \leq 10^{-10}$ are necessary. To simulate this, regular Monte-Carlo (MC) simulations require too much computing time. Importance Sampling (IS) [1] is a more advanced technique that provides sufficiently accurate results and is relatively easy to implement. A speed up of several orders can be achieved when compared to regular Monte Carlo methods.

2 Regular Monte Carlo

Let Y be a real-valued random variable with probability density function f. We assume that N independent random observations Y_i $(i = 1, \ldots, N)$ of Y are taken. We define $X_i = I_A(Y_i)$ for a given set $A = (-\infty, x)$ where $I_A(Y_i) = 1$ if $Y_i \in A$ and 0 otherwise. Then $p_f^{\text{MC}}(A) = \frac{1}{N}\sum_{i=1}^{N} X_i$ estimates $p = \int_{-\infty}^{x} f(z)dz = P(Y \in A)$. The X_i are Bernoulli distributed, hence $Np_f^{\text{MC}} \sim \text{Bin}(N, p)$, $\text{E}(p_f^{\text{MC}}) = \frac{1}{N}Np = p$, and $\sigma^2(p_f^{\text{MC}}) = \frac{p(1-p)}{N}$. Let $\Phi(x) = \int_{-\infty}^{x} \text{e}^{-z^2/2}dz$ and define z_α by $\Phi(-z_\alpha) = \alpha$. From the Central Limit Theorem (CLT) we derive

$$P(|p_f^{\text{MC}} - p| > \varepsilon) = P\left(\frac{|p_f^{\text{MC}} - p|}{\sigma(p_f^{\text{MC}})} > z\right) \overset{N_{\text{MC}}\to\infty}{\longrightarrow} 2\Phi(-z) \leq 2\Phi(-z_{\alpha/2}) = \alpha,$$

where $z = \varepsilon/\sqrt{p(1-p)/N_{\text{MC}}}$ and $N_{\text{MC}} = N$. Hence, if $z \geq z_{\alpha/2}$ we deduce

$$N_{\text{MC}} \geq p(1-p)\left(\frac{z_{\alpha/2}}{\varepsilon}\right)^2 = \frac{1-p}{p}\left(\frac{z_{\alpha/2}}{\nu}\right)^2, \tag{1}$$

for $\varepsilon = \nu p$. We take $\nu = 0.1$ and $p = 10^{-10}$. Now let $\alpha = 0.02$, then $z_{\alpha/2} \approx 2$. Then $N_{\text{MC}} \geq 4\ 10^{12}$. If we do not know p, we can use $p(1-p) \geq 1/4$ yielding $N_{\text{MC}} \geq \frac{1}{4}\left(\frac{z_{\alpha/2}}{\varepsilon}\right)^2 = 10^{22}$. And if N_{MC} is not large enough to apply the CLT, Chebyshev's inequality even results to $N_{\text{MC}} \geq 10^{24}$. These general bounds are much too pessimistic. Large Deviations Theory (LDT) [1,4] results in a sharp upper bound [6]

$$P(|p_f^{\text{MC}} - p| > \nu p) \leq \exp\left(-\frac{N_{\text{MC}}}{2}\frac{p}{1-p}\nu^2\right). \tag{2}$$

For $\nu = 0.1$, $p = 10^{-10}$ and $\alpha = 0.02$, as above, we find: $N_{\text{MC}} \geq 8\ 10^{12}$ (which is a sharp result – see at the end of the next proof). Note that an extra k-th decimal in ν increases N_{MC} with a factor k^2.

Proof of (2) [6]. The sequence of the Monte Carlo results $P_N(A) := p_f^{\mathrm{MC}}$ satisfies a *Large-Deviation Principle* [1, 4, 5], meaning that there is some *'rate function'* $I : \mathbb{R} \to \mathbb{R} \cup \{-\infty, +\infty\}$ such that

(i) $\limsup_{N\to\infty} \frac{1}{N} \ln P_N(C) \le -\inf_{x\in C} I(x)$ for all closed subsets $C \subset \mathbb{R}$,
(ii) $\liminf_{N\to\infty} \frac{1}{N} \ln P_N(G) \ge -\inf_{x\in G} I(x)$ for all open subsets $G \subset \mathbb{R}$.

Let X be a Bernoulli variable with success probability p. The *logarithmic moment generating function* for X is given by $\ln\left(\mathrm{E}\left[e^{\lambda X}\right]\right) = \ln\left(q + e^{\lambda} p\right)$, where as usual $q = 1 - p$. We define the following function [5]

$$J(x,\lambda) = \lambda x - \ln\left(\mathrm{E}\left[e^{\lambda X}\right]\right) = \lambda x - \ln(q + e^{\lambda} p), \tag{3}$$

where $x, \lambda \in \mathbb{R}$. We note that an optimum value λ^* must satisfy

$$\frac{\partial J}{\partial \lambda} = x - \frac{pe^{\lambda^*}}{q + pe^{\lambda^*}} \quad = \quad 0, \quad \text{hence}$$

$$\lambda^* = \ln\left(\frac{qx}{p(1-x)}\right), \quad \text{and} \quad pe^{\lambda^*} = \frac{qx}{1-x}, \quad \text{and} \quad q + pe^{\lambda^*} = \frac{q}{1-x}. \tag{4}$$

In our case, the rate function can be shown to be equal to

$$I(x) = \sup_{\lambda\in\mathbb{R}} J(x,\lambda) = J(x,\lambda^*) = x \ln\left(\frac{qx}{p(1-x)}\right) - \ln\left(\frac{q}{1-x}\right), \tag{5}$$

a function which is continuous on the interval $(0, 1)$. With $C = [p - \nu p, p + \nu p] \subset (0, 1)$ and $G = \mathbb{R} \setminus C$, the Large-Deviation Principle above implies

$$\lim_{N\to\infty} \frac{1}{N} \ln P\left(\left|\frac{1}{N}\sum_{k=1}^{N} X_k - p\right| \ge \nu p\right) = -\inf_{|x-p|\ge\nu p} I(x).$$

From (5) we can calculate $I'(x)$ and $I''(x)$ explicitly. For $x \in (0, 1)$ we have $I''(x) > 0$, which implies that I' is increasing and that I is *convex*. Also $I(0^+) = -\ln(q) > 0$ and $I(1^-) = \ln(q/p) \in \mathbb{R}$. Clearly I can be extended continuously at both $x = 0$ and $x = 1$. Furthermore $I(p) = 0$ and $I'(p) = 0$. Hence $I(p) = 0$ is a global minimum. This implies that actually the infimum of I on $\{x : |x - p| > \nu p\}$ is assumed at $x = p \pm \nu p$. This can be analyzed further using Taylor expansion [6]. Thus from part (i) of the Large Deviation Principle, we obtain (2) for all N with a possible exception of finitely many. Part (ii) implies that the exponential bound in (2) is also valid from below and thus is sharp. □

3 Importance Sampling

With Importance Sampling we sample the Y_i according to a different distribution function g and observe that $p_f(A) = \int_{-\infty}^{x} f(z)dz = \int_{-\infty}^{x} \frac{f(z)}{g(z)} g(z)dz$. Define $V_i = I_A(Y_i) f(Y_i)/g(Y_i)$ and $V = V(A) = I_A(Y) f(Y)/g(Y)$. Let $p_f^{\mathrm{IS}}(A) = \frac{1}{N}\sum_{i=1}^{N} V_i$. Then $\mathrm{E}_g\left(p_f^{\mathrm{IS}}\right) = \frac{1}{N}\sum_{i=1}^{N} \mathrm{E}_g\left(V_i\right) = p_f(A)$. When $\frac{f(z)}{g(z)} \leq 1$ on A we have $\mathrm{Var}_g\left(p_f^{\mathrm{IS}}\right) \leq \mathrm{Var}_f\left(p_f^{\mathrm{MC}}\right)$ (variance reduction, using the same number of samples). This does not yet imply more efficiency. However, similar to (2), we derive (in which $N_{\mathrm{IS}} = N$) [6]

$$P\left(\left|p_f^{\mathrm{IS}} - p\right| > \nu p\right) \leq \exp\left(-\frac{N_{\mathrm{IS}}\, p^2}{2\mathrm{Var}_g(V)}\, \nu^2\right). \tag{6}$$

Assuming the same upper bounds, comparing (2) and (6) gives $\frac{N_{IS}}{N_{MC}} = \frac{\mathrm{Var}_g(V)}{p(1-p)} = \frac{\mathrm{E}_g(V^2)-p^2}{p(1-p)}$. Suppose $\frac{f(z)}{g(z)} \leq \kappa < 1$ on A and $p \leq \kappa$, then, with $q = 1-p$,

$$\frac{N_{IS}}{N_{MC}} = \frac{\mathrm{E}_g(V^2)}{pq} - \frac{p}{q} \leq \frac{\kappa}{q} - \frac{p}{q} \leq \kappa(1+\zeta) \tag{7}$$

for $|(1-\frac{1}{\kappa})p + \mathcal{O}(p^2)| \leq \zeta$, which for $\kappa = 0.1$ and $p = 10^{-10}$ means that $\zeta \leq 10^{-9}$. Hence for $\kappa = 0.1$ we can take an order less samples with Importance Sampling to get the same accuracy as with Monte Carlo. This even becomes better with smaller κ. Efficiency is the main message. Indeed the asymptotic accuracy also improves, but less: $\mathrm{Var}_g\left(p_f^{\mathrm{IS}}\right) \leq \kappa\, \mathrm{Var}_f\left(p_f^{\mathrm{MC}}\right) - \frac{1-\kappa}{N} p^2$ and thus $\sigma_g\left(p_f^{\mathrm{IS}}\right) \leq \sqrt{\kappa}\, \sigma_f\left(p_f^{\mathrm{MC}}\right)$, which for $\kappa = 0.1$ means that here not an order is gained, but a factor $\sqrt{\kappa} \approx 0.316$.

Proof of (6) [6]. Let Y be distributed according to g, $V = I_{(-\infty,x)}(Y) f(Y)/g(Y)$ and $v(y) = I_{(-\infty,x)}(y) f(y)/g(y)$. Then

$$\mathrm{E}_g\left[e^{\lambda V}\right] = \int_{-\infty}^{\infty} g(y)\, e^{\lambda I_{(-\infty,x)} f(y)/g(y)}\, dy = \int_{-\infty}^{x} g(y)\, e^{\lambda f(y)/g(y)}\, dy + 1 - G(x),$$

where $G(x) = \int_{-\infty}^{t} g(y)\, dy$. We will restrict ourselves to simple *sufficient* conditions and we will not strive for full generality. We assume:

1. There is no $y \in \mathbb{R}$ such that $P(Y = y) = 1$ (Y is not supported by a single point),
2. $0 < \mathrm{E}_g\left[e^{\lambda V}\right] < \infty$ for all $\lambda \in \mathbb{R}$,
3. Introduce the density function $\rho_\lambda(y)$

$$\rho_\lambda(y) = \frac{e^{\lambda v(y)} g(y)}{\mathrm{E}_g\left[e^{\lambda V}\right]} \quad \text{(thus } \int \rho_\lambda(y)\,dy = 1)$$

(which is well-defined for all $\lambda \in \mathbb{R}$) and let Y_λ be a random variable distributed according to ρ_λ. We assume that for all $\lambda \in \mathbb{R}$

$$\mathrm{E}_{\rho_\lambda}(Y_\lambda) = \int y_\lambda \rho_\lambda(y_\lambda)\,dy_\lambda = \int y \frac{e^{\lambda v(y)} g(y)}{\mathrm{E}_g\left[e^{\lambda V}\right]}\,dy < \infty$$

and

$$\mathrm{Var}_{\rho_\lambda}(Y_\lambda) = \mathrm{E}\left[Y_\lambda^2\right] - \mathrm{E}^2_{\rho_\lambda}(Y_\lambda) < \infty.$$

Now let $\varphi(\lambda) = \ln \mathrm{E}_g\left[e^{\lambda V}\right]$. Then, $\varphi(\lambda)$ is a well-defined, two times differentiable, real function with derivatives

$$\varphi'(\lambda) = \frac{\mathrm{E}_g\left[V\,e^{\lambda V}\right]}{\mathrm{E}_g\left[e^{\lambda V}\right]} = \mathrm{E}_{\rho_\lambda}(Y_\lambda), \quad \varphi''(\lambda) = \frac{\mathrm{E}_g\left[V^2\,e^{\lambda V}\right]}{\mathrm{E}_g\left[e^{\lambda V}\right]} - \frac{\mathrm{E}_g^2\left[V\,e^{\lambda V}\right]}{\mathrm{E}_g^2\left[e^{\lambda V}\right]} = \mathrm{Var}_{\rho_\lambda}(Y_\lambda).$$

Clearly, $\mathrm{Var}(Y_\lambda) > 0$ and φ is therefore *strictly convex*. Let $J(x,\lambda) = \lambda x - \varphi(\lambda)$. As in Sect. 2 we again consider the function $I(x) = \sup_{\lambda\in\mathbb{R}} J(x,\lambda)$ [5]. Clearly $I(x) \geq J(x,0) = -\varphi(0) = -\ln e^0 = 0$. To compute the supremum in $I(x)$, we consider

$$\frac{d}{d\lambda} J(x,\lambda) = x - \frac{d}{d\lambda}\varphi(\lambda) = x - \frac{\mathrm{E}_g\left[Ve^{\lambda V}\right]}{\mathrm{E}_g\left[e^{\lambda V}\right]}. \tag{8}$$

We observe that

$$\frac{d}{d\lambda} J(x,\lambda) = 0 \Longrightarrow x = \Psi(\lambda), \quad \text{where} \quad \Psi(\lambda) = \frac{\int y\, e^{\lambda\, v(y)} g(y)\,dy}{\int e^{\lambda\, v(y)} g(y)\,dy}. \tag{9}$$

Here we note that

$$\Psi'(\lambda) = \frac{\int e^{\lambda\, v(y)} g(y)\,dy \int y^2 e^{\lambda\, v(y)} g(y)\,dy \quad - \quad [\int y e^{\lambda\, v(y)} g(y)\,dy]^2}{[\int e^{\lambda\, v(y)} g(y)\,dy]^2}. \tag{10}$$

At the right-handside we can recognize a weighted inner-product (using weight function $e^{\lambda\, v(y)}$): $< 1, y > \equiv \int 1 \cdot ye^{\lambda v(y)} g(y)\,dy$. By the Cauchy-Schwarz inequality, $< 1, y > \leq \sqrt{< 1, 1 >)}\sqrt{< y, y >}$ we obtain $\Psi'(\lambda) > 0$ because $y \neq 1$. This implies that Ψ is invertible and hence (9) defines $\lambda = \lambda(x) = \Psi^{-1}(x)$. Hence

$$I(x) = J(x, \lambda(x)) \tag{11}$$

and we can write $x = \Psi(\lambda) = \mathrm{E}_{\rho_\lambda}[Y]$. Clearly $\rho_{\lambda=0}(y) = g(y)$. Further, to calculate the first (total) derivative of $I(x)$, we differentiate (11) with respect to x and substitute (9) to obtain $I'(x) = \lambda(x)$ and $I''(x) = \lambda'(x) = 1/\frac{\partial x}{\partial \lambda} = 1/\mathrm{Var}_{\rho_\lambda}(V)$ [6]. By [5, Lemma I.4, p. 8], $I(x)$ is strictly (proper) convex which means that the minimizer of I is unique. Now let p be as in Sect. 2. Then $I(p) = 0$, since the Strong Law of Large Numbers implies that the empirical measure of every neighbourhood of p tends to one. Hence, p is the unique minimizer of I and $I'(p) = 0$. Since p is also an internal point, we obtain that $0 = I'(p) = \lambda(p)$. Hence,

$$I''(p) = \frac{1}{\mathrm{Var}_{\rho_{\lambda(p)}}(V)} = \frac{1}{\mathrm{Var}_{\rho_{\lambda=0}}(V)} = \frac{1}{\mathrm{Var}_g\,(V)}. \tag{12}$$

Finally, by Taylor expansion, $I(p \pm \nu p) = \frac{1}{2}\nu^2 p^2 I''(p) + \mathcal{O}(\nu^3 p^3) = \frac{1}{2}\frac{\nu^2 p^2}{\mathrm{Var}_g\,(V)}$. Thus, after applying the *Large-Deviation Principle* [1,4,5], as in Sect. 2,

$$P\left(\left|\frac{1}{N}\sum_{k=1}^{N} V_k - p\right| > \nu p\right) \leq \exp\left(-N \inf_{|x-p|>\nu p} I(x)\right) \approx \exp\left(-\frac{Np^2}{2\mathrm{Var}_g(V)}\,\nu^2\right), \tag{13}$$

for all sufficiently large N. This implies (6), which completes the proof.
We finally note that, if $g(x) \equiv 1$, as in Sect. 2, we have $\mathrm{Var}_g\,(V) = \frac{1}{pq}$, see (2). □

4 Accurate Estimation of SRAM Yield

The threshold voltages V_t of the six transistors in an SRAM cell are the most important parameters causing variations of the characteristic quantities of an SRAM cell [2] like Static Noise Margin (SNM) and Read Current (I_{read}). In [2, 6] Importance Sampling (IS) was used to accurately and efficiently estimate low failure probabilities for SNM and I_{read}. SNM $=$ min(SNM$_h$, SNM$_l$) is a measure for the read stability of the cell. SNM$_h$ and SNM$_l$ are identically Gaussian distributed. The min() function is a non-linear operation by which the distribution of SNM is no longer Gaussian. Figure 1-left, shows the cumulative distribution function (CDF) of the SNM, using 50k trials, both for regular MC (solid) and IS (dotted). Regular MC can only simulate down to $P_{\mathrm{fail}} \leq 10^{-5}$. Statistical noise becomes apparent below $P_{\mathrm{fail}} \leq 10^{-4}$. With IS (using a broad uniform distribution g), $P_{\mathrm{fail}} \leq 10^{-10}$ is easily simulated (we checked this with more samples). The correspondence between regular MC and IS is very good down to $P_{\mathrm{fail}} \leq 10^{-5}$. Figure 1-left clearly shows that using extrapolated MC leads to overestimating the SNM at $P_{\mathrm{fail}} = 10^{-10}$. The Read Current I_{read} is a measure for the speed of the memory cell. It has a non-Gaussian distribution. Figure 1-right shows that extrapolated MC (dashed) can result in serious underestimation of I_{read}. This can lead to over-design of the memory cell. Also here IS is essentially needed for sampling I_{read} appropriately.

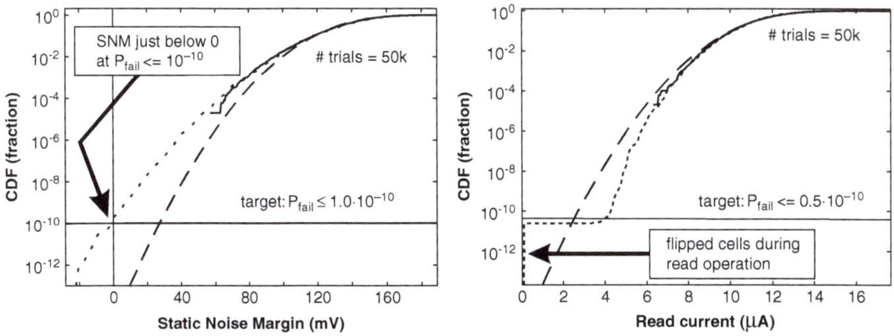

Fig. 1 SNM (left) and I_{read} (right) cumulative distribution function for extrapolated MC (dashed), regular MC (solid) and IS (dotted). Extrapolation assumes a normal distribution

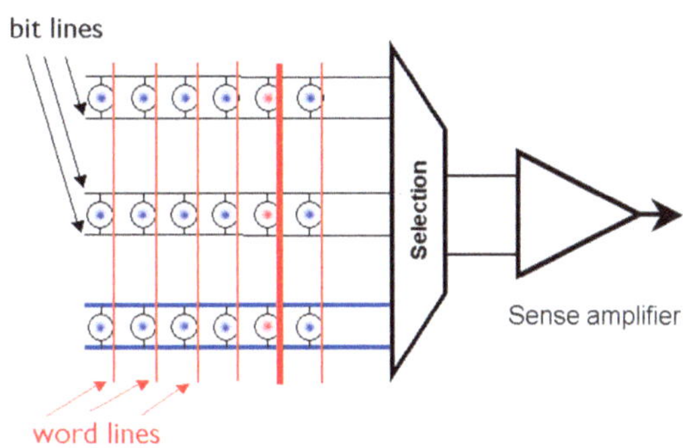

Fig. 2 Block of SRAMs (rotated 90°)

5 Optimization of SRAM Block

The block in Fig. 2 (rotated 90°) contains a Sense Amplifier (SA), a selector, and a number of SRAM cells. The selector chooses one "column" of cells. Then the voltage difference is $\Delta V_{\text{cell}} = \Delta V_{\text{k}}$. A block B works if $\min_{\text{k}}(\Delta V_{\text{k}}) \geq \Delta V_{\text{SA}}$. With m blocks B and n cells per block we define Yield Loss by $YL = P(\#B \geq 1) \leq m\, P(B)$, where the fail probability $P(B) = P_{\text{fail}}(B)$ of one block is (accurately) approximated by the lower bound $P(B) \approx \frac{YL}{m} = \frac{n\,YL}{N}$, where $N = n\,m$. For $YL = 10^{-3}$, $m = 10^4$ blocks, $n = 1000$ we find $P(B) \leq 10^{-7}$.
For $X = \min_k(\Delta V_k)$, and $Y = \Delta V_{\text{SA}}$ we have

$$P(B) = P(X < Y) = \int\int_{-\infty \leq x < y \leq \infty} f_{X,Y}(x,y)dx\,dy = \int_{-\infty}^{\infty} f_Y(y)\,F_X(y)dy.$$

Thus we need the pdf $f_Y(y)$ and the cdf $F_X(y)$ (probability and cumulative density functions of Y and X). Note that

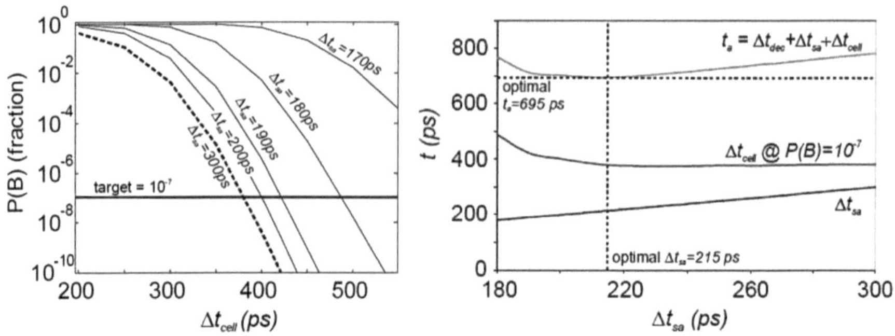

Fig. 3 Left: $P(B)$ as function of Δt_{cell} and Δt_{SA}. Right: Delay time t as function of Δt_{SA}

$$
\begin{aligned}
F_X(y) &= P(X < y) = P(\min_k \Delta V_k < y) \\
&= 1 - [1 - P(\Delta V_k < y)]^n \leq n\, P(\Delta V_k < y).
\end{aligned}
$$

For each simulation of the block we can determine the access times Δt_{cell} and Δt_{SA}. We come down to an optimization problem with a statistical constraint:

Minimize $\Delta t_{\text{cell}} + \Delta t_{\text{SA}}$ such that $P(B) \leq 10^{-7}$.

This has led to the following algorithm. We only give a sketch; for details see [3].

- By **Importance Sampling** sample ΔV_k. Collect ΔV_k at same Δt_{cell}.
- By **Monte Carlo** sample ΔV_{SA}. Collect ΔV_{SA} at same Δt_{SA}.
- For given Δt_{cell}:
  - Estimate pdf $f_{\Delta V_k}$ and cdf $P(\Delta V_k < y)$.
  - From this calculate $F_X(y) = F_X(y; \Delta t_{\text{cell}})$. Note that $\frac{\partial F_X(y;\Delta t_{\text{cell}})}{\partial \Delta t_{\text{cell}}} \leq 0$.
- For given Δt_{SA}:
  - Estimate pdf of ΔV_{SA}: $f_Y(y)$.
- Calculate (numerical integration)
  - $P(B) = \int_{-\infty}^{\infty} f_Y(y)\, F_X(y) dy$.

Hence $P(B) = G(\Delta t_{\text{cell}}, \Delta t_{\text{SA}})$ for some function G. For given Δt_{SA} $G_1(\Delta t_{\text{cell}}; \Delta t_{\text{SA}}) = G(\Delta t_{\text{cell}}, \Delta t_{\text{SA}})$ is monotonically decreasing in Δt_{cell}, see Fig. 3. Hence we *Minimize* $G_1^{-1}(10^{-k}; \Delta t_{\text{SA}}) + \Delta t_{\text{SA}}$. The optimization with the statistical constraint on $P(B)$ led to a reduction of 6% of the access time of an already optimized SA while simultaneously reducing the silicon area [3].

6 Conclusions

Large Deviation Theory allows to derive sharp lower and upper bounds for estimating accuracy of tail probabilities of quantities that have a non-Gaussian distribution. For Monte Carlo this leads to a realistic number of samples that should be taken. We extended this to Importance Sampling (IS). IS was applied to estimate fail probabilities $P_{\text{fail}} \leq 10^{-10}$ of SRAM characteristics like Static Noise Margin (SNM) and Read Current (I_{read}). We also applied IS to minimise the access time of an SRAM block while guaranteeing that the fail probability of one block is small enough.

In our experiments we used a fixed distribution g in the parameter space. In [6] ideas with an adaptively determined distribution g can be found.

References

1. Bucklew, J.A.: Introduction to Rare Event Simulation. Springer, Berlin (2004)
2. Doorn, T.S., ter Maten, E.J.W., Croon, J.A., Di Bucchianico, A., Wittich, O.: Importance Sampling Monte Carlo simulation for accurate estimation of SRAM yield. In: Proceedings of the IEEE ESSCIRC'08, 34th European Solid-State Circuits Conference, Edinburgh, Scotland, pp. 230–233 (2008)
3. Doorn, T.S., Croon, J.A., ter Maten, E.J.W., Di Bucchianico, A.: A yield statistical centric design method for optimization of the SRAM active column. In: Proceedings of the IEEE ESSCIRC'09, 35th European Solid-State Circuits Conference, Athens, Greece, pp. 352–355 (2009)
4. de Haan, L., Ferreira, A.: Extreme Value Theory. Springer, Berlin (2006)
5. den Hollander, F.: Large Deviations. Fields Institute Monographs 14, The Fields Institute for Research in Math. Sc. and AMS, Providence, RI (2000)
6. ter Maten, E.J.W., Doorn, T.S., Croon, J.A., Bargagli, A., Di Bucchianico, A., Wittich, O.: Importance Sampling for high speed statistical Monte-Carlo simulations – Designing very high yield SRAM for nanometer technologies with high variability. TUE-CASA 2009-37, TU Eindhoven (2009), http://www.win.tue.nl/analysis/reports/rana09-37.pdf

Effective Numerical Computation of Parameter Dependent Problems

Lennart Jansen and Caren Tischendorf

Abstract We analyse parameter dependent differential-algebraic-equations (DAEs)

$$Ad'(x,t,p) + b(x,t,p) = 0.$$

For these systems one is interested in the relation between the numerical solutions x and some associated parameters p. The standard approach is to discretise the equations with respect to the parameters and solve the parameter independent equations afterwards. This approach forces a calculation of the differential equations multiple times (for a huge number of parameter values p). This may lead to high computational costs. By using the already computed solutions to calculate the remaining ones and thus exploiting the smoothness of the solution with respect to the parameters, it is possible to save the majority of the computational cost.

1 Introduction

Nowadays parameter dependent problems have their applications in various fields. In the electric circuit simulation a circuit is modeled by a differential-algebraic-equation obtain via a modified-nodal-analysis (MNA) like in [1]. These DAEs depend on the parameters of the many electric parts in the circuit. Since these parts are afflicted with a manufacturing error, one is interested in the relation between small variation in these parameters and the behavior of the circuit.

In modern medicine the effect of the drugs used during a chemo therapy can be described by an ordinary-differential-equation (ODE), on of the models is presented in [2]. To obtain the optimal dosing of the drugs, simulating the therapy many

L. Jansen (✉) and C. Tischendorf
Mathematical Institute, University of Cologne, Eyestalk Weyertal 86-90, 50931 Köln
e-mail: lejansen@math.uni-koeln.de; tischendorf@math.uni-koeln.de

B. Michielsen and J.-R. Poirier (eds.), *Scientific Computing in Electrical Engineering SCEE 2010*, Mathematics in Industry 16, DOI 10.1007/978-3-642-22453-9_6,

times is necessary. Therefore the associated ODE has to be solved for many sets of parameters.

In meteorology the reliability of a weather forecast suffers because of the huge amount of the needed data and the chaotic behavior of the weather. These two problems force the calculation of a big partial-differential-equation(PDE) for many different parameters during the weather forecast.

These problems can be described mathematically by a parameter dependent differential-equation and a set of parameters. The general task is to efficiently solve the equations at all parameters. The main objective of this article is to present a new approach to accelerate this calculation. We will restrict the analysis to DAEs, even so the idea can be transferred to PDEs as well. The improvement will be demonstrated through the comparison of the convergence estimate and an example in the circuit simulation.

This paper is organized as follows. First we describe parameter dependent DAE and state the well-known convergence estimate for a Backward Differentiation Formula Method (BDF). Section 3 is devoted to the presentation of the new solving approach and its numerical results. In Sect. 4 these results are exemplified by a numerical example.

2 Parameter Dependent DAEs

In this chapter we want to introduce the problem in a general setup. Therefore we need to define the structure of the DAE and its properties. Furthermore an associate set of parameters is needed. To combine these two things define a parameter dependent DAE:

Definition 1. Define a semi-linear parameter dependent DAE as followed:

$$Ad'(x,t,p) + b(x,t,p) = 0, \tag{1}$$

with

$$A \in \mathbb{R}^{n \times k},\ d(x,t,p) \in \mathbb{R}^k,\ b(x,t,p) \in \mathbb{R}^n,\ x \in \mathbb{D} \subset \mathbb{R}^n,\ t \in \mathbb{I} \subset \mathbb{R},\ p \in \mathbb{P} \subset \mathbb{R}^l.$$

Recall that $(.)'$ means the total derivative with respect to time, i.e. $\frac{\mathrm{d}}{\mathrm{d}t}(.)$. Furthermore d and b with their partial derivatives d_y, d_x, b_x and b_t are continuous. We call this DAE properly stated, if:

- $\forall p \in \mathbb{P}$: $\ker A$ and $\operatorname{im} d_x$ are $C^1 - subspaces$
- $\forall p \in \mathbb{P}$: $\ker A \oplus \operatorname{im} d_x(x,t,p) = \mathbb{R}^n \quad \forall\ y \in \mathbb{R}^n,\ x \in \mathbb{D},\ t \in \mathbb{I}.$

In the following we assume that the DAE is properly stated. The concept of DAEs with properly stated leading terms is described in detail in [5, 6]. Also the index of the DAE is limited by two. An DAE index describes the complexity of the structure of a DAE. There are many ways to define an index, since in our applications there is often little smoothness we refer here to the tractability index.

To solve the parameter problem we need to calculate the numerical solution of the DAE at every single parameter point respectively. For the computation we use a BDF-method and achieve the well-known convergence estimate for the numerical solution of the DAE.

Theorem 1. *Discretise the DAE at a fixed parameter point p_0 with a BDF-method and achieve the following system:*

$$A(\frac{1}{h}\sum_{i=0}^{K}\alpha_i d(x_{n-i}, t_{n-i}, p_0)) + b(x_n, t_n, p_0) = 0. \quad (2)$$

Let h be the constant step size in time. Let the initial steps be sufficiently accurate. Then the error of the numerical solution obtained by solving the BDF-discretised system can be bounded by:

$$\max_{n>\mu K} ||x(t_n) - x_n|| \le c(h^K + \frac{1}{h^\mu}\max_{0<j<\mu K} ||\delta_{n-j}||) \quad (3)$$

with K the order of the BDF-method, $\mu + 1$ the index DAE and c being a bounded constant.

The proof can be found in [3, 7]. Notice that the error depends on the index. With an index bigger than one the computational error δ of the linear solvers must be guarantied to be small enough in relation to the step size h. The order K of the

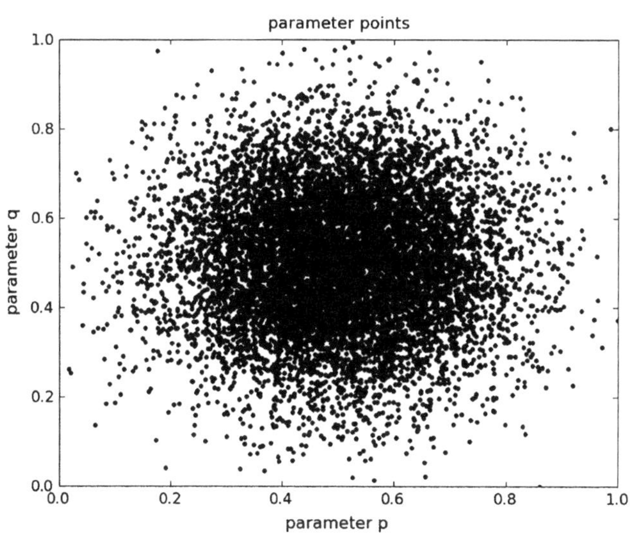

Fig. 1 A set of 10000 random parameter points

BDF-method mostly has to be chosen very small because of the stiffness of the applications.

For better understanding of the next chapter visualize a set of parameter points: Let for example $\mathbb{P} \subset \mathbb{R}^2$ be a set of 10000 random parameter points (Fig. 1).

Due to the applications we are interested in solving the DAE in every of these points to achieve the sensitivity of the solution with respect to the parameters. This may lead to high computational cost. So we need an approach that take advantage of the situation.

3 Parameter-Time-Integrator

This section will be used to present a more efficient approach to calculate the numerical solutions of the DAE at every point of the parameter set. To accelerate the computation of the solutions of the parameter dependent equation we want to take advantage of the solutions which are already calculated. Since the solutions won't be similar to other solutions for big enough differences in the parameters, one has to ensure that the change in the parameters will be sufficiently small. Therefore split the parameter points in sufficiently small packages and observe every package separately. The splitting of the points can depend on external data or it can be implemented in a adaptive way (Fig. 2). Solve the parameter dependent DAE at some parameter points in every single package to obtain interpolation nodes with respect to the parameters. With this nodes it is possible to improve the performance

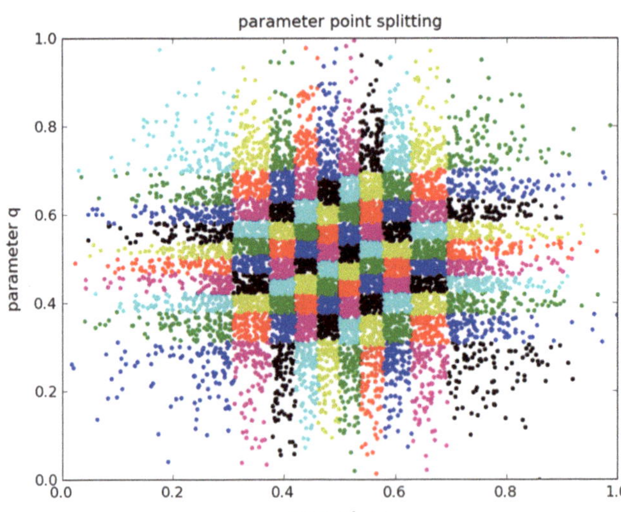

Fig. 2 A set of 10000 random parameter points splitted in parameter packages regarding their distribution

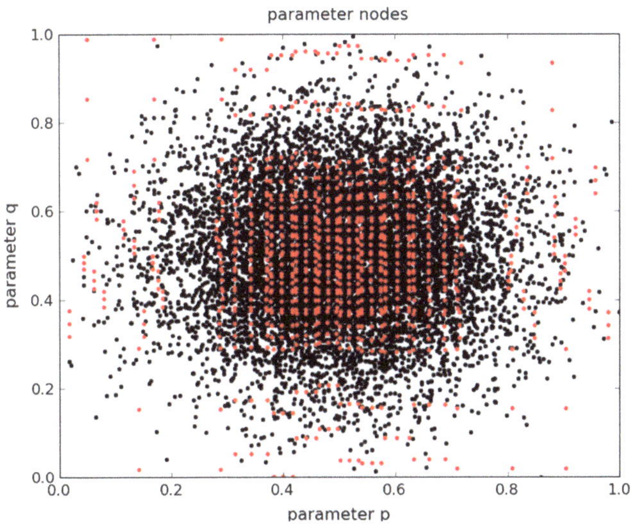

Fig. 3 A set of 10000 random parameter points and the associated set of 100 parameter nodes

of the calculation of the numerical solutions. For this reason formulate a modified system of DAEs involving the parameter interpolation nodes (Fig. 3).

Definition 2. Define a new system of DAEs by modifying the right hand side of a giving DAE with the help of some solutions x_{p_j}:

$$Ad'(x,t,p) + b(x,t,p) - \sum_{j=1}^{(m+1)^d} c_j(p)(Ad'(x_{p_j},t,p_j) + b(x_{p_j},t,p_j)) = 0 \tag{4}$$

with $m \in \mathbb{N}$ the order of the parameter interpolation, $p_j \in \mathbb{P}$ the nodes of the parameter interpolation, $x_{p_j} \in \mathbb{R}^m$ the solutions at the interpolation nodes and $c_j(p) \in \mathbb{R}$ the weighting functions of an multidimensional polynomal interpolation.

Notice that the exact solution x_{p_j} fulfills $Ad'(x_{p_j},t,p_j) + b(x_{p_j},t,p_j) = 0$, therefore the modified DAE will be solved by x_p and can be called equivalent to the original formula (1). In praxis one only has the numerical solution $x_{p_j,n}$ at a given point t_n and $Ad'(x_{p_j},t,p_j) + b(x_{p_j},t,p_j)$ will not be exactly zero. For the interpolation in the parameter space a polynomial interpolation is used. This will force the interpolation nodes p_j to be on a tensor-product grid in the parameter space. In praxis this is objectionable and can be avoided by an interpolation with radial basis functions or other interpolation methods. But assuming interpolation nodes p_j to be on a tensor-product grid and using an multidimensional polynomial interpolation does make the proofs and formulas much easier. Now again use a BDF-method to solve the modified system of DAEs.

Theorem 2. *Discretise the modified DAE for a fixed parameter point p_0 with a BDF-method. Discretise the parameter interpolation part as well with the same BDF-method and achieve:*

$$A\frac{1}{h}\sum_{i=0}^{K}\alpha_i d(x_{n-i},t_{n-i},p_0)+b(x_n,t_n,p_0)$$
$$-\sum_{j=1}^{(m+1)^d} c_j(p_0)(A\frac{1}{h}\sum_{i=0}^{K}\alpha_i d(x_{p_j,n-i},t_{n-i},p_j)+b(x_{p_j,n},t_n,p_j))=0 \quad (5)$$

Let h be the constant step size in time. Let the initial steps be sufficiently accurate. Then the error of the numerical solution obtained by solving the BDF-discretised modified system can be bounded by:

$$\max_{n>\mu K}||x(t_n)-x_n|| \le c^*(diam(\mathbb{P})^{m+1}(h^K+\frac{1}{h^\mu}\max_{0<j<\mu K}||\delta_{n-j}||)+h_0^K) \quad (6)$$

with K the order of the BDF-method, $\mu+1$ the index DAE and c being a bounded constant. Furthermore $\mathbb{P}$ is the parameter domain in one package and h_0 is the step size used in the calculation of the solutions at the interpolation nodes.

Compare this result with [4]. At this point notice the h_0^K term in the error estimate, because of this term the solutions computed with the help of the modified system cannot be more accurate than the solutions at the parameter nodes. Therefore the step size h_0 must be as small as the step size we would have chosen without this new approach. The source of the improvement of this error estimation is the parameter interpolation witch yields to the term $diam(\mathbb{P})^{m+1}$ in front of the normal relation of the error to the step size of the BDF-method h. You could say that we can accelerate the computation of the numerical solution because we have a good guess of the solution before the calculation itself starts. At this point there are three different cases to be observed. First the interpolation guess is as accurate as the given tolerance. In that case we don't have to calculate a new solution. In this trivial case we don't need a Parameter-Time-Integration since the parameter interpolation is already good enough. Second the interpolation guess is not as accurate as the given tolerance but accurate enough to accelerate the calculation of the new numerical solution. That means again that we can chose a bigger step size but still achieve the given tolerance. And third the accuracy of the guess is to low to improve the computation, which means we have to solve the original system. With an adaptive time step solver these three cases can switch on every timestep depending of the smoothness of the solution at the given time point regarding the parameters.

Think again of the example parameter set, so let $\mathbb{P}\subset\mathbb{R}^2$ again be a set of 10000 random parameter points. In this example one has to solve only 100 DAE of the original system, if the new Parameter-Time-Integrator is be used. The remaining solutions can be obtained by solving the modified system with the improved convergence estimate. In the example the parameter point splitting is based on external information in that case the distribution of the points, therefore we can assume that the parameter smoothness is big enough in relation to the size of the parameter packages.

4 Example

As an example for the applications in the circuit simulation observe the following circuit (Fig. 4):

The linear time-varying DAE

$$q_1' + G_1(t)e_1 + j_V = 0, \quad q_2' + G_2(t)e_2 + \eta t j_V = 0$$
$$e_1 = -\frac{1}{1-\eta t}e_2, \quad q_1 = C_1 e_1, \quad q_2 = C_2 e_2$$

simulates the electric circuit.

Here $G_1(t) = (1-\eta t - \lambda)$ and $G_2(t) = (\frac{\eta}{1-\eta t} - \lambda - \eta t)$ are the resistor functions which can be changed by the parameters η and λ to simulate the circuit with different resistors. Let $\lambda = -5$ be constant and $p = \eta$ be the variating parameter of our system. So this is a one-dimensional parameter space.

Furthermore e_1, e_2 describe the voltages at the nodes 1 and 2 with respect to the mass node. j_V is the current through the voltage source and q_1, q_2 represent the charges of the capacitances C_1 and C_2. For simplicity $C_1 = C_2 = 1$ is assumed. With

$$x = \begin{pmatrix} q_1 \\ q_2 \\ e_1 \\ e_2 \\ j_V \end{pmatrix}, \; A = \begin{pmatrix} 1\,0 \\ 0\,1 \\ 0\,0 \\ 0\,0 \\ 0\,0 \end{pmatrix}, \; b(x,t,p) = \begin{pmatrix} (6-pt)x_2 + x_4 \\ (5 + \frac{p}{1-pt} - pt)x_3 + ptx_4 \\ x_2 + \frac{1}{1-pt}x_3 \\ x_0 - x_2 \\ x_1 - x_3 \end{pmatrix}$$

$$d(x,t,p) = \begin{pmatrix} 1\,0\,0\,0\,0 \\ 0\,1\,0\,0\,0 \end{pmatrix} x$$

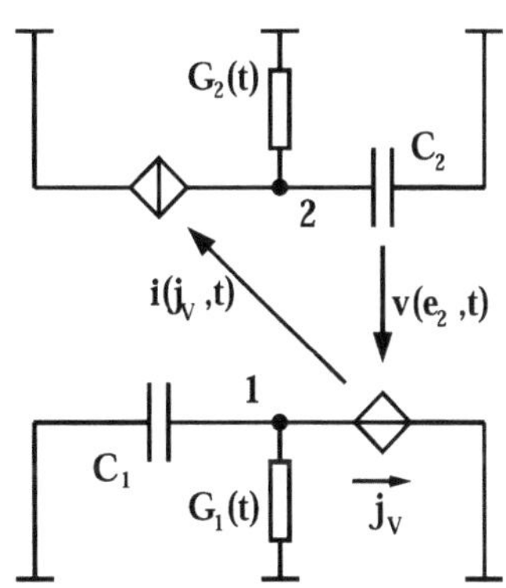

Fig. 4 The circuit diagram shows two small separated circuit whose sources are being controlled by the current or the voltage of the respective subcircuit. Apart from that the subcircuit consists of a resistor and a capacitor only

The circuit describing DAE can be written as:

$$Ad'(x,t,p) + b(x,t,p) = 0.$$

The parameter domain is defined through $p \in \mathbb{P} := [-20, -19]$. Solve the normal system at two parameter points $p_1 = -19.15$ and $p_2 = -19.85$ with an implicit Euler method with a constant step size $h = 10^{-4}$ in a time interval $T = [0, 1]$. This means 10000 steps have to be calculated with one Newton step each. At each parameter point a numerical solution x_{p_1} and x_{p_2} with an error $||x_{p_i}(t_n) - x_{p_i,n}|| \leq 2 \cdot 10^{-2}$ is calculated. Choose a random parameter in $\mathbb{P}$, for example $p_3 = -19.6$ and just solve it normally with a constant step size $h = 10^{-3}$ with an implicit Euler. Then the third component of the solution j_v is calculated with an error $||(j_v)_{p_3}(t_n) - (j_v)_{p_3,n}|| \leq 2 \cdot 10^{-1}$ (Fig. 5).

Now again solve the DAE with a constant step size $h = 10^{-3}$ and with an implicit Euler at $p_3 = -19.6$, but use the changed system. Again 1000 time and Newton steps are required, but an error $||(j_v)_{p_3}(t_n) - (j_v)_{p_3,n}|| \leq 2.5 \cdot 10^{-2}$ is achieved (Fig. 6).

5 Conclusion

In this paper we have seen a new approach to solve parameter dependent problems. The main idea was to approximate the solution at a parameter point with the already calculated solution of other points to have a good guess of the solution before calculating it. With this guess it is possible to accelerate the computation of the

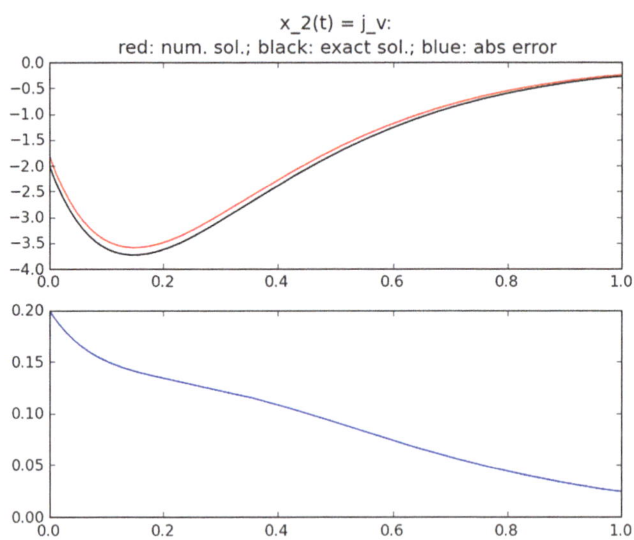

Fig. 5 Solution j_v with time on the x-axis and absolute error or solution values on the y-axis. The initial system is used

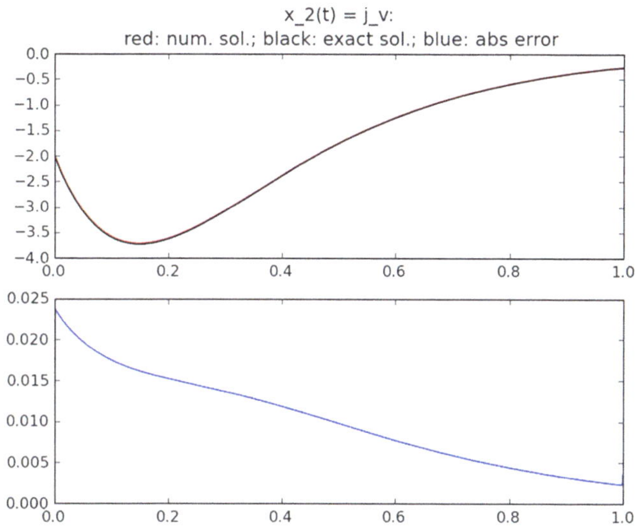

Fig. 6 Solution j_v with time on the x-axis and absolute error or solution values on the y-axis. The modified system is used

numerical solution. The degree of acceleration depends strongly on the smoothness of the solution regarding the parameter, since it wont be possible to obtain a good guess before the calculation, if there is nearly no connection between the parameters and the solution of our problem. But since we can decide at every single parameter point whether we use the parameter-time-integration or we solve the equation of the problem without it, one can exploit the smoothness of the parameter as long as there is some.

References

1. Schwarz, D.E., Tischendorf, C.: Structural analysis of electrical circuits and consequences for MNA. Int. J. Circuit Theory Appl. **28**, 131–162 (2000)
2. Scholz, M., Engel, C., Loeffler, M.: Modelling Human Granulopoiesis under Polychemotherapy with G-CSF Support. Math. Biol. **50**(4), 397–439 (2004)
3. Tischendorf, C.: Coupled Systems of Differential Algebraic and Partial Differential Equations in Circuit and Device Simulation, Habitilationsschrift, pp. 6–31. Humboldt University of Berlin (2004), http://www.mi.uni-koeln.de/~ctischen/publications.html
4. Jansen, L.: Effektive Numerische Simulation Elektrischer Schaltungen in Bezug auf Parameterschwankungen, Master's thesis, University of Cologne (2009)
5. März, R.: Nonlinear Differential-Algebraic Equations with Properly Formulated Leading Term, Preprint 3, Humboldt University of Berlin, Institute of Mathematics (2001)
6. Higueras, I., März, R.: Differential-algebraic equations with properly stated leading terms. Comput. Math. Appl. **48**, 215–235 (2004)
7. Voigtmann, S.: General Linear Methods for Integrated Circuit Design. Dissertation, pp. 111–173 (2004)

Analytical Properties of Circuits with Memristors

Ricardo Riaza

Abstract The memristor is a new lumped circuit element defined by a nonlinear charge-flux characteristic. The recent design of such a device has motivated a lot of research on this topic. In this communication we address certain analytical properties of semistate models of memristive circuits formulated in terms of differential-algebraic equations (DAEs). Specifically, we focus on the characterization of the *geometric index* of the DAEs arising in so-called branch-oriented analysis methods, which cover in particular tree-based techniques. Some related results involving nodal models and non-passive problems are discussed in less detail.

1 Introduction

The recent appearance of a nanometer-scale device displaying a memristive characteristic [24] has had a great impact in the electrical and electronic engineering communities; cf. [5, 12, 16, 20, 23] and references therein. Although their existence was already postulated by Chua in 1971 [2], the actual appearance of memristors in nanoscale electronics, reported in [24], has raised a renewed interest in this device. The memristor is considered as the fourth basic circuit element (besides the resistor, the inductor and the capacitor) and is defined by a nonlinear charge-flux characteristic, which may have either a charge-controlled or a flux-controlled form. This device is likely to play a relevant role in electronics in the near future, especially at the nanometer scale. Many applications are already reported (see e.g. [5, 11, 12, 15]).

R. Riaza (✉)
Depto. Matemática Aplicada TTI, ETSI Telecomunicación, Universidad Politécnica de Madrid, Ciudad Universitaria s/n, 28040 Madrid, Spain
e-mail: ricardo.riaza@upm.es

B. Michielsen and J.-R. Poirier (eds.), *Scientific Computing in Electrical Engineering SCEE 2010*, Mathematics in Industry 16, DOI 10.1007/978-3-642-22453-9_7,

The characteristic of a charge-controlled memristor has the form $\varphi = \phi(q)$, the incremental *memristance* being defined as

$$M(q) = \frac{d\phi(q)}{dq}.$$

Using the relations $v(t) = (\phi(q))'(t)$, $i(t) = q'(t)$ and the chain rule we get the voltage-current characteristic $v(t) = M(q(t))i(t)$. The device behaves as a resistor in which the resistance depends on $q(t) = \int_{-\infty}^{t} i(\tau)d\tau$, hence the *memory-resistor* (or *memristor*) name. In turn, a flux-controlled memristor is governed by $q = \sigma(\varphi)$, the current-voltage relation being $i(t) = W(\varphi(t))v(t)$ with incremental *memductance*

$$W(\varphi) = \frac{d\sigma(\varphi)}{d\varphi}.$$

A charge-controlled (resp. flux-controlled) memristor is *strictly locally passive* if $M(q) > 0$ (resp. $W(\varphi) > 0$) for all q (resp. φ). In the presence of coupling effects, this requirement should be restated by asking the full memristance or memductance matrices to be positive definite.

The analysis of memristive circuits involves the use of models based on differential-algebraic equations (DAEs). This stems from the fact that most circuit simulation programs set up the circuit equations in DAE form; this is the case in SPICE and its commercial variants [6, 9, 27]. A major problem in the study of DAE circuit models is the characterization of their index [1, 17, 19]: in this paper we characterize, in terms of the circuit topology, the so-called *geometric index* of memristive circuits, focusing the attention mainly on branch-oriented and tree-based models. The geometric index, which supports reduction methods, displays several advantages in the study of different analytical aspects of DAEs, involving e.g. stability issues or bifurcations; detailed discussions about this and other related index notions can be found in the above-mentioned references [1, 17, 19].

2 Memristive Circuit Models, DAEs, and the Geometric Index

So-called branch-oriented circuit models are based on the use of the loop and cutset matrices B, Q to describe Kirchhoff laws, being closely related to hybrid models [10, 19, 26]. The entries (b_{ij}) (resp. (q_{ij})) of the loop matrix B (resp. cutset matrix Q) are set to $+1$ or -1 if branch j belongs to loop i (resp. cutset i) with the same or opposite orientation, respectively, being 0 otherwise. Branch-oriented models avoid the use of the node potentials as circuit variables and are well-suited regarding analytical aspects such as the state space formulation problem; with respect to this problem, DAE-reduction methods and the geometric index notion [17, 19] arise naturally. Branch-oriented models have the form

$$
\begin{aligned}
C(v_c)v_c' &= i_c && \text{(1a)}\\
L(i_l)i_l' &= v_l && \text{(1b)}\\
q_m' &= i_m && \text{(1c)}\\
\varphi_w' &= v_w && \text{(1d)}\\
0 &= v_r - \gamma_r(i_r) && \text{(1e)}\\
0 &= i_g - \gamma_g(v_g) && \text{(1f)}\\
0 &= v_m - M(q_m)i_m && \text{(1g)}\\
0 &= i_w - W(\varphi_w)v_w && \text{(1h)}\\
0 &= B_c v_c + B_l v_l + B_r v_r + B_g v_g + B_m v_m + B_w v_w + B_u v_s(t) + B_j v_j && \text{(1i)}\\
0 &= Q_c i_c + Q_l i_l + Q_r i_r + Q_g i_g + Q_m i_m + Q_w i_w + Q_u i_u + Q_j i_s(t), && \text{(1j)}
\end{aligned}
$$

where we accommodate both current- and voltage-controlled resistors (cf. (1e) and (1f)) and, analogously, charge- and flux-controlled memristors in (1g) and (1h). For later use, denote by R and G the incremental resistance and conductance matrices $\gamma_r'(i_r)$, $\gamma_g'(v_g)$. The subscripts c, l correspond to capacitors and inductors; r, g to current- and voltage-controlled resistors; m, w to charge- and flux-controlled memristors and, finally, u, j to voltage and current sources, respectively. Sometimes it is more convenient to use a so-called *proper formulation* [13, 14, 19] and write the left-hand sides of (1a) and (1b) as $(q_c(v_c))'(t)$ and $(\varphi_l(i_l))'(t)$, respectively. In this paper we will not make use of this type of models, though.

In particular, when the loop and cutset matrices arise from the choice of a spanning tree (see e.g. [19]), then (1i) and (1j) take the form

$$
\begin{aligned}
0 &= v_{\mathrm{co}} + K v_{\mathrm{tr}}\\
0 &= i_{\mathrm{tr}} - K^T i_{\mathrm{co}},
\end{aligned}
$$

for a certain matrix K; with the subscripts $_{\mathrm{tr}}$ and $_{\mathrm{co}}$ we refer to tree and cotree branches, respectively. Proceeding as in [19], it is not difficult to show that the index is not affected by the specific choice of the matrices B and Q, and therefore we may assume w.l.o.g. that they arise from the choice of a given spanning tree.

Differential-algebraic equations and the geometric index. If the capacitance and inductance matrices $C(v_c)$, $L(i_l)$ are non-singular, (1) defines a semi-explicit differential-algebraic equation (DAE) [1, 19] of the form

$$x' = f(x, y, t) \tag{2a}$$

$$0 = g(x, y, t), \tag{2b}$$

where (2a) comprises (1a)–(1d) and (2b) stands for (1e)–(1j). The DAE is said to have geometric index one if the matrix of the partial derivatives of g with respect to y is non-singular. Index one DAEs display many advantageous features from both analytical and numerical standpoints [1, 17, 19]; in particular, a local *reduction* of the form $x' = f(x, h(x,t), t)$ follows from the implicit function theorem.

The definition of DAEs with geometric index two is more involved, and the reader is referred to [17, 19] for details. The key aspect is that *two* reduction steps are now necessary to describe the dynamics of the DAE. In particular, for so-called *Hessenberg DAEs* $x' = f(x, y, t)$, $0 = g(x, t)$, in which g does not depend on y, the index is two when the product $g_x f_y$ defines a non-singular matrix.

3 Geometric Index Characterization

Our main result (stated in Theorem 1 below) extends to the geometric index framework, and in terms of branch-oriented models, previous characterizations of the tractability index of nodal models [6, 23, 27].

Theorem 1. *Assume that the matrices C, L, R, G, M, W are positive definite. The model (1) has (I) geometric index one in the absence of VC-loops and IL-cutsets, and (II) geometric index two in the presence of at least one VC-loop including capacitors and/or at least one IL-cutset including inductors.*

Proof. (I) Let us first assume that the circuit exhibits neither VC-loops nor IL-cutsets. In this situation there is no loss of generality in assuming that (1) arises from the choice of a *proper tree*, including all voltage sources and capacitors and neither current sources nor inductors. This assumption gives (1i) and (1j) the form

$$\begin{pmatrix} v_{mw_{\text{co}}} \\ v_{rg_{\text{co}}} \\ v_j \\ v_l \end{pmatrix} = - \begin{pmatrix} K_{11} & K_{12} & K_{13} & K_{14} \\ K_{21} & K_{22} & K_{23} & K_{24} \\ K_{31} & K_{32} & K_{33} & K_{34} \\ K_{41} & K_{42} & K_{43} & K_{44} \end{pmatrix} \begin{pmatrix} v_{mw_{\text{tr}}} \\ v_{rg_{\text{tr}}} \\ v_s(t) \\ v_c \end{pmatrix}$$

and

$$\begin{pmatrix} i_{mw_{\text{tr}}} \\ i_{rg_{\text{tr}}} \\ i_u \\ i_c \end{pmatrix} = \begin{pmatrix} K_{11}^T & K_{21}^T & K_{31}^T & K_{41}^T \\ K_{12}^T & K_{22}^T & K_{32}^T & K_{42}^T \\ K_{13}^T & K_{23}^T & K_{33}^T & K_{43}^T \\ K_{14}^T & K_{24}^T & K_{34}^T & K_{44}^T \end{pmatrix} \begin{pmatrix} i_{mw_{\text{co}}} \\ i_{rg_{\text{co}}} \\ i_s(t) \\ i_l \end{pmatrix},$$

where we have joined together the entries corresponding to charge- and flux-controlled memristors, on the one hand, and current- and voltage-controlled resistors, on the other.

Additionally, proceeding as it is done in [19] for RLC circuits one can show that the positive definiteness assumption on R and G makes it possible to recast the resistors' multiport equations (1e)–(1f) in a local hybrid form

$$\begin{aligned} v_{r_{\mathrm{tr}}} &= h_{r_1}(i_{r_{\mathrm{tr}}}, v_{r_{\mathrm{co}}}) \\ i_{r_{\mathrm{co}}} &= h_{r_2}(i_{r_{\mathrm{tr}}}, v_{r_{\mathrm{co}}}) \\ v_{g_{\mathrm{tr}}} &= h_{g_1}(i_{g_{\mathrm{tr}}}, v_{g_{\mathrm{co}}}) \\ i_{g_{\mathrm{co}}} &= h_{g_2}(i_{g_{\mathrm{tr}}}, v_{g_{\mathrm{co}}}), \end{aligned}$$

with positive definite Jacobian matrices

$$H_r = \begin{pmatrix} \frac{\partial h_{r_1}}{\partial i_{r_{\mathrm{tr}}}} & \frac{\partial h_{r_1}}{\partial v_{r_{\mathrm{co}}}} \\ \frac{\partial h_{r_2}}{\partial i_{r_{\mathrm{tr}}}} & \frac{\partial h_{r_2}}{\partial v_{r_{\mathrm{co}}}} \end{pmatrix} = \begin{pmatrix} H_{r_{11}} & H_{r_{12}} \\ H_{r_{21}} & H_{r_{22}} \end{pmatrix}, \; H_g = \begin{pmatrix} \frac{\partial h_{g_1}}{\partial i_{g_{\mathrm{tr}}}} & \frac{\partial h_{g_1}}{\partial v_{g_{\mathrm{co}}}} \\ \frac{\partial h_{g_2}}{\partial i_{g_{\mathrm{tr}}}} & \frac{\partial h_{g_2}}{\partial v_{g_{\mathrm{co}}}} \end{pmatrix} = \begin{pmatrix} H_{g_{11}} & H_{g_{12}} \\ H_{g_{21}} & H_{g_{22}} \end{pmatrix}.$$

Similarly, the memristors' characteristics (1g)-(1h) can be rewritten as

$$\begin{pmatrix} v_{m_{\mathrm{tr}}} \\ i_{m_{\mathrm{co}}} \end{pmatrix} = \begin{pmatrix} H_{m_{11}} & H_{m_{12}} \\ H_{m_{21}} & H_{m_{22}} \end{pmatrix} \begin{pmatrix} i_{m_{\mathrm{tr}}} \\ v_{m_{\mathrm{co}}} \end{pmatrix}, \; \begin{pmatrix} v_{w_{\mathrm{tr}}} \\ i_{w_{\mathrm{co}}} \end{pmatrix} = \begin{pmatrix} H_{w_{11}} & H_{w_{12}} \\ H_{w_{21}} & H_{w_{22}} \end{pmatrix} \begin{pmatrix} i_{w_{\mathrm{tr}}} \\ v_{w_{\mathrm{co}}} \end{pmatrix},$$

being

$$H_m(q_m) = \begin{pmatrix} H_{m_{11}}(q_m) & H_{m_{12}}(q_m) \\ H_{m_{21}}(q_m) & H_{m_{22}}(q_m) \end{pmatrix}, \; H_w(\varphi_w) = \begin{pmatrix} H_{w_{11}}(\varphi_w) & H_{w_{12}}(\varphi_w) \\ H_{w_{21}}(\varphi_w) & H_{w_{22}}(\varphi_w) \end{pmatrix}$$

positive definite.

The index one condition for (1) can be easily checked to rely on the singularity of the matrix of derivatives of (1e)–(1j) with respect to the variables $v_{mw_{\mathrm{tr}}}$, $v_{rg_{\mathrm{tr}}}$, $i_{mw_{\mathrm{co}}}$, $i_{rg_{\mathrm{co}}}$, $i_{mw_{\mathrm{tr}}}$, $i_{rg_{\mathrm{tr}}}$, $v_{mv_{\mathrm{co}}}$ and $v_{rg_{\mathrm{co}}}$. This matrix has the form

$$J = \begin{pmatrix} I & -H \\ \tilde{K} & I \end{pmatrix}$$

with

$$H = \begin{pmatrix} H_{mw_{11}} & 0 & H_{mw_{12}} & 0 \\ 0 & H_{rg_{11}} & 0 & H_{rg_{12}} \\ H_{mw_{21}} & 0 & H_{mw_{22}} & 0 \\ 0 & H_{rg_{21}} & 0 & H_{rg_{22}} \end{pmatrix} \tag{3}$$

and

$$\tilde{K} = \begin{pmatrix} 0 & 0 & -K_{11}^T & -K_{21}^T \\ 0 & 0 & -K_{21}^T & -K_{22}^T \\ K_{11} & K_{12} & 0 & 0 \\ K_{21} & K_{22} & 0 & 0 \end{pmatrix} = \begin{pmatrix} 0 & 0 & -I & 0 \\ 0 & 0 & 0 & -I \\ I & 0 & 0 & 0 \\ 0 & I & 0 & 0 \end{pmatrix} \begin{pmatrix} K_{11} & K_{12} & 0 & 0 \\ K_{21} & K_{22} & 0 & 0 \\ 0 & 0 & K_{11}^T & K_{21}^T \\ 0 & 0 & K_{21}^T & K_{22}^T \end{pmatrix}.$$

Note that in (3) we have joined together, again, the entries corresponding to charge- and flux-controlled memristors, and also to current- and voltage-controlled resistors.

Assume that $(u, v) \in \ker J$. It follows that $v = -\tilde{K}u$ and $(I + H\tilde{K})u = 0$. Premultiply this relation by $u^T \tilde{K}^T$ to derive

$$u^T \tilde{K}^T u + u^T \tilde{K}^T H \tilde{K} u = 0.$$

The block-structure of $\tilde{K}$ implies $\tilde{K}^T = -\tilde{K}$, so that $u^T \tilde{K}^T u = 0$. From the resulting relation $u^T \tilde{K}^T H \tilde{K} u = 0$ and the positive definiteness of H we get $\tilde{K}u = 0$ and, since u verifies $(I + H\tilde{K})u = 0$ we get $u = 0$ and, in turn, $v = -\tilde{K}u = 0$, proving that the kernel of J is trivial. Hence, in the absence of VC-loops and IL-cutsets J is non-singular and therefore (1) has geometric index one. This completes the proof of (I).

(II) The index two analysis associated with the presence of VC-loops and/or IL-cutsets is more involved. Now we assume that the spanning tree is a *normal* one, including all voltage sources, the maximum possible number of capacitors, the minimum possible number of inductors and no current source. The relations (1i) and (1j) now take the form

$$\begin{pmatrix} v_{mw_{\mathrm{co}}} \\ v_{rg_{\mathrm{co}}} \\ v_j \\ v_{l_{\mathrm{co}}} \\ v_{c_{\mathrm{co}}} \end{pmatrix} = - \begin{pmatrix} K_{11} & K_{12} & K_{13} & K_{14} & 0 \\ K_{21} & K_{22} & K_{23} & K_{24} & 0 \\ K_{31} & K_{32} & K_{33} & K_{34} & K_{35} \\ K_{41} & K_{42} & K_{43} & K_{44} & K_{45} \\ 0 & 0 & K_{53} & K_{54} & 0 \end{pmatrix} \begin{pmatrix} v_{mw_{\mathrm{tr}}} \\ v_{rg_{\mathrm{tr}}} \\ v_s(t) \\ v_{c_{\mathrm{tr}}} \\ v_{l_{\mathrm{tr}}} \end{pmatrix}$$

and

$$\begin{pmatrix} i_{mw_{\mathrm{tr}}} \\ i_{rg_{\mathrm{tr}}} \\ i_u \\ i_{c_{\mathrm{tr}}} \\ i_{l_{\mathrm{tr}}} \end{pmatrix} = \begin{pmatrix} K_{11}^T & K_{21}^T & K_{31}^T & K_{41}^T & 0 \\ K_{12}^T & K_{22}^T & K_{32}^T & K_{42}^T & 0 \\ K_{13}^T & K_{23}^T & K_{33}^T & K_{43}^T & K_{53}^T \\ K_{14}^T & K_{24}^T & K_{34}^T & K_{44}^T & K_{54}^T \\ 0 & 0 & K_{35}^T & K_{45}^T & 0 \end{pmatrix} \begin{pmatrix} i_{mw_{\mathrm{co}}} \\ i_{rg_{\mathrm{co}}} \\ i_s(t) \\ i_{l_{\mathrm{co}}} \\ i_{c_{\mathrm{co}}} \end{pmatrix}$$

where K_{51}, K_{52}, K_{55}, K_{15} and K_{25} do vanish because of the choice of a normal tree. Now the circuit model can be reduced to a Hessenberg DAE of the form

$$
\begin{aligned}
C(v_c)\begin{pmatrix} v_{c_{\mathrm{tr}}} \\ v_{c_{\mathrm{co}}} \end{pmatrix}' &= \begin{pmatrix} \alpha(q_m, \varphi_w, v_{c_{\mathrm{tr}}}, i_{l_{\mathrm{co}}}, i_{c_{\mathrm{co}}}, v_s(t), i_s(t)) \\ i_{c_{\mathrm{co}}} \end{pmatrix} \\
L(i_l)\begin{pmatrix} i_{l_{\mathrm{tr}}} \\ i_{l_{\mathrm{co}}} \end{pmatrix}' &= \begin{pmatrix} v_{l_{\mathrm{tr}}} \\ \beta(q_m, \varphi_w, v_{c_{\mathrm{tr}}}, i_{l_{\mathrm{co}}}, v_{l_{\mathrm{tr}}}, v_s(t), i_s(t)) \end{pmatrix} \\
q_m' &= \eta(q_m, \varphi_w, v_{c_{\mathrm{tr}}}, i_{l_{\mathrm{co}}}, v_s(t), i_s(t)) \\
\varphi_w' &= \zeta(q_m, \varphi_w, v_{c_{\mathrm{tr}}}, i_{l_{\mathrm{co}}}, v_s(t), i_s(t)) \\
v_{c_{\mathrm{co}}} &= -K_{53}v_s(t) - K_{54}v_{c_{\mathrm{tr}}} \\
i_{l_{\mathrm{tr}}} &= K_{35}^T i_s(t) + K_{45}^T i_{l_{\mathrm{co}}},
\end{aligned}
$$

for suitable functions α, β, η, ζ, with $\partial\alpha/\partial i_{c_{\mathrm{co}}} = K_{54}^T$ and $\partial\beta/\partial v_{l_{\mathrm{tr}}} = -K_{45}$. The index two condition for this Hessenberg DAE then relies on the non-singularity of

$$
(K_{54} \;\; I\,)(C(v_c))^{-1}\begin{pmatrix} K_{54}^T \\ I \end{pmatrix}, \quad (I \;\; -K_{45}^T)(L(i_l))^{-1}\begin{pmatrix} I \\ -K_{45} \end{pmatrix}.
$$

Simple computations show that their non-singularity follows from the definiteness assumption on C and L, which implies that of C^{-1} and L^{-1}; details are left to the reader. In this setting the geometric index of (1) is two and the proof is complete. □

The geometric index characterization presented in Theorem 1 is a novel result. Noteworthy, it applies in particular to models constructed from the choice of a spanning tree, very often used in practice. Below, we survey some related properties involving other circuit models and other working scenarios.

4 Nodal Models and Non-passive Circuits

Nodal analysis. Most circuit simulation programs, such as SPICE, set up circuit equations using nodal analysis [6, 9, 19, 27]. Nodal models, which are well-suited from a computational point of view, are based on the so-called reduced incidence matrix A and are formulated in terms of the node potentials e and some branch currents.

After choosing a reference node, the reduced incidence matrix of a connected circuit is defined as (a_{ij}), where a_{ij} is $+1$ (resp. -1) if branch j leaves (resp. enters) node i, or 0 otherwise. In terms of the reduced incidence matrix, the nodal equations can be written in the form

$$C(v_c)v_c' = i_c \tag{4a}$$

$$L(i_l)i_l' = A_l^T e \tag{4b}$$

$$\varphi_w' = A_w^T e \tag{4c}$$

$$0 = A_g \gamma(A_g^T e) + A_w i_w + A_c i_c + A_l i_l + A_u i_u + A_j i_s(t) \tag{4d}$$

$$0 = v_c - A_c^T e \tag{4e}$$

$$0 = v_s(t) - A_u^T e \tag{4f}$$

$$0 = i_w - W(\varphi_w) A_w^T e. \tag{4g}$$

Note that Kirchhoff voltage law is now stated in the form $v = A^T e$, making it possible to eliminate several branch voltages from the model. We assume the memristors to be flux-controlled, and the resistors to have a voltage-controlled characteristic $i_g = \gamma(v_g)$. As detailed in [23], under strict passivity assumptions the index characterization stated in Theorem 1 also holds for the tractability index [13, 14] of the nodal model (4).

Non-passive problems. The positive definiteness assumptions in Theorem 1 are not met in many practical situations. Using the approach discussed in [22], it is possible to derive a characterization of index one problems without this passivity requirement, as shown in [20]. Specifically, assuming that the matrices C, L are non-singular, and that neither resistors nor memristors exhibit coupling effects, the models (1) and (4) are still index one if and only if a) the circuit has neither VC-loops nor IL-cutsets, and b) the sum of conductance-memductance products in proper trees does not vanish. The proof of this assertion can be found in [20].

5 Concluding Remarks

The memristor and related devices are likely to play a very relevant role in electronics in the near future. The characterization of analytical properties such as the index of DAE models is important in the study of dynamical and numerical properties of circuits with memristive devices. The results here discussed should be useful in future analyses of nonlinear aspects involving e.g. stability, bifurcations, oscillations, chaotic effects or impasse phenomena in memristive circuits, extending to this context related results proved for circuits without memristors [3, 4, 7, 8, 10, 18, 21, 25].

Acknowledgements The author acknowledges support by Project MTM2007-62064, Ministerio de Educación y Ciencia, Spain.

References

1. Brenan, K.E., Campbell, S.L., Petzold, L.R.: Numerical Solution of Initial-Value Problems in Differential-Algebraic Equations. SIAM, Philadelphia (1996)

2. Chua, L.O.: Memristor – The missing circuit element. IEEE Trans. Circuit Theory. **18**, 507–519 (1971)
3. Chua, L.O., Deng, A.D.: Impasse points, I: Numerical aspects. Internat. J. Circuit Theory Appl. **17**, 213–235 (1989)
4. Demir, A.: Floquet theory and non-linear perturbation analysis for oscillators with differential-algebraic equations. Internat. J. Circuit Theory Appl. **28**, 163–185 (2000)
5. Di Ventra, M., Pershin, Y.V., Chua, L.O.: Circuit elements with memory: memristors, memcapacitors and meminductors. Proc. IEEE **97**, 1717–1724 (2009)
6. Estévez-Schwarz, D., Tischendorf, C.: Structural analysis of electric circuits and consequences for MNA. Internat. J. Circuit Theory Appl. **28**, 131–162 (2000)
7. Fosséprez, M.: Non-Linear Circuits: Qualitative Analysis of Non-linear, Non-Reciprocal Circuits. Wiley, New York, (1992)
8. Green, M.M., Willson Jr, A.N.: An algorithm for identifying unstable operating points using SPICE. IEEE Trans. Computer-Aided Des. Circ. Sys. **14**, 360–370 (1995)
9. Günther, M., Feldmann, U.: CAD-based electric-circuit modeling in industry. Surv. Math. Ind. **8**, 97–129, 131–157 (1999)
10. Hasler, M., Neirynck, J.: Nonlinear Circuits. Artech House, Boston, (1986)
11. Itoh, M., Chua, l.O.: Memristor oscillators. Intl. J. Bifurcation Chaos **18**, 3183–3206 (2008)
12. Itoh, M., Chua, L.O.: Memristor cellular automata and memristor discrete-time cellular neural networks. Internat. J. Bifurcation Chaos **19**, 3605–3656 (2009)
13. März, R.: Differential algebraic equations anew. Appl. Numer. Math. **42**, 315–335 (2002)
14. März, R.: The index of linear differential algebraic equations with properly stated leading terms. Results Math. **42**, 308–338 (2002)
15. Muthuswamy, B., Kokate, P.P.: Memristor-based chaotic circuits. IETE Tech. Rev. **26**, 417–429 (2009)
16. Pershin, Y.V., Di Ventra, M.: Spin memristive systems: Spin memory effects in semiconductor spintronics. Phys. Rev. B **78**, 113309 (2008)
17. Rabier, P.J., Rheinboldt, W.C.: Theoretical and numerical analysis of differential-algebraic equations. Handbook of Numerical Analysis, vol. VIII, pp. 183–540. North-Holland, Amsterdam (2002)
18. Riaza, R.: On the singularity-induced bifurcation theorem. IEEE Trans. Aut. Contr. 47, 1520–1523 (2002)
19. Riaza, R.: Differential-Algebraic Systems. Analytical Aspects and Circuit Applications, World Scientific, Singapore (2008)
20. Riaza, R.: Nondegeneracy conditions for active memristive circuits. IEEE Trans. Circuits Systems – II. **57**, 223–227 (2010)
21. Riaza, R.: Graph-theoretic characterization of bifurcation phenomena in electrical circuit dynamics. Internat. J. Bifurcation Chaos **20**, 451–465 (2010)
22. Riaza, R., Encinas, A.: Augmented nodal matrices and normal trees. Discrete Appl. Math. **158**, 44–61 (2010)
23. Riaza, R., Tischendorf C.: Semistate models of electrical circuits including memristors. Internat. J. Circuit Theory Appl. **39**(6), 607–627 (June 2011)
24. Strukov, D.B., Snider, G.S., Stewart, D.R., Williams, R.S.: The missing memristor found. Nature **453**, 80–83 (2008)
25. Tadeusiewicz, M.: Global and local stability of circuits containing MOS transistors. IEEE Trans. Circuits Systems Part I **48**, 957–966 (2001)
26. Takamatsu, M., Iwata S.: Index characterization of differential-algebraic equations in hybrid analysis for circuit simulation. Internat. J. Circuit Theory Appl. **38**, 419–440 (2010)
27. Tischendorf, C.: Topological index calculation of DAEs in circuit simulation. Surv. Math. Ind. **8** 187–199 (1999)

Scattering Problems in Periodic Media with Local Perturbations

Therese Pollok, Lin Zschiedrich, and Frank Schmidt

Abstract Within this paper we consider scattering problems with periodic exterior domains, modeled by the Helmholtz equation. Most current works on this subject make specific assumptions on the geometry of the periodic cell, e.g. special symmetries or shapes, and cannot be generalized to higher space dimensions in an easy way. In contrast our goal is the realization of an easy dimension independent concept which is valid for all kinds of periodic structures with local defects. We will first give a general analytical formulation and then present an algorithmic realization. At the end of the paper we will also depict a 1D and 2D example.

1 Introduction and Problem Setting

Periodic structures such as photonic crystals or metamaterials have many applications in modern optic devices due to their optical properties, as for example the occurrence of band gaps, i.e. forbidden frequency ranges, or negative refractive indices. Particularly defects within the periodicity of band gap materials are of special interest since they can be used to manipulate the flow of light efficiently. By disturbing a whole line of unit cells for example one can produce a waveguide for frequencies within the band gap whereas local perturbations yield optical cavities. For more details see for example [1,4]. We will confine our considerations to local defects, i.e., the perturbation is restricted to a bounded region Ω as illustrated in Fig. 1. Without loss of generality we will further assume, that Ω is contained in one single unit cell C_0.

In real applications photonic crystals often consist of a very large number of unit cells. We will therefore assume that the crystal is of infinite size.

T. Pollok (✉) · L. Zschiedrich · F. Schmidt
Zuse Institute Berlin, Takustr. 7, 14195 Berlin, Germany
e-mail: pollok@zib.de; zschiedrich@zib.de; frank.schmidt@zib.de

B. Michielsen and J.-R. Poirier (eds.), *Scientific Computing in Electrical Engineering SCEE 2010*, Mathematics in Industry 16, DOI 10.1007/978-3-642-22453-9_8,

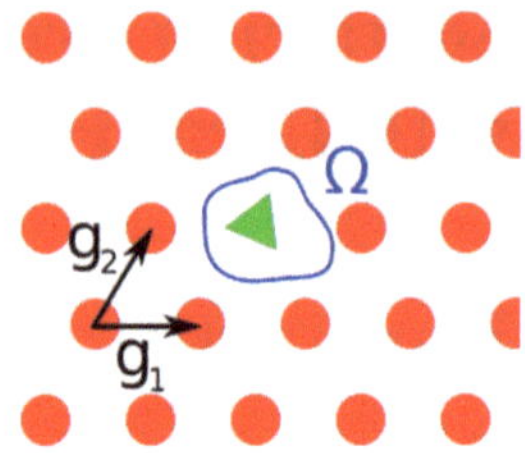

Fig. 1 2D periodic structure with a local perturbation

Furthermore we assume a non-conducting and charge-free medium and a time-harmonic incoming wave and model the scattering problem by the scalar Helmholtz equation $\Delta u(\mathbf{r}) + k^2(\mathbf{r})u(\mathbf{r}) = f(\mathbf{r})$. The material properties thereby are contained in $k^2(\mathbf{r}) = \varepsilon(\mathbf{r})\mu(\mathbf{r})\omega^2$, where ω is the frequency of the incoming wave. The right hand term $f(\mathbf{r})$ is the source term, resulting from the incoming wave (see (5)).

One of the most important works on this subject is the work of P. Joly [3] in which for 2D structures a coupled operator equation system for four operators is derived. It can be decoupled in the special case of so called *double symmetric* refractive indices but is much more involved for the general unsymmetric case. In [2], which is a second important work in this field, only 2D structures are considered which consist of cylinders of refractive index n_i in an elsewise homogeneous medium of refractive index n_e. In view of the development of an universal tool we are interested in an easier and general concept which is independent of the spacial dimension.

2 Scattering Problems

Let u_{in} be an incident wave which satisfies the Helmholtz equation in the exterior domain $\Omega_{\text{ext}} = \mathbb{R}^d \setminus \Omega$:

$$\Delta u_{\text{in}}(\mathbf{r}) + k_{\text{per}}^2(\mathbf{r})u_{\text{in}}(\mathbf{r}) = 0, \tag{1}$$

where k_{per} is an (undisturbed) periodic function with lattice vectors $\mathbf{g}_j$ for $j = 1, \dots, d$, i.e.,

$$k_{\text{per}}(\mathbf{r} + \mathbf{g}_j) = k_{\text{per}}(\mathbf{r}) \quad \forall \mathbf{r} \in \mathbb{R}^d. \tag{2}$$

Find the scattered wave u_{sc} such that for the total field $u_{\text{tot}} = u_{\text{in}} + u_{\text{sc}}$,

$$\Delta u_{\text{tot}}(\mathbf{r}) + k^2(\mathbf{r})u_{\text{tot}}(\mathbf{r}) = 0 \quad \text{in } \mathbb{R}^d, \tag{3}$$

where

$$k^2(\mathbf{r}) = k_{\text{per}}^2(\mathbf{r}) + k_{\Delta}^2(\mathbf{r}) \tag{4}$$

for a function k_{Δ} with support in Ω. Without loss of generality we assume that Ω is contained in a unit cell of the periodic structure.

By inserting (1) and (4) into (3) one obtains

$$\Delta u_{\mathrm{sc}} + k^2 u_{\mathrm{sc}} = -k_\Delta^2 u_{\mathrm{in}} =: f \quad \text{in } \mathbb{R}^d, \tag{5}$$

with supp $f \subseteq \Omega$.

In order to obtain the physically correct solution we require in addition a radiation condition that ensures that u_{sc} is purely outgoing. For the mathematical formulation of this radiation condition we will use the limiting absorption principle which is part of the next section.

For the sake of simplicity we will omit the subscript of the scattered field u_{sc} and denote it in the following as u.

3 Bloch-Floquet Transform and Limiting Absorption Principle

Due to backscattering off the periodic configuration of materials, distinguishing between incoming and outgoing waves is much more involved than in the homogeneous case. The standard approach to overcome this difficulty is to introduce artificial damping by replacing $k \to k(1 + \mathrm{i}\sigma)$, where $\sigma \in \mathbb{R}_+$ is the damping parameter. The outgoing waves of the damped problem are exponentially decaying for $|\mathbf{r}| \to \infty$ and thus can be distinguished from the exponentially growing incoming waves. This is known as *limiting absorption principle* and was first introduced by Joly [5] and reads:

Find $u \in L^2(\mathbb{R}^d)$ such that

$$\Delta u + k^2(1 + \mathrm{i}\sigma)^2 u = f. \tag{6}$$

Next, we introduce the so-called Bloch-Floquet transform, a second standard technique for the treatment of periodic problems. Its application to the Helmholtz equation leads to boundary value problems with finite computational domains, as detailed in the following.

Let $\mathbf{G} := (\mathbf{g}_1, \ldots, \mathbf{g}_d)$ the matrix consisting of the lattice vectors $\mathbf{g}_j$ (see (2) and Fig. 1) and

$$\Gamma := \left\{ \mathbf{Gn} | \mathbf{n} \in \mathbb{Z}^d \right\}.$$

For exponentially decaying u one can define the Bloch-Floquet transform $\mathrm{Fl}(u) =: \hat{u}$ by

$$\hat{u}(\mathbf{k}_B, \mathbf{r}) := \sum_{\mathbf{d} \in \Gamma} u(\mathbf{r} + \mathbf{d}) \exp(-\mathrm{i}\mathbf{k}_B \cdot \mathbf{d}), \tag{7}$$

where $\mathbf{k}_B \in \mathbb{R}^d$ is called *Bloch vector*. (See [6].)

It can easily be shown that $\hat{u}$ is periodic with respect to $\mathbf{k}_B$: Let $\tilde{\mathbf{g}}_1, \ldots, \tilde{\mathbf{g}}_d$ the *reciprocal lattice vectors*, i.e.

$$\mathbf{g}_i \cdot \tilde{\mathbf{g}}_j = 2\pi\delta_{ij}.$$

Then

$$\hat{u}(\mathbf{k}_B + \tilde{\mathbf{g}}_j, \mathbf{r}) = \hat{u}(\mathbf{k}_B, \mathbf{r}).$$

Hence it is sufficient to consider $\mathbf{k}_B \in \mathrm{BZ}$, where BZ is the so called *first Brillouin zone* which is defined as the primitive unit cell of the reciprocal lattice. The inverse of the Bloch-Floquet transform reads

$$u(\mathbf{r}) = \frac{1}{|BZ|} \int_{\mathrm{BZ}} \hat{u}(\mathbf{k}_B, \mathbf{r}) \mathrm{d}\mathbf{k}_B. \tag{8}$$

Let us now apply the Bloch-Floquet transform to the damped Helmholtz equation (6). It can be easily shown that for the periodic part k_{per} of k,

$$\mathrm{Fl}(k_{\mathrm{per}}^2 u) = k_{\mathrm{per}}^2 \,\mathrm{Fl}(u). \tag{9}$$

Thus, by using (4) and (9), we obtain

$$\Delta\hat{u} + k_{\mathrm{per}}^2 (1 + \mathrm{i}\sigma)^2 \hat{u} + \mathrm{Fl}\left(k_\Delta^2 (1 + \mathrm{i}\sigma)^2 u\right) = \mathrm{Fl}(f). \tag{10}$$

Without loss of generality we can assume $\Omega \subset C_0$, where C_0 is a unit cell of the periodic structure (otherwise define C_0 as a sufficient large cell). Then for every function g with support in Ω holds $\mathrm{Fl}(g)(\mathbf{k}_B, \mathbf{r}) = g(\mathbf{r})$ for all $\mathbf{r} \in C_0$. Thus we may write

$$\Delta\hat{u} + k_{\mathrm{per}}^2 (1 + \mathrm{i}\sigma)^2 \hat{u} + k_\Delta^2 (1 + \mathrm{i}\sigma)^2 u = f \quad \text{in } \mathrm{BZ} \times C_0. \tag{11}$$

We can obtain an equation for $\hat{u}$ from (10) by applying the inverse transform (8):

$$\Delta\hat{u} + k_{\mathrm{per}}^2 (1 + \mathrm{i}\sigma)^2 \hat{u} + k_\Delta^2 (1 + \mathrm{i}\sigma)^2 \frac{1}{|\mathrm{BZ}|} \int_{\mathrm{BZ}} \hat{u} \mathrm{d}\mathbf{k}_B = f \quad \text{in } \mathrm{BZ} \times C_0. \tag{12}$$

By adding a lattice vector $\mathbf{g}_j$ of the periodic structure to the space argument $\mathbf{r}$ in the definition (7) of the Bloch-Floquet transform we see that $\hat{u}(\mathbf{k}_B, \mathbf{r})$ is *quasi-periodic*:

$$\hat{u}(\mathbf{k}_B, \mathbf{r} + \mathbf{g}_j) = \hat{u}(\mathbf{k}_B, \mathbf{r}) \exp(\mathrm{i}\mathbf{k}_B \cdot \mathbf{g}_j). \tag{13}$$

Equation (13) yields the boundary conditions that enable us to restrict the spacial computational domain to the unit cell C_0. It also tells us how to extend the solution to $\mathbb{R}^d$.

Remark 1. Since any bounded solution $\hat{u}$ of (12) is extended by (13) it remains bounded on $\mathbb{R}^d$. Therefore the solution u of (6) which can be obtained by inverting $\hat{u}$ with (8) is also bounded and thus contains no exponentially growing part. This means that the obtained solution must be outward radiating as demanded.

4 Algorithmic Solution

Equation (12) is a $2d$-dimensional problem since $\hat{u}$ depends on $\mathbf{k}_B$ and $\mathbf{r}$. In the case of $k_\Delta = 0$ (12) would decouple into one d-dimensional boundary value problem

$$\Delta\hat{u}(\mathbf{k}_B, \bullet) + k_{\text{per}}^2(1 + \mathrm{i}\sigma)^2\hat{u}(\mathbf{k}_B, \bullet) = f \qquad \text{in } C_0 \tag{14}$$

$$\hat{u}(\mathbf{k}_B, \mathbf{r} + \mathbf{g}_j) = \hat{u}(\mathbf{k}_B, \mathbf{r}) \exp(\mathrm{i}\mathbf{k}_B \cdot \mathbf{g}_j)$$

for each $\mathbf{k}_B \in \text{BZ}$, which we can solve with standard methods.

Let us now assume we already knew the solution u of the original problem with $k_\Delta \neq 0$. Then we would get from (10)

$$\Delta\hat{u}(\mathbf{k}_B, \bullet) + k_{\text{per}}^2(1 + \mathrm{i}\sigma)^2\hat{u}(\mathbf{k}_B, \bullet) = \tilde{f} \qquad \text{in } C_0 \tag{15}$$

with the new right hand side $\tilde{f} = f - k_\Delta^2(1 + \mathrm{i}\sigma)^2 u$ and we would have again a problem of the same type as (14). This motivates the following iteration:

$u_n = u_0$;
for $n = 0$: max_iteration_steps **do**
  $\hat{u} \leftarrow$ solve (15) with $\tilde{f} = f - k_\Delta^2(1 + \mathrm{i}\sigma)^2 u_n$;
  $u_{n+1} \leftarrow$ FloquetInvert($\hat{u}$);
end for

As initial value for the iteration we choose $u_0 = 0$. This implies that in the first iteration step the algorithm solves the unperturbed periodic problem.

To solve (15) for one specific value of $\mathbf{k}_B$ we use a standard finite element method with quasi-periodic ansatz functions

$$\phi(\mathbf{r} + \mathbf{g}_j) = \phi(\mathbf{r}) \exp(\mathrm{i}\mathbf{k}_B \cdot \mathbf{g}_j) \text{ for } j = 1, \ldots, d. \tag{16}$$

This special choice of ansatz functions eliminates the boundary integral from the variational formulation of (15). The numerical integration in (8) requires the evaluation of (15) for several values $\mathbf{k}_B$ which means computing several finite element solutions per iteration step. To keep the computational effort low we used an adaptive integration formula that uses an unstructured grid. The choice of the damping parameter σ influences the effort required for the numerical integration, since $\sigma = 0$ would yield a solution with singularities which are smoothed when increasing σ. Therefore the choice of σ is a tradeoff between low integration costs and perturbation of the original problem and its solution.

Remark 2. To get the solution of the undamped case $\sigma = 0$ one can use extrapolation methods.

Convergence

The convergence of the iteration depends on the magnitude of the perturbation function k_Δ^2. The more the periodicity is disturbed, the worse is the convergence. In the case of divergence one may help oneself by applying the following trick: Let $k_\Delta^2 = k_{\Delta,1}^2 + k_{\Delta,2}^2$, where $k_{\Delta,1}^2$ is small enough so that the problem $\Delta u + (k_{\text{per}}^2 + k_{\Delta,1}^2)(1 + \mathrm{i}\sigma)^2 u = f$ leads to a convergent iteration and is thus solvable with the above algorithm. By shifting the troubling term $k_{\Delta,2}^2(1+\mathrm{i}\sigma)^2 u$ to the right hand side one obtains again an iteration problem:

$$\Delta u_{n+1} + (k_{\text{per}}^2 + k_{\Delta,1}^2)(1 + \mathrm{i}\sigma)^2 u_{n+1} = f - k_{\Delta,2}^2(1 + \mathrm{i}\sigma)^2 u_n$$

This way one may reduce the problem recursively to solvable problems.

Perturbation of Multiple Cells

In the case $\Omega \not\subset C_0$ it is not necessary, to solve (15) on a larger lattice cell, since (13) and the left hand side of (15) remain unaffected. We only have to take into account more terms of the Bloch-Floquet transform of the right hand side of (15):

$$\tilde{f} \rightarrow \sum_{\mathbf{d}\in\Gamma_0} \tilde{f}(\mathbf{r} + \mathbf{d}) \exp(-\mathrm{i}\mathbf{k}_B \cdot \mathbf{d})$$

where $\Gamma_0 = \left\{\mathbf{d} \in \Gamma \,|\, \operatorname{supp} \tilde{f} \cap (\mathbf{d} + C_0) \neq \emptyset\right\}$.

5 Examples

We implemented this algorithm for the one- and two-dimensional case. In the following we present a 1D as well as a 2D example. Figures 2 and 3 show the geometry of the 1D and 2D media which each consist of two materials with $k_1 = 6$ and $k_2 = 8$. The source term is $f(\mathbf{r}) = \exp(-25\mathbf{r} \cdot \mathbf{r})$ in Ω and $f = 0$ outside Ω, where $\Omega = [-0.5; 0.5]$ and $\Omega = [-0.5; 0.5]^2$, respectively. The damping parameter is in both cases $\sigma = 10^{-2}$.

Figure 4 shows the intensity $|u(\mathbf{r})|^2$ of the computed scattered field in the inner part of the infinite domain for the two-dimensional case.

6 Conclusions

Within the current work we demonstrated a dimensionally independent formulation for scattering problems with periodic exterior domains, that is valid for all kinds of infinite periodic structures with local defects. By means of a simple iterative scheme

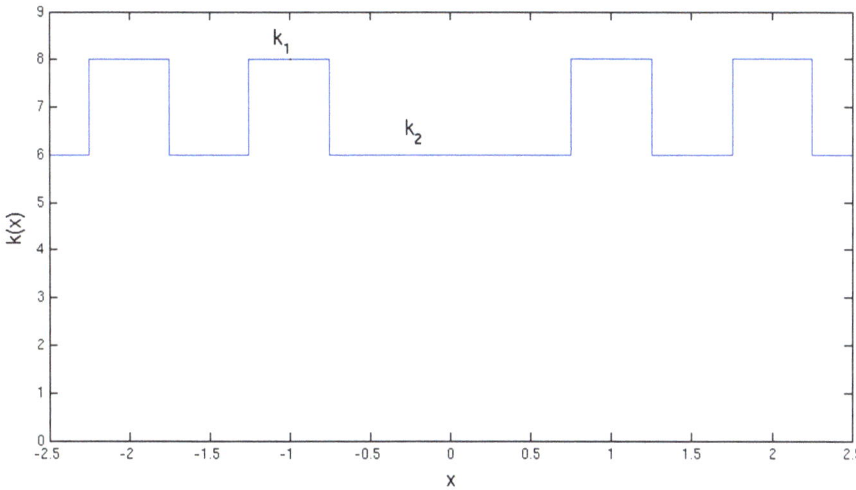

Fig. 2 1D geometry consisting of two materials with difference refractive indices

Fig. 3 2D geometry with square lattice. The red dashed line marks the domain which is depicted in Fig. 4

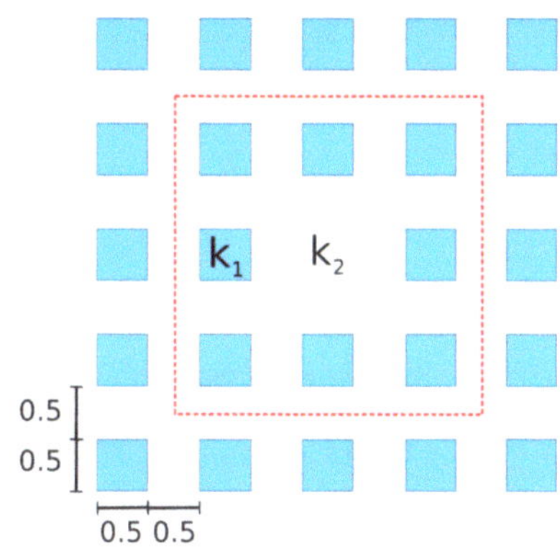

the original problem with local defects is reduced to a series of exact periodic problems which can be solved efficiently with standard numerical techniques. For each of the individual exact periodic subproblems the computational domain can be restricted to a finite region by applying the limiting absorption principle and the Bloch-Floquet transform. We demonstrated the feasibility of our formulation by implementing it for the one-dimensional as well as for the two-dimensional case. For both cases a practical test case has been shown to yield meaningful results with a good convergence behaviour (Fig. 5). Our investigations indicate that the convergence behaviour of the iteration depends on the magnitude of the perturbation. Consequently, for strongly perturbed geometries, the problem might not converge at all. However, for such cases, convergence can be re-established by employing a nested iteration.

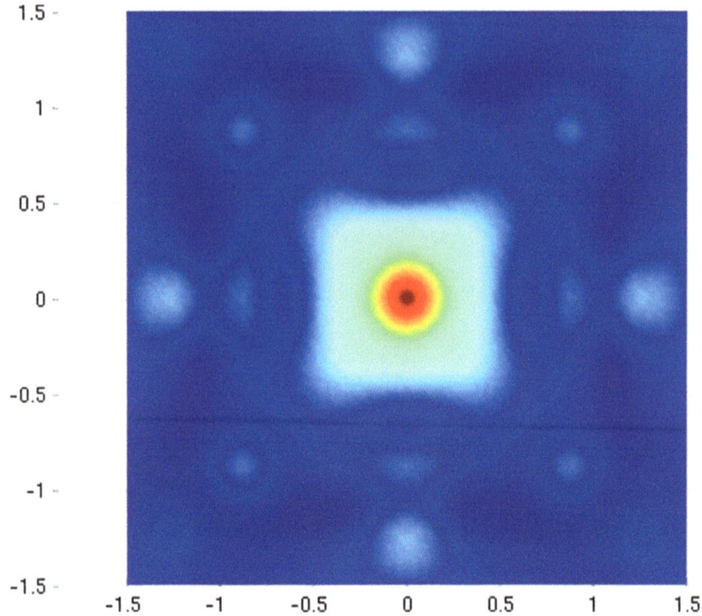

Fig. 4 Intensity of the computed 2d-solution for the infinite periodic scattering problem

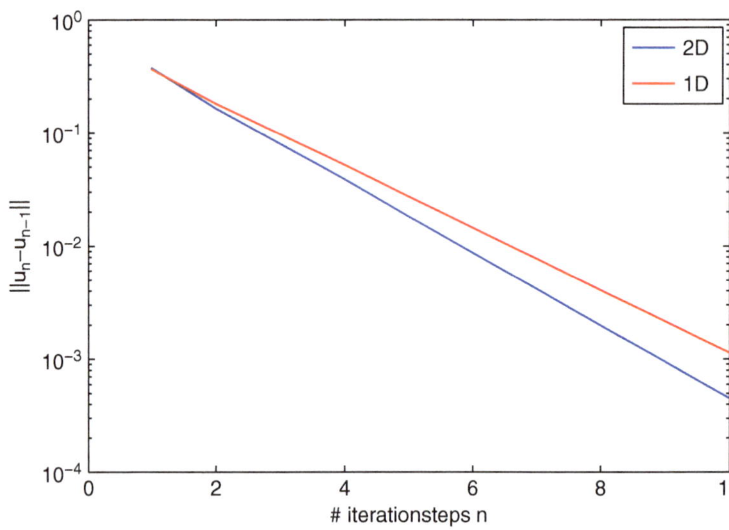

Fig. 5 Convergence of the iteration

References

1. Benisty, H., Berger, V., Gérard, J.M., Maystre, D., Tchelnokov, A.: Photonic Crystals. Towards Nanoscale Photonic Devices. Springer, Berlin (2008)
2. Botten, L.C., Dossou, K., Wilcox, S., McPhedran, R.C., de Sterke, C.M., Nicorovici, N.A., Asatryan, A.A.: Accurate defect mode modelling in photonic crystals using the generalised fictitious source superposition method. PIERS Online **3**(3), 320–324 (2007)
3. Fliss, S., Joly, P.: Exact boundary conditions for time-harmonic wave propagation in locally perturbed periodic media. Appl. Numer. Math. **59**, 2155–2178 (2009)
4. Joannopoulos, J.D., Johnson, S.G., Winn, J.N., Meade, R.D.: Photonic Crystals. Molding the Flow of Light. Princeton University Press, Princeton, NJ (2008)
5. Joly, P., Li, J., Fliss, S.: Exact boundary conditions for periodic waveguides containing a local perturbation. Comm. Comput. Phys. **1**(6), 945–973 (2006)
6. Kuchment, P.: The mathematics of photonic crystals. In: G. Bao, L. Cowsar, W. Masters (eds.) Mathematical Modeling in Optical Science, Frontiers in Applied Mathematics, vol. 22, pp. 207–272. SIAM, Philadelphia (2001)

Part II
Computational Electromagnetics

Introduction

This second part is concerned with methods for electromagnetic computation. The first paper by G. Sylvand (an invited speaker at the conference) sets the industrial scene of electromagnetic field computations by showing some examples of impressive large-scale computations made at the European Aeronautic Defence and Space company (EADS). Modern implementations of the fast-multipole methods on parallel computers are presently used to solve frequency domain electromagnetic scattering problems with boundary integral equation methods with upto forty million degrees of freedom.

The contribution by J. Ostrowski (an invited speaker at the conference) et al. deals with a finite element solution of the full linear Maxwell equations for slow processes. The authors propose a novel stabilisation technique that allows for the use of very large time steps in an explicit Euler scheme. This is of great importance for the efficient simulation of slow processes in order to keep the number of time steps reasonably small. The improved robustness is demonstrated through a numerical experiment on the lightning impulse test of a transformer.

The contribution by M. Kolmbauer and U. Langer proposes a preconditioned minimum residue solver for finite element discretisation of the frequency domain eddy current problem. The method is shown to be robust with regard to both the mesh size and the frequency. N. Sajjad, A. Khenchaf, and A. Coatanhay study the depolarisation of electromagnetic waves in scattering at soil surfaces. To include the effects of surface roughness, the authors add second order scattering effects at a small scale and develop an improved two-scale method. The performance of the new method is assessed by comparing backscattering simulation results with measured data and integral equation computation results. The contribution by J. Trommler, S. Koch, and T. Weiland, discusses two finite-element approaches to handle the inclusion of thin sheets in three-dimensional electro-quasistatic problems. The different methods, i.e. either using a high order approximation in the direction orthogonal to the surface or using an analytical model for the thin sheet jump relations, are then compared to a conventional discretisation method. In the following contribution, B. Bandlow and R. Schuhmann present new mode selecting eigensolvers for three-dimensional problems which, upon discretisation, may lead to unsymmetrical matrices. For the computation of interior eigen functions, one usually needs an estimate of the eigenvalues and the eigenvectors in order to obtain good convergence to the desired result. The authors propose an extended selection strategy for the Ritz pairs appearing in the Jacobi-Davidson eigensolver algorithm.

The following four papers rely on a quasistatic approximation to the Maxwell equations. The paper by P. Dular (an invited speaker at the conference) et al. discusses various methods of using model refinement in magnetic circuit computations. Transitions from, for example, one-dimensional to three-dimensional problems, from statics to dynamics or from perfect to real materials, can all be treated as model refinements. This paper shows a general method for making the mentioned transitions (and more) using finite element methods. In the paper by D. Ioan, G. Ciuprina, and A. Lazar a new modelling approach appropriate for substrate modelling is proposed. The main idea is to perform a hierarchical modelling based on an exponential partitioning scheme, which leads to a circuit model of linear complexity. In the paper by A. Fröhlcke, E. Gjonaj and T. Weiland a boundary conformal high-order Discontinuous Galerkin method on Cartesian grids is proposed for solving three-dimensional electro-quasistatic problems. Material boundary surfaces are handled using an accurate cut-cell approach. Two numerical examples show the optimal convergence rate of the method for arbitrary geometries. The paper by F. Muntean et al. analyses optimisation approaches for smoothing certain edge-effects in electro-deposited layers obtained in electrochemical reactors.

The next two papers study the movement of electrical charges in electromagnetic fields. The paper by T. Christen et al. introduces procedures for an improved prediction of streamer paths in complex geometries. Although the method is still based on the electric background field, it generalizes conventional models and is able to explain both streamer inception points other than at field maxima as well as streamer paths deviating from field lines. In the following paper, L. Pebernet et al. present a Particle-In-Cell (PIC) method based on a Discontinuous Galerkin scheme for the solution of the Maxwell-Vlasov equations in the time domain. Comparisons with a two-dimensional finite difference result found in the literature are made in order to validate the method.

The last two papers are examples of model coupling in which the electromagnetic field equations are coupled to the charge dynamic equations. This coupling is rather natural, though, because the charges are already present in the Maxwell equations. In the next part, we will see more examples of coupled problems, and in some cases the nature of the coupled problems is rather different.

From Quasi-static to High Frequencies : An Overview of Numerical Simulation at EADS

Guillaume Sylvand

Abstract EADS IW produces mathematical methods, numerical schemes and softwares in the field of electromagnetic simulation for the various needs of all EADS Business Units. Hence, we have produced over the years a wide range of tools for time domain and frequency domain EM problems. The aim of this talk is to give an overview of this work, underlining its most remarkable aspects, the recent developments and future perspectives.

1 Context

Innovation Works is, inside EADS, an entity devoted to research and development for the usage of EADS Business Units (Airbus, Eurocopter, MBDA, etc.). The numerical analysis team has been working for now more than 20 years on various methods for wave propagation simulations, first in electromagnetism, later in acoustics, both in frequency and time domain. The aim of this talk is to present the various tools currently developed and used inside EADS (with a focus on BEM, which is the main field of your author).

2 Boundary Element Method

The boundary element method has been a subject of work at EADS Innovation Works since the middle of the 1980s. Originally, the application was electromagnetism and, more specifically, stealth applications. The integral equations, solved

G. Sylvand (✉)
EADS Innovation Works, 18 rue Marius Terce, TOULOUSE, France
e-mail: guillaume.sylvand@eads.net

B. Michielsen and J.-R. Poirier (eds.), *Scientific Computing in Electrical Engineering SCEE 2010*, Mathematics in Industry 16, DOI 10.1007/978-3-642-22453-9_9,

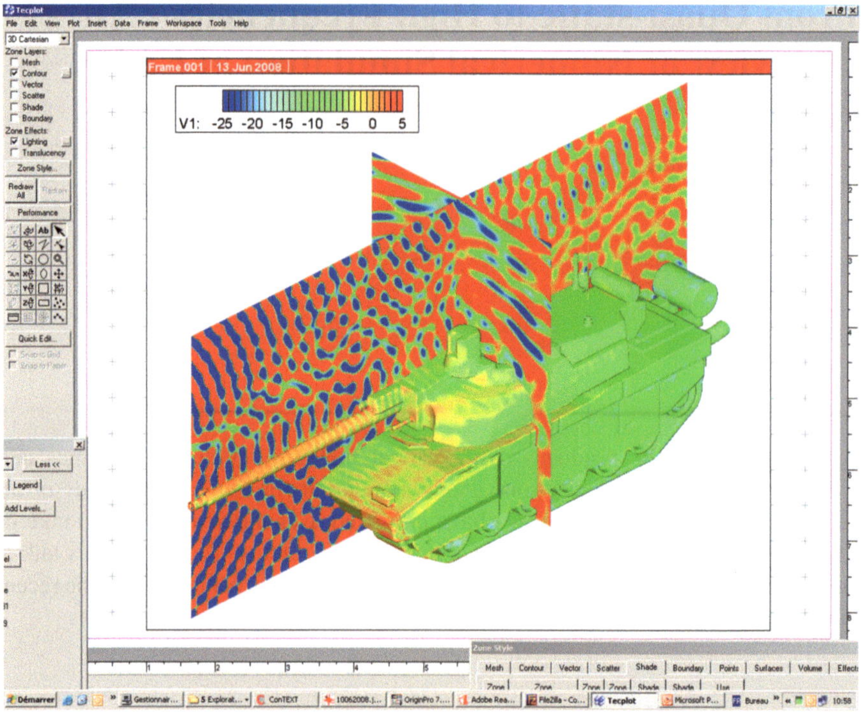

Fig. 1 Examples of BEM computation in frequency domain : Near field around a structure

with a boundary element method, were chosen because of their high accuracy. Another advantage of this method is that they only require a surface mesh of the object to compute, which is much simpler to realize, especially for complex "real world" (that is to say : not sphere !) objects such as planes or engines (Fig. 1).

2.1 Frequency Domain

Integral equations and BEM require manipulating more complex mathematical formulae (with singularities coming from the Green kernel), and they lead to dense symmetric matrices. For solving these linear systems, one can use either direct solvers (such as those based on an L.D.Lt factorization of the matrix). The cost of these methods grows like the cube of number of unknowns, which make them very expensive in CPU time for large problems (with several millions of unknowns). EADS IW has therefore developed an out-of-core parallel direct solver called SPIDO to handle the resolution. The other class of solver is iterative solvers, which can be very efficient when used in conjunction with a multipole method for speeding up the matrix-vector product.

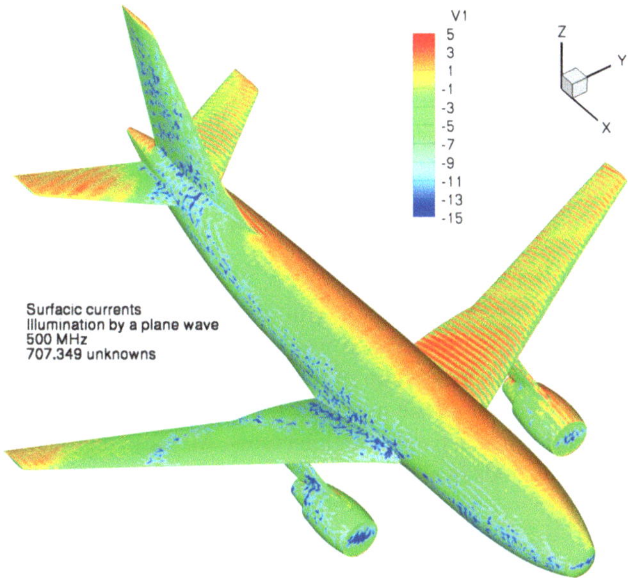

Fig. 2 Examples of BEM computation in frequency domain: surfacic currents on an aircraft

Iterative solvers allow solving a linear system by computing matrix-vector products. These products need O(n2) operations (where n is the number of unknowns) if they are written classically. The FMM (Fast Multipole Method) allows to compute them in O(n.log n) operations, hence leading to a much lower overall computation time. The algorithm is based on recursive decomposition of the object through a tree-like 3D-structure called octree.

Using this decomposition, one can separate interactions to compute between close interactions (between domain that touch each others) and distant interactions (the remaining ones). The latter are then treated in an approximate but faster way by the FMM algorithm. The FMM can accelerate the resolution of problems that can already be solved with direct solvers, but it also gives the possibility to handle very large problems out-of-reach until now (up to forty millions unknowns).

This method has been implemented at EADS IW (through a collaboration with CERMICS) since 1997. The software is parallel, out-of-core, and now widely used among the Business Units of EADS for acoustics (Airbus) or electromagnetism (Astrium, Airbus, Eurocopter, ...) applications (Fig. 2).

2.2 *Time Domain*

In time domain, BEM produce sparse matrices, solved with a marching-on-in-time algorithm. In this case, the numerical scheme is inconditionnaly stable, without any

CFL-like condition, thus allowing to very wide band computation on a given mesh (which is a precious advantage, from an industrial point of view).

3 Other Methods

3.1 FDTD

The finite difference time domain is still very widely used. Until the recent development of time domain BEM, it was the time domain of choice for a very wide range of computations. On top of that, it allows to represent realistic shapes and materials, adapted to real life needs. Its (relatively) simple scheme is also well suited for high performance parallel architectures. All these reasons make FDTD a very widely used methods in all EADS (Figs. 3 and 4).

Fig. 3 Examples of FDTD applications : NH90 helicopter mesh

Fig. 4 Examples of FDTD applications : A380 computation

Fig. 5 Examples of HF applications : aircraft case

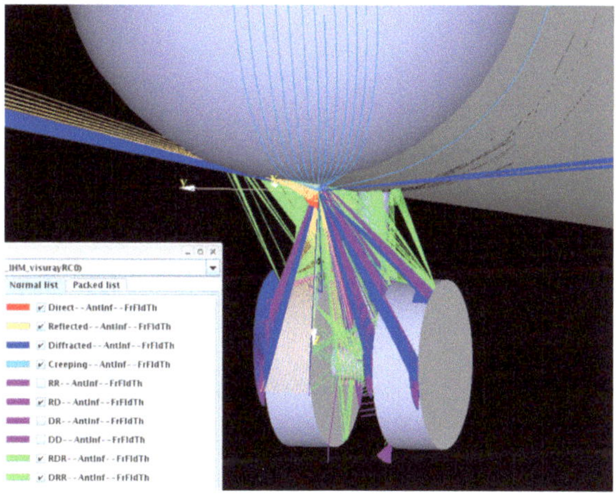

Fig. 6 Examples of HF applications : zoom on the landing gear area

3.2 High Frequency Method

For frequency domain computation, high frequency scheme based on ray launching techniques produce satisfactory results with very low computation time (Figs. 5 and 6).

Transient Full Maxwell Computation of Slow Processes

J. Ostrowski, R. Hiptmair, F. Krämer, J. Smajic, and T. Steinmetz

Abstract This article deals with finite element solution of the full linear Maxwell's equations. The focus lies on the transient simulation of slow processes, i.e. of processes, where wave propagation does not play a role. We employ an implicit Euler method for time discretization of the $\mathbf{A}, \varphi$-based Galerkin-formulation with Coulomb-gauge. We propose a novel stabilization technique that makes possible the use of very large timesteps. This is of supreme importance for efficient simulation of slow processes in order to keep the number of timesteps reasonably small. The greatly improved robustness in comparison with a standard formulation is demonstrated through numerical experiments. As an example we simulate the *lightning impulse test* of an industrial dry-type transformer.

1 Motivation

Electromagnetic field simulations of slow processes, i.e. of processes in the so-called *low frequency range*, where wave propagation does not play a role, are normally carried out by using either [8]

- A static model, i.e. *electrostatics* or *magnetostatics*, if all variations in time can be neglected.
- Or a quasi-static model, i.e. *electro-quasistatics* or *magneto-quasistatics*.

J. Ostrowski (✉) · J. Smajic · T. Steinmetz
ABB Switzerland Ltd., Corporate Research, Segelhofstr. 1 K, CH-5405 Baden, Switzerland
e-mail: joerg.ostrowski@ch.abb.com; jasmin.smajic@ch.abb.com; thorsten.steinmetz@ch.abb.com

R. Hiptmair · F. Krämer
Seminar for Applied Mathematics, ETH Zürich, Rämistr. 101, 8092 Zurich, Switzerland
e-mail: hiptmair@sam.math.ethz.ch; floriankra@gmail.com

B. Michielsen and J.-R. Poirier (eds.), *Scientific Computing in Electrical Engineering SCEE 2010*, Mathematics in Industry 16, DOI 10.1007/978-3-642-22453-9_10,

The static models are just special cases of the full Maxwell's equations, whereas the quasi-static models are approximations that are only valid in special situations [10,11]. If capacitive effects are dominant and the magnetic field energy is negligible against the electric field energy, then the electro-quasistatic model can be used, but induction is neglected. On the contrary, if inductive effects matter and the electric field energy is negligible versus the magnetic field energy, then the magneto-quasistatic model (or eddy-current model) can be used. The displacement-current is neglected in the magneto-quasistatic model.

This zoo of models forces the computational engineer to acquire and learn several simulation modules to cover the wide range of industrial applications. Some expertise in electromagnetics is also required in order to select the appropriate model. This is not desirable, because it limits the possible users of electromagnetic field simulation to a circle of experts. Moreover, the quasi-static models do not allow the simulation of configurations with coupled inductive/capacitive effects. For these reasons, we propose a generally applicable full Maxwell solver that unifies the four models of the low frequency range. However, standard full Maxwell formulations lack stability in this range. Therefore we describe a remedy in the form of a particular stabilization. Through this we achieve a robust Maxwell formulation.

We structured our article like this: first we analyze the reason for the instability of the standard full Maxwell formulation. Next we add the stabilization. Since this technique has already been introduced in frequency domain [3], we focus here on its realization in time domain. We demonstrate the strongly improved robustness by numerical experiments. At the end we show an industrial application of a transient simulation.

2 Instability of the Full Maxwell Model

We assume that the bounded computational domain $\Omega = \Omega_c \cup \Omega_n$ consists of a conductive domain Ω_c and a non-conductive domain Ω_n. For completeness we include possible prescribed solenoidal currents $\mathbf{j}^s$ and prescribed charges $\rho^s := -\text{div}\mathbf{j}^{sp}$. We assume stationary, i.e. non-moving, and non-deforming ohmic conductors. Thus the current is $\mathbf{j} = \sigma \cdot \mathbf{E} + \mathbf{j}^s$. We use an implicit Euler scheme for time discretization, because we deal with an essentially dissipative regime. With these definitions the standard Coulomb gauged $\mathbf{A}$, φ-based full Maxwell formulation that has to be solved in each timestep (k) writes

$$\mathbf{curl}\,\frac{1}{\mu}\,\mathbf{curl}\,\mathbf{A}_k + \left(\frac{\varepsilon}{\Delta t^2} + \frac{\sigma}{\Delta t}\right)\mathbf{A}_k + \left(\frac{\varepsilon}{\Delta t} + \sigma\right)\mathbf{grad}\,\varphi_k \tag{1}$$

$$= \left(\frac{2\varepsilon}{\Delta t^2} + \frac{\sigma}{\Delta t}\right)\mathbf{A}_{k-1} - \frac{\varepsilon}{\Delta t^2}\mathbf{A}_{k-2} + \frac{\varepsilon}{\Delta t}\,\mathbf{grad}\,\varphi_{k-1} + \mathbf{j}_k^s + \frac{\mathbf{j}_k^{sp} - \mathbf{j}_{k-1}^{sp}}{\Delta t} \text{ in } \Omega,$$

$$\text{div}(\varepsilon\mathbf{A}_k) = 0 \text{ in } \Omega\,. \tag{2}$$

Herein Δt is the size of the timestep and μ, ε and σ are material coefficients. The boundary conditions are chosen such that they model the contacts, see [3,4]. In [3] it was explained that in frequency domain this standard formulation lacks stability in the stationary limit, i.e. for vanishing angular frequency $\omega \to 0$. The same instability occurs in the time-domain for large timesteps Δt due to the equivalence of $i\omega$ in the frequency domain with $\frac{1}{\Delta t}$ in the time domain. The reason for the instability is the fact that for $\frac{1}{\Delta t} \to 0$ the electric scalar potential φ is not controlled by (1) and (2) in the non-conducting domain Ω_n (where $\sigma = 0$). As a consequence, φ becomes undetermined locally, and the electric field cannot be recovered in Ω_n. Theoretically, this is only the case at $\frac{1}{\Delta t} = 0$, but in computations one observes severe ill-conditioning already for positive but small $\frac{1}{\Delta t}$. This is caused by the very small parameter ε in the crucial term $\frac{\varepsilon}{\Delta t}\,\mathbf{grad}\,\varphi_k$ of (1).

This instability also haunts other standard formulations. If, for example, *temporal gauge* is used, where the electric scalar potential is set to zero, then we have to solve the system

$$\mathbf{curl}\,\frac{1}{\mu}\,\mathbf{curl}\,\mathbf{A}_k + \left(\frac{\varepsilon}{\Delta t^2} + \frac{\sigma}{\Delta t}\right)\mathbf{A}_k \tag{3}$$

$$= \left(\frac{2\varepsilon}{\Delta t^2} + \frac{\sigma}{\Delta t}\right)\mathbf{A}_{k-1} - \frac{\varepsilon}{\Delta t^2}\mathbf{A}_{k-2} + \mathbf{j}_k^s + \frac{\mathbf{j}_k^{sp} - \mathbf{j}_{k-1}^{sp}}{\Delta t} \quad \text{in } \Omega,$$

$$\varphi = 0 \quad \text{in } \Omega\,. \tag{4}$$

In this formulation we lose uniqueness of $\mathbf{A}_k$ in the non-conducting domain for large timesteps $\Delta t \to \infty$. In the limit, any gradient may be added to the solution of $\mathbf{A}$ in the non-conducting domain. Consequently the electric field is also poorly controlled for large Δt in temporal gauge, as is strikingly illustrated in Sect. 4. The same holds for the equivalent **E**-*based formulation*. If the gauge is removed in the *ungauged formulation* (i.e. only (1)), then one loses control of both φ and $\mathbf{A}$ in Ω_n. Again, the electric field cannot be recovered in Ω_n.

3 Stabilization

Stabilization, i.e. control of the electric field in the non-conductive Ω_n, is achieved according to the recipe of [3] by incorporating the charge neutrality of Ω_n aside from prescribed charges (i.e. Gauss' law). This extra condition is balanced by an extra unknown that results from the non-direct splitting of the electric scalar potential (φ) into two parts, $\varphi = \tilde{\varphi} + \psi$, with $\psi =$ constant in the conducting domain Ω_c. The final stable formulation in Coulomb gauge is then given by

$$\mathbf{curl}\,\frac{1}{\mu}\,\mathbf{curl}\,\mathbf{A}_k + \left(\frac{\varepsilon}{\Delta t^2} + \frac{\sigma}{\Delta t}\right)\mathbf{A}_k + \left(\frac{\varepsilon}{\Delta t} + \sigma\right)\mathbf{grad}(\tilde{\varphi}_k + \psi_k) \tag{5}$$

$$= \left(\frac{2\varepsilon}{\Delta t^2} + \frac{\sigma}{\Delta t}\right)\mathbf{A}_{k-1} - \frac{\varepsilon \mathbf{A}_{k-2}}{\Delta t^2} + \frac{\varepsilon\,\mathbf{grad}\,(\tilde{\varphi}_{k-1} + \psi_{k-1})}{\Delta t} + \mathbf{j}_k^s + \frac{\dot{\mathbf{j}}_k^{sp} - \dot{\mathbf{j}}_{k-1}^{sp}}{\Delta t} \text{ in } \Omega,$$

$$\operatorname{div}(\varepsilon \mathbf{A}_k) = 0 \text{ in } \Omega, \tag{6}$$

$$\operatorname{div}(\varepsilon\,\mathbf{grad}(\tilde{\varphi}_k + \psi_k)) = \operatorname{div}\dot{\mathbf{j}}_k^{sp} \text{ in } \Omega_n. \tag{7}$$

Note that the additional third equation (7) is independent of the timestep Δt, which achieves the stabilization. We chose the Coulomb gauge as typical gauge for the low frequency range, but this is not mandatory.

To cast (5) and (7) into weak form, we have to introduce an appropriate function space for the extra unknown ψ_k: $H_n^1 := \{\psi \in H_0^1(\Omega) \colon \psi_{|\Omega_c} \equiv \text{const}\}$. The function spaces for the other unknowns follow from standard choices, see [2, Sect. 5] for notations and details. Then the variational formulation reads: seek $\mathbf{A}_k \in \boldsymbol{H}_0(\mathbf{curl}, \Omega)$, $\tilde{\varphi}_k \in H_0^1(\Omega)$, $\psi_k \in H_n^1(\Omega)$ such that

$$\left\langle \frac{1}{\mu}\,\mathbf{curl}\,\mathbf{A}_k,\, \mathbf{curl}\,\mathbf{A}' \right\rangle + \left\langle \left(\frac{\varepsilon}{\Delta t^2} + \frac{\sigma}{\Delta t}\right)\mathbf{A}_k,\, \mathbf{A}' \right\rangle \tag{8}$$

$$+ \left\langle \left(\frac{\varepsilon}{\Delta t} + \sigma\right)\mathbf{grad}(\tilde{\varphi}_k + \psi_k),\, \mathbf{A}' \right\rangle = \left\langle \left(\frac{2\varepsilon}{\Delta t^2} + \frac{\sigma}{\Delta t}\right)\mathbf{A}_{k-1} - \frac{\varepsilon}{\Delta t^2}\mathbf{A}_{k-2},\, \mathbf{A}' \right\rangle$$

$$+ \left\langle \frac{\varepsilon}{\Delta t}\,\mathbf{grad}(\tilde{\varphi}_{k-1} + \psi_{k-1}),\, \mathbf{A}' \right\rangle + \left\langle \mathbf{j}_k^s,\, \mathbf{A}' \right\rangle + \left\langle \frac{\dot{\mathbf{j}}_k^{sp} - \dot{\mathbf{j}}_{k-1}^{sp}}{\Delta t},\, \mathbf{A}' \right\rangle \text{ in } \Omega,$$

$$\left\langle \varepsilon \mathbf{A}_k\,,\, \mathbf{grad}\,\tilde{\varphi}' \right\rangle = 0 \text{ in } \Omega, \tag{9}$$

$$\left\langle (\varepsilon\,\mathbf{grad}(\tilde{\varphi}_k + \psi_k))\,,\, \mathbf{grad}\,\psi' \right\rangle = \left\langle \operatorname{div}\dot{\mathbf{j}}_k^{sp}\,,\, \psi' \right\rangle \text{ in } \Omega_n. \tag{10}$$

for all $\mathbf{A}' \in \boldsymbol{H}_0(\mathbf{curl}, \Omega)$, $\varphi' \in H_0^1(\Omega)$, and $\psi' \in H_n^1(\Omega)$.

Remark: We point out that the solution cannot be unique, because (10) can be obtained by testing (8) with $\mathbf{A}' := \mathbf{grad}\,\psi'$, $\psi' \in H_n^1(\Omega)$.

4 Numerical Experiments

We employ a conformal Galerkin finite element discretization of (8)–(10) based on first order edge elements for the vector potential and first order nodal elements for the scalar potentials [2, Sect. 3]. This was implemented in an in-house simulation framework at ABB.

According to the above remark, we face a *singular* linear system of equations with consistent right hand side in each timestep. Iterative solvers can tackle this kind of problem and we used a preconditioned BiCGstab method to solve the system. We constructed a preconditioner by using the direct solver Pardiso [7] for solving the regularized system that results from applying a lower conductivity bound of

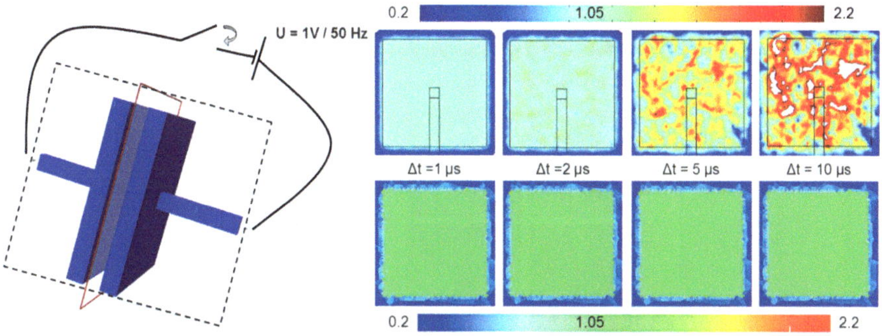

Fig. 1 *Left:* Plate capacitor. *Right:* Electric field in [V/m] after 100 μs in the center between the capacitor plates (solid red plane) for different timesteps. The upper row shows the solutions of the system (3)–(4), and the lower row shows the solutions of the stabilized system (8)–(10). Please note the different color-scales due to the different visualization software. The expected value is 1.05 V/m. A mesh of 200,000 elements was used in both cases

1 $(\Omega\mathrm{m})^{-1}$ in (8). Note that this expensive preconditioner is almost identical to a direct solver, because the regularized system differs only slightly from (8)–(10). An alternative is probably the cheaper preconditioner that was introduced for frequency domain in [6], but that has not yet been transferred to time domain.

In order to compare formulations, we used the RF module of the commerical software COMSOL [1] (with the direct solver Pardiso) for the solution of the standard non-stabilized formulation in temporal gauge (3) and (4). A simple rectangular plate capacitor with plate distance of 3 cm and plate diameter of 43 cm was computed. We switched on a sinusoidal voltage of 1 V/50 Hz. Figure 1 shows the greatly improved robustness of the stabilized system.

For the standard system (3) and (4) one encounters a severe stability constraint on the timestep, despite the use of implicit timestepping, and the electric field is disturbed for timesteps larger than 1 μs. We observed that the disturbance started even earlier, at timesteps of 0.5 μs for a larger mesh with one Million elements. This timestep constraint is much more relaxed for the stabilized system (8)–(10), where we could use three orders of magnitude larger timesteps of 1 ms. This is confirmed by a comparison with a computation in the frequency domain, see Fig. 2.

5 Lightning Impulse Test Simulation

As a practical example we simulated the lightning impulse test of an ABB dry-type transformer.

Power and distribution transformers are not only exposed to the rated voltage over their life time; occasionally, transformers can experience transient voltage surges produced by network switching operations or atmospheric overvoltages. The insulation between the windings has to be very carefully designed to ensure

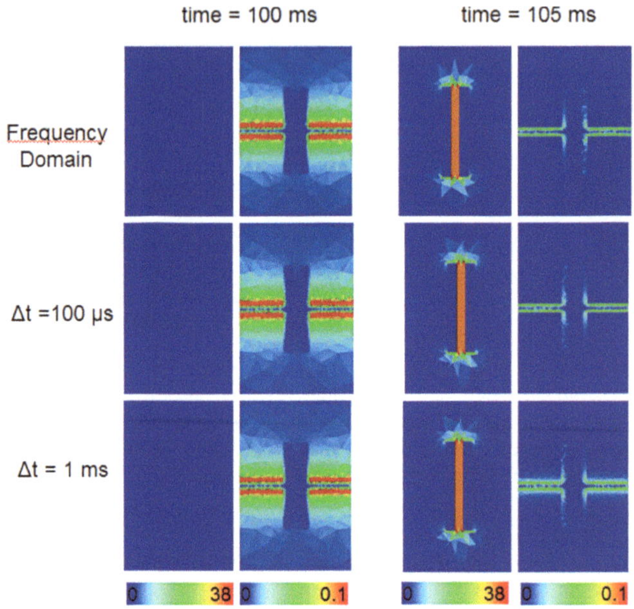

Fig. 2 Comparison of the electric field (0 − 38 V/m) and the magnetic field (0 − 0.1 pT) between the solution in frequency domain, and the solution in time domain for timesteps of 1 ms and 100 μs. The fields are shown after a steady oscillation is reached (i.e. here after five periods of 20 ms) at the zero-voltage-intersect after 100 ms and at the peak voltage after 105 ms. The pictures are in the dashed black plane of the capacitor in Fig. 1

reliable operation even if a voltage surge occurs. This is tested experimentally by the lightning impulse test which is precisely defined by the IEC standard [5]. Due to the lack of insulating oil in dry-type transformers, more sophisticated dielectric design is required compared to the oil-immersed counterparts. Thus the dielectric design of dry-type transformers can be strongly supported by electromagnetic field simulations of the lightning impulse test. An accurate simulation of the electric field between the winding sections is therefore of paramount importance.

The configuration for the lightning impulse test is shown in Fig. 3. The peak value of the applied impulse voltage is roughly five times the nominal voltage. The 1.2 μs rise time and 50 μs decay time of the voltage in the lightning impulse test are specified to mimic the real nature of the surge, see [5]. The Fourier spectrum of the applied signal comprises waves with wavelengths comparable to the size of the windings. The propagation of the waves along the windings can produce local field enhancement regions that are caused by both multiple reflections of the electromagnetic waves and internal resonance effects by the interaction of the capacitances and inductances of the windings. Due to the complex winding geometries, it is practically impossible to predict critical field regions of the windings without performing transient full Maxwell simulations. Taking into account only static simulations may strongly falsify the estimation of a possible dielectric breakdown.

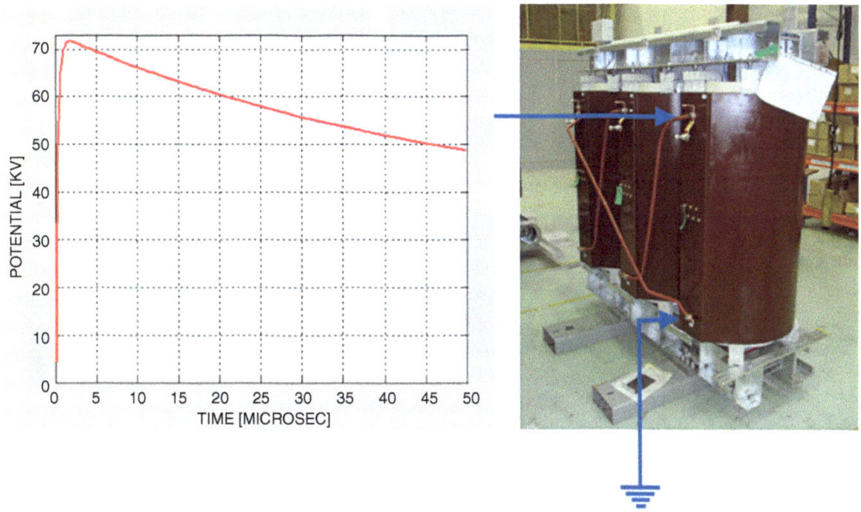

Fig. 3 Dry-Transformer and applied test voltage pulse

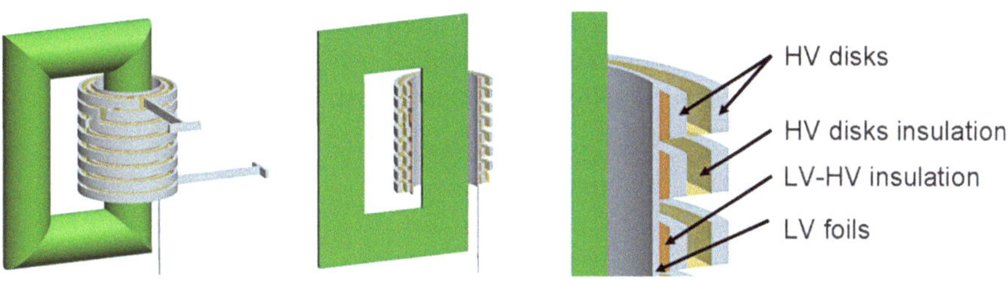

Fig. 4 Transformer model showing windings and core (*left*), its cross-section (*center*), and the details of the winding disks (*right*)

However, the windings in their full complexity cannot be modeled in 3D: Each disk consists of several tens of turns of conductive foil with a thickness in the range of some hundred microns. The foils are insulated against each other. The thickness of the insulation is even smaller than the thickness of the conductive foil. This yields a huge aspect ratio compared to the height of about 2 m of the transformer. Therefore, the internal structure of the winding sections is simplified. Effective values of the dielectric permittivity and magnetic permeability are used to model the internal capacitances and inductances of the real winding sections. The regions of the effective material parameters are shown in Fig. 4. They are placed in the regions where the corresponding capacitances (turn-to-turn capacitance of a single winding disk) and inductances (stray field in the cooling channel) are confined. The effective material parameters are derived from 2D computations. An alternative treatment of the windings could be to use the homogenization technique, see [9].

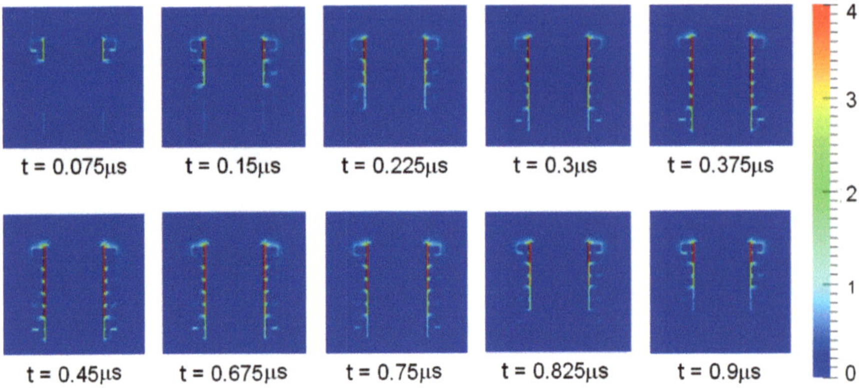

Fig. 5 Electric field distribution (in kV/mm) over the axial symmetry slice (Fig. 4, *center*) shown at different time instants. The propagation and the reflection of the electromagnetic wave is visible. Note that this oscillation is much faster than the applied voltage pulse of Fig. 3

The transient electric field between the windings as computed by an impulse lightning simulation is shown in Fig. 5. One recognizes a reflection during the rise time of 1 μs of the voltage. These reflections continue during the first 10 μs of the simulation. This agrees with the measured oscillations of the electric field between the windings in the experimental lightning impulse test.

6 Conclusion

In this article we introduced a *robust full Maxwell formulation* in time domain. This formulation is stable for large timesteps in simulations of slow processes. We demonstrated the improved stability, compared to standard formulations, by numerical experiments. A three order of magnitude improvement of the timestep was achieved. As an example, we simulated a transient industrial application with coupled capacitive and inductive effects. We conclude that the stabilized formulation (8)–(10) represents a unified Maxwell model that is robust even with large timesteps.

References

1. Comsol website, URL http://www.comsol.com
2. Hiptmair, R.: Finite elements in computational electromagnetism. Acta Numerica **11**, 237–339 (2002)
3. Hiptmair, R., Krämer, F., Ostrowski, J.: A robust Maxwell formulation for all frequencies. IEEE Trans. Magn. **44**(6), 682–685 (2008)

4. Hiptmair, R., Sterz, O.: Current and voltage excitations for the eddy current model. Int. J. Numer. Model **18**(1), 1–21 (2005)
5. International standard IEC 60076-11: Dry-type transformers. International Electrotechnical Commission, Geneve, Switzerland (2004)
6. Ostrowski, J., Bebendorf, M., Hiptmair, R., Krämer, F.: H-matrix based operator preconditioning for full maxwell at low frequencies. IEEE Trans. magn. **46**(8), 3193–3196 (2010)
7. Pardiso website, www.pardiso-project.org
8. van Rienen, U.: Numerical methods in computational electrodynamics. In: Lecture Notes in Computational Science and Engineering, vol. 12. Springer, Berlin (2001)
9. Sabariego, R., Dular, P., Gyselinck, J.: Time-domain homogenization of windings in 3-d finite element models. IEEE Trans. magn. **44**(6), 1302–1305 (2008)
10. Schmidt, K., Sterz, O., Hiptmair, R.: Estimating the eddy-current modelling error. IEEE Trans. Magn. **44**(6), 686–689 (2008)
11. Steinmetz, T., Kurz, S., Clemens, M.: Domains of validity of quasistatic and quasistationary field approximations. Submitted to COMPEL **30**(4), 1237–1247 (2011)

A Frequency-Robust Solver for the Time-Harmonic Eddy Current Problem

Michael Kolmbauer and Ulrich Langer

Abstract This work is devoted to fast and parameter-robust iterative solvers for frequency domain finite element equations, approximating the eddy current problem with harmonic excitation. We construct a preconditioned MinRes solver for the frequency domain equations, that is robust (= parameter-independent) in both the discretization parameter h and the frequency ω.

1 Introduction

In many practical applications, the excitation is time-harmonic. Switching from the time domain to the frequency domain allows us to replace expensive time-integration procedures by the solution of a system of partial differential equations for the amplitudes belonging to the sine- and to the cosine-excitation. Following this strategy Copeland et al. [7, 8] and Bachinger et al. [5, 6] applied harmonic and multiharmonic approaches to parabolic initial-boundary value problems and the eddy current problem, respectively. Indeed, in [7] a MinRes solver for the solution of parabolic initial-boundary value problems is constructed, that is robust with respect to both the discretization parameter h and the frequency ω. The aim of this work is

M. Kolmbauer (✉)
Institute of Computational Mathematics, Johannes Kepler University, Altenbergerstr. 69, A-4040 Linz, Austria
e-mail: kolmbauer@numa.uni-linz.ac.at

U. Langer
Institute of Computational Mathematics, Johannes Kepler University, Altenbergerstr. 69, A-4040 Linz, Austria

RICAM, Johann Radon Institute for Computational and Applied Mathematics, Austrian Academy of Sciences, Altenbergerstr. 69, A-4040 Linz, Austria
e-mail: ulrich.langer@assoc.oeaw.ac.at;ulanger@numa.uni-linz.ac.at

B. Michielsen and J.-R. Poirier (eds.), *Scientific Computing in Electrical Engineering SCEE 2010*, Mathematics in Industry 16, DOI 10.1007/978-3-642-22453-9_11,

to generalize these ideas to the eddy current problem. Due to the non-trivial kernel of the **curl**-operator, the generalization of this solver is not straight forward. In order to achieve a positive definite reformulation of the frequency domain equations, we perform a regularization in terms of an additional gauging term. The regularized problem can be solved in a MinRes setting, applying a preconditioning technique proposed by Schöberl and Zulehner [19].

2 Frequency Domain FEM

As a model problem we consider the eddy current problem with homogeneous Dirichlet boundary condition and an inhomogeneous initial condition.

$$\left\{\begin{array}{rcll} \sigma \dfrac{\partial \mathbf{u}}{\partial t} + \mathbf{curl}\,(\nu\, \mathbf{curl}\, \mathbf{u}) & = & \mathbf{f} & \text{in } \Omega \times (0,T] \\ \mathbf{u}(\mathbf{x},0) & = & \mathbf{0} & \text{in } \bar{\Omega} \\ \mathbf{u}(\mathbf{x},t) & = & \mathbf{0} & \text{on } \partial\Omega \times [0,T] \end{array}\right. \tag{1}$$

We assume, that $\Omega \subset \mathbb{R}^3$ is a bounded Lipschitz domain. The reluctivity $\nu = \nu(\mathbf{x})$ is supposed to be independent of $|\mathbf{curl}\ \mathbf{u}|$, i.e. we assume that the eddy current problem (1) is linear. The conductivity σ is piecewise constant and zero in non-conducting regions. We assume that the source $\mathbf{f}$ is weakly divergence free. Bachinger et al. [5] provide existence und uniqueness results for linear and non-linear eddy current problems in appropriate gauged spaces.

Furthermore we assume that $\mathbf{f}$ is given by a time-harmonic excitation with frequency $\omega > 0$ and amplitudes $\mathbf{f}^{\mathrm{c}}$ and $\mathbf{f}^{\mathrm{s}}$, i.e. $\mathbf{f}(\mathbf{x},t) = \mathbf{f}^{\mathrm{c}}(\mathbf{x})\cos(\omega t) + \mathbf{f}^{\mathrm{s}}(\mathbf{x})\sin(\omega t)$ Therefore the solution $\mathbf{u}$ is time-harmonic as well, with the same base frequency ω:

$$\mathbf{u}(\mathbf{x},t) = \mathbf{u}^{\mathrm{c}}(\mathbf{x})\cos(\omega t) + \mathbf{u}^{\mathrm{s}}(\mathbf{x})\sin(\omega t). \tag{2}$$

In fact, (2) is the real reformulation of a complex time-harmonic approach $\mathbf{u}(\mathbf{x},t) = \hat{\mathbf{u}}(\mathbf{x})e^{i\omega t}$ with the complex-valued amplitude $\hat{\mathbf{u}} = \mathbf{u}^{\mathrm{c}} - i\mathbf{u}^{\mathrm{s}}$. Using the real-valued time-harmonic representation of the solution (2), we can state the eddy current problem (1) in the frequency domain as follows:

$$\text{Find } \mathbf{u} = (\mathbf{u}^{\mathrm{c}}, \mathbf{u}^{\mathrm{s}}) : \quad \left\{\begin{array}{l} \mathbf{curl}\,(\nu\, \mathbf{curl}\, \mathbf{u}^{\mathrm{c}}) + \omega\, \sigma\, \mathbf{u}^{\mathrm{s}} = \mathbf{f}^{\mathrm{c}} \\ \mathbf{curl}\,(\nu\, \mathbf{curl}\, \mathbf{u}^{\mathrm{s}}) - \omega\, \sigma\, \mathbf{u}^{\mathrm{c}} = \mathbf{f}^{\mathrm{s}}, \end{array}\right. \tag{3}$$

with the corresponding boundary conditions from (1).

Remark 1. Having in mind applications to problems with non-linear reluctivity ν, we prefer to use the real reformulation (3) instead of a complex approach (see [3, Sect. 3.4]).

The finite element discretization of the variational formulation of (3) with lowest order edge elements, introduced by Nédélec in [13], yields the following system of linear equations

$$\begin{pmatrix} \mathbf{A_h} & \omega \mathbf{M}_{\sigma,\mathbf{h}} \\ -\omega \mathbf{M}_{\sigma,\mathbf{h}} & \mathbf{A_h} \end{pmatrix} \begin{pmatrix} \mathbf{u}_\mathbf{h}^\mathrm{c} \\ \mathbf{u}_\mathbf{h}^\mathrm{s} \end{pmatrix} = \begin{pmatrix} \mathbf{f}_\mathbf{h}^\mathrm{c} \\ \mathbf{f}_\mathbf{h}^\mathrm{s} \end{pmatrix} \tag{4}$$

with stiffness matrix $\mathbf{A_h}$ and mass matrix $\mathbf{M}_{\sigma,\mathbf{h}}$.

3 Exact Regularization

Eddy current problems are essentially different for conducting ($\sigma > 0$) and non-conducting regions ($\sigma = 0$). In order to gain uniqueness in the non-conducting regions, we persue an exact regularization strategy.

Due to the non-trivial kernel of the **curl**-operator, the resulting stiffness matrix $\mathbf{A_h}$ is only positive semi-definite. However, for later preconditioning purposes, we require that the sum of certain blocks of the system matrix (4) is positive definite. In order to achieve that, we follow a gauging strategy proposed by Kuhn [12]. The regularized variational problem reads as

$$\text{Find } \mathbf{u} = (\mathbf{u}^\mathrm{c}, \mathbf{u}^\mathrm{s}) \in H_0(\mathbf{curl})^2 : \quad a_Q(\mathbf{u}, \mathbf{v}) = \langle F, \mathbf{v} \rangle, \quad \forall \mathbf{v} \in H_0(\mathbf{curl})^2 \tag{5}$$

with the regularized bilinear form

$$a_Q(\mathbf{u}, \mathbf{v}) := \sum_{j \in \{c,s\}} \int_\Omega \nu \,\mathbf{curl}\, \mathbf{u}^\mathrm{j} \mathbf{curl}\, \mathbf{v}^\mathrm{j} + \omega \nabla P_D \mathbf{u}^\mathrm{j} \nabla P_D \mathbf{v}^\mathrm{j} \mathrm{d}\mathbf{x} + \omega \int_\Omega \sigma \left[\mathbf{u}^\mathrm{c} \mathbf{v}^\mathrm{s} - \mathbf{u}^\mathrm{s} \mathbf{v}^\mathrm{c}\right] \mathrm{d}\mathbf{x}. \tag{6}$$

Here $P_D : H_0(\mathbf{curl}) \to H_0^1(\Omega)$ is the Helmholtz projection (see e.g. [12]). For any $\mathbf{v} \in H_0(\mathbf{curl})$, $P_D \mathbf{v} := p$ is defined by the unique solution of the variational problem

$$(\nabla p, \nabla q)_{L_2(\Omega)} = (\mathbf{v}, \nabla q)_{L_2(\Omega)}, \quad \forall q \in H_0^1(\Omega). \tag{7}$$

Hence we replace $\mathbf{A_h}$ by the sum of $\mathbf{A_h}$ and a regularization term $\omega \mathbf{Q_h}$, i.e. $\mathbf{A_h} + \omega \mathbf{Q_h}$. Here $\mathbf{Q_h}$ is the discretization of the operator $\mathbf{Q}$, defined by $(\mathbf{Q}\,\mathbf{u}, \mathbf{v})_{L_2(\Omega)} := \int_\Omega \nabla P_D \mathbf{u} \nabla P_D \mathbf{v} \mathrm{d}\mathbf{x}$, by Nédélec finite elements of lowest order.

$$\begin{pmatrix} \mathbf{A_h} + \omega \mathbf{Q_h} & \omega \mathbf{M}_{\sigma,\mathbf{h}} \\ -\omega \mathbf{M}_{\sigma,\mathbf{h}} & \mathbf{A_h} + \omega \mathbf{Q_h} \end{pmatrix} \begin{pmatrix} \mathbf{u}_\mathbf{h}^\mathrm{c} \\ \mathbf{u}_\mathbf{h}^\mathrm{s} \end{pmatrix} = \begin{pmatrix} \mathbf{f}_\mathbf{h}^\mathrm{c} \\ \mathbf{f}_\mathbf{h}^\mathrm{s} \end{pmatrix}. \tag{8}$$

The operator P_D and hence the matrix $\mathbf{Q_h}$ are chosen in such a way, that on the one hand it ensures the positive definiteness of the block $\mathbf{A_h} + \omega \mathbf{Q_h}$ and on the other hand $\mathbf{Q_h} \mathbf{u}_\mathbf{h}^{\mathrm{c/s}}$ vanishes at the solution, i.e. the regularized system (8) and the original system (4) have one and the same solution. The proof of the equivalence of the original and exact regularized problem (5) follows the same steps as in [12].

4 MinRes Preconditioner

For preconditioning purpose we have to reformulate the system (8) with a positive definite but block skew-symmetric system matrix, as a symmetric but indefinite one. This system can be solved by a preconditioned MinRes method [14]. The key points for the construction of a parameter robust preconditioner are the introduction of a non-standard norm in $H_0(\mathbf{curl})$ and the theorem of Babuška-Aziz [2].

The symmetric and indefinite reformulation of the variational formulation with a positive definite but skew-symmetric bilinear form (5) is given by:

$$\text{Find } (\mathbf{x},\mathbf{y}) \in H_0(\mathbf{curl})^2 : \quad \mathscr{A}_M((\mathbf{x},\mathbf{y}),(\mathbf{v},\mathbf{w})) = \int_\Omega \left[\frac{1}{\omega}\mathbf{f}^c\mathbf{v} + \mathbf{f}^s\mathbf{w}\right] d\mathbf{x} \tag{9}$$

for all $(\mathbf{v},\mathbf{w}) \in H_0(\mathbf{curl})^2$, with the scaled vectors $(\mathbf{x},\mathbf{y}) = (\mathbf{u}^s, \frac{1}{\omega}\mathbf{u}^c)$ and $(\mathbf{v},\mathbf{w}) = (\omega\mathbf{v}^c, \mathbf{v}^s)$ and the symmetrised bilinear form $\mathscr{A}_M(\cdot,\cdot)$, given by

$$\begin{aligned}\mathscr{A}_M((\mathbf{x},\mathbf{y}),(\mathbf{v},\mathbf{w})) = {} & (\sigma\ \mathbf{x},\mathbf{v})_{L_2(\Omega)} - \omega^2(\sigma\ \mathbf{y},\mathbf{w})_{L_2(\Omega)} \\ & + (\nu\ \mathbf{curl}\ \mathbf{y}, \mathbf{curl}\ \mathbf{v})_{L_2(\Omega)} + \omega(\nabla P_D\mathbf{y}, \nabla P_D\mathbf{v})_{L_2(\Omega)} \\ & + (\nu\ \mathbf{curl}\ \mathbf{x}, \mathbf{curl}\ \mathbf{w})_{L_2(\Omega)} + \omega(\nabla P_D\mathbf{x}, \nabla P_D\mathbf{w})_{L_2(\Omega)}.\end{aligned}$$

Hence we can reformulate the block skew-symmetric and positive definite system (8) as a symmetric but indefinite system (10) with system matrix $\mathbf{D_h}$:

$$\begin{pmatrix} \mathbf{M}_{\sigma,\mathbf{h}} & \mathbf{A_h} + \omega\mathbf{Q_h} \\ \mathbf{A_h} + \omega\mathbf{Q_h} & -\omega^2\mathbf{M}_{\sigma,\mathbf{h}} \end{pmatrix} \begin{pmatrix} \mathbf{u_h^s} \\ \frac{1}{\omega}\mathbf{u_h^c} \end{pmatrix} = \begin{pmatrix} \frac{1}{\omega}\mathbf{f_h^c} \\ \mathbf{f_h^s} \end{pmatrix}. \tag{10}$$

Next we construct a block-diagonal preconditioner according to the preconditioning technique proposed by Schöberl and Zulehner [19]. We introduce the non-standard norm $\|\cdot\|_{V_M}$ in $H_0(\mathbf{curl})$

$$\|\mathbf{y}\|_{V_M}^2 = \frac{1}{\omega}\left[(\nu\ \mathbf{curl}\ \mathbf{y}, \mathbf{curl}\ \mathbf{y})_{L_2(\Omega)} + \omega\|\nabla P_D\mathbf{y}\|_{L_2(\Omega)}^2 + \omega(\sigma\ \mathbf{y},\mathbf{y})_{L_2(\Omega)}\right]. \tag{11}$$

Note, that the regularization term ensures, that this norm is well defined even in non-conducting regions. This definition gives rise to a non-standard norm $\|\cdot\|_{Q_M}$ in the product space $H_0(\mathbf{curl})^2$

$$\|(\mathbf{x},\mathbf{y})\|_{Q_M}^2 = \|\mathbf{x}\|_{V_M}^2 + \omega^2\|\mathbf{y}\|_{V_M}^2. \tag{12}$$

Lemma 1. *We have*

$$\frac{1}{\sqrt{2}}\|(\mathbf{x},\mathbf{y})\|_{Q_M} \leq \sup_{0\neq(\mathbf{v},\mathbf{w})\in H_0(\mathbf{curl})^2} \frac{\mathscr{A}_M((\mathbf{x},\mathbf{y}),(\mathbf{v},\mathbf{w}))}{\|(\mathbf{v},\mathbf{w})\|_{Q_M}} \leq \|(\mathbf{x},\mathbf{y})\|_{Q_M}. \tag{13}$$

Proof. Boundedness follows from reapplication of Cauchy's inequality. The lower estimate can be attained by choosing $\mathbf{v} = \omega\mathbf{y} + \mathbf{x}$ and $\mathbf{w} = \frac{1}{\omega}\mathbf{x} - \mathbf{y}$. □

Since we are dealing with conforming finite elements, the estimate (13) is also valid in the Nédélec finite element subspace. Hence, it follows by the theorem of Babuška-Aziz, that there exists a unique solution of the corresponding variational problem (9), and that the solution continuously depends on the data, uniformly on ω and σ. Hence we conclude that the block-diagonal preconditioner

$$\mathbf{C_h} = \frac{1}{\omega}\begin{pmatrix} \tilde{\mathbf{C}}_\mathbf{h} & \mathbf{0} \\ \mathbf{0} & \omega^2\tilde{\mathbf{C}}_\mathbf{h} \end{pmatrix}, \tag{14}$$

with $\tilde{\mathbf{C}}_\mathbf{h} = \omega(\mathbf{M}_\sigma,\mathbf{h} + \mathbf{Q_h}) + \mathbf{A_h}$, is robust with respect to both the discretization parameter h and the parameters ω and σ. Thus the spectral condition number (16) of the preconditioned system

$$\mathbf{C_h}^{-1}\mathbf{D_h}\mathbf{u_h} = \mathbf{C_h}^{-1}\mathbf{f_h} \tag{15}$$

can be estimated by a constant c that is independent of h, ω and σ i.e.

$$\kappa(\mathbf{C_h}^{-1}\mathbf{D_h}) := \|\mathbf{C_h}^{-1}\mathbf{D_h}\|_{\mathbf{C_h}}\|\mathbf{D_h}^{-1}\mathbf{C_h}\|_{\mathbf{C_h}} \leq c \neq c(\omega,h,\sigma). \tag{16}$$

Therefore the number of MinRes iterations required for reducing the initial error by some fixed factor $\varepsilon \in (0,1)$ is independent of the discretization parameter h and the frequency ω.

In practice, the diagonal blocks $\tilde{\mathbf{C}}_\mathbf{h}$ in (14) are replaced by some appropriate preconditioners, e.g. by robust multigrid preconditioners as proposed in [1].

Theorem 1 (Entire robust and optimal solver). *The MinRes method applied to the preconditioned system (15) converges. At the m-th iteration, the preconditioned residual* $\mathbf{r_h^m} = \mathbf{C_h}^{-1}\mathbf{f_h} - \mathbf{C_h}^{-1}\mathbf{D_h}\mathbf{u_h^m}$ *is bounded as*

$$\|\mathbf{r_h^{2m}}\|_{\mathbf{C_h}} \leq \frac{2q^m}{1+q^{2m}}\|\mathbf{r_h^0}\|_{\mathbf{C_h}} \quad \textit{where} \quad q = \frac{\kappa(\mathbf{C_h}^{-1}\mathbf{D_h}) - 1}{\kappa(\mathbf{C_h}^{-1}\mathbf{D_h}) + 1}. \tag{17}$$

If we additionally apply the Arnold/Falk/Winther multigrid preconditioner [1] to the diagonal blocks, the whole convergence rate q is independent of ω and h.

Proof. The convergencerate of the MinRes method [14] can be found in [10]. Combining this result with the estimate of the condition number (16) and the multigrid convergence [1], yields the desired result. □

Table 1 Number of MinRes iterations for reducing the initial residual by 10^{-6}

DOF	$\log_{10}\omega$	−4	−3	−2	−1	0	1	2	3	4	5	6	7	8	CPU time
1,208	$h = 0.25$	3	3	3	5	7	14	15	16	14	8	6	4	4	< 0.48 s
8,368	$h = 0.125$	3	3	3	5	7	13	15	16	16	12	6	4	4	< 2.48 s
62,048	$h = 0.0625$	3	3	3	5	7	13	15	16	16	14	8	6	4	< 29.79 s
477,376	$h = 0.03125$	3	3	3	5	7	8	16	16	16	13	12		4	< 477.55 s
Skin depth $\sqrt{2\nu/(\omega\sigma)}$		141.4	44.6	14.1	4.5	1.4	0.4	0.14	0.044	< 0.03125					

Table 2 Number of MinRes iterations for reducing the initial residual by 10^{-6}

DOF	$\log_{10}\sigma_2$	−4	−3	−2	−1	1	2	3	4	5	6	7	8
196	$h = 0.5$	7	7	7	7	13	15	14	8	8	8	7	7
1,208	$h = 0.25$	6	6	6	7	11	15	16	12	8	8	7	7
8,368	$h = 0.125$	5	5	6	6	11	15	16	16	10	8	7	7
62,048	$h = 0.0625$	5	5	5	6	9	15	17	18	14	8	8	7

Since σ is not constant in general, we loose robustness with respect to σ in the multigrid procedure. Note that for constant σ, we additionally get robustness with respect to σ.

5 Numerical Results

Finally, we report two numerical tests for an academic three dimensional eddy current problem. The numerical results presented in this section were attained using ParMax [16]. First, we demonstrate the robustness of the block-diagonal preconditioner with respect to the frequency ω. Therefore, for the inversion of the diagonal blocks we use the exact solver PARDISO [17, 18]. Table 1 provides the number of MinRes iterations needed for reducing the initial residual by a factor of 10^{-6} for different ω and h. These numerical experiment was performed for a three-dimensional linear problem on the unit-cube, discretized by tetrahedra for the case $\nu = \sigma = 1$. These experiment demonstrates the independence of the frequency and the meshsize as the number of iterations is bounded by 16. Next we repeat the numerical experiment for piecewise constant conductivity σ, i.e.

$$\sigma = \begin{cases} \sigma_1 \text{ in } \Omega_1 = \{(x,y,z)^T \in [0,1]^3 : z > 0.5\} \\ \sigma_2 \text{ in } \Omega_2 = \{(x,y,z)^T \in [0,1]^3 : z \leq 0.5\} \end{cases}. \tag{18}$$

In Table 2 we give the number of iterations for fixed $\omega = 1$ and $\sigma_1 = 1$ and various σ_2. We observe, that the number of iterations is bounded by 18. Both experiments demonstrate the robustness of the block-diagonal preconditioner with respect to the involved parameters. Moreover, this theory-based parameter-robust block-diagonal preconditioner is appropriate to be incorporated in a Newton-based multiharmonic solver for solving (nonlinear) shielding and welding problems (see [5,6]).

6 Further Applications

The presented preconditioning technique provides a robust tool for solving linear eddy current problems with time-harmonic excitation. The theory can be extended to multiharmonic excitations and even to problems with non-harmonic excitation of the right-hand side. The theory in this paper is presented for exact regularized problems. Furthermore we can develop this preconditioning technique also for inexact regularized problems.

6.1 Non-harmonic Excitation

By approximating any non-harmonic right-hand side by a multiharmonic excitation in terms of a truncated Fourier series, it follows, that the solution $\mathbf{u_N}$ has the structure:

$$\mathbf{u_N}(\mathbf{x}, t) = \sum_{k=0}^{N} \mathbf{u_k^c}(\mathbf{x}) \cos(k\omega t) + \mathbf{u_k^s}(\mathbf{x}) \sin(k\omega t). \tag{19}$$

Using the truncated Fourier approximation (19), the corresponding system matrix in the frequency domain decouples into a block-diagonal matrix of the form

$$\operatorname{diag}\left\{\begin{pmatrix} \mathbf{A_h} & k\omega \mathbf{M_{\sigma,h}} \\ -k\omega \mathbf{M_{\sigma,h}} & \mathbf{A_h} \end{pmatrix}\right\}_{k=0,\dots,N}, \tag{20}$$

where each block has almost the same structure as the two-by-two system matrix in (4). Hence we can apply either the exact or the inexact regularization technique and precondition each block robustly with respect to the frequency ω, the mode k and the meshsize h. By approximating a general right-hand side $\mathbf{f}$ by a finite Fourier series with N summands, we introduce an additional truncation error of order N^{-1}.

$$\|\mathbf{u} - \mathbf{u_N}\|_{L_2((0,T),H_0(\mathbf{curl}))} = \mathcal{O}(N^{-1}). \tag{21}$$

6.2 Inexact Regularization (Conductivity Regularization)

Instead of the exact regularization an inexact regularization, as for example in [5], can also be applied by introducing a regularized conductivity σ_ε, defined as $\max\{\sigma, \varepsilon\}$ with the regularization parameter $\varepsilon > 0$. In this case the same strategy can be used to construct a block diagonal preconditioner, that is robust with respect to ω, h and σ_ε, leading to the system matrix $\mathbf{D_h}$ and the preconditioner $\mathbf{C_h}$.

$$\mathbf{D_h} = \begin{pmatrix} \mathbf{M}_{\sigma_\varepsilon,\mathbf{h}} & \mathbf{A_h} \\ \mathbf{A_h} & -\omega^2 \mathbf{M}_{\sigma_\varepsilon,\mathbf{h}} \end{pmatrix} \quad \mathbf{C_h} = \frac{1}{\omega} \begin{pmatrix} \omega \mathbf{M}_{\sigma_\varepsilon,\mathbf{h}} + \mathbf{A_h} & \mathbf{0} \\ \mathbf{0} & \omega^2(\omega \mathbf{M}_{\sigma_\varepsilon,\mathbf{h}} + \mathbf{A_h}) \end{pmatrix} \tag{22}$$

In contrast to the exact regularization, where no additional regularization error is introduced, in the case of inexact regularization, we have to deal with an additional error of order $\mathcal{O}(\varepsilon)$ (see [5]).

7 Conclusion and Outlook

The method developed in this work shows great potential for solving both, time-harmonic and non harmonic eddy current problems in a very efficient and robust way, in the linear case. Up to now, theory only guarantees robustness in the case of constant coefficients ω and σ, but currently we are working on the extension also to the piecewise constant case. Indeed, based on the results in [11], we are working on a domain decomposition preconditioner for the inversion of the diagonal blocks, that guarantees robustness also for piecewise constant conductivity σ.

In the non-linear case, i.e. $\nu = \nu(\mathbf{x}, |\mathbf{curl\ u}|)$, it turns out, that even for harmonic excitation of the right-hand side, we have to take all frequencies $k\omega$ into account. For earlier works see e.g. [4, 9, 15]. Additionally, due to the nonlinearity, we lose the advantageous block-diagonal structure and therefore have to deal with a fully-coupled system of non-linear equations in the Fourier coefficients. Since the Fréchet derivative of the non-linear frequency domain equations is explicitly computable, the nonlinearity can easily be overcome by applying Newton's method. Anyhow, at each step of Newton's iteration, a huge and fully block-coupled Jacobi system with sparse blocks has to be solved. The applicableness of the parameter-robust MinRes solver to the Jacobi system is not clear at the first glance.

Acknowledgements The authors gratefully acknowledge the financial support of the Austrian Science Fund (FWF) research project P19255 and DK W1214.

References

1. Arnold, D.N., Falk, R.S., Winther, R.: Multigrid in H(div) and H(curl). Numer. Math. **85**(2), 197–217 (2000)
2. Babuška, I.: Error-bounds for finite element method. Numer. Math. **16**(4), 322–333 (1971)
3. Bachinger, F.: Multigrid solvers for 3D multiharmonic nonlinear magnetic field computations. Master's thesis, Johannes Kepler University, Linz (2003)
4. Bachinger, F., Kaltenbacher, M., Reitzinger, S.: An Efficient Solution Strategy for the HBFE Method. In: Proceedings of the IGTE '02 Symposium Graz, Austria, pp. 385–389 (2002)

5. Bachinger, F., Langer, U., Schöberl, J.: Numerical analysis of nonlinear multiharmonic eddy current problems. Numer. Math. **100**(4), 593–616 (2005)
6. Bachinger, F., Langer, U., Schöberl, J.: Efficient solvers for nonlinear time-periodic eddy current problems. Comput. Vis. Sci. **9**(4), 197–207 (2006)
7. Copeland, D., Kolmbauer, M., Langer, U.: Domain decomposition solvers for frequency-domain finite element equation. In: Domain Decomposition Methods in Science and Engineering XIX, *LNCSE*, vol. 78, pp. 301–308. Springer, Heidelberg (2011)
8. Copeland, D., Langer, U.: Domain decomposition solvers for nonlinear multiharmonic finite element equations. J. Numer. Math. **18**(3), 157–175 (2010)
9. Gersem., H.D., Sande, H.V., Hameyer, K.: Strong coupled multi-harmonic finite element simulation package. COMPEL **20**, 535–546 (2001)
10. Greenbaum, A.: Iterative methods for solving linear systems, Frontiers in Applied Mathematics, vol. 17. Society for Industrial and Applied Mathematics (SIAM), Philadelphia, PA (1997)
11. Hu, Q., Zou, J.: A nonoverlapping domain decomposition method for Maxwell's equations in three dimensions. SIAM J. Numer. Anal. **41**(5), 1682–1708 (electronic) (2003)
12. Kuhn, M.: Efficient parallel numerical simulation of magnetic field problems. Ph.D. thesis, Johannes Kepler University, Linz (1998)
13. Nédélec, J.C.: Mixed finite elements in $\mathbf{R}^3$. Numer. Math. **35**(3), 315–341 (1980)
14. Paige, C.C., Saunders, M.A.: Solutions of sparse indefinite systems of linear equations. SIAM J. Numer. Anal. **12**(4), 617–629 (1975)
15. Paoli, G., Biro, O., Buchgraber, G.: Complex representation in nonlinear time harmonic eddy current problems. IEEE Trans. Magn. **34**(5), 2625–2628 (1998)
16. Pechstein, C.: ParMax (2010). http://www.numa.uni-linz.ac.at/P19255/software.shtml
17. Schenk, O., Bollhöfer, M., Römer, R.A.: On large scale diagonalization techniques for the Anderson model of localization. SIAM Rev. **50**(1), 91–112 (2008). SIGEST Paper
18. Schenk, O., Wächter, A., Hagemann, M.: Matching-based preprocessing algorithms to the solution of saddle-point problems in large-scale nonconvex interior-point optimization. Comput. Optim. Appl. **36**(2-3), 321–341 (2007)
19. Schöberl, J., Zulehner, W.: Symmetric indefinite preconditioners for saddle point problems with applications to PDE-constrained optimization problems. SIAM J. Matrix Anal. Appl. **29**(3), 752–773 (2007)

Depolarization of Electromagnetic Waves from Bare Soil Surfaces

Naheed Sajjad, Ali Khenchaf, and Arnaud Coatanhay

Abstract An improved Two Scale Model (TSM) has been investigated for the depolarization of electromagnetic waves from bare soil surfaces. Classical TSM produces depolarized results due to the tilt of reflecting plane. To include the contribution of actual phenomenon, we add the second order scattering effects at small scale and develop an improved TSM. The performance of the new TSM is assessed by comparing the simulation results in backscattering configuration with the measured data, Advanced Integral Equation Model and Second order Small Slope Approximation at L-, S-, C- and X-band frequencies for a variety of roughness conditions. Finally, we use the new TSM to predict the bistatic scattering and compare the results with classical TSM.

1 Introduction

Depolarization in a radar return results in corruption of the received signal. It is an undesired effect for a given transmitter, limiting the useful radar coverage distance. However, the cross-polarization in conjunction with co-polarization information can be used to retrieve the surface roughness parameters, the geometrical configuration of scatterers while giving important clues to the electrical properties of surfaces etc. Hence the study of depolarization cannot be used only to discriminate the unwanted reflections but it is also used for the identification and optimization purposes since it permits a deeper insight into physical phenomena. Due to this reason, the cross-polarized (or depolarized) radar returns are of interest to some EMC engineers, hydrologists, meteorologists and agriculturists.

N. Sajjad (✉) · A. Khenchaf · A. Coatanhay
ENSTA Bretagne/E^{3}I^2 – EA3876 Laboratory, 2 rue Francois Verny, 29806, Brest Cedex 09, France
e-mail: naheed.sajjad@ensta-bretagne.fr; ali.khenchaf@ensta-bretagne.fr; arnaud.coatanhay@ensta-bretagne.fr

B. Michielsen and J.-R. Poirier (eds.), *Scientific Computing in Electrical Engineering SCEE 2010*, Mathematics in Industry 16, DOI 10.1007/978-3-642-22453-9_12,

Cross polarization in a radar return from a rough surface has been observed experimentally [1, 2]. First order Small Perturbation Method (SPM1) [3] and Kirchhoff Approximation (KA) [4] do not predict this phenomenon. In order to account for the observed cross polarization, most theoreticians have used the methods of Advanced Integral Equation Model (AIEM) [5], second order Small Slope Approximation (SSA2) [6], second order Small Perturbation Method (SPM2) [1, 2, 7], Two Scale Model (TSM) [8–10] and empirical models [11] etc. In the classical TSM (TSM1) [9, 10] it is assumed that the short wavelength waves are riding on the longer waves and thus tilted with respect to the horizontal surface. It uses SPM1 at small scale i.e. for short wavelength waves and the effect of long wavelength part is taken into account by averaging over the tilt angles. Hence by using TSM1 based on first order theory, depolarization is basically due to the tilt of reflecting plane. Due to this reason, the simple TSM needs to be improved.

Since the mechanism of multi-scattering due to target surface roughness also causes depolarization [1], this observation motivates us to develop an improved TSM (TSM2) by taking into account the contribution of higher order scattering (up to second order) at small scale. The purpose of this paper is to present the mathematical development of TSM2. In addition, we assume that the bare soil surface can be modeled as having two average sizes of roughness, this model is then applied to depolarization case. In backscattering configuration, we assess the performance of TSM2 by comparing the numerical results with the measured data [11], AIEM [12] and SSA2. Finally, the simulation results are presented for bistatic case.

2 Scattering Models

This section contains a brief review of SPM up to second order and TSM1. The development of TSM2 is then presented.

2.1 Small Perturbation Method (SPM)

The scattering of electromagnetic waves from a slightly rough surface can be studied by using SPM. In this method it is assumed that the surface variations are much smaller than the incident wavelength and the slope of the rough surface is relatively small.

The first and second order bistatic scattering coefficients and correlation products can be written as [7]

$$\sigma_{pq}^{(1)} = 4\pi k^2 \cos^2\theta_s \left|\alpha_{pq}^{(1)}\right|^2 W\left(\overline{k}_{s\perp} - \overline{k}_{i\perp}\right) \tag{1}$$

$$\sigma_{pqmn}^{(1)} = 4\pi k^2 \cos^2\theta_s \alpha_{pq}^{(1)} \alpha_{mn}^{*(1)} W\left(\overline{k}_{s\perp} - \overline{k}_{i\perp}\right) \tag{2}$$

$$\sigma_{pq}^{(2)} = 4\pi k^2 \cos^2\theta_s \int W\left(\overline{k}_{s\perp} - \overline{k}_{\perp}\right) W\left(\overline{k}_{\perp} - \overline{k}_{i\perp}\right) \alpha_{pq}^{(2)} \left[\alpha_{pq}^{*(2)} + \beta_{pq}^{*(2)}\right] d\overline{k}_{\perp} \tag{3}$$

$$\sigma_{pqmn}^{(2)} = 4\pi k^2 \cos^2\theta_s \int W\left(\overline{k}_{s\perp} - \overline{k}_{\perp}\right) W\left(\overline{k}_{\perp} - \overline{k}_{i\perp}\right) \alpha_{pq}^{(2)} \left[\alpha_{mn}^{*(2)} + \beta_{mn}^{*(2)}\right] d\overline{k}_{\perp} \tag{4}$$

where k is the wave number, $\alpha_{pq}^{(1)}$ and $\alpha_{pq}^{(2)}$, $\beta_{pq}^{(2)}$ are the first and second order polarization-dependent factors respectively, $W(.)$ is the roughness spectrum and $\overline{k}_{\perp}$ denotes vector $k_x\widehat{x} + k_y\widehat{y}$ in $x - y$ plane.

SPM2 can be used for longer correlation lengths and large values of rms height and has larger domain of validity as compared to SPM1 [13]. In backscattering direction, using SPM2 one is able to calculate the non-zero cross-polarized (depolarized) scattering coefficients (which become zero by SPM1) in conjunction with co-polarized scattering coefficients. Hence depolarization is a second order effect in a plane of incidence. Moreover, the expression for the depolarization scattering cross section is of the form obtained in multiple scattering studies [1]. It is therefore hypothesized that depolarization is due to multiple scattering and the inclusion of this effect in all polarizations can be useful for the study of depolarization. This observation motivates us to include the second order corrections at small scale in TSM and develop an improved TSM.

2.2 Two-Scale Model

TSM1 [10] approximates the rough surface as a two-scale surface with small-scale waves riding on the top of large-scale waves. The scattering coefficients are then estimated in two steps. Firstly, TSM1 uses SPM1 on a small scale waves and then determine the diffuse component in the global reference by a tilting process.

Assume the incident wave $\mathbf{E}^i$ to be $\mathbf{E}^i = \widehat{\mathbf{a}}E_0$ with $E_0 = |E_0| \exp\{-jk(\widehat{\mathbf{n}}_i.\mathbf{r})\}$, where $\widehat{\mathbf{a}}$ is the unit polarization vector and $\widehat{\mathbf{n}}_i$ is the unit vector in the incident direction. In the local reference frame, the unit polarized incident wave will appear as a horizontal and a vertical incident wave given by

$$\mathbf{E}^i = E_{h'}^i\widehat{\mathbf{h}}' + E_{v'}^i\widehat{\mathbf{v}}' = \left[\left(\widehat{\mathbf{h}}'.\widehat{\mathbf{a}}\right)\widehat{\mathbf{h}}' + \left(\widehat{\mathbf{v}}'.\widehat{\mathbf{a}}\right)\widehat{\mathbf{v}}'\right]E_0 \tag{5}$$

and the locally scattered fields due to incident waves are:

$$\mathbf{E}^s = E_{h_s'}^s\widehat{\mathbf{h}}_s' + E_{v_s'}^s\widehat{\mathbf{v}}_s' = [S]\,\mathbf{E}^i = \begin{bmatrix} S_{h_s'h'}E_{h'}^i + S_{h_s'v'}E_{v'}^i \\ S_{v_s'h'}E_{h'}^i + S_{v_s'v'}E_{v'}^i \end{bmatrix} \tag{6}$$

where $S_{p'q'}$ is the scattered field for unit incident fields. In the global frame of reference, the scattering matrix is given by

$$S = \begin{bmatrix} \widehat{\mathbf{h}}_s.\widehat{\mathbf{h}}'_s & \widehat{\mathbf{h}}_s.\widehat{\mathbf{v}}'_s \\ \widehat{\mathbf{v}}_s.\widehat{\mathbf{h}}'_s & \widehat{\mathbf{v}}_s.\widehat{\mathbf{v}}'_s \end{bmatrix} \begin{bmatrix} S_{h'_sh'} & S_{h'_sv'} \\ S_{v'_sh'} & S_{v'_sv'} \end{bmatrix} \begin{bmatrix} \widehat{\mathbf{h}}'.\widehat{\mathbf{h}} & \widehat{\mathbf{h}}'.\widehat{\mathbf{v}} \\ \widehat{\mathbf{v}}'.\widehat{\mathbf{h}} & \widehat{\mathbf{v}}'.\widehat{\mathbf{v}} \end{bmatrix} \tag{7}$$

For the received polarization p ($\widehat{\mathbf{h}}_s$ or $\widehat{\mathbf{v}}_s$) and the transmitted polarization q ($\widehat{\mathbf{h}}$ or $\widehat{\mathbf{v}}$), the scattered polarized and depolarized fields are obtained from

$$E^s_{pq} = \left[\left(\mathbf{p}.\widehat{\mathbf{h}}'_s\right)\left\{\left(\widehat{\mathbf{h}}'.\mathbf{q}\right) S_{h'_sh'} + (\widehat{\mathbf{v}}'.\mathbf{q})\, S_{h'_sv'}\right\} + (\mathbf{p}.\widehat{\mathbf{v}}'_s)\left\{\left(\widehat{\mathbf{h}}'.\mathbf{q}\right) S_{v'_sh'} + (\widehat{\mathbf{v}}'.\mathbf{q})\, S_{v'_sv'}\right\}\right] E_0 \tag{8}$$

The correlation product $\left\langle E^s_{pq} E^{s*}_{pq} \right\rangle$ with respect to the large-scale roughness can be calculated and rewritten in terms of the scattering coefficients σ^s_{pq} as a function of the transmitter polarization q and the receiver polarization p [10]. The average $\langle . \rangle$ in the scattering coefficients may then be calculated by using any model of the surface slopes distribution.

2.3 Improved Two-Scale Model

To include the contribution of second order scattering effects, we add the second order scattered field $S^{(2)}_{pq}$ to $S^{(1)}_{pq}$, for unit incident field, in local domain. Hence the local scattering matrix will become

$$S' = \begin{bmatrix} S^{(1)}_{h'_sh'} + S^{(2)}_{h'_sh'} & S^{(1)}_{h'_sv'} + S^{(2)}_{h'_sv'} \\ S^{(1)}_{v'_sh'} + S^{(2)}_{v'_sh'} & S^{(1)}_{v'_sv'} + S^{(2)}_{v'_sv'} \end{bmatrix} \tag{9}$$

and in global frame of reference, the scattering matrix S is given by

$$S = \begin{bmatrix} \widehat{\mathbf{h}}_s.\widehat{\mathbf{h}}'_s & \widehat{\mathbf{h}}_s.\widehat{\mathbf{v}}'_s \\ \widehat{\mathbf{v}}_s.\widehat{\mathbf{h}}'_s & \widehat{\mathbf{v}}_s.\widehat{\mathbf{v}}'_s \end{bmatrix} \begin{bmatrix} S^{(1)}_{h'_sh'} + S^{(2)}_{h'_sh'} & S^{(1)}_{h'_sv'} + S^{(2)}_{h'_sv'} \\ S^{(1)}_{v'_sh'} + S^{(2)}_{v'_sh'} & S^{(1)}_{v'_sv'} + S^{(2)}_{v'_sv'} \end{bmatrix} \begin{bmatrix} \widehat{\mathbf{h}}'.\widehat{\mathbf{h}} & \widehat{\mathbf{h}}'.\widehat{\mathbf{v}} \\ \widehat{\mathbf{v}}'.\widehat{\mathbf{h}} & \widehat{\mathbf{v}}'.\widehat{\mathbf{v}} \end{bmatrix} \tag{10}$$

Now the components of scattered field, in global frame of reference, are given as

$$\begin{aligned} E^s_{pq} = &\left[\left(\mathbf{p}.\widehat{\mathbf{h}}'_s\right)\left\{\left(\widehat{\mathbf{h}}'.\mathbf{q}\right)\left(S^{(1)}_{h'_sh'} + S^{(2)}_{h'_sh'}\right) + (\widehat{\mathbf{v}}'.\mathbf{q})\left(S^{(1)}_{h'_sv'} + S^{(2)}_{h'_sv'}\right)\right\}\right. \\ &\left. + (\mathbf{p}.\widehat{\mathbf{v}}'_s)\left\{\left(\widehat{\mathbf{h}}'.\mathbf{q}\right)\left(S^{(1)}_{v'_sh'} + S^{(2)}_{v'_sh'}\right) + (\widehat{\mathbf{v}}'.\mathbf{q})\left(S^{(1)}_{v'_sv'} + S^{(2)}_{v'_sv'}\right)\right\}\right] E_0 \end{aligned} \tag{11}$$

The improved scattering coefficient σ^s_{pq} is given by [14]

$$
\begin{aligned}
\sigma_{pq}^{s} = \Big\langle & \left(\mathbf{p}.\widehat{\mathbf{h}}_s'\right)^2 \left(\widehat{\mathbf{h}}'.\mathbf{q}\right)^2 \left(\sigma_{h_s'h'}^{(1)} + \sigma_{h_s'h'}^{(2)}\right) + \left(\mathbf{p}.\widehat{\mathbf{h}}_s'\right)^2 \left(\widehat{\mathbf{v}}'.\mathbf{q}\right)^2 \left(\sigma_{h_s'v'}^{(1)} + \sigma_{h_s'v'}^{(2)}\right) \\
& + \left(\mathbf{p}.\widehat{\mathbf{v}}_s'\right)^2 \left(\widehat{\mathbf{h}}'.\mathbf{q}\right)^2 \left(\sigma_{v_s'h'}^{(1)} + \sigma_{v_s'h'}^{(2)}\right) + \left(\mathbf{p}.\widehat{\mathbf{v}}_s'\right)^2 \left(\widehat{\mathbf{v}}'.\mathbf{q}\right)^2 \left(\sigma_{v_s'v'}^{(1)} + \sigma_{v_s'v'}^{(2)}\right) + \left(\mathbf{p}.\widehat{\mathbf{v}}_s'\right)^2 \\
& \times \left(\widehat{\mathbf{h}}'.\mathbf{q}\right) \left(\widehat{\mathbf{v}}'.\mathbf{q}\right) \left(\sigma_{v_s'h'v_s'v'}^{(1)} + \sigma_{v_s'h'v_s'v'}^{(2)}\right) + \left(\mathbf{p}.\widehat{\mathbf{v}}_s'\right)^2 \left(\widehat{\mathbf{h}}'.\mathbf{q}\right) \left(\widehat{\mathbf{v}}'.\mathbf{q}\right) \left(\sigma_{v_s'v'v_s'h'}^{(1)}\right. \\
& \left. + \sigma_{v_s'v'v_s'h'}^{(2)}\right) + \left(\mathbf{p}.\widehat{\mathbf{h}}_s'\right) \left(\mathbf{p}.\widehat{\mathbf{v}}_s'\right) \left(\widehat{\mathbf{v}}'.\mathbf{q}\right)^2 \left(\sigma_{h_s'v'v_s'v'}^{(1)} + \sigma_{h_s'v'v_s'v'}^{(2)}\right) + \left(\mathbf{p}.\widehat{\mathbf{h}}_s'\right) \left(\mathbf{p}.\widehat{\mathbf{v}}_s'\right) \\
& \times \left(\widehat{\mathbf{v}}'.\mathbf{q}\right)^2 \left(\sigma_{v_s'v'h_s'v'}^{(1)} + \sigma_{v_s'v'h_s'v'}^{(2)}\right) + \left(\mathbf{p}.\widehat{\mathbf{h}}_s'\right)^2 \left(\widehat{\mathbf{h}}'.\mathbf{q}\right) \left(\widehat{\mathbf{v}}'.\mathbf{q}\right) \left(\sigma_{h_s'h'h_s'v'}^{(1)} + \sigma_{h_s'h'h_s'v'}^{(2)}\right) \\
& + \left(\mathbf{p}.\widehat{\mathbf{h}}_s'\right)^2 \left(\widehat{\mathbf{h}}'.\mathbf{q}\right) \left(\widehat{\mathbf{v}}'.\mathbf{q}\right) \left(\sigma_{h_s'v'h_s'h''}^{(1)} + \sigma_{h_s'v'h_s'h'}^{(2)}\right) + \left(\mathbf{p}.\widehat{\mathbf{h}}_s'\right) \left(\mathbf{p}.\widehat{\mathbf{v}}_s'\right) \left(\widehat{\mathbf{h}}'.\mathbf{q}\right)^2 \\
& \times \left(\sigma_{h_s'h'v_s'h'}^{(1)} + \sigma_{h_s'h'v_s'h'}^{(2)}\right) + \left(\mathbf{p}.\widehat{\mathbf{h}}_s'\right) \left(\mathbf{p}.\widehat{\mathbf{v}}_s'\right) \left(\widehat{\mathbf{v}}'.\mathbf{q}\right) \left(\widehat{\mathbf{h}}'.\mathbf{q}\right) \left(\sigma_{h_s'v'v_s'h'}^{(1)} + \sigma_{h_s'v'v_s'h'}^{(2)}\right) \\
& \times \left(\widehat{\mathbf{v}}'.\mathbf{q}\right) \left(\sigma_{h_s'h'v_s'v'}^{(1)} + \sigma_{h_s'h'v_s'v'}^{(2)}\right) + \left(\mathbf{p}.\widehat{\mathbf{h}}_s'\right) \left(\mathbf{p}.\widehat{\mathbf{v}}_s'\right) \left(\widehat{\mathbf{h}}'.\mathbf{q}\right) \left(\widehat{\mathbf{v}}'.\mathbf{q}\right) \left(\sigma_{v_s'v'h_s'h'}^{(1)}\right. \\
& \left. + \sigma_{v_s'v'h_s'h'}^{(2)}\right) + \left(\mathbf{p}.\widehat{\mathbf{h}}_s'\right) \left(\mathbf{q}.\widehat{\mathbf{v}}_s'\right) \left(\widehat{\mathbf{h}}'.\mathbf{q}\right)^2 \left(\sigma_{v_s'h'h_s'h'}^{(1)} + \sigma_{v_s'h'h_s'h'}^{(2)}\right) \Big\rangle
\end{aligned}
\tag{12}
$$

where $\sigma_{p'q'}^{(1)}$,$\sigma_{p'q'm'n'}^{(1)}$, $\sigma_{p'q'}^{(2)}$ and $\sigma_{p'q'm'n'}^{(2)}$ are obtained from (1)–(4) and calculated at local angles.

Note that in (12) we ignore the terms involving the product of first and second order fields i.e., $\sigma_{p'q'}^{(12)}$ and $\sigma_{p'q'm'n'}^{(12)}$ for the sake of simplicity.

3 Numerical Results

In this section, initially we illustrate the numerical simulation results of the cross polarized backscattering coefficient (σ_{hv}). Applications of the developed model for co-polarization case are discussed in [15]. The bistatic case is represented at the end of this section.

Figures 1 and 2 show the angular dependence of σ_{hv} for a bare soil surface with rms height (h) of 0.40 cm and correlation length (l) of 8.4 cm. For all three plots, the values of frequencies and relative dielectric constants are taken as 1.5 GHz,15.57 (L1); 4.75 GHz, 15.42 (X1) and 9.5 GHz, 12.31 (X1), respectively. The simulation results of TSM2 are compared with SPM2, TSM1 and measured data [11]. It is observed that the TSM2 give enhanced results which are in good agreement with the measured data.

We carry on the comparison between TSM1 and TSM2 for a relatively rough surface with $h = 1.12$ cm and $l = 8.4$ cm at L2 ($f = 1.5$ GHz, $\varepsilon_r = 15.34$), C2 ($f = 4.75$ GHz, $\varepsilon_r = 15.23$) and X2 ($f = 9.5$ GHz, $\varepsilon_r = 13.14$) and found that the difference between two models increases as the roughness of the surface

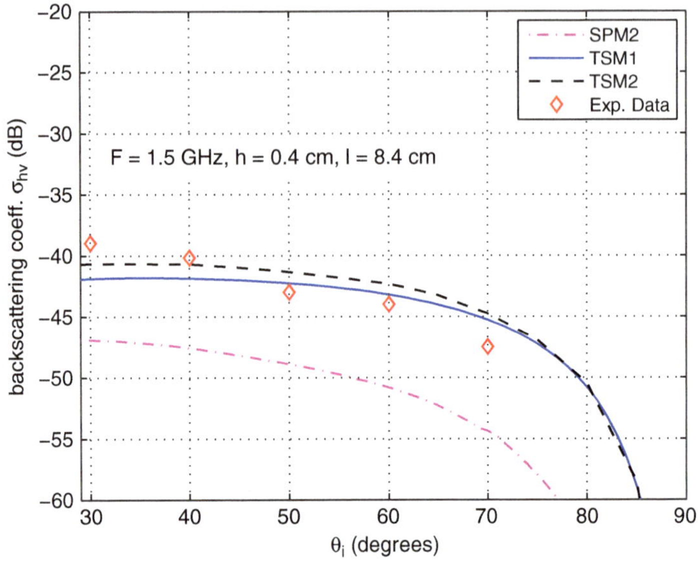

Fig. 1 TSM2 compared to the measured data [11], SPM2 and TSM1 for $h = 0.40$ cm and $l = 8.4$ cm at L-band

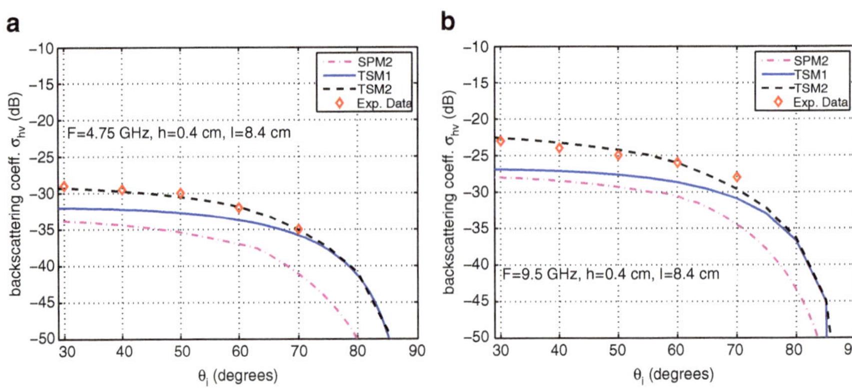

Fig. 2 TSM2 compared to the measured data [11], SPM2 and TSM1 for $h = 0.40$ cm and $l = 8.4$ cm at (**a**) C-band, (**b**) X-band

increases. In Fig. 3 the comparison with measured data is given for a very rough surface with $h = 3.02$ cm and $l = 8.8$ cm at L3 ($f = 1.5$ GHz, $\varepsilon_r = 8.92$), C3 ($f = 4.75$ GHz, $\varepsilon_r = 9.64$) and X3 ($f = 9.5$ GHz, $\varepsilon_r = 7.57$). For L3 we are not so far from the measured data but for the other two frequencies (i.e., C3 and X3) TSM2 over-estimates.

Next, to study further the consistency and validity of TSM2, we compare our results with AIEM and experimental data [12] in Fig. 4a which is plotted at S-band

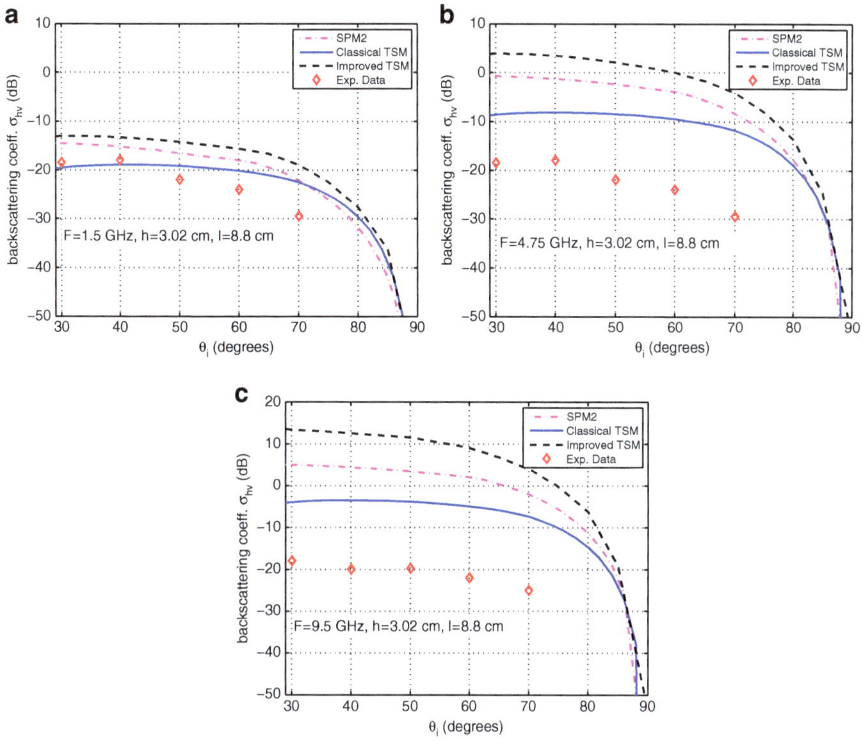

Fig. 3 TSM2 compared to the measured data [11], SPM2 and TSM1 for $h = 3.02$ cm and $l = 8.8$ cm at (**a**) L-band, (**b**) C-band, (**c**) X-band

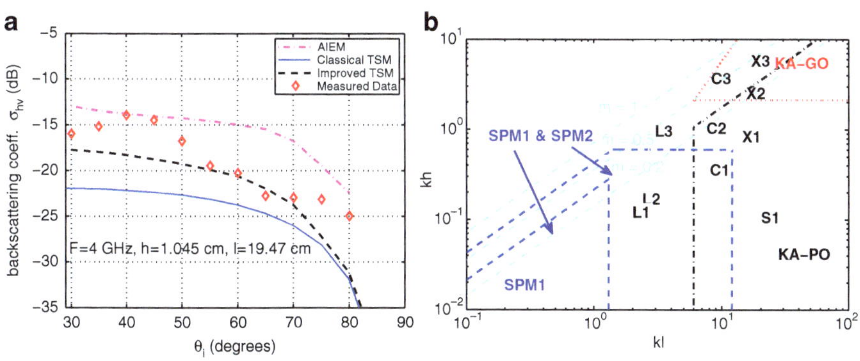

Fig. 4 (**a**) TSM2 compared with AIEM, measured data [12] and TSM1 for $h = 3.02$ cm and $l = 8.8$ cm at S-band; (**b**) roughness parameters and the qualitative region of validity of SPM1, SPM2, PO and GO models

(f = 3 GHz) with h = 1.045 cm, l = 19.47 cm and ε_r = 11.5. Again the predictions by TSM2 are in better agreement with the measured data and AIEM.

To evaluate the applicability of TSM2, the locations of considered points are identified in $ks - kl$ space in Fig. 4b along with the qualitative regions of validity of SPM1, SPM2 [13], PO (Physical Optics) and GO (Geometrical Optics) models. It is observed that TSM2 predictions are good/reasonable as far as $kh \leq 1$ and it overestimates otherwise. It is quite logical because actual estimation of cross polarized coefficients are due to the inclusion of second order scattering by SPM2 which is valid till $kh < 0.6$ [13], for moderate incident angles. On the other hand TSM2 can be used successfully for longer correlation lengths.

Furthermore, the comparison of TSM2 with SSA2 is also presented in Figs. 5a,b for $kh = 0.5$, $kl = 3$ and $kh = 1$, $kl = 6$. TSM2 gives enhanced results as compared to SSA2 which is due to the fact that TSM also includes the averaging effects over slope distribution for long scale waves along with the scattering coefficient calculations by SPM for small scale waves.

Finally, Figs. 6a,b show the angular responses of the hv-polarized bistatic scattering coefficient (σ_{hv}). The incident angle is fixed at 45° while the received one

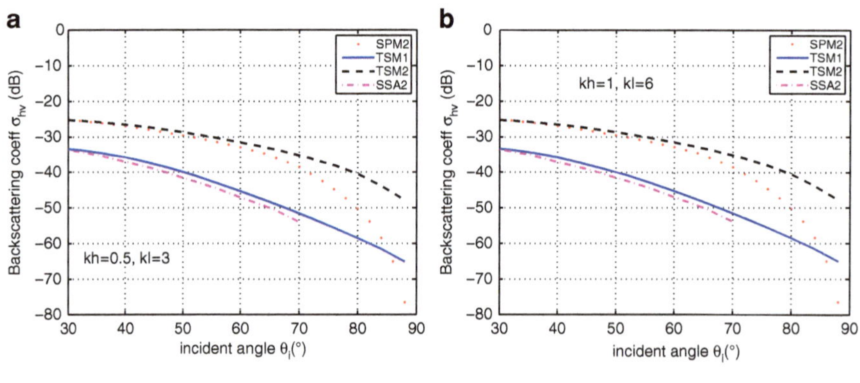

Fig. 5 Comp. of SPM2, TSM1 and TSM2 with SSA2 (**a**) $kh = 0.5, kl = 3$; (**b**) $kh = 1, kl = 6$

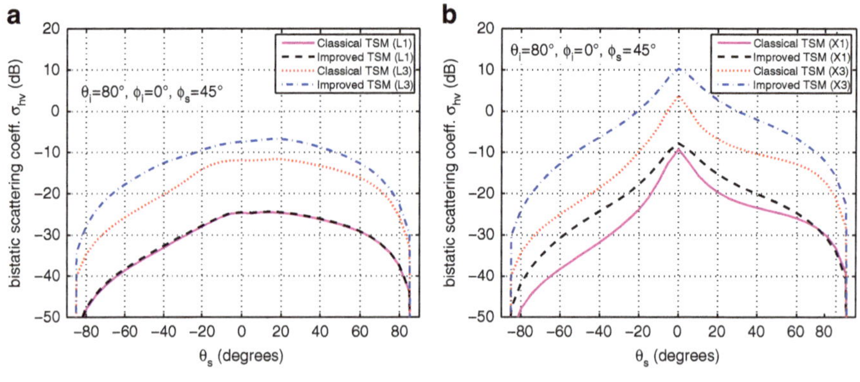

Fig. 6 Bistatic scattering coefficient (σ_{hv}): Comp. of TSM1 and TSM2 (**a**) L-band (**b**) X-band

varies from -90° to 90° and received azimuth is set at 45°.The numerical results are given for L1, L3, X1 and X3. It can be observed that the difference between two models increases with the increase in frequency and roughness level.

4 Conclusion

By taking into account the contribution of second order scattering effects at small scale, the development of an improved two-scale model is presented. Comparisons of numerical results with measured data and with other scattering models shows that TSM2 gives better predictions of depolarized components and can be used adequately as far as the value of kh remains less than or nearly equal to one. The new model may have promising applications for electromagnetic scattering from the ocean surface at grazing angles due to the inclusion of actual scattering mechanism of multiple/higher order scattering, which are under investigation and will be reported later on.

References

1. Valenzuela, G.R.: Depolarization of EM waves by slightly rough surfaces. IEEE Trans. Ant. Prop., **15**, 552–557 (1967)
2. Raemer, H.R., Preis, D.D.: Aspects of Parallel-Polarized and Cross-Polarized Radar Returns from a Rough Sea Surface. IEEE Trans. EM Compat. **22**(1), 29–44 (1980)
3. Rayleigh, J.W.S.: The theory of sound. vol. 2, dover, New York (1945)
4. Beckmann, P.: The Scattering of Electromagnetic Waves from Rough Surfaces. Macmillan Co., New York (1963)
5. Wu, T.D., Chen, K.S., Shi, J.C., Lee, H.W., Fung, A.K.: A study of AIEM Model for Bistatic Scattering from Randomly Surfaces. IEEE TGRS **46**(9), 2584–2598 (2008)
6. Gilbert, M.S., Johnson, J.T.: A study of the higher-order small-slope approximation for scattering from a Gaussian rough surface. Waves Random Media **13**, 137–149 (2003)
7. Tsang, L. and Kong, J.A.: Scattering of Electromagnetic Waves, Vol. 3. Advanced Topics, Wiley Interscience (2001)
8. Wright, J.W.: A new model for sea clutter. IEEE Trans. Ant. Pro. **16**, 217–223 (1968)
9. Bass, F.G., Fuks, I.M.: Wave Scattering from Statistically Rough Surfaces. Pergamon Press Oxford, New York (1979)
10. Khenchaf, A.: Bistatic scattering and depolarization by randomly rough surface: application to the natural rough surface in X-band. Waves Random Media **11**, 61–87 (2001)
11. Oh, Y., Sarabandi, K., Ulaby, F.T.: An Empirical Model and an Inversion Technique for Radar Scattering from Bare Soil Surfaces. IEEE TGRS **30**, 370–381 (1992)
12. Liu et al.: Study on the backscattering characteristic of typical earth substances in northwest of Chaina. IGARSS 09, Cape Town, South Africa, july 12–17 (2009)
13. Thorsos, E.I., Jackson, D.R.: The validity of the perturbation approximation for rough surface scattering using a Gaussian roughness spectrum. J. Acoust. Soc. Am. **86**(1) (1989)
14. Sajjad, N., Khenchaf, A., Coatanhay, A., Awada, A.: An improved two scale model for the ocean surface bistatic scattering. IGARSS, Boston, USA, 6-11 july, 2008
15. Sajjad, N., Khenchaf, A., Coatanhay, A.: Electromagnetic Wave Scattering From Sea and Bare Soil Surfaces Based On An Improved Two-Scale Model. invited paper, IEEE Radar, Bordeaux, France, 2009

Two Finite-Element Thin-Sheet Approaches in the Electro-Quasistatic Formulation

Jens Trommler, Stephan Koch, and Thomas Weiland

Abstract Two finite-element approaches to cope with thin sheets in the electro-quasistatic formulation are presented. Both rely on the well-known strategy to reduce the sheet volume to a surface. In the first approach, polynomials in the lateral direction are used to allow for a field variation across the sheet. Using the second method, the presence of the thin sheet is modeled by a modification of the local discretization. In contrast to the first approach, here, no additional degrees of freedom are introduced. The different methods are compared to a conventional solution based on simple test examples.

1 Introduction

In case of models containing objects of very small extension in one direction, the application of a volume-based discretization, such as the finite element method, is cumbersome. Often, because of local refinement, this procedure leads to a very large number of mesh cells, whereas the elements related to the thin sheet exhibit a bad aspect ratio. This leads to ill-conditioned matrices and, as a consequence, to a high computational effort when solving the related system of equations using iterative solvers. In order to avoid these difficulties, an object of thickness δ, where δ is very small compared to the extension in the remaining directions, is commonly considered during the mesh generation as a surface layer Γ_s instead of a volume Ω_s.

This modeling technique was initially introduced for the simulation of air gaps in transformer cores as well as for eddy-current shielding [1]. It is commonly applied for applications in the magneto-static and magneto-quasistatic (MQS) regime [2,3].

J. Trommler (✉) · S. Koch · T. Weiland
Technische Universität Darmstadt, Institut für Theorie Elektromagnetischer Felder (TEMF), Schloßgartenstraße 8, 64289 Darmstadt, Deutschland
e-mail: trommler@temf.tu-darmstadt.de; koch@temf.tu-darmstadt.de; thomas.weiland@temf.tu-darmstadt.de

B. Michielsen and J.-R. Poirier (eds.), *Scientific Computing in Electrical Engineering SCEE 2010*, Mathematics in Industry 16, DOI 10.1007/978-3-642-22453-9_13,

In many cases, the field variation across a thin sheet can be neglected, e.g., if the skin depth is large compared to the thickness δ. As long as this assumption holds, the method is appropriate as it avoids both meshing difficulties and the deterioration of the condition number of the system matrix. If this assumption no longer holds, the field variation can be taken into account by means of a higher-order discretization inside the sheet as, e.g., in [4, 5]. This approach is based on virtual or degenerated prismatic elements for the thin sheet [6, 7].

This technique is also applicable for electro-quasistatic (EQS) models as shown, e.g., in [8] using constant elements across the sheet. However, for specific examples within this application range, a field variation across the thin sheet can be relevant. Therefore, in Sect. 2.1, the approach using a higher-order discretization perpendicular to the sheet surface is transferred from MQS to EQS. While the difficulties during the mesh generation are avoided, the system matrices resulting from the according thin-sheet discretization still exhibit a dependence on the thickness δ, even if only a linear approximation perpendicular to the sheet surface is applied. Moreover, for many EQS applications, considering a linear discretization perpendicular to the sheet leads to a sufficient approximation. Thus, a different approach, which is able to overcome the undesired effect, is introduced in Sect. 2.2. Both methods are compared in terms of the numerical results as well as regarding the condition number of the respective system matrices based on simple test cases in Sect. 3.

2 Finite-Element Discretization for Thin Sheets

In the electro-quasistatic limit, the electric field strength $\mathbf{E}$ is irrotational. Therefore, the electric scalar potential φ with $\mathbf{E} = -\nabla\varphi$ can be introduced. The resulting partial differential equation derived from the full set of the Maxwell equations reads

$$\nabla \cdot (\kappa \nabla \varphi) + j\omega \nabla \cdot (\epsilon \nabla \varphi) = 0 \tag{1}$$

in frequency domain. Here κ denotes the electric conductivity, ϵ the permittivity and $\omega = 2\pi f$ the angular frequency. Discretizing the weak form of (1) by means of nodal Whitney basis functions $w_i(\mathbf{x})$ and element-wise constant ϵ and κ leads to

$$\sum_{k=1}^{K} (\kappa_k + j\omega\epsilon_k) \int_{\Omega_k} \nabla\varphi(\mathbf{x}) \cdot \nabla w_j(\mathbf{x})\, dV = 0 \qquad \text{with} \qquad \varphi(\mathbf{x}) = \sum_{i=1}^{N} \phi_i w_i(\mathbf{x}), \tag{2}$$

assuming homogenous Dirichlet boundary conditions for φ at $\partial\Omega$. The domain Ω is sub-divided into K elements and N denotes the number of nodes in the mesh. Based on the volume discretization, the surface elements of the reduced sheet are included during the assembly of the final system matrices for the two different methods described in the following.

2.1 Method 1: Thin-Sheet Bases

The first method is based on prismatic elements as described in [6,7]. Following the notation in [5], the different sets of basis functions defined on the resulting prisms are referred to as thin-sheet bases. As in [4], basis functions of arbitrary order can be considered. During the assembly of the system matrix, the contribution of the thin sheet is neglected, as its volume is zero at mesh level. Subsequently, it is included by defining specific thin-sheet bases in order to represent the volumetic nature of the sheet at the physical level. These modified bases are obtained by multiplying the triangular basis function defined at the sheet surface by a polynomial $p : \mathbb{R} \to \mathbb{R}$. As a consequence, the variation of the fields perpendicular to the surface element can be considered. A Lagrangian base of order o is chosen for p. This implies the need of o additional coefficients for each volume-based degree of freedom (DoF) located at the sheet surface. Considering a three-dimensional domain with $\mathbf{e}_z$ perpendicular to the sheet element k, the according thin-sheet base reads

$$\tilde{w}_{k,i,j}(x, y, z) = w^s_{k,i}(x, y) p_j(z) , \tag{3}$$

where $w^s_{k,i}$ is the i-th surface basis function of the k-th sheet element and $p_j(z)$ the j-th base of the polynomial $p(z)$ defined for $z \in [-\delta/2, \delta/2]$. For the lowest order ($o = 0$, $p(z) = 1$), no additional DoFs are required. In this case, the volume integrals in (2) reduce to the related surface integrals multiplied by the thickness δ, namely,

$$\delta \int_{\Gamma_k} \nabla w^s_{k,i}(x, y) \cdot \nabla w^s_{k,j}(x, y)\, dA . \tag{4}$$

For linear or higher-order variation the two outermost polynomial degrees of freedom of $p(z)$ at $z = -\delta/2$ and $z = \delta/2$ have to be coupled to the domain above and below the sheet, respectively. All other DoFs are internal DoFs for the polynomial p as indicated in Fig. 1. The coupling terms between all DoFs of the sheet element are

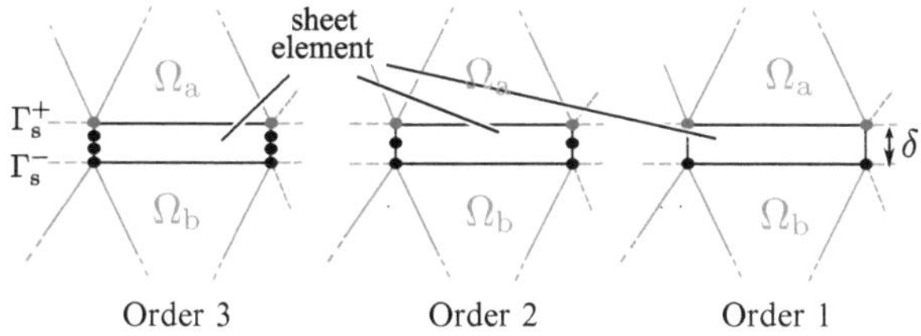

Fig. 1 A sheet element is shown in its surrounding mesh for polynomials p of different order. Geometrically, the sheet is a surface Γ_s connecting the meshes below (Ω_b) and above (Ω_a) the sheet. *Black dots* denote the additional degrees of freedom (DoFs) used for the polynomial p

$$\int_{\Omega_k} \nabla \tilde{w}_{k,i,j} \cdot \nabla \tilde{w}_{k,g,h} \, dV = \int_{-\delta/2}^{\delta/2} \nabla p_j(z) \cdot \nabla p_h(z) \, dz \int_{\Gamma_k} w^s_{k,i}(x,y) w^s_{k,g}(x,y) \, dA$$
$$+ \int_{-\delta/2}^{\delta/2} p_j(z) p_h(z) \, dz \int_{\Gamma_k} \nabla w^s_{k,i}(x,y) \cdot \nabla w^s_{k,g}(x,y) \, dA \,, \tag{5}$$

where Γ_k is the area of the sheet element. These terms can be evaluated for arbitrary order of both the polynomial p and the surface basis function w^s_k. Each sheet element can be interpreted as a prismatic element inserted to the tetrahedral mesh at the position of the thin sheet. The additional DoFs are allocated along the height δ of the inserted prisms.

2.2 Method 2: Semi-analytic Sheet

In order to avoid the unfavorable dependence of the system matrix entries on the factor $1/\delta$, a second method to deal with thin sheets is developed. Moreover, no additional DoFs are introduced, while providing linear order of approximation across the sheet. The information about the voltage drop between the domain above and below the sheet is included in the basis functions of the elements that are connected to the sheet.

Figure 2 shows one Lagrangian element k which is adjacent to the sheet. Here, each DoF of its corresponding sheet element s represents the electric scalar potential at the centerline of the sheet Γ_s. Assuming a linear variation of the scalar potential across the sheet, which is very often the case in EQS, the related voltage drop between the centerline and the sheet boundary evaluates to

$$\int_{\mathbf{x}_i}^{\mathbf{x}_i^*} \mathbf{E}_\mathrm{s} \cdot d\mathbf{s} = -\frac{\delta}{2} \nabla \varphi_s(\mathbf{x}_i) \cdot \mathbf{n}_k \,, \tag{6}$$

where $\mathbf{n}$ is the vector normal to the sheet pointing from the connected element towards the sheet and $\mathbf{x}_i$ is the position vector corresponding to the node i in the

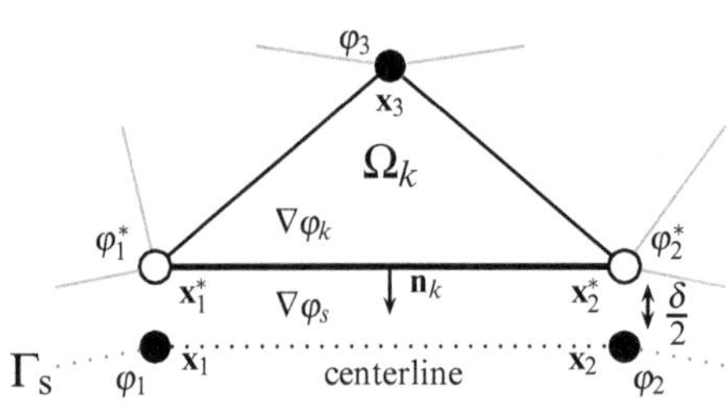

Fig. 2 Element k (in 2D a triangle) is one of the elements connected to the sheet. The *black dots* at the bottom denote the DoFs defined at the sheet centerline. The *empty circles* denote the potential at the border of Ω_k which is apriori unknown but can be evaluated

mesh. The gradient of the electric potential within the sheet $\nabla\varphi_s$ can be obtained by the evaluation of the continuity equation (1) at the sheet boundary of element k, namely

$$(\kappa_k + j\omega\epsilon_k)\nabla\varphi_k(\mathbf{x}_i)\cdot\mathbf{n}_k = (\kappa_s + j\omega\epsilon_s)\nabla\varphi_s(\mathbf{x}_i)\cdot\mathbf{n}_k\ . \tag{7}$$

Using (6) and (7), the electric potential ϕ_i^* at the sheet boundary is given by

$$\phi_i^* = \phi_i - \frac{\delta}{2}\alpha_k\nabla\varphi_k(\mathbf{x}_i)\cdot\mathbf{n}_k \qquad \text{with} \qquad \alpha_k = (\kappa_k + j\omega\epsilon_k)/(\kappa_s + j\omega\epsilon_s)\ , \tag{8}$$

where ϕ_i is the known potential at the sheet centerline.

The set of all element DoFs N_k consists of a set C_k of all DoFs that are connected to the sheet and a set U_k of all DoFs that are not. The discrete form of the normal electric field then reads

$$\nabla\varphi_k(\mathbf{x}_i)\cdot\mathbf{n}_k = \sum_{j\in U_k}\phi_j\nabla w_j(\mathbf{x}_i)\cdot\mathbf{n}_k + \sum_{j\in C_k}\phi_j^*\nabla w_j(\mathbf{x}_i)\cdot\mathbf{n}_k\ . \tag{9}$$

Inserting (9) in (8) for all C_k results in a small local system of equations. Solving this system for all ϕ_i^* in (8) and replacing the solution in the discrete gradient of element k leads to a modified gradient $\tilde{\nabla}\varphi_k(\mathbf{x})$ of the electric potential in the connected element k. For first-order elements, this gradient reads

$$\tilde{\nabla}\varphi_k(\mathbf{x}) = \sum_{i\in N_k}\phi_i\left(\nabla w_i(\mathbf{x}) - \frac{\delta}{2}\alpha_k\beta_k\zeta_k\nabla w_i\cdot\mathbf{n}_k\right) \tag{10}$$

with

$$\beta_k = \frac{1}{1+\frac{\delta}{2}\alpha_k\sum_{j\in C_k}\nabla w_j\cdot\mathbf{n}_k} \qquad \text{and} \qquad \zeta_k = \sum_{j\in C_k}\nabla w_j(\mathbf{x}). \tag{11}$$

and is directly used in the standard FEM matrix assembly (2). As the gradient of the test functions remains unchanged, the system matrix becomes unsymmetric.

To consider also the tangential variation of the normal component in the sheet, additionally, the coupling terms of the lowest-order thin-sheet bases (first method with constant p) are added for each sheet element. This is possible, because the sheet DoFs are defined at the centerline of the sheet, which is approximately the mean-value and therefore representative for constant elements.

3 Comparison

A simple example is chosen to show the difference between both methods, in particular regarding the condition number of the related system matrix A. Here, the spectral conditioning number $\mathrm{cond}_2(A) = ||A||_2||A^{-1}||_2$ is chosen in order to

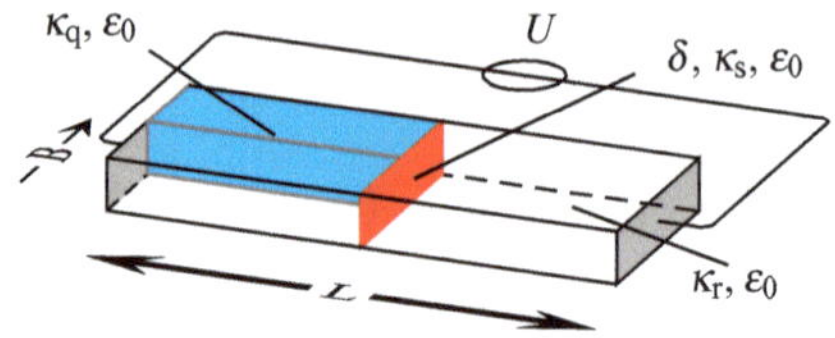

Fig. 3 Resistor with a thin crack (thickness δ, κ_s, ϵ_0) at its center. The resistor is completely filled with one material (κ_r, ϵ_0) except of one quarter which is filled with a different material (κ_q, ϵ_0)

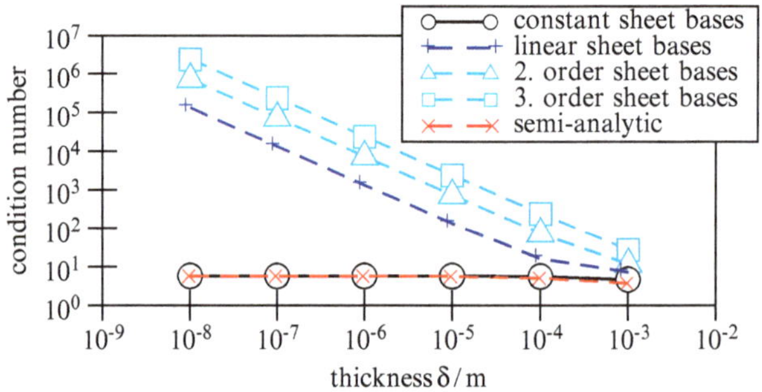

Fig. 4 Condition number of system matrix A versus the sheet thickness δ for the 1D example

obtain a quantitative statement. The selected test mode is shown in Fig. 3. It consists of a resistor with a thin crack which is excited harmonically by a voltage source ($U = 2\,\text{V}$, $\omega = 2\pi 50\,\text{Hz}$, $L = 1\,\text{m} + \delta$, $B = 1\,\text{m}$). At the remaining boundaries, homogeneous Neumann boundary conditions are applied. The comparison in the following subsections is carried out based on this example geometry.

3.1 1D Example

The example in Fig. 3 reduces to a 1D model for the choice of equal electric conductivity $\kappa_q = \kappa_r = 1{,}000\,\text{Sm}^{-1}$ in the different parts, while $\kappa_s = 1\,\text{Sm}^{-1}$. In this case, an analytical solution is available. Here, the variation of the electric scalar potential is piece-wise linear with respect to the only remaining coordinate direction. As a consequence, two elements are sufficient to resemble the exact solution. Nevertheless, for verification purpose, a mesh consisting of four equally-sized elements is selected. For, e.g. $\delta = 10^{-4}\,\text{m}$, the voltage drop within the sheet

$$\Delta\varphi = U\delta \frac{\omega\epsilon_0 - j\kappa_r}{\omega\epsilon_0 L - j\,(\delta\kappa_r + (L-\delta)\kappa_s)} \approx 0.18182\,\text{V} + j4.59313 \cdot 10^{-10}\,\text{V}$$

is correct for both methods, except for lowest-order thin-sheet bases as the assumption to be constant across the sheet is not valid. However, Fig. 4 shows, that for thin-sheet bases with linear or higher-order polynomial p the condition number of the system matrix, which is a measure for the computational effort, increases strongly with decreasing thickness δ. In contrast, for the semi-analytic method, it never exceeds the value for the lowest-order thin-sheet bases (where p is one), even though the latter method is not suitable in this case. Note that the thin-sheet bases with linear polynomials are equivalent to a first-order volume-based FEM with the sheet resolved in the mesh.

3.2 2D Example

For, e.g., the choice of $\kappa_r = 20\,\mathrm{Sm}^{-1}$ and $\kappa_q = 40\,\mathrm{Sm}^{-1}$, the variation of the electric potential in the example is no longer linear. The graph for the condition number of the system matrix with respect to the thickness δ, although not plotted, is similar to Fig. 4. Figure 5 shows the relative error of the electric losses P_{el} in all regions except the sheet for different levels of discretization. Both methods are compared to the respective values without the application of any thin-sheet modeling technique. For all methods, the biconjugate gradient method is used to solve the system of linear equations. The reference value is always the best solution obtained with the finest mesh ($P_{el} \approx 47.10083\,\mathrm{W}$ at about 10^5 elements). Both methods are comparable in the convergence order to the standard volume-based, first-order FEM where the sheet is resolved as a volume in the mesh. For some levels of discretization also the condition number of the system matrix is compared, shown in Fig. 6. Again, the condition number in the semi-analytic method never exceeds the condition of

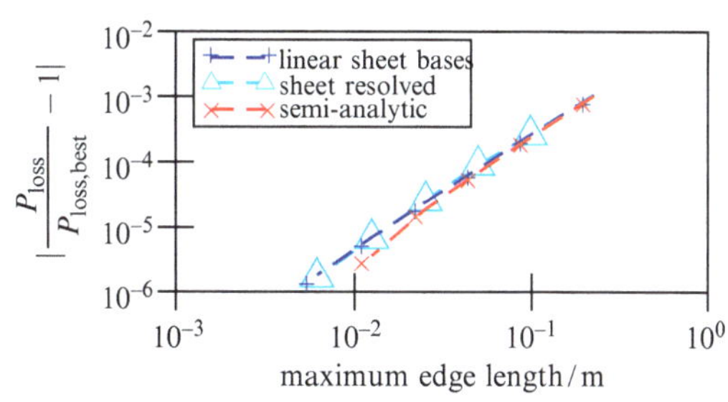

Fig. 5 Relative error of electric losses outside sheet versus the maximum edge length in the mesh

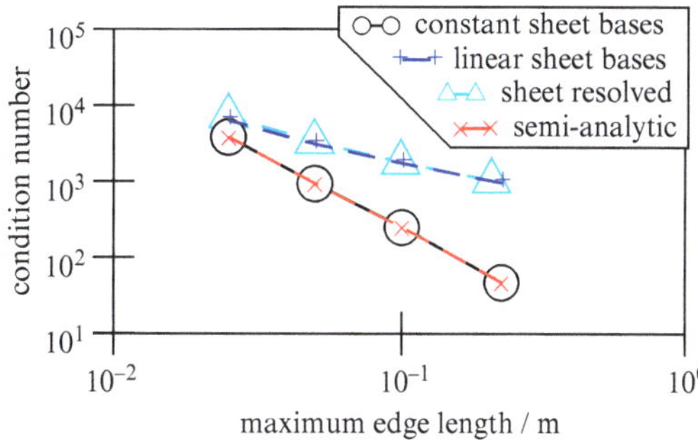

Fig. 6 Condition number of system matrix versus the maximum edge length in mesh

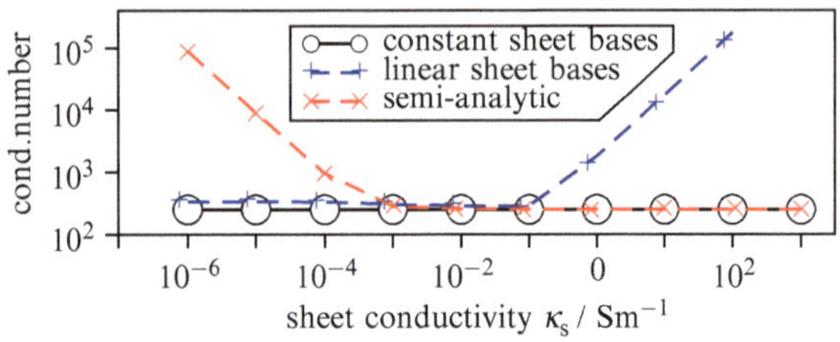

Fig. 7 Condition number of system matrix A versus sheet conductivity κ_s (2D example)

the method with constant thin-sheet bases. However, the length ratio of the smallest sheet edge to smallest edge outside the sheet decreases with finer meshes. So, of course, the advantage regarding the condition number decreases as well.

The benefit of the different thin-sheet modeling techniques depends on the ratio of the conductivity κ_s to κ_r. Figure 7 shows the condition number of the system matrix for a constant sheet thickness of $\delta = 10^{-4}\,m$ while κ_s is varied in the range $10^{-6} < \kappa_s < 10^3$. For $\kappa_s > 10^5$ the voltage drop within the sheet is negligible. Therefore, the constant sheet bases are applicable and provide, due to their simplicity, the best choice.

On the other hand, for $\kappa_s < 0.1$, the thin sheet has no significant effect on the matrix condition. As a consequence, no thin-sheet approach is actually required. In the range $10^{-1} < \kappa_s < 10^4$ the semi-analytic method is appropriate, as it leads to condition numbers comparable to the respective values for lowest-order thin-sheet bases while considering the voltage drop within the sheet.

4 Conclusion

Two finite-element thin-sheet approaches are investigated in EQS. The well-known method of using higher-order prismatic elements for the discretization of thin sheets introduces additional degrees of freedom. While difficulties in mesh generation are avoided, the condition number of the resulting system matrices deteriorates for higher-order approximations. For many cases, however, allowing for a linear field variation across the sheet already leads to a sufficient accuracy. In this case, the proposed method yields superior properties such as the independence of the condition number of the system matrices from the thickness of the sheet. The new method is compared to the common approach and is found to deliver identical results. Furthermore, the concepts can be used in combination with higher-order finite-element discretizations.

References

1. Nakata, T., Takahashi, N., Fujiwara, K., Shiraki, Y.: 3-D magnetic field analysis using special elements. IEEE Trans. Magn. **26**(5), 2379–2381 (1990)

2. Krähenbühl, L., Muller, D.: Thin layers in electrical engineering-example of shell models in Analysing Eddy-Currents by boundary and finite element methods. IEEE Trans. Magn. **29**(2), 1450–1455 (1993)
3. Dular, P., Geuzaine, C.: Modeling of thin insulating layers with dual 3-D magnetodynamic formulations. IEEE Trans. Magn. **39**(3), 1139–1142 (2003)
4. Gyselinck, J., Sabariego, R., Dular, P., Geuzaine, C.: Time-domain finite-element modeling of thin electromagnetic shells. IEEE Trans. Magn. **44**(6), 742–745 (2008)
5. Schmidt, K.: High-order numerical modeling of highly conductive thin sheets. Ph.D. thesis, ETH Zürich (2008)
6. Ren, Z.: Degenerated Whitney prism elements-general nodal and edge shell elements for field computation in thin structures. IEEE Trans. Magn. **34**(5), 2547–2550 (1998)
7. Abenius, E., Edelvik, F.: Thin sheet modeling using shell elements in the finite-element time-domain method. IEEE Trans. Antennas Propag. **54**(1), 28–34 (2006)
8. Weida, D., Steinmetz, T., Clemens, M.: Electro-quasistatic high voltage field simulations of large scale insulator structures including 2-D models for nonlinear field-grading material layers. IEEE Trans. Magn. **45**(3) (2009)

Mode Selecting Eigensolvers for 3D Computational Models

Bastian Bandlow and Rolf Schuhmann

Abstract For the computation of interior eigenpairs an educated initial guess on the eigenvalue is mandatory in general. The convergence behavior of eigensolvers can be improved by using a starting vector, which should be a reasonable approximation of the searched eigenvector. However, these two provisions do not lead necessarily to the searched eigenpair. We propose an extended selection strategy for the Ritz pairs occurring within the Jacobi-Davidson eigensolver algorithm and compare its performance with the Rayleigh quotient iteration. A complex unsymmetric standard eigenvalue problem resulting from a finite integration discretization of a dielectric disk in free-space serves for numerical experiments.

1 Introduction

The computation of interior eigenvalues of complex unsymmetric matrices arising in numerical models from engineering problems is still a challenging task. At least an educated guess on the eigenvalue location within the spectrum has to be available, in order to succeed. In many cases there is some additional information on the eigenvector available that can be used as a starting vector for iterative eigensolvers. However, using a priori knowledge as a starting vector does not necessarily cause convergence towards the searched eigenvector, if the guessed eigenvalue is too far away from the actual solution. Inside the Jacobi-Davidson eigensolver algorithm [9] one of several eigenpair approximations – the most promising one – has to be selected for the use in subsequent iterations. In this paper we focus on the extension of that systematic selection process, in order to compute only one single

B. Bandlow (✉) · R. Schuhmann
FG Theoretische Elektrotechnik, Universität Paderborn, Warburger Str. 100, D-33098 Paderborn, Germany,
e-mail: bandlow@tet.upb.de, schuhmann@tet.upb.de

B. Michielsen and J.-R. Poirier (eds.), *Scientific Computing in Electrical Engineering SCEE 2010*, Mathematics in Industry 16, DOI 10.1007/978-3-642-22453-9_14,

Algorithm 3 Jacobi-Davidson-Method

Require: matrix $\mathbf{A} \in \mathbb{C}^{n \times n}$ (normalized properly), target τ, initial vector $\mathbf{v}_0 \in \mathbb{C}^n$, initial subspace $\mathbf{V}_m \in \mathbb{C}^{n \times m}$, tolerance ϵ
Ensure: Eigenpair $(\lambda, \mathbf{x})$

$\mathbf{t} \leftarrow \mathbf{v}_0$
$\theta \leftarrow \tau$
loop
  Expand and orthonormalize $\mathbf{V}_m$ by $\mathbf{t}$
  Project $\mathbf{A}$ on subspace $\mathbf{V}_m$
  Solve projected eigenvalue problem
  Select Ritz pair $(\theta, \mathbf{u})$ with θ next to τ
  $\mathbf{r} \leftarrow (\mathbf{A} - \theta \mathbf{I})\mathbf{u}$
  Break if $\|\mathbf{r}\| < \epsilon$
  Solve JD correction equation for $\mathbf{t}$
end loop
$\lambda \leftarrow \theta, \mathbf{x} \leftarrow \mathbf{u}$

Algorithm 4 Rayleigh-Quotient-Iteration

Require: matrix $\mathbf{A} \in \mathbb{C}^{n \times n}$ (normalized properly), initial guess σ, initial vector $\mathbf{v}_0 \in \mathbb{C}^n$, threshold ι, tolerance ϵ
Ensure: Eigenpair $(\lambda, \mathbf{x})$

$\mathbf{v}_0 \leftarrow \mathbf{v}_0 / \|\mathbf{v}_0\|$
for $i = 1, \ldots, i_{max}$ **do**
  if $i < \iota$ **then**
    $\mathbf{v}_1 \leftarrow (\mathbf{A} - \sigma \mathbf{I})^{-1} \mathbf{v}_0$
  else
    $\mathbf{v}_1 \leftarrow (\mathbf{A} - \sigma \mathbf{I})^{-1} \mathbf{v}_1$
  end if
  $\mathbf{v}_1 \leftarrow \mathbf{v}_1 / \|\mathbf{v}_1\|$
  $\sigma \leftarrow \mathbf{v}_1^H \mathbf{A} \mathbf{v}_1$
  Break if $\|(\mathbf{A} - \sigma \mathbf{I})\mathbf{v}_1\| < \epsilon$
end for
$\lambda \leftarrow \sigma$ und $\mathbf{x} \leftarrow \mathbf{v}_1$

eigensolution – especially for the case that some specific features of the eigenvector are known in advance, but the initial guess on the eigenvalue is still insufficient.

2 Jacobi-Davidson Method

The outline of the Jacobi-Davidson[1] method (JD) for a standard eigenvalue problem $\mathbf{Ax} = \lambda \mathbf{x}$, $\mathbf{A} \in \mathbb{C}^{n \times n}$ is given in Algorithm 3. The JD is feasible for the computation of a modest number of interior or exterior eigenvalues λ of the spectrum. It is important to note, that the JD algorithm generates the eigenpairs one by one and not a whole block of eigenvalues simultaneously. According to Algorithm 3 the original full-dimensional eigenvalue problem is projected and solved on a low-dimensional subspace $\mathbf{V}_m \in \mathbb{C}^{n \times m}, m \ll n$, which is gradually extended by the solution of a correction equation. An important feature of the JD is that the correction equation may be solved inexactly. Usually, an end of the spectrum or an arbitrary interior value of the spectrum can be specified as a so-called target value τ, and the eigenvalues λ next to the target value τ are computed during the solution process. See Algorithm 3 and [5, 9] for further details on the single steps.

However, the solution of the low-dimensional, projected eigenvalue problem does not only yield Ritz values θ_i, but we can also obtain approximations of the corresponding eigenvectors (Ritz vectors $\mathbf{u}_i$) by expanding them back to full dimension. It is important to note that the residual norm of the Ritz vectors $\|\mathbf{r}_i\| := \|(\mathbf{A} - \theta_i \mathbf{I})\mathbf{u}_i\|$ is rather large during the first JD iterations. The goal is to have a simple measure beyond the target value τ, in order to be able to decide which of the Ritz

[1] A Matlab implementation of the JD as well as further bibliography on the JD is available from [5].

vectors $\mathbf{u}_i$ incorporates suitable information on the eigenvector we are searching for. Preliminary results for the computation of selected 2D eigenmodes in waveguide cross-sections have been shown in [2].

2.1 Selection of Appropriate Ritz Pairs

The basis for the selection of the actual Ritz pair is the solution of the projected, low-dimensional eigenvalue problem. In the JDQR method the *standard* extraction of Ritz values leads to a low-dimensional standard eigenvalue problem with matrix $\mathbf{M} = \mathbf{V}_m^H \mathbf{A} \mathbf{V}_m$. The *harmonic* extraction is based on subspaces $\mathbf{V}_m$ and $\mathbf{W}_m$ and therefore Ritz values are obtained from the matrix pencil $(\mathbf{W}_m^H \mathbf{A} \mathbf{V}_m, \mathbf{W}_m^H \mathbf{V}_m)$ with $\mathbf{W}_m = \mathbf{A}\mathbf{V}_m - \sigma \mathbf{V}_m$ and some shift σ (JDQZ method). Usually, the selection of the Ritz pair is done by sorting the Ritz values with respect to the target value τ [4].

The JDQR solver is based on a Schur decomposition rather than a complete eigen decomposition of the low-dimensional matrix $\mathbf{M} = \mathbf{Q}\mathbf{R}\mathbf{Q}^H$, since the complete eigen decomposition is known not to be computationally most efficient [4]. The Schur decomposition consists of orthonormal $\mathbf{Q}$ and upper triangular $\mathbf{R}$ which are extended gradually in every iteration. The diagonal of $\mathbf{R}$ contains the Ritz values θ_i and the first column of $\mathbf{Q}$ (left-multiplied by $\mathbf{V}_m$) together with the first diagonal entry of $\mathbf{R}$ yields a Ritz pair $(\theta_1, \mathbf{u}_1)$, while all other columns of $\mathbf{Q}$ will lead only to Ritz vectors, if the strict upper triangular part of $\mathbf{R}$ vanishes. Yet for sufficiently diagonal dominant $\mathbf{R}$ all columns of $\mathbf{V}_m \mathbf{Q}$ may be regarded as reasonable approximations of Ritz vectors $\mathbf{u}_i$.

The JDQZ solver leads to a generalized Schur form (QZ decomposition) of the low-dimensional pencil $(\mathbf{W}_m^H \mathbf{A} \mathbf{V}_m, \mathbf{W}_m^H \mathbf{V}_m)$. The special choice of search and test subspace for the *harmonic* extraction has been found to be beneficial for the computation of interior eigenpairs [4]. The Ritz values θ_i can be obtained by the generalized Schur form but the Ritz vectors are not – besides the first one. The additional computation of the solution of the low-dimensional generalized eigenvalue problem yields the Ritz vectors and they can be assigned to the generalized Schur form by using the Ritz values.

So far we have shown ways to get Ritz vectors (or approximations) from the low-dimensional eigenvalue problem occurring inside the JDQR and JDQZ algorithms. Now, the selection process of an appropriate Ritz pair $(\theta, \mathbf{u})$ can be extended by a criterion which uses a scalar product of some weighting vector $\mathbf{f}$, which contains the a priori knowledge and a specific Ritz vector $\mathbf{u}_i$

$$\mathbf{f}^H \mathbf{u}_i > \alpha. \tag{1}$$

The weighting vector $\mathbf{f}$ may be established from previous calculations or analytical considerations, and may be restricted to *some* components of $\mathbf{u}_i$. A good Ritz vector $\mathbf{u}_i$ for further JD iterations yields a value larger than the predefined threshold α. The selection process finally reads:

1. Sort Ritz values with respect to target τ.
2. Evaluate (1) for all Ritz vectors.
3. If and only if there are Ritz pairs satisfying (1) place them at the top of order.
4. Resume JD iteration with first Ritz pair.

Note that no change in sort order occurs if there is no Ritz vector satisfying (1).

3 Rayleigh Quotient Iteration

The Rayleigh quotient iteration (RQI) of Algorithm 4 is feasible for the fast computation of a single eigenpair if the occurring linear system can be solved exactly [7]. Its singularity has to be checked in advance and the RQI may fail if $\|\mathbf{v}_1\| = 0$ occurs. The RQI shows local cubic convergence behavior for well chosen initial vectors. In Algorithm 4 we keep the initial vector as a (constant) right hand side for $\iota \geq 1$ iterations and study the convergence behavior for different thresholds ι. Within the RQI there is no subspace projection and the JD can be regarded as a subspace accelerated (inexact) RQI. More details on the relationship of JD and RQI can be found in the literature [1, 6].

4 Numerical Example

For numerical experiments we compute higher-order eigenmodes of a pierced dielectric disk in free-space, cf. Fig. 1a. We consider a disk having a refractive index of 3.3, a radius of 1 μm and a thickness of 375 nm. In [10] an analytical eigenfrequency estimation is proposed, which is applicable when a specific eigenmode has nearly no interaction with the pierced hole. An example obtained by the analytical approximation for the E_z electric field of the $\mathrm{TM}_{1,12}$ eigenmode is given in Fig. 1b. This is an approximation since the behavior of the electric field at the edges is not properly modeled and it is valid in the $z = 0$ plane only.

The discrete eigenvalue problem for the electric grid voltages is formulated using the finite integration technique (FIT) [11]. The FIT is based on a spatial segmentation of the computational domain by a computational grid pair, the normal grid G and the dual grid $\tilde{G}$. The degrees of freedom of the method are the so-called integral state variables, defined as integrals of the electric and magnetic field vectors over edges L_i, $\tilde{L}_i$ of the normal grid G and the dual grid $\tilde{G}$, respectively:

$$\widehat{\mathbf{e}}_i = \int_{L_i} \mathbf{E} \cdot \mathrm{d}\mathbf{s}, \qquad \widehat{\mathbf{h}}_j = \int_{\tilde{L}_j} \mathbf{H} \cdot \mathrm{d}\mathbf{s}. \tag{2}$$

Maxwell's grid equations can be written down in frequency domain, neglecting sources, as

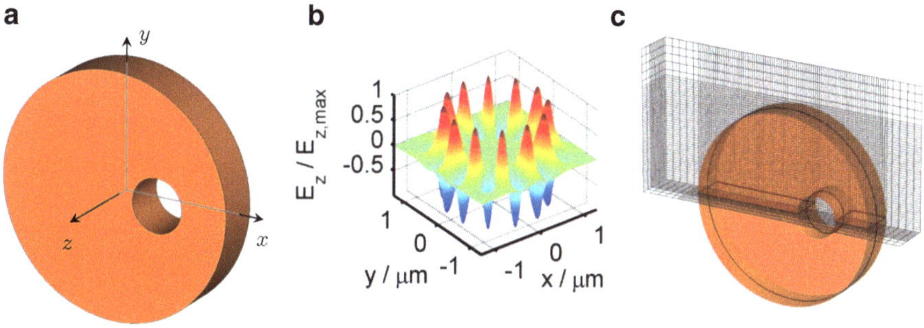

Fig. 1 (**a**) Structure of a pierced dielectric disk. (**b**) Analytical approximation in the $z = 0$ plane of the $TM_{1,12}$ eigenmode obtained by the approach of [10]. (**c**) Mesh (without PML) for the pierced disk: 1/4 of the computational domain has to be discretized

$$\mathbf{C}\widehat{\mathbf{e}} = -\mathrm{i}\omega\mathbf{M}_{\mu}\widehat{\mathbf{h}}, \qquad \widetilde{\mathbf{C}}\widehat{\mathbf{h}} = \mathrm{i}\omega\mathbf{M}_{\varepsilon}\widehat{\mathbf{e}}. \tag{3}$$

$\mathbf{C}$ and $\widetilde{\mathbf{C}}$ are the topological curl-operators containing entries $\{-1;0;1\}$ only. Perfectly matched layers (PML) [12] as an absorbing boundary condition based on complex metric stretching can be introduced in FIT in a straight-forward manner, which causes the diagonal material matrices to become complex and frequency dependent $\underline{\mathbf{M}}_{\varepsilon}(\omega)$ and $\underline{\mathbf{M}}_{\mu}(\omega)$. The PML is only theoretically *perfectly matched*, since the discretization of the damping conductivities introduces a remaining reflection error that can be controlled by the number of layers. In frequency domain we solve the curl-curl eigenmode equation for resonance frequencies $\underline{\omega}^2$, which can be derived from (3) easily as

$$\underline{\mathbf{A}}\widehat{\mathbf{e}} = \underline{\omega}^2\widehat{\mathbf{e}}, \qquad \underline{\mathbf{A}} = \underline{\mathbf{M}}_{\varepsilon^{-1}}(\underline{\omega})\mathbf{C}^T\underline{\mathbf{M}}_{\mu^{-1}}(\underline{\omega})\mathbf{C}. \tag{4}$$

The frequency dependent material matrices are linearized at the estimation frequency and the radiation losses introduced by the PML lead to a complex, non-symmetrizable system matrix $\underline{\mathbf{A}}$ with complex eigenvalues $\underline{\omega}^2$. Its solution can be computationally expensive, but yields the modal fields as well as their resonance frequency and quality factor $Q = \Re\{\underline{\omega}\}/2\Im\{\underline{\omega}\}$. Moreover, the spectrum is spoilt by some undesired modes, which are trapped within the PML and occur at similar frequencies like the desired modes.

For discretization we use CST MICROWAVE STUDIO [3] and a mesh with 14 lines per wavelength. Two symmetry conditions reduce the number of unknowns to one quarter of the original problem (see Fig. 1c). Four PML layers are attached to the mesh using some laboratory code in Matlab. The final algebraic system matrix has 270,936 complex unknowns. The system matrix is reordered and its spectrum is shifted and normalized at first. For the JD we opt to solve the correction equation directly and the tolerance for the residual norm of the Ritz pairs has to drop below 10^{-9} within at most 50 JD iterations. The estimation frequency obtained from the analytical approximation is 242.9 THz and is used as target value τ within the JD

Table 1 Results for the original JDQR and the JDQR and JDQZ with extended selection criterion (1) in dependency on different starting vectors $\mathbf{v}_0$ which include between 0 and 7 layers of the analytical approximation from [10]

	Original JDQR from [5]			JDQR with selection (1)			JDQZ with selection (1)		
$\mathbf{v}_{ana}$ in $\mathbf{v}_0$	# JD iterations	Add. modes	$TM_{1,12}$ f / THz	# JD iterations	Add. modes	$TM_{1,12}$ f / THz	# JD iterations	Add. modes	$TM_{1,12}$ f / THz
0	40	11	253.6	21	–	$\neq$253.6	5	–	$\neq$253.6
1	31	11	253.6	6	0	253.6	4	0	253.6
3	31	12	253.6	4	0	253.6	5	0	253.6
4	31	11	253.6	4	0	253.6	3	0	253.6
7	32	11	253.6	5	0	253.6	4	0	253.6

and RQI. The analytical approximation in the $z = 0$ plane from Fig. 1b is used as a weighting vector for the extended selection criterion in (1). Moreover, different starting vectors $\mathbf{v}_0$ can be built from the analytical approximation by plugging it also into planes with $z > 0$, since that might be a better choice than random values.

5 Results

In Table 1 the results of the JD solvers with different starting vectors are shown for $TM_{1,12}$ eigenmode. The first column denotes the number of mesh planes which contain the analytical approximation. For the original JDQR solver the required number of iterations depends on whether the analytical approximation is considered or not. 31–40 iterations are needed and 11–12 additional eigenmodes are computed before the $TM_{1,12}$ eigenmode occurs. The computed part of the eigenfrequencies is shown with their Q factors in Fig. 2a: The dashed line denotes the estimation obtained by the analytical approximation, the black square is the final result for the $TM_{1,12}$ eigenmode with a resonance frequency of 253.6 THz and the blue circles and the red diamond are the additional modes, which are also computed by the JDQR. The upper inset in Fig. 2a shows the E_z field distribution in plane $z = 0$ of the $TM_{1,12}$ eigenmode. The JDQR with selection (columns 5–7 in Table 1) fails as expected when no analytical aproximation is included in the starting vector $\mathbf{v}_0$, since no extended selection takes place, and (1) is never fulfilled. Better starting vectors which include the analytical approximation in one or more layers cause convergence in 4–6 iterations, and no undesired modes are computed. We choose $\alpha = 0.2$ for this study. Nearly the same holds for the JDQZ method with extended selection and $\alpha = 0.2$ (columns 8–10 in Table 1) with a similar number of JD iterations. In most cases the JDQZ converges a bit faster than the JDQR to interior eigenvalues as expected [4].

The influence of the threshold value α in (1) is studied when $\mathbf{v}_0$ includes a single layer of the analytical approximation. It turns out that the JDQR with selection

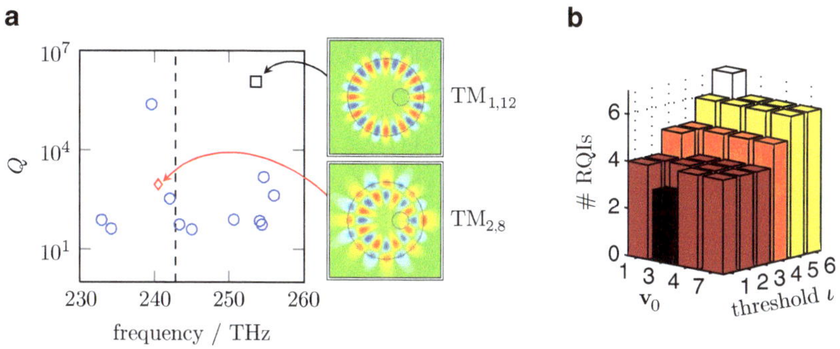

Fig. 2 (**a**) Eigenfrequencies and Q: Real estimation of the analytical approach (*dashed line*) for the searched $TM_{1,12}$ eigenmode (*square*) and additional modes (○) of the discrete model. The $TM_{2,8}$ eigenmode (*diamond*) can be computed with the JDQR with selection. (**b**) Number of iterations of the RQI for different starting vectors $\mathbf{v}_0$ and thresholds ι when computing the $TM_{1,12}$ eigenmode

converges after six iterations when $\alpha \in [0.2, 0.95]$. Criterion (1) can not be satisfied any more if $\alpha = 0.97$ or larger and so undesired eigenpairs are retrieved.

The correction equation within the JD may also be solved inexactly e.g. by an iterative solver. In that case sometimes Ritz vectors may occur which satisfy (1) but belong to Ritz values that are far away from τ. In order to prevent the JD from *hopping* through the spectrum we suggest not to use too low values of α. Setting $\alpha = 0.8$ and using bicgstab for the inexact solution of the JD correction equation leads to convergence after 80 iterations.

When the combination of problem size and available computing or memory capabilities allow the exact solution of the occurring linear systems, the RQI is an alternative to the JD. The results of the RQI are given in the diagram of Fig. 2b. Again the number of layers of the analytical approximation, which is included in $\mathbf{v}_0$ is under study. Moreover, the threshold value ι defines for how many iterations of the RQI the right hand side of the equation to be solved is kept fixed. At least the first solve has to be performed with $\mathbf{v}_0$ as right hand side and the smallest number of iterations are achieved for $\iota = 1, 2, 3$. But also for permanent use of $\mathbf{v}_0$ as right hand side ($\iota = 6$) the residual norm finally drops below 10^{-9} as required.

For the $TM_{2,8}$ eigenmode (red diamond and lower inset in Fig. 2a the electric field interacts with the disk's pierced hole. Therefore only a very crude analytical approximation at 405.1 THz is obtained by the approach from [10]. The RQI does not converge towards the desired $TM_{2,8}$ eigenmode for arbitrary starting vectors in that case. However, for the selection criterion (1) it is possible to use only that part of the approximate field distribution, which is on the opposite side of the hole (i.e. $x < 0$ cf. Fig. 1a. The JDQR with extended selection criterion is reliably able to calculate $TM_{2,8}$ eigenmode of the pierced disk at a frequency of 240.5 THz in 7–10 iterations when the starting vector contains three or more z-planes of analytical approximation for the region having $x < 0$.

The JD and RQI are implemented in Matlab. As linear solver the PARDISO [8] included in Intel's MKL is used. The computing node consists of four Intel Xeon 7350 CPUs with in total 16 cores and 128 GByte memory running a Microsoft OS. Parameter combinations from Table 1 and Fig. 2b which need four iterations to converge lead to a computation time of 328 s for the JDQR, 328 s for the JDQZ and 334 s for the RQI. The small deviations show that most time is needed by the linear solver.

6 Conclusions

We have shown how to include a priori known features of searched eigenvectors in the selection process of the Ritz pair within the Jacobi-Davidson method using *standard* or *harmonic* extraction. The extended selection is based on a reordering of potential Ritz pairs, which are weighted according to the a priori known features. It has been shown that the exact choice of the introduced threshold value is not very critical for the success of the extended selection process. This holds for exact solutions of JD correction equation as well as for inexact solutions where typically a larger number of JD iterations is necessary until convergence is reached.

For the possibility of direct solutions of the occurring linear systems the Rayleigh quotient iteration is a good alternative, when a single eigenpair is searched only. The number of iterations needed is comparably low as in the JD method. The most computation time is spent for solving the linear system in both approaches.

Acknowledgements The authors wish to thank Dr. J. Rommes for bringing the potential application of the RQI to their attention.

References

1. Arbenz, P., Hochstenbach, M.E.: A Jacobi–Davidson method for solving complex symmetric eigenvalue problems. SIAM J. Sci. Comput. **25**(5), 1655–1673 (2004)
2. Bandlow, B., Sievers, D., Schuhmann, R.: An improved Jacobi-Davidson method for the computation of selected eigenmodes in waveguide cross sections. IEEE Trans. Magn. **46**(8), 3461 -3464 (2010)
3. Computer Simulation Technology AG (CST): CST Studio Suite. http://www.cst.com
4. Fokkema, D.R., Sleijpen, G.L.G., van der Vorst, H.A.: Jacobi-Davidson style QR and QZ algorithms for the reduction of matrixpencils. SIAM J. Sci. Comput. **20**(1), 94–125 (1998)
5. Hochstenbach, M.E.: Jacobi-Davidson gateway. http://www.win.tue.nl/casa/research/topics/jd
6. Notay, Y.: Convergence analysis of inexact Rayleigh quotient iteration. SIAM J. Matrix Anal. Appl. **24**, 627–644 (2002)
7. Ostrowski, A.M.: On the convergence of the Rayleigh quotient iteration for the computation of the characteristic roots and vectors V. Arch. Ration. Mech. Anal. **3**, 472–481 (1959)
8. Schenk, O., Gärtner, K.: Solving unsymmetric sparse systems of linear equations with pardiso. Future Gener. Comput. Syst. **20**(3), 475–487 (2004)

9. Sleijpen, G.L.G., Van der Vorst, H.A.: A Jacobi-Davidson iteration method for linear eigenvalue problems. SIAM J. Matrix Anal. Appl. **17**(2), 401–425 (1996)
10. Smotrova, E., Nosich, A., Benson, T., Sewell, P.: Cold-cavity thresholds of microdisks with uniform and nonuniform gain: quasi-3-D modeling with accurate 2-D analysis. IEEE J. Sel. Top. Quant. Electron. **11**(5), 1135 – 1142 (2005)
11. Weiland, T.: Eine Methode zur Lösung der Maxwellschen Gleichungen für sechskomponentige Felder auf diskreter Basis. AEÜ **31**, 116–120 (1977)
12. Zhao, L., Cangellaris, A.: GT-PML: generalized theory of perfectly matched layers and its application to the reflectionless truncation of finite-difference time-domain grids. IEEE Trans. Microw. Theor. Tech. **44**(12), 2555 –2563 (1996)

Magnetic Model Refinement via a Coupling of Finite Element Subproblems

Patrick Dular, Ruth V. Sabariego, Laurent Krähenbühl, and Christophe Geuzaine

Abstract Model refinements of magnetic circuits are performed via a subdomain finite element method. A complete problem is split into subproblems with overlapping meshes, to allow a progression from source to reaction fields, ideal to real flux tubes, 1-D to 3-D models, perfect to real materials, statics to dynamics, with any coupling of these changes. Its solution is then the sum of the subproblem solutions. The procedure simplifies both meshing and solving processes, and quantifies the gain given by each refinement on both local fields and global quantities.

1 Introduction

The perturbation of finite element (FE) solutions provides clear advantages in repetitive analyses and helps improving the solution accuracy [1–6]. It allows to benefit from previous computations instead of starting a new complete FE solution for any geometrical, physical or model variation. It also allows different problem-adapted meshes and computational efficiency due to the reduced size of each subproblem.

A general framework allowing a wide variety of refinements is herein developed. It is defined as a subproblem FE approach based on canonical magnetostatic and magnetodynamic problems solved in a sequence, with at each step volume sources (VSs) and surface sources (SSs) originated from previous solutions. VSs express

P. Dular (✉) · R.V. Sabariego · C. Geuzaine
University of Liège, Department of Electrical Engineering and Computer Science, ACE, B-4000 Liège, Belgium
F.R.S.-FNRS, Fonds de la Recherche Scientifique, Belgium
e-mail: Patrick.Dular@ulg.ac.be

L. Krähenbühl
Université de Lyon, Ampère (UMR CNRS 5005), École Centrale de Lyon, F-69134 Écully Cedex, France

B. Michielsen and J.-R. Poirier (eds.), *Scientific Computing in Electrical Engineering SCEE 2010*, Mathematics in Industry 16, DOI 10.1007/978-3-642-22453-9_15,

changes of material properties. SSs express changes of boundary conditions (BCs) or interface conditions (ICs). Common and useful changes from source to reaction fields, ideal to real flux tubes (with leakage flux), 1-D to 3-D models, perfect to real materials, and statics to dynamics, can all be defined through combinations of VSs and SSs.

The developments are performed for the magnetic vector potential FE formulation, paying special attention to the proper discretization of the constraints involved in each subproblem. The method will be illustrated and validated on various problems.

2 Series of Coupled Subproblems

Instead of solving a complete problem, with all its details, it is proposed to split it into a sequence of subproblems, some with approximated geometrical or physical data, including model simplifications, and others performing adequate corrections. The complete solution is then the sum of the subproblem solutions. Each subproblem p is defined in a domain Ω_p, with boundary $\partial\Omega_p = \Gamma_p$. It is governed by magnetostatic or magnetodynamic equations and constrained with VSs and SSs, of which some components originate from previous problems q. The involved fields are the magnetic field $\mathbf{h}_p$, the magnetic flux density $\mathbf{b}_p$ and the electric field $\mathbf{e}_p$.

Classical VSs fix remnant inductions in magnetic materials and current densities in stranded inductors. Similar VSs can also express changes of permeability μ and conductivity σ from a problem q to a problem p [4, 5]. For changes from μ_q to μ_p and from σ_q to σ_p, the magnetic and electric material relations for problem p are $\mathbf{h}_p = \mu_p^{-1}\mathbf{b}_p + \mathbf{h}_{s,p}$ and $\mathbf{j}_p = \sigma_p\mathbf{e}_p + \mathbf{j}_{s,p}$, with VSs $\mathbf{h}_{s,p} = (\mu_p^{-1} - \mu_q^{-1})\,\mathbf{b}_q$ and $\mathbf{j}_{s,p} = (\sigma_p - \sigma_q)\,\mathbf{e}_q$ limited to the modified regions.

The usually homogeneous SSs, i.e. BCs or ICs for the traces $\mathbf{n} \times \mathbf{h}_p|_{\gamma_p}$, $\mathbf{n} \cdot \mathbf{b}_p|_{\gamma_p}$ and $\mathbf{n} \times \mathbf{e}_p|_{\gamma_p}$, with $\mathbf{n}$ the unit exterior normal and $\gamma_p \subset \Gamma_p$, can be extended to non-zero constraints. The resulting ICs, i.e. the discontinuities $[\mathbf{n} \times \mathbf{h}_p]_{\gamma_p} = \mathbf{j}_{f,p}$, $[\mathbf{n} \cdot \mathbf{b}_p]_{\gamma_p} = \mathbf{b}_{f,p}$ and $[\mathbf{n} \times \mathbf{e}_p]_{\gamma_p} = \mathbf{f}_{f,p}$ through an interface γ_p, involve SSs $\mathbf{j}_{f,p}$, $\mathbf{b}_{f,p}$ and $\mathbf{f}_{f,p}$ obtained from previous problems. Usually, free forced discontinuities in a problem q, allowing some simplifications with idealized thin regions [2–5], can be removed in a problem p via opposed SSs, i.e. $\mathbf{j}_{f,p} = -[\mathbf{n} \times \mathbf{h}_q]_{\gamma_p}$, $\mathbf{b}_{f,p} = -[\mathbf{n} \cdot \mathbf{b}_q]_{\gamma_p}$ and $\mathbf{f}_{f,p} = -[\mathbf{n} \times \mathbf{e}_q]_{\gamma_p}$ (γ_p and γ_q only differ at the discrete level by their meshes). For the weakly defined ICs, a post-treatment of the FE weak formulation is done to naturally express the SSs via a volume integration limited to a layer of FEs surrounding the interface [2–5]. VSs and SSs involve previous solutions in subdomains of the current problem p. At the discrete level, this means these solutions have to be expressed in portions of the mesh of problem p, while initially given in the mesh of problem q. This is done via an L^2-projection [2–6].

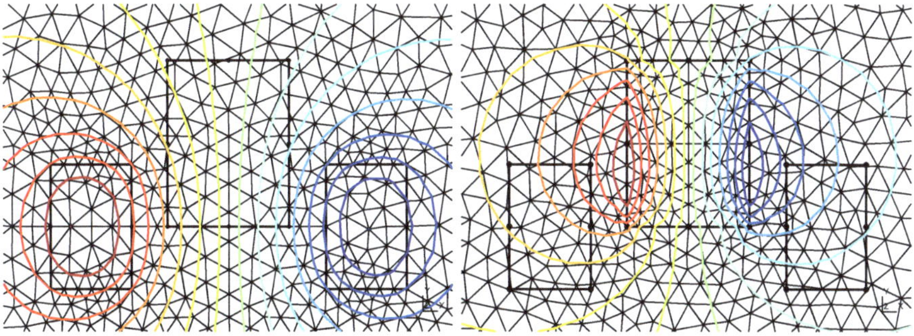

Fig. 1 Field lines for an inductor alone ($\mathbf{b}_1$, *left*) and for an added core ($\mathbf{b}_2$, $\mu_{r,core} = 100$) (*right*); distinct meshes are used for problems 1 and 2

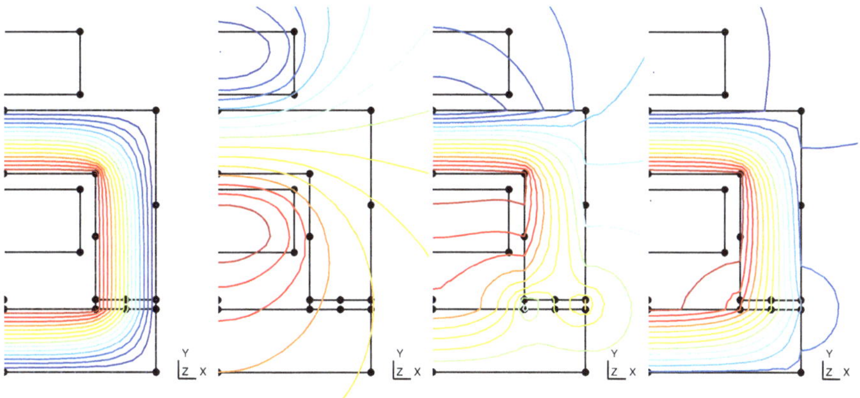

Fig. 2 An electromagnet: field lines in an ideal flux tube ($\mathbf{b}_1$, $\mu_{r,core} = 100$), for the inductor alone ($\mathbf{b}_2$), for the leakage flux ($\mathbf{b}_3$) and for the total field ($\mathbf{b}$) (*left* to *right*)

3 Various Correction Procedures

Various correction schemes, appropriate to practical magnetic system analyses, can benefit from the developed subproblem approach. These are summarized below and will be discussed in details.

(1) *Change of material properties* (Fig. 1) – A typical problem is that of a region put in an initially calculated source field $\mathbf{b}_1$. The associated subproblem 2 is solved in its proper mesh, with the added core and its surrounding region, and VSs limited to this core, where μ and/or σ are modified. Such changes can occur when adding or suppressing materials or portions of those, in, e.g., shape optimization, non-destructive testing [1, 6], moving systems.

(2) *Change from ideal to real flux tubes* (Fig. 2) [3, 4] – A problem q can first consider ideal tubes, i.e. surrounded by perfect flux walls through which $\mathbf{n}\cdot\mathbf{b}_q|_{\gamma_q}$

is zero and $\mathbf{b}_q$ and $\mathbf{h}_q$ outside are zero. The complementary trace $\mathbf{n} \times \mathbf{h}_q|_{\gamma_q}$ is unknown and non-zero. Consequently, a change to a permeable flux wall defines a problem p with SSs opposed to this non-zero trace. This change (2) can be done simultaneously with change (1), which is the case in Fig. 2: the leakage flux $\mathbf{b}_3$ completes the ideal distribution $\mathbf{b}_1$ while knowing the source $\mathbf{b}_2$ proper to the inductor; this allows independent overlapping meshes for both source and reaction fields.

(3) *Change from 1-D to 3-D* [5] – Change (2) can be extended to allow a dimension change, e.g. from 2-D to 3-D: a 2-D solution is first considered as limited to a certain thickness in the third dimension, with a zero field outside; on the other side, another independent problem is solved. Changes of ICs on each side of this portion, via SSs, then allow the calculation of 3-D end effects.

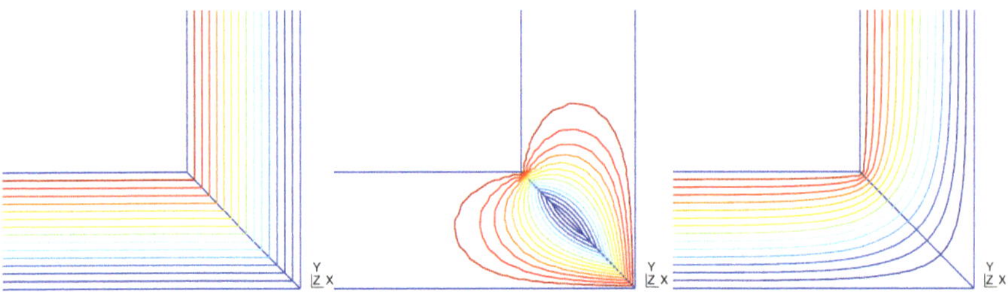

Fig. 3 Series connection of two flux tubes: field lines in ideal flux tubes ($\mathbf{b}_1$, *left*), local correction at the junction ($\mathbf{b}_2$, *middle*) and complete solution ($\mathbf{b}$, *right*)

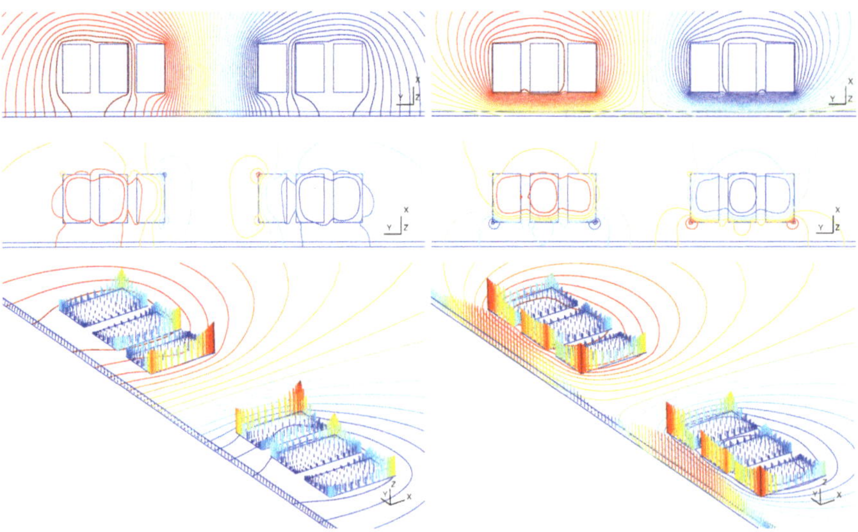

Fig. 4 A 3-turn inductor over a half plate, with perpendicular flux horizontal symmetry axis below); low (*left*) and high (*right*) plate conductivity; (*top*) flux lines for $\mathbf{b}_1$ with a $\sigma \to \infty$ inductor; (*middle*) the correction solution $\mathbf{b}_2$ and (*bottom*) the total $\mathbf{b}$ and current density modulus

Series connections of flux tubes use a similar procedure: a violation of ICs when connecting two flux tubes can be corrected via an opposed SS, e.g. which allows changes from 1-D to 2-D (Fig. 3).

(4) *Change from perfect to real materials* (Fig. 4) [2] – A problem q can first consider perfect conducting (resp. magnetic) materials, with $\sigma_q \to \infty$ (resp. $\mu_q \to \infty$), in which case the trace $\mathbf{n} \cdot \mathbf{b}_q|_{\gamma_q}$ (resp. $\mathbf{n} \times \mathbf{h}_q|_{\gamma_q}$) on its boundary is zero and $\mathbf{b}_q$ (resp. $\mathbf{h}_q$) inside is zero. The complementary trace $\mathbf{n} \times \mathbf{h}_q|_{\gamma_q}$ (resp. $\mathbf{n} \cdot \mathbf{b}_q|_{\gamma_q}$) is unknown and non-zero. Consequently, a change to a finite σ_p (resp. μ_p) defines a problem p with SSs opposed to this non-zero trace.

References

1. Badics, Z., Matsumoto, Y., Aoki, K., Nakayasu, F., Uesaka, M., Miya, K.: An effective 3-D finite element scheme for computing electromagnetic field distorsions due to defects in eddy-current nondestructive evaluation. IEEE Trans. Magn. **33**(2), 1012–1020 (1997)
2. Dular, P., Sabariego, R.V., Gyselinck, J., Krähenbühl, L.: Sub-domain finite element method for efficiently considering strong skin and proximity effects. COMPEL **26**(4), 974–985 (2007)
3. Dular, P., et al.: Perturbation finite element method for magnetic model refinement of air gaps and leakage fluxes. IEEE Trans. Magn. **45**(3), 1400–1403 (2009)
4. Dular, P., et al.: Perturbation finite-element method for magnetic circuits. IET Sci. Meas. Technol. **2**(6), 440–446 (2008)
5. Dular, P., Sabariego, R.V., Krähenbühl, L.: Magnetic model refinement via a perturbation finite element method – From 1-D to 3-D. COMPEL **28**(4), 974–988 (2009)
6. Dular, P., Sabariego, R.V.: A perturbation method for computing field distortions due to conductive regions with h-conform magnetodynamic finite element formulations. IEEE Trans. Magn. **43**(4), 1293–1296 (2007)

Substrate Modeling Based on Hierarchical Sparse Circuits

Daniel Ioan, Gabriela Ciuprina, and Ioan-Alexandru Lazăr

Abstract In this paper, a new modeling approach appropriate for the substrate modeling is proposed. More generally, this technique can be applied for any homogeneous layer for which an exponential decay of the field variation can be assumed. The main idea is to perform a hierarchical modeling based on an exponential partitioning scheme conducing to a circuit model of linear complexity which is extracted with a low computational effort. The model obtained is further coupled with the models of the other parts in which the integrated circuit is decomposed or its sparse matrix is used as a boundary condition for field in SiO2 domain.

1 Introduction

With the continuous downscaling of CMOS devices analog, RF and digital circuitry are integrated on a single chip. However, due to the conducting nature of the common substrate, noise generated by the digital circuitry can be easily injected into and propagate through the silicon substrate. Accurate and efficient modeling of the electromagnetic effects in the semiconductor substrate is an important still open problem for the EDA community [1, 2].

The IC substrate is a semiconductor body represented by computational domains of rectangular shapes. It is usually structured in homogeneous layers, with constant material parameters. The traces of the circuit devices on the top surface of the substrate are called *connectors* or *contacts*. The bottom surface of the substrate is the *backplane contact*, usually a grounded or a floating metallic layer [3]. The top surface of the computational domain and its lateral surfaces have a virtual character, being conventional cuts in the real semiconductor substrate body. The contacts are

D. Ioan · G. Ciuprina (✉) · I.-A. Lazăr
Electrical Engineering Faculty, Numerical Methods Lab., "Politehnica" University of Bucharest, Spl. Independenţei 313, 060042, Bucharest, Romania
e-mail: daniel@lmn.pub.ro; gabriela@lmn.pub.ro; alexlz@lmn.pub.ro

B. Michielsen and J.-R. Poirier (eds.), *Scientific Computing in Electrical Engineering SCEE 2010*, Mathematics in Industry 16, DOI 10.1007/978-3-642-22453-9_16,

also conventional surfaces [1, 3]. The number, shapes and sizes of the contacts are very much dependent on the actual circuit layer as well as on the modeling approach. Inhomogeneous, high-conductivity layers and structures such as the epi-layer, wells, diffusion gradients, and buried layers are usually included in the substrate models, but a simpler solution we will consider in our approach is the one in which the modeled substrate contains only the homogeneous Silicon bulk. The top contacts are placed on an orthogonal, regularly structured grid. They can be clustered to match the actual circuit layers.

The substrate models are based on electromagnetic (EM) field modeling. The choice of the most appropriate EM field regime for a particular model of the substrate depends on the values of the material constants and the required operating frequency range. At low frequencies, the substrate behavior is well described by static regimes, the most appropriate model being obtained by using, in conjunction, electrostatics (ES), electric conduction (EC) and magnetostatics (MS), aiming to model capacitive, conductive losses and inductive effects of the integrated circuit.

Numerical studies in [4] show that the electroquasistatic (EQS) assumptions are valid for high-resistivity substrates at frequencies below 20 GHz. In the case of low-resistivity substrates EQS can be used at least for frequencies up to 100 GHz.

Even in the simplest static regimes, the complexity of the extracted model with n connectors is O(n^2), since the number of lumped circuit elements linking the connectors is given by $n(n-1)/2$. For instance, two millions of R, L or C elements are required to model 2,000 connectors in EC, MS or ES regimes. RC "equivalent" circuits are extracted from the EQS field solution. Fortunately, not all these elements have a similar importance in the model, as many of them describe weak interactions. Typical examples are links between far connectors or connectors screened by other connectors. That is why, a hierarchical modeling approach for the substrate is necessary. Several hierarchical approaches are described in [1, 5].

2 Hierarchical Approach

2.1 Main Idea

The substrate modeling approach we propose is valid at frequencies where the EQS regime may be considered valid. In order to also model the magnetic/inductive field effects in the substrate, we consider the EQS field in conjunction with the MS one. Thus, two independent models are extracted, to be connected in the global model of the IC. For this, we use the domain partitioning (DP) technique as described in [6]. The IC devices and the substrate interact by means of EM *hooks* [6]. The hierarchical sparsification we propose is based on an exponential partitioning scheme of the substrate (Fig. 1). Virtual contacts (hooks) are buried in the substrate at different depths (according to their levels), thus realizing a domain-partitioning of the substrate in horizontal layers structured in rectangular

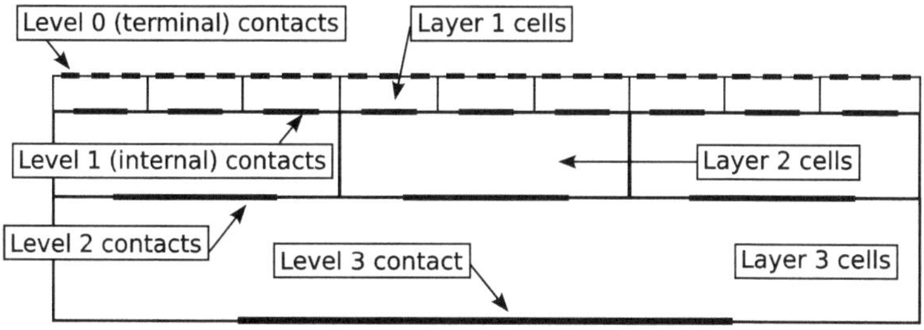

Fig. 1 Partitioning of the substrates in macro-cells

super-elements (macro-cells). The cell-walls thus generate an adapted *discretization macro-grid*, progressively refined from bottom to top. Unlike the literature, in our approach, the equivalent contacts have a physical meaning, being the terminals of the macro-cells in which the domain is partitioned. Thus, sparse hierarchical circuit-models with a reduced number of lumped elements are generated.

2.2 Theoretical Basis

The main reason which makes our hierarchical modeling approach valid is the exponential decay of the field variation on deeper horizontal planes. For instance, in EC, ES and MS field regimes, the scalar potential satisfies in homogeneous media the Laplace equation $\Delta V = 0$.

For the sake of simplicity, let's consider a 2D domain $\mathscr{D} = [0, a] \times [0, b]$, which represents the homogeneous substrate we want to model, with $V(x, 0) = 0$ for $x \in [0, a]$, $\partial V/\partial x(0, y) = 0$ and $\partial V/\partial x(a, y) = 0$ for $y \in [0, b]$. The top segment, corresponding to $y = b$ and $x \in [0, a]$ ensures the link with the upper part, so that a certain non-zero Dirichlet boundary condition has to be imposed for it: $V(x, b) = f(x)$ for $x \in [0, a]$. This problem can be solved analytically, the solution obtained after imposing three out of the four boundary conditions being:

$$V(x, y) = C_0 y + \sum_{i=1}^{\infty} C_i \cos(\lambda_i x) \sinh(\lambda_i y), \tag{1}$$

where $\lambda_i = \pi i/a$.

The constants C_i in (1) can be obtained by imposing the Dirichlet condition on the top horizontal segment $y = b$.

$$C_0 b = F_0/2 = \frac{1}{a} \int_0^a f(x)\,\mathrm{d}x,$$

$$C_i \sinh(\lambda_i b) = F_i = \frac{2}{a} \int_0^a f(x) \cos(\lambda_i x)\,\mathrm{d}x. \tag{2}$$

The values of the potential on the top segment depend on the device placed on the substrate. However, any possible variation can be approximated accurately as a piecewise linear function, which tends to the real variation when the fineness of the discretization tends to infinity. Such a function can be expressed as a linear combination of "hat functions". If N is the number of contacts (terminals) equidistantly placed on the top side and m is the index of the contact that is excited with a $V_0 = 1$ V potential, then the Fourier coefficients given by (2) are

$$F_0 = 2V_0/(N-1),$$
$$F_i = \frac{2V_0(N-1)}{\pi^2 i^2}\left[2\cos\frac{\pi i(m-1)}{N-1} - \cos\frac{\pi i(m-2)}{N-1} - \cos\frac{\pi i m}{N-1}\right]. \quad (3)$$

Finally, the coefficients in (1) are $C_0 = V_0/b(N-1)$ and $C_i = F_i/\sinh(\lambda_i b)$ where F_i is given by (3).

The Fourier series can be truncated, but, according to [7], in order to keep a minimal accuracy on the top segment (given by $y = b$ and also called "level 0"), the number of retained terms (spatial harmonics along the Ox direction) should be at least twice the total number of contacts in that direction: $n_0 = 2N$. The error due to the truncation of the series (1) at the M-th term is

$$|V_M(x,y) - V(x,y| \le \sum_{i=M+1}^{\infty} |C_i| \sinh(\lambda_i y)$$
$$\le \frac{8V_0(N-1)}{\pi^2} \sum_{i=M+1}^{\infty} \frac{\sinh(\lambda_i y)}{i^2 \sinh(\lambda_i b)} \le A \int_M^{\infty} \frac{\sinh(\pi x y/a)}{x^2 \sinh(\pi x b/a)} \mathrm{d}x$$
$$\approx A \int_M^{\infty} \frac{\exp(\pi(y-b)x/a)}{x^2} \mathrm{d}x$$
$$= A\left(\frac{\exp(\pi(y-b)M/a)}{M} + \frac{\pi(y-b)}{a} E_1(M)\right)$$
$$\le A\left(\frac{\exp(\pi(y-b)M/a)}{M} + \frac{\pi(y-b)}{a} e^{-M} \log(1+\frac{1}{M})\right), \quad (4)$$

where $A = 8V_0(N-1)/\pi^2$ and $E_1(M) = \int_M^{\infty} \exp(-t)/t \, \mathrm{d}t$ is the exponential integral and its margins are well known [8].

It is obvious from Fig. 2 that on deeper levels $y_k < b$, for the same accuracy, only a lower number of terms need to be retained from the Fourier series given by (1). For instance, if we would like that level 1 be accurately described by $N_1 = N/3$ contacts, the number of terms that have to be summed is $n_1 = 2N/3$. By imposing

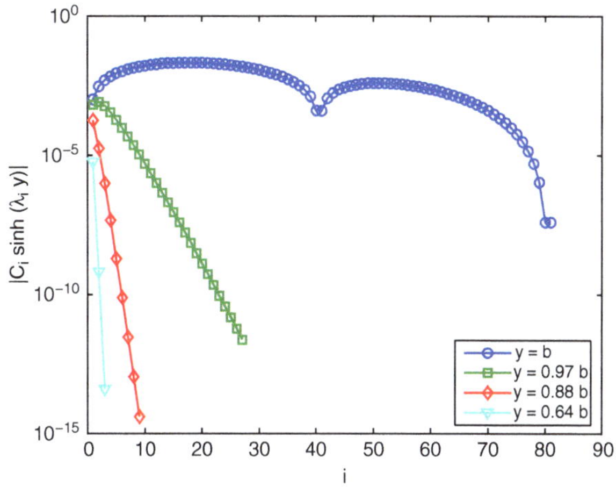

Fig. 2 Fourier coefficients for several depths

the condition that the first neglected term at level 1 be equal to the first neglected term at level 0, it follows that level 1 has to be placed as given by

$$y_1 \leq b - \frac{a}{\pi(2N/3+1)} \log \frac{|F_{2N/3+1}|}{|F_{2N+1}|}. \tag{5}$$

For the example above, it follows that $y_1 \leq 0.97b$. This kind of restriction gives a guidance about how deep a reduced number of contacts can be buried while keeping the modeling accuracy. By applying this procedure recursively, the substrate is partitioned in layers having an exponentially decreasing number of contacts. To simplify the presentation, we will assume that N is a power of 3, $N = 3^L$. Thus, in order to decrease the number of degrees of freedom (dofs) by 3 for each level, we will need $L+1$ levels, a level j having $N_j = 3^{L-j}$ contacts. A layer j is placed between level j and level $j-1$ (for $j = 1, \ldots, L$) and it will have a certain thickness Δy_j. We will place the layers according to a geometric progression of ratio $r > 1$: $\Delta y_2 = r\Delta y_1$, $\Delta y_3 = r^2 \Delta y_1, \ldots, \Delta y_L = r^{L-1}\Delta y_1$. It follows that $\Delta y_1(r^L - 1)/(r-1) = b$. If we chose $r = N_0/N_1 = 3$, the first level has a thickness of $\Delta y_1 = 2b/(3^L - 1)$ and its corresponding position $y_1 = b - \Delta y_1$ satisfies relation (5) only if the number of layers L is less than 4. In this case the cells that are used to discretize the substrate will have the same input-output behavior. Thus, only one cell, called *reference cell* has to be solved in order to find its input-output relationship and thus an equivalent circuit for it (Fig. 3).

Going down in the substrate, the number of dofs necessary to describe the solution decreases exponentially (level j has 3^{L-j} dofs). A lower number of dofs means a lower number of hooks (contacts) on deeper layers. Hence, the grid necessary to describe the field may be coarser, deeper in substrate. This is the main conclusion of the above study. Going down, the field distribution is smother and requires a lower number of spatial harmonics (samples) to be represented accurately.

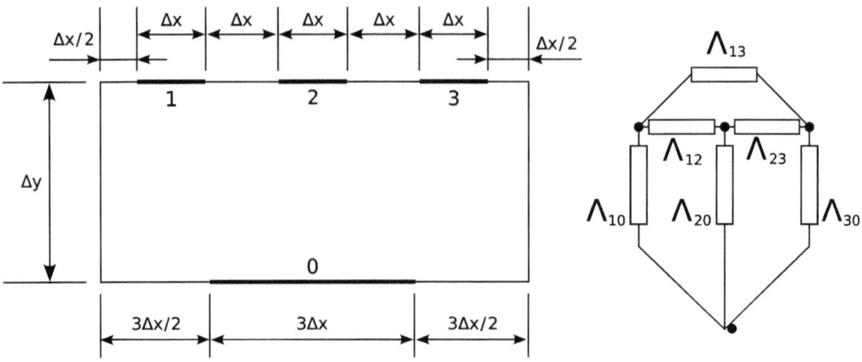

Fig. 3 Layout of a standard cell and its equivalent circuit

Even if the above explanations were given for the 2D case, they can be generalized without difficulty for the 3D case.

It is easy to show that the complexity of the equivalent circuit increases linearly. In 2D, a standard cell will have four terminals, three on the top segment and one on the bottom. The number of layers is $L = \ln(N)/\ln(3)$, the total number of cells is $(3^L - 1)/2$, and the number of branches of the equivalent circuit is $O(3(N-1)) = O(N)$. A model with 13 layers can handle about 1.6 million top-connectors, using almost 4.8 million lumped elements. In 3D, a standard cell will have ten terminals, nine on the top face and one on bottom, the number of layers is $L = \ln(N)/\ln(9)$, the total number of cells is $(9^L - 1)/8$, and the complexity of the equivalent circuit is $O(45(N-1)/8) = O(N)$. A model with seven layers can handle about 4.8 million top-connectors, about 600,000 cells and, consequently, using about twenty seven millions lumped elements, each cell having 45 lumped elements. The linear order of the extracted model is another great advantage of this approach.

2.3 Algorithm

The algorithm we propose has the following steps:

Step 1: Chose appropriate EMCE formulation for the upper part and simulate it. This implies the setting of the appropriate shape and position of terminals on the boundary part that will be connected to the substrate. This setting depends on the actual configuration of the device.

Step 2: Compute the number N of equidistant terminals necessary for the level 0 of the substrate. This depends on the minimum discretization step used at step 1.

Step 3: Compute the number of necessary layers in the substrate L and their heights.

Step 4: Simulate the reference cell (2D case shown in Fig. 3 left), and compute the geometric permeances (Fig. 3 right).

In this step, by solving the reference cell (e.g. with a BEM or FIT solver), the nodal permeances matrix is obtained. In 2D this matrix has 3×3 entries, but due to reciprocity and geometric symmetry, only four values are different:

$$\Lambda_{\text{ref.cell}}^{(n)} = \begin{bmatrix} \Lambda_{11}^{(n)} & \Lambda_{12}^{(n)} & \Lambda_{13}^{(n)} \\ \Lambda_{12}^{(n)} & \Lambda_{22}^{(n)} & \Lambda_{12}^{(n)} \\ \Lambda_{13}^{(n)} & \Lambda_{12}^{(n)} & \Lambda_{11}^{(n)} \end{bmatrix}. \tag{6}$$

These values are dimensionless and depend solely on the cell and contact sizes. The values of the permeances of the lumped elements are:

$$\begin{aligned} \Lambda_{12} &= -\Lambda_{12}^{(n)}, \quad \Lambda_{10} = \Lambda_{11}^{(n)} + \Lambda_{12}^{(n)} + \Lambda_{13}^{(n)}, \\ \Lambda_{13} &= -\Lambda_{13}^{(n)}, \quad \Lambda_{20} = \Lambda_{22}^{(n)} + 2\Lambda_{12}^{(n)}. \end{aligned} \tag{7}$$

Step 5: Assemble the nodal permeances matrix for the hierarchical sparsified circuit that models the substrate.

In this step, the nodal permeance matrix $\Lambda_{HSS}^{(n)}$ for the whole hierarchical circuit (Fig. 4) is derived. This is a sparse, symmetric matrix, having four non-zero elements on each row (Fig. 5). Its pattern depends on the node numbering. For instance, if the numbering is carried out from left to right and from bottom to top, the last N lines and columns correspond to the terminals. It is useful to partition the nodal permeance matrix of the HSS circuit according to this numbering as:

$$\Lambda_{HSS}^{n} = \begin{bmatrix} \Lambda_{HSS,11}^{n} & \Lambda_{HSS,12}^{n} \\ \Lambda_{HSS,21}^{n} & \Lambda_{HSS,22}^{n} \end{bmatrix} \tag{8}$$

Step 6: Compute the terminal admittance matrix of the top level with (9).

By eliminating the internal nodes of the model, the terminal permeance matrix Λ_T of the sparsified model can be obtained:

$$\Lambda_T = \Lambda_{HSS,11}^{n} - \Lambda_{HSS,12}^{n} \left(\Lambda_{HSS,22}^{n}\right)^{-1} \Lambda_{HSS,21}^{n}. \tag{9}$$

Algebraically, this means the computation of the Schur complement of the lower right block. The terminal admittance matrix is $\mathbf{Y}_{HSS} = (\sigma + j\omega\varepsilon)\Lambda_T$, where σ and ε are the conductivity and the permeability of the substrate. The same geometric permeances are used to compute the magnetic reluctances and, based on them, the fundamental loop inductances of the circuit placed above the substrate can be extracted.

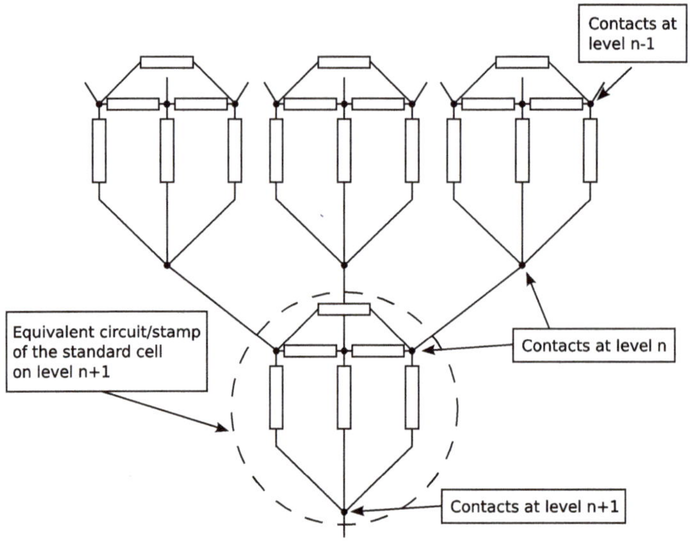

Fig. 4 Hierarchical sparse circuit of the substrate

Fig. 5 Structure of the nodal matrix Λ^n_{HSS}

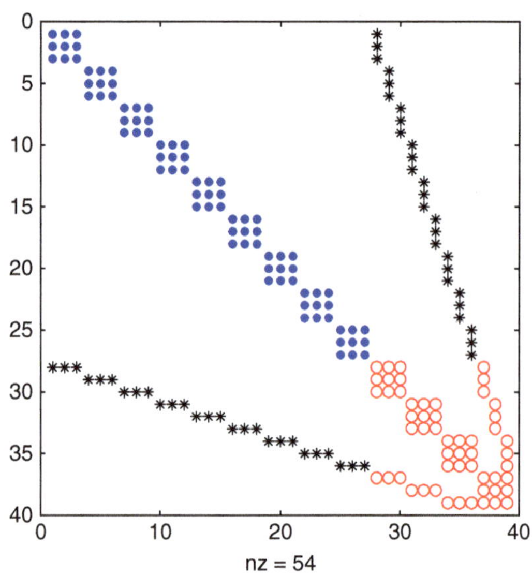

Step 7: Clusterise the terminals of the substrate according to the terminals of the upper part.

Step 8: Couple the models with relation (10).

In the last step, the models are coupled by means of their contacts. Assuming that the top model has the terminals numbered so that the hooks that are connected to the substrate are numbered at the end, its transfer matrix is partitioned as $\mathbf{Y}_{top} = [\mathbf{Y}_{11}\mathbf{Y}_{12}; \mathbf{Y}_{21}\mathbf{Y}_{22}]$ then, by imposing the coupling conditions, the admittance matrix of the whole model is

$$\mathbf{Y} = \mathbf{Y}_{11} - \mathbf{Y}_{12}\left(\mathbf{Y}_{HSS} + \mathbf{Y}_{22}\right)^{-1}\mathbf{Y}_{21}. \tag{10}$$

3 Results and Conclusions

In order to verify the proposed approach, a simple study case of a micro-strip dual conductor line in SiO2 over a lossy Si substrate was considered. The line admittance was computed by FIT, using DP with conform grids (Fig. 6) and with the hierarchical modeling for the substrate. The result shown in Fig. 7 validates the latter approach. The substrate was decomposed into five layers having progressive increasing thicknesses with a constant rate. The CPU time needed to extract the admittance matrix by hierarchical sparsification was 0.093 s, whereas the same time when using DP with conform grids was 20 s. This illustrates the important reduction of the extraction time. The proposed approach allows a fast extraction of parasitic GLC parameters in complex integrated circuits with over a million of components, modeling the EM coupling and noise propagation.

Acknowledgements The support of the projects: FP6/Chameleon-RF, FP5/Codestar, UEFISCSU/IDEI 609/2009 is gratefully acknowledged. This work has been co-funded as well by the Sectoral Operational Programme Human Resources Development 2007–2013 of the Romanian Ministry of Labour, Family and Social Protection through the Financial Agreement POSDRU/89/1.5/S/62557.

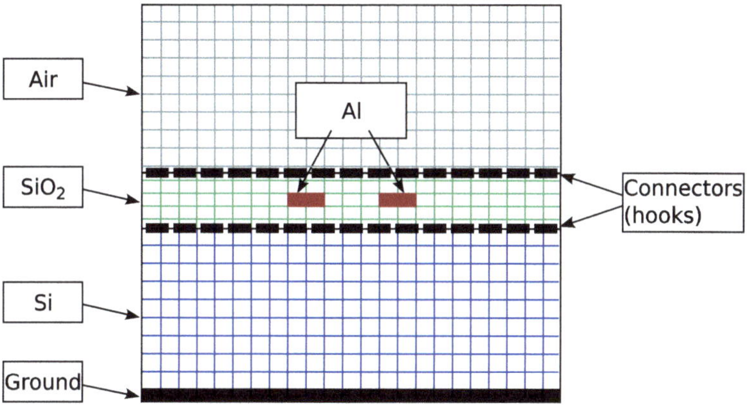

Fig. 6 Computational domain partitioned in three parts. Conform grids are shown

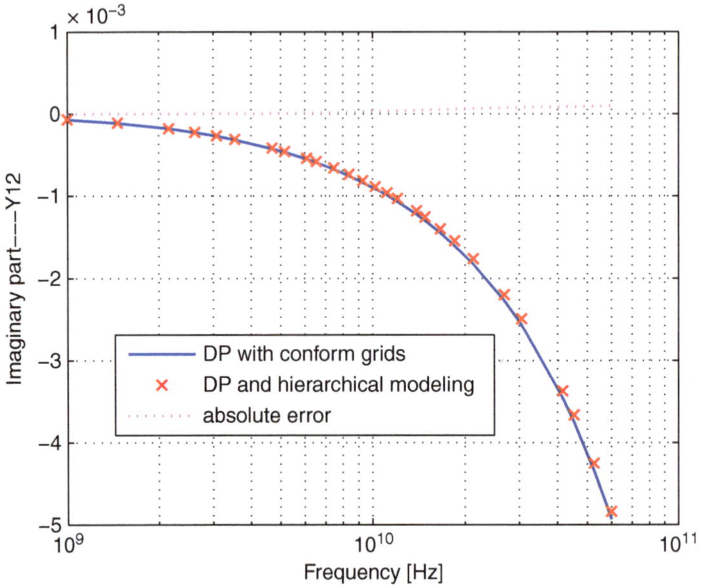

Fig. 7 Hierarchical approach is as accurate as DP with conform grids

References

1. Gharpurey, R.: Modeling and analysis of substrate coupling in integrated circuits. Ph.D. thesis, Univ. of California, Berkeley (1995)
2. Lan, H.: Synthesized compact models for substrate noise coupling in mixedsignal ICs. Ph.D. thesis, Stanford University (2006)
3. Kristiansson, S., Ingvarson, F., Kagganti, S., Simic, N., Zgrda, M., Jeppson, K.: A surface potential model for predicting substrate noise coupling in integrated circuits. IEEE J. Solid State Circ. **40**(9), 1797–1803 (2005)
4. Veronis, G., Lu, Y.C., Dutton, R.W.: Modeling of wave behavior of substrate noise coupling for mixed-signal IC design. In: ISQED '04: Proc. of the IEEE International Symposium on Quality Electonic Design, pp. 303–308. San Jose, CA, USA (2004)
5. Phillips, J.R., Silveira, L.M.: Simulation approaches for strongly coupled interconnect systems. In: ICCAD '01: Proceedings of the Int. Conf. on Computer-Aided Design, pp. 430–437. San Jose, California (2001)
6. Ioan, D., Ciuprina, G., Silveira, L.: Effective domain partitioning with electric and magnetic hooks. IEEE Trans. Magn. **45**(3), 1328–1331 (2009)
7. Unser, M.: Sampling—50 Years after Shannon. Proc. IEEE **88**(4), 569–587 (2000)
8. Abramovitz, M., Stegun, I.: Handbook of Mathematical Functions with Formulas, Graphs, and Mathematical Tables. Dover, New York (1964). http://www.math.sfu.ca/~cbm/aands

A Boundary Conformal DG Approach for Electro-Quasistatics Problems

A. Fröhlcke, E. Gjonaj, and T. Weiland

Abstract A boundary conformal technique for solving three dimensional electro-quasistatic problems with a high order Discontinuous Galerkin method on Cartesian grids is proposed. The method is based on a cut-cell approach which is applied only on elements intersected by curved material boundaries. A particular numerical quadrature technique is applied which allows for an accurate integration of the finite element operators taking into account the exact geometry of the cut-cells. Two numerical examples are presented which demonstrate the optimal convergence rate of the method for arbitrary geometry.

1 Introduction

Staircase discretization errors for Finite Difference (FD) type discretizations on Cartesian grids represent a serious limitation on the accuracy of numerical simulations. Major efforts have been made by several authors to overcome this difficulty. Among others, the Partially Filled Cell approach for the Finite Integration Technique [1] and the Dey-Mittra conformal boundary algorithm for the Finite Difference Time Domain method [2] have been proposed. These techniques can reduce staircasing errors at curved material boundaries by incorporating explicit information on the boundary geometry into the numerical scheme. Unfortunately,

A. Fröhlcke
Graduate School of Computational Engineering, Technische Universität Darmstadt, Dolivostr. 15, 64293 Darmstadt, Germany
e-mail: froehlcke@gsc.tu-darmstadt.de

E. Gjonaj (✉) · T. Weiland
Computational Electromagnetics Laboratory, Technische Universität Darmstadt, Schloßgartenstr. 8, D-64289 Darmstadt, Germany
e-mail: gjonaj@temf.tu-darmstadt.de; thomas.weiland@temf.tu-darmstadt.de

B. Michielsen and J.-R. Poirier (eds.), *Scientific Computing in Electrical Engineering SCEE 2010*, Mathematics in Industry 16, DOI 10.1007/978-3-642-22453-9_17,

these techniques are designed specifically for low order discretizations. Indeed, high order FD methods rely on a large spatial stencil which makes the implementation of conformal boundary conditions cumbersome and numerically inefficient.

Finite Element Methods (FEM) on unstructured boundary fitted grids, on the other hand, are free of staircasing errors. These methods do provide an improved geometrical flexibility compared to FD methods. In addition, compact stencil and high order accuracy FEM can be easily formulated for a variety of electromagnetic field problems. The price due for this flexibility is a reduced numerical efficiency compared to simple FD schemes. This is directly related to the use of unstructured grids which leads to a more complicated data storage and access pattern in FEM-based computations. Furthermore, the numerical effort for generating boundary fitted unstructured grids for complex geometries can be extremely high.

In this paper, we propose a discrete formulation which combines the accuracy of high order approximations with the simple implementation and numerical efficiency of Cartesian grids. The basic idea is illustrated in Fig. 1 where a computational domain containing a single material block is discretized by a regular Cartesian grid. The material boundary subdivides several grid cells into sub-cells which are associated with (at least) two different sets of material parameters. In the following, we will refer to them as cut-cells. The challenge consists in deriving an appropriate numerical approximation within these cells. Since no general set of basis functions satisfying continuity conditions can be defined for an arbitrarily shaped cut-cell, the standard FEM formulation cannot be applied. Instead, we propose a formulation based on the high order Discontinuous Galerkin (DG) method.

As implied by the figure, the paper refers primarily to electro-quasistatics problems characterized by material parameters such as the dielectric permittivity ϵ and the electrical conductivity κ. However, the proposed discretization approach can be easily extended to other types of electromagnetic field problems. The structure of the paper is as follows. In Sect. 2, the high order DG formulation for the time-harmonic electro-quasistatics equations is introduced. In Sect. 3 the application of the boundary conformal approach with cut-cells within the framework of DG

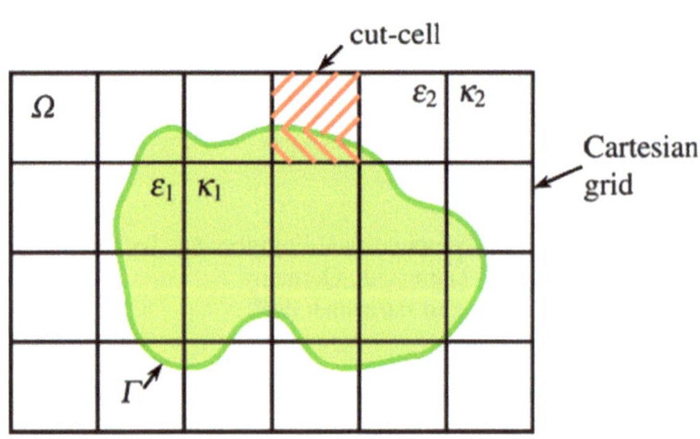

Fig. 1 Exemplary Cartesian-grid domain containing an arbitrarily shaped material block. The shaded area represents a cut-cell intersected by the material boundary

is described. The numerical accuracy and the practicability of the method are demonstrated is Sect. 4 where a simple validation example as well as the fully 3D simulation of a low frequency heating module are presented.

2 DG Formulation for Electro-Quasistatics

The time-harmonic Maxwell's equations for electro-quasistatic fields are written as:

$$\frac{1}{\epsilon}\mathbf{D}(\mathbf{x},t) = -\nabla\phi(\mathbf{x},t), \tag{1}$$

$$\mathbf{i}\omega\nabla\cdot\mathbf{D}(\mathbf{x},t) = -\nabla\cdot\left[\frac{\kappa(\mathbf{x})}{\epsilon(\mathbf{x})}\mathbf{D}(\mathbf{x},t)\right], \tag{2}$$

where ω is the angular frequency, ϕ is the electric potential, $\mathbf{D}$ is the electric flux density; ϵ and κ denote the permittivity and electric conductivity, respectively.

Given a partition of the computational domain Ω into Cartesian grid cells Ω_i, $i = 1\ldots N$ (see, e.g. Fig. 1) we introduce a discrete approximation for (1) and (2) by employing a mixed DG approach. Denoting the approximations of the electric potential and flux density by $\phi_h(\mathbf{x},t)$ and $\mathbf{D}_h(\mathbf{x},t)$, respectively, the weak problem for electro-quasistatics in the DG formulation reads: Find $\mathbf{D}_h$, ϕ_h such that

$$\int_{\Omega_i}\boldsymbol{\psi}^D_{i,q}\cdot\frac{1}{\epsilon_i}\mathbf{D}_h\,\mathrm{d}^3\mathbf{x} = -\int_{\Omega_i}\boldsymbol{\psi}^D_{i,q}\cdot\nabla\phi_h\,\mathrm{d}^3\mathbf{x}, \tag{3}$$

$$\mathbf{i}\omega\int_{\Omega_i}\psi^\phi_{i,q}\nabla\cdot\mathbf{D}_h\,\mathrm{d}^3\mathbf{x} = -\int_{\Omega_i}\psi^\phi_{i,q}\nabla\cdot\left[\frac{\kappa_i}{\epsilon_i}\mathbf{D}_h\right]\mathrm{d}^3\mathbf{x}, \tag{4}$$

$\forall i = 1\ldots N$ and $\forall q = 1\ldots P$, where P is the highest polynomial order used. In (3) and (4), $\psi^\phi_{i,q}$ and $\boldsymbol{\psi}^D_{i,q}$ represent two sets of scalar and vectorial polynomial basis functions for the electric potential and flux density, respectively. Note the index i running over all grid cells for every polynomial order q. It indicates the cell-wise definition of the DG basis functions. Thus, in contrast to the conventional FEM, the approximations obtained are, generally, discontinuous at grid cell interfaces.

Due to the discontinuous DG approximation, the evaluation of element integrals requires special attention. Considering, e.g., (3), the volume integral containing derivatives of the discontinuous electric potential is transformed as:

$$\int_{\Omega_i}\boldsymbol{\psi}^D_{i,q}\cdot\frac{1}{\epsilon_i}\mathbf{D}_h\,\mathrm{d}^3\mathbf{x} = -\int_{\Omega_i}\phi_h\nabla\cdot\boldsymbol{\psi}^D_{i,q}\,\mathrm{d}^3\mathbf{x} + \int_{\partial\Omega_i}\tilde{\boldsymbol{\phi}}_h\boldsymbol{\psi}^D_{i,q}\cdot\mathbf{n}\,\mathrm{d}^2\mathbf{x}. \tag{5}$$

In (5), $\tilde{\phi}_h$ denotes the *numerical flux* for the electric potential defined at the cell interface and $\mathbf{n}$ is the outward pointing interface normal. In order to complete the

DG formulation (3)–(5), a numerically consistent relation for these fluxes must be provided. Several possibilities exist for defining them (see, e.g., [3] for a complete review of choices). In the numerical examples presented below, the so called central flux scheme for the electric potential as well as for the flux density is applied.

Expressing the field approximations ϕ_h and $\mathbf{D}_h$ by means of the basis functions $\psi^{\phi}_{i,q}$ and $\boldsymbol{\psi}^{D}_{i,q}$, respectively, and evaluating the integrals (3) and (4) using numerical fluxes as in (5), yields the set of matrix equations:

$$\mathbf{M}\mathbf{I}_{1/\epsilon}\mathbf{d} = -\mathbf{G}\boldsymbol{\phi} + \mathbf{f}_{\phi}, \tag{6}$$

$$\mathrm{i}\omega\mathbf{G}^T\mathbf{d} = -\mathbf{G}^T\mathbf{I}_{\kappa/\epsilon}\mathbf{d} + \mathbf{f}_d, \tag{7}$$

where $\mathbf{G}$ is the discrete gradient operator, $\mathbf{M}$ is the mass matrix, $\mathbf{I}_{1/\epsilon}$ and $\mathbf{I}_{\kappa/\epsilon}$ are diagonal matrices containing the cell-wise constant material parameters and $\mathbf{f}_{\phi}$ and $\mathbf{f}_d$ are vectors of boundary conditions. Equations (6) and (7) can be further reduced by a Schur complement approach resulting in

$$-\mathbf{G}^T\left(\mathrm{i}\omega\mathbf{I}_{\epsilon} + \mathbf{I}_{\kappa}\right)\mathbf{M}^{-1}\mathbf{G}\boldsymbol{\phi} = \mathbf{f}_d - \mathbf{G}^T\left(\mathrm{i}\omega\mathbf{I}_{\epsilon} + \mathbf{I}_{\kappa}\right)\mathbf{M}^{-1}\mathbf{f}_{\phi}. \tag{8}$$

The above equation can be solved for the potential degrees of freedom $\boldsymbol{\phi}$ using an iterative or direct solver for complex symmetric systems. The Schur complement reduction in (8) can be trivially applied since the mass matrix $\mathbf{M}$ in the DG formulation is block-diagonal. The choice of the basis functions $\psi^{\phi}_{i,q}$ and $\boldsymbol{\psi}^{D}_{i,q}$ is, generally, uncritical for DG-type discretizations. In this work, the high-order hierarchical basis functions proposed in [4] is used. The definition of the electric potential basis functions in the reference element is identical with that employed in H^1-conforming FEM. Correspondingly, the flux density within each element is approximated using a set of high order basis functions which coincides with that used in $H(div)$-conforming FEM (cf. [4]). The reason for this choice is to maintain some degree of equivalence with the standard FEM for comparison and (possibly) hybridization purposes. Note, however, that DG can be neither H^1- nor $H(div)$-conforming, since the global approximation is generally discontinuous.

3 Cut-Cell Approach

The basic observation is that the above derivation does not depend on cell (element) geometry. In particular, it can be applied on the cut-cells of a Cartesian grid as shown in Fig. 1. The latter can be considered as independent grid cells characterized by a unique material. The weak DG equations for the cut-cells can be formally written as in (3) and (4) for the standard (Cartesian) cells provided that, for each cut-cell, a set of independent approximation functions, $\psi^{\phi}_{c,q}$ and $\boldsymbol{\psi}^{D}_{c,q}$, is specified. Thus, the cut-cell approach can be interpreted as a modification of the original Cartesian grid to include additional cells of arbitrary curved geometry. This modification, however, is applied only in the vicinity of material boundaries corresponding to the splitting of the original Cartesian grid cells into several cut-cells with different material content.

In the present implementation, the approximation functions within the cut-cells are chosen to be identical with those in the parent Cartesian cell. This choice is independent from the geometry of the cut-cell, since the DG formulation does not impose conformity constrains on these functions; not even for the regular grid cells away from material boundaries. The field discontinuity at the boundary surface between two neighboring cut-cells is treated naturally within the DG framework by introducing numerical fluxes as in (5).

The numerical evaluation of the DG integrals, however, needs an appropriate description for the cut-cell geometry. For this purpose, the Open CASCADE geometry kernel [5] is used. It enables a geometrical representation of the cut-cells based on parametrized Bezier and B-Spline surfaces. Furthermore, high order Gauss quadrature rules for evaluating surface integrals are provided. Internal integral terms require a separate treatment. Referring again to the weak equation (3) for a cut-cell volume Ω_c, the following transformations are performed:

$$\begin{aligned} \int_{\Omega_c} \boldsymbol{\psi}^D_{c,q} \cdot \frac{1}{\epsilon_c} \mathbf{D}_h \, \mathrm{d}^3\mathbf{x} &= - \int_{\Omega_c} \phi_h \nabla \cdot \boldsymbol{\psi}^D_{c,q} \, \mathrm{d}^3\mathbf{x} + \int_{\partial\Omega_c} \tilde{\boldsymbol{\phi}}_h \boldsymbol{\psi}^D_{c,q} \cdot \mathbf{n} \, \mathrm{d}^2\mathbf{x} \\ &= \int_{\partial\Omega_c} \left(-\mathbf{S}^D_{c,q} + \tilde{\boldsymbol{\phi}}_h \boldsymbol{\psi}^D_{c,q} \right) \cdot \mathbf{n} \, \mathrm{d}^2\mathbf{x}, \end{aligned} \tag{9}$$

where $\mathbf{S}^D_{c,q}$ is a primitive function of the integrand in the first integral term defined by the relation, $\nabla \cdot \mathbf{S}^D_{c,q} = \phi_h \nabla \cdot \boldsymbol{\psi}^D_{c,q}$. Since a polynomial basis approximation is assumed, $\mathbf{S}^D_{c,q}$ can be determined analytically for arbitrarily high orders. Thus, the weak formulation integrals (3) and (4) can be fully reduced to surface integrals along the cut-cell faces which can be further evaluated by the numerical quadrature rules provided by the geometry kernel.

4 Numerical Examples

4.1 Validation

The simple model of a cylindrical capacitor filled with an electrically conducting material is considered (see Fig. 2). For simplicity, a time independent setup is assumed. It consists in a constant voltage excitation applied between the inner and outer electrodes of the capacitor. Thus, the problem reduces to a stationary current flow problem with exact analytical solution which can be used for investigating the accuracy of the method.

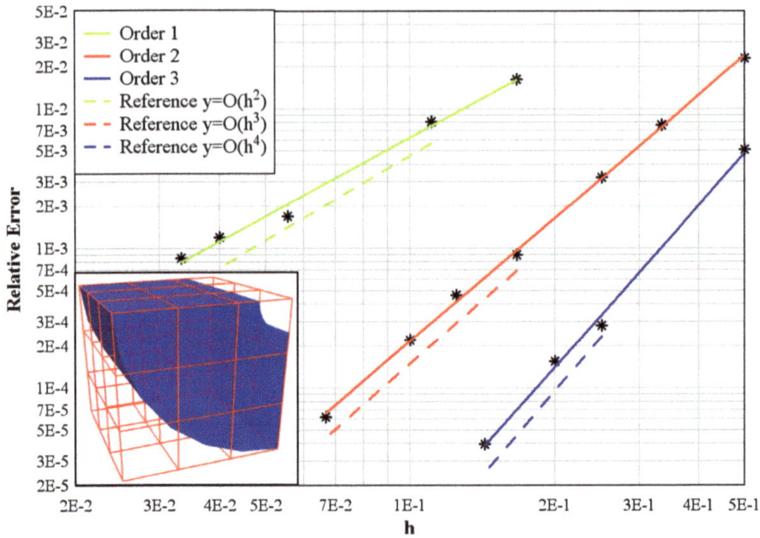

Fig. 2 Relative error measured in the L2-norm of electric potential vs. mesh parameter for different DG approximation orders

Figure 2 shows the numerical error for the electric potential vs. grid resolution for different DG approximation orders. Obviously, the numerical result converges with the optimal convergence order, $P + 1$, where P is the highest degree of polynomials used in the approximation. The cut-cell approach is, thus, exact in the sense that, for arbitrarily curved geometry, it does not introduce additional numerical errors (like staircasing errors) apart for the usual approximation error of DG. In the simulations, uniform and comparatively sparse Cartesian grids with 2–30 cells along the side of the computational domain were used.

4.2 *Simulation of a Heating Module*

As a real world example, the simulation of a heating module is considered (see Fig. 3). The device is commonly used in the food processing industry to improve the shelf life of liquid products such as milk or juice [6]. It consists of two steel electrodes embedded in a teflon case and operated at 250 kHz. The model dimensions are $20 \times 20 \times 20$ cm with rectangular electrodes of side length 13.5 cm. The fluid flowing between the electrodes is assumed to be orange juice with an electrical conductivity of 0.5 S / m and a relative permittivity of 80. The conductivity and relative permittivity of teflon are assumed to 10^{-12} S / m and 3, respectively.

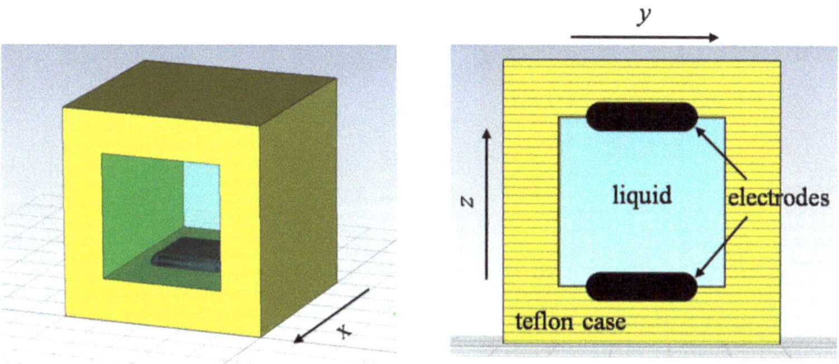

Fig. 3 *Left*: Geometry of the heating module. *Right*: Cross sectional view of the device

The rectangular problem geometry suggests the use of a Cartesian grid. Exceptions make the two electrodes having rounded edges to avoid local peaks in the electric field distribution. The situation can be well handled by the cut-cell approach, since only a small number of cut-cells along the electrode surfaces needs to be considered. In the present simulation, a uniform Cartesian grid with $10 \times 10 \times 10$ cells is used. For the numerical field solution the high order cut-cell DG approach with quadratic basis functions is applied.

Figure 4 shows some of the field distributions obtained by simulation on several cross-sections of the heating module. Note the high resolution of the electric potential and current density obtained in the vicinity of the electrodes, although, an extremely sparse regular grid is used. This accuracy is due to the high order approximation of the DG formulation combined with the cut-cell approach presented in the paper. The heating module example demonstrates the capability of the method to handle practical problems efficiently on simple Cartesian grids by completely avoiding staircasing errors which are typical for FD based methods.

5 Conclusions

A cut-cell approach for the high order DG method is proposed. The method is derived for the case of time-harmonic electro-quasistatics problems, although, it can be easily applied for the solution of other types of static or time dependent electromagnetic field problems. The strength of this approach consists in its capability to obtain high order accuracy solutions on trivial meshes. The discrete problem formulation is simple and easy to implement. This is because the cut-cell approach can be naturally embedded within the DG framework which does not impose conformity conditions on the approximation spaces. The validation example presented in the paper shows that this approach converges at optimal rate for any approximation order.

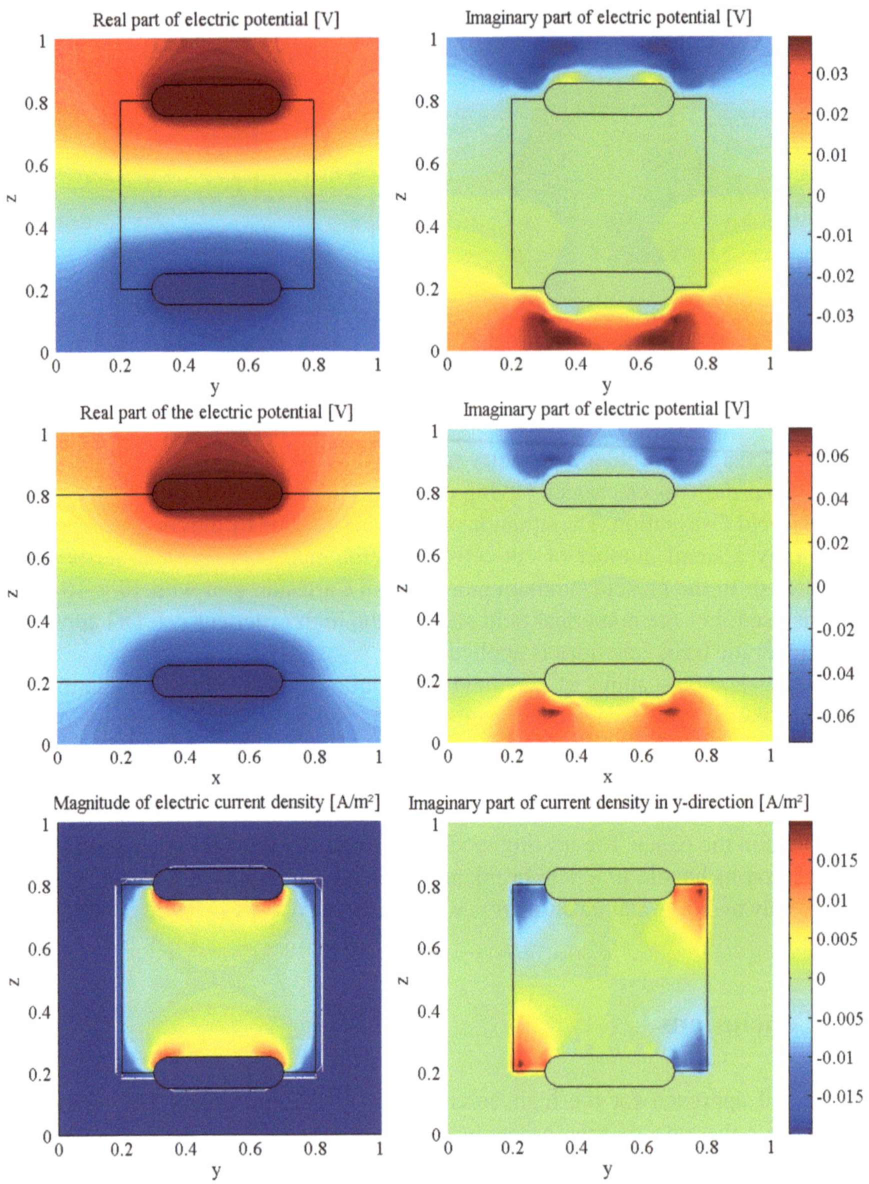

Fig. 4 *Top*: Real and imaginary part of the normalized electric potential on the yz-plane. *Middle*: Real and imaginary parts of the normalized electric potential on the xz-plane. *Bottom*: Magnitude (*left*) and imaginary part of the y-component of the current density on the yz-plane

References

1. Thoma, P.: Zur numerischen Lösung der Maxwellschen Gleichungen im Zeitbereich. PhD Dissertation, TU Darmstadt (1997)
2. Dey, S., Mittra, R.: A locally conformal finite-difference time-domain (FDTD) algorithm modeling modeling three-dimensional perfectly conducting objects IEEE Microw. Guid. Wave Lett. **7**, 273–275 (1997)
3. Arnold, D.N., Brezzi, F., Cockburn, B., Marini, D.: Unified analysis of discontinuous Galerkin methods for elliptic problems. SIAM J. Numer. Anal. **39**, 1749–1779 (2002)
4. Schöberl, J., Zaglmayr, S.: High order Nedelec elements with local complete sequence properties. COMPEL **24**, 374–384 (2005)
5. OpenCascade 4.0, Open-Source Toolkit for 3D modeling (2001). URL: http://www.opencascade.com
6. Scholler, C., et al.: Numerical simulation of thermally coupled electromagnetic fields and fluid flow. In: Proceedings of Computational Methods for Coupled Problems in Science and Engineering, Papadrakakis, M., Onate, E., Schreffler, B. (eds.) CIMNE, Barcelona (2005), Santorini, Greece, May 2005

Optimization of the Current Density Distribution in Electrochemical Reactors

Florin Muntean, Alexandru Avram, Johan Deconinck, Marius Purcar, Vasile Topa, Calin Munteanu, Laura Grindei, and Ovidiu Garvasuc

Abstract This paper proposes to investigate, analyze and compare two practical optimization approaches for smoothing the side effects of electrodeposited layers in electrochemical reactors. The study case consists in a hydraulic component protected by a thin chromium (Cr) layer. Both optimization approaches are investigated by using a 3D finite element software for solving the Laplace equation. The obtained results using these approaches are compared with the numerical results for an electrodepositing process without any additional thief current systems. The uniformity of the chromium deposition on the test component is greatly improved.

1 Introduction

The design of an electroplating rack requires many preliminary steps such as the choice of the electrolyte and the location, the shape and number of electrodes, masks and currents thieves. These parameters affect deposit thickness and plating distribution. Preliminary steps taken to optimize a plating process might be very time consuming if they are performed in a trial-and-error fashion, i.e. plating parts, measuring thickness, plating again etc. If those trial-and-error steps can be simulated accurately, large gains can be made in overall plating cost reduction and the time-to-market of new part designs [1, 2].

F. Muntean · A. Avram · M. Purcar · V. Topa (✉) · C. Munteanu · L. Grindei · O. Garvasuc
Faculty of Electrical Engineering, Technical University of Cluj-Napoca, G. Baritiu 24-26, 400020, Cluj-Napoca, Romania
e-mail: Vasile.Topa@et.utcluj.ro

J. Deconinck
Faculty of Engineering, Vrije Universiteit Brussel, Pleinlaan 2, B-1050 Brussels, Belgium

B. Michielsen and J.-R. Poirier (eds.), *Scientific Computing in Electrical Engineering SCEE 2010*, Mathematics in Industry 16, DOI 10.1007/978-3-642-22453-9_18,

Therefore, effective electrolytic plating thickness simulation helps plating industries to design the most appropriate rack and tools to produce the best deposit uniformity.

2 Electrochemical Models

The nature of the involved electrochemical processes is generally very complex. However, several assumptions and simplifications of limited validity can be made in order to tackle the main aspects of the problem. For example, if the electrode reactions take place at low rates, such that the concentration gradients are neglected, the potential distribution may be found using Laplace's equation. As a consequence, the resulting model describes the ohmic effects in the electrolyte [2]. This model is referred to as the *P*otential *M*odel (PM). An early interest for modeling these kinds of topics has been shown in a number of works. Several authors applied the PM to compute the current density distribution for electroplating applications. Alkire and Bergh [3] applied the Finite Element Method (FEM) to solve the resulting Laplace equation, with nonlinear boundary conditions to account for the electrode charge transfer reactions. Deconinck [4] discretized the equations of the PM using the Boundary Element Method (BEM), in order to compute the changes of the electrode profile for nonlinear boundary conditions. In order to deal with concentration gradients near electrodes, Nernst [5] proposed to decouple the total volume of the electrochemical cell into the bulk solution, where the convective motion takes place, and a thin boundary layer called the Nernst or diffusion layer, near the electrode surface(s). This model is referred to as the *N*ernst's *M*odel (NM). A more general class of models is based on a complete description of the dilute solution theory referred to as the *M*ulti-*I*on *T*ransport and electrode *Re*actions *M*odel (MITReM), [6].

The dilute solution model considers mass transfer diffusion, convection, migration and homogenous reactions together, with electroneutrality conditions and nonlinear concentrations, depending on polarization relations at the electrolyte electrode interface.

Most of the above mentioned papers deal with the mathematical and numerical formulation of the electrochemical models, but do not treat the aspects of optimization of the layer thickness and current density distribution in the electrochemical reactors.

2.1 The Potential Model

If the electrode reactions take place at low rates, the concentration gradients are neglected and the potential distribution may be found using Laplace's equation [2]:

$$\nabla \cdot (-\sigma \cdot \nabla U) = 0 \tag{1}$$

where: U represents the electrolyte's electric potential in $[V]$ and σ the electric conductivity of the solution in $[\Omega^{-1} \cdot \text{m}^{-1}]$. The current density $\mathbf{J}$ in $[A \cdot \text{m}^{-2}]$ according Ohm's law is given by:

$$\mathbf{J} = -\sigma \cdot \nabla U \tag{2}$$

Note that the conductivity σ does not need to be constant. Indeed, it is possible to couple domains with a different conductivity or systems with a local varying conductivity (e.g. function of the temperature T). The reactor's walls, as well as the gaseous medium in contact with the electrolyte, may be seen as insulators. No current flows through them and therefore the normal current density is zero [7]:

$$J_n = \mathbf{J} \cdot \mathbf{l_n} = -\sigma \cdot \nabla U \cdot \mathbf{l_n} = -\sigma \cdot \frac{\partial U}{\partial n} = 0 \tag{3}$$

where: the subscript **n** refers to the normal direction. The same boundary conditions can be applied to the symmetry planes. Depending on the working conditions, the current density distribution may be presented using different expressions. One option is to use a linear relation [4, 7]:

$$J_n = A \cdot (V - U - E_0) + B \tag{4}$$

with: A $[\text{A} \cdot \text{m}^{-2} \cdot \text{V}^{-1}]$ and B $[\text{A} \cdot \text{m}^{-2}]$ the polarization constants, V the metal potential and E_0 the equilibrium potential, all these electric quantities in $[V]$. For single metal deposition processes, the current density distribution is accurately described by a Butler-Volmer relation [4, 7]:

$$J_n = J_0 \cdot (e^{\frac{\alpha_a \cdot F}{R \cdot T} \cdot \eta} - e^{\frac{\alpha_c \cdot F}{R \cdot T} \cdot \eta}) \tag{5}$$

where: J_0 in $[\text{A} \cdot \text{m}^{-2}]$ is the exchange current density, α_a and α_c the anodic and cathodic charge transfer coefficients for the deposition reaction, η the overpotential in [V], R the gas constant in $[\text{J} \cdot \text{mol}^{-1} \cdot \text{K}^{-1}]$, F the Faraday constant in $[\text{C} \cdot \text{mol}^{-1}]$ and T the temperature of the electrolyte in $[K]$.

2.2 Numerical Study Case

The PM model was applied for a study case consisting in a hydraulic component. Being part of a complex mechanical system, this hydraulic part is usually under stress due to friction and/or other forces. For these reasons some parts of the surface must be protected and strengthened by a thin Cr layer, of around 10–30 μm.

The numerical computation of the proposed study case, using the PM model, was done using a 3D FEM simulation tool, specially tailored for electroplating process [8]. In realistic conditions, the electrochemical process takes place in large

Fig. 1 CAD of the electrochemical reactor with a single hydraulic component

tanks with a lot of pieces (electrochemical reactors). Due to the fact that the purpose of our study was to optimize the Cr thickness deposition of a single component, but with two different practical optimization approaches, a small square tank is considered, as in Fig. 1. In order to obtain a uniform layer, very small currents are used. The main disadvantage in this case is a very long process time. For this reason, in the real life, higher currents are used in order to increase the efficiency of the process, by decreasing the total process time. Unfortunately, this change comes with some risks. If higher currents are used, side effects like hydrogen evolution, Cr over burn and porous deposit may appear. In order to overcome these types of problems, we used two different practical optimization approaches for the optimal design of the Cr layer thickness, respectively a current robber and a shielding system.

The following phenomena are taken into account during the optimization process: the ohmic drop in the electrolyte solution; the anodic polarization; the cathode shape changes over different time steps; the reactor configuration, including anode positioning, screens and current thieves; the work piece shape and dimensions; the selective insulation of work piece surfaces; the total current injected and the anode work piece contacting method. After the numerical analysis and electroplating simulations, the obtained results are compared with the first simulation results, which correspond to the initial study case without any additional optimization system.

2.3 Dry Run Simulation

The first simulation is the so called "dry run simulation". The following parameters are used for the numerical computation in this case, respectively the plating time 30 min, the average current density 4,000 A/m^2, the main current 110 A and the main voltage source 4.50 V. Using the PM mathematical model and the 3D FEM software [8] with the above mentioned parameters, the gain in the Cr weight is 3.33 g. The thickness of the Cr deposition on the active zone is between 9 μm and 35 μm, while the current density is between 2,300 $\mathrm{A/m^2}$ and 6,000 $\mathrm{A/m^2}$.

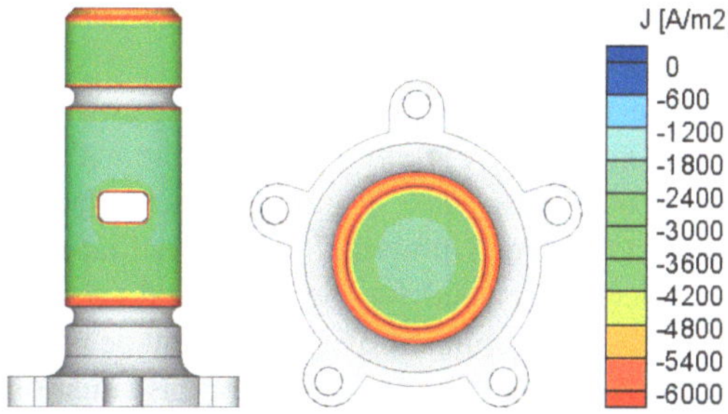

Fig. 2 Current density distribution on the hydraulic component, for the dry run simulation

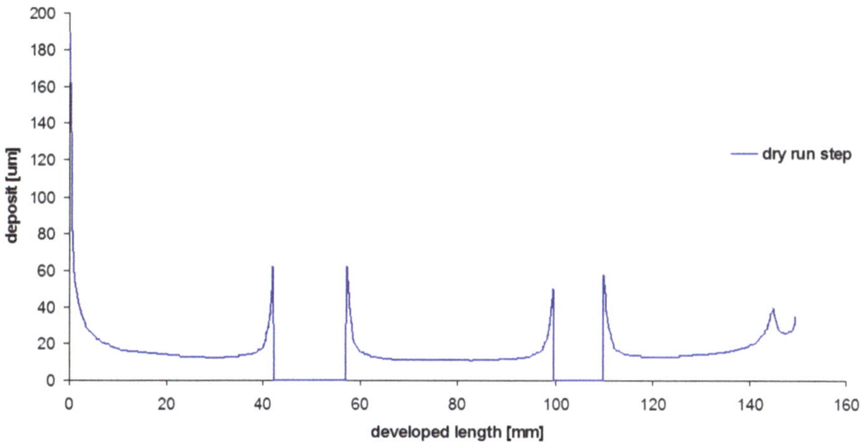

Fig. 3 Cr deposition over the plating surface for the "first run" simulation

Figure 2 indicates the risk for the occurrence of Cr burn or other high current densities related defects. It may easily be seen that, some zones are exposed to risks and a big grinding effort has to be made in order to obtain a smooth surface.

The corresponding Cr deposition thickness for the whole studied hydraulic component is given in Fig. 3. The edge effects are obvious.

3 Practical Optimization Approach

3.1 Current Robbers Approach

The current robber system, used by the authors for the optimization of the Cr deposit layer, consists in one ring robber near the top of the work piece, a lead tape for the recessed area and a lead tape for the hole's edges of the part, mounted on a plastic support, as in Fig. 4. The whole system is short circuited to the mass, through the work pieces. The lead tape system is perfectly joined with the plating area, providing a cylindrical flat surface.

The optimization process consists in three steps by modifying the distance between the ring current robber and the width of the lead tape as following: 20 mm width of the lead tape, 2 mm near each edge; 15 mm width of the lead tape, 4 mm near each edge; 8 mm width of the lead tape, 5 mm near each edge. The following parameters are used for the numerical computation in this case, respectively the plating time 30 min, the average current density 4,000 A/m^2, the main current 138 A, the main voltage source 5.15 V. In this situation the gain in Cr weight is 3.37 g. It's easy to observe that there are no more ï£¡red zonesï£¡ to indicate problems for the Cr deposit, as in Fig. 4.

Since the current robbers are part of the electrode system, during the optimization process the main power source must be changed in order to keep the current density on the work piece on constant 4,000 A/m^2 value. The evolution of the Cr deposition thickness on the plating surface during the optimization process is given in Fig. 5.

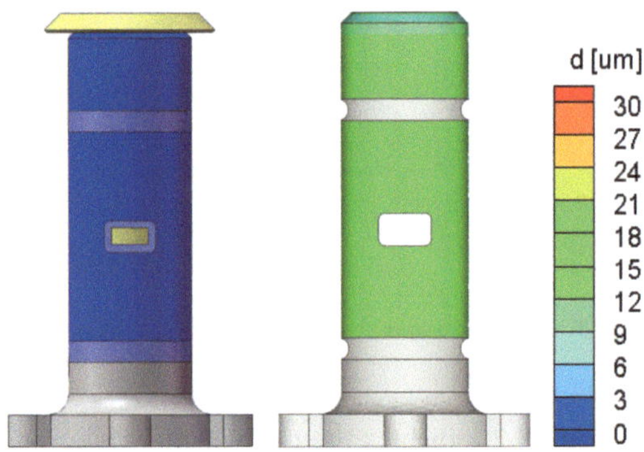

Fig. 4 The current robbers system and the Cr deposition thickness

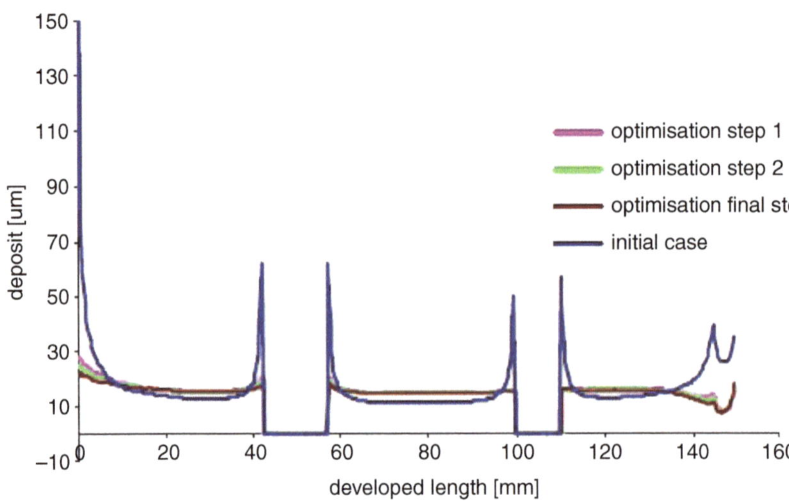

Fig. 5 Cr deposition over the plating surface for each optimization step

Table 1 Current densities and Cr thickness during the optimization process (case 1)

Case	Current source		Min value	Max value	Average value
Intial case	110 A	d[μm]	10	307.3	16.63
		J[A/m^2]	2,822	4,236	3,948
Step 1	127 A	d[μm]	6.65	41.63	15.53
		J[A/m^2]	2,182	8,159	4,017
Step 2	133 A	d[μm]	7.1	28.85	15.22
		J[A/m^2]	2,287	6,243	3,971
Final step	138 A	d[μm]	7.33	23.65	15.07
		J[A/m^2]	2,336	5,429	3,948

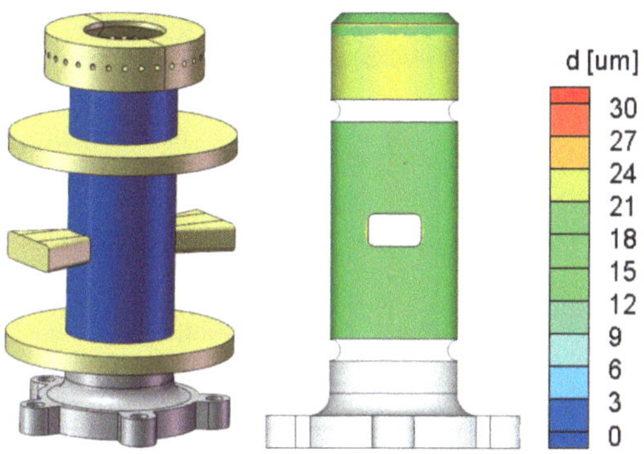

Fig. 6 The shielding system and the Cr deposition thickness

The edge effects are significantly reduced and the thickness of the layer deposit is kept constant (15μm) on the active zone.

In Table 1 there are given the values of the obtained thickness of the Cr depositions and the current densities (min, max and average values) for each optimization step.

3.2 *Shielding System Approach*

The shielding system, used by the authors for the optimization of the Cr deposit layer consists in two collar flat screens with 125 mm diameter for the recessed area, two conical shields for the work piece hole and one L-shape screen with holes on the top of the piece, as in Fig. 6.

The optimization process consists in three steps, by variating the collar flat screens diameter, the length of the conical shield from the surface and the L-shaped distance between the pieces, as following: diameter 70 mm, length 10 mm, L-shape distance 25 mm; diameter 100 mm, length 20 mm from the surface and the L-shape distance 12 mm; diameter 125 mm, length 12 mm and the L-shape distance 12 mm, with holes.

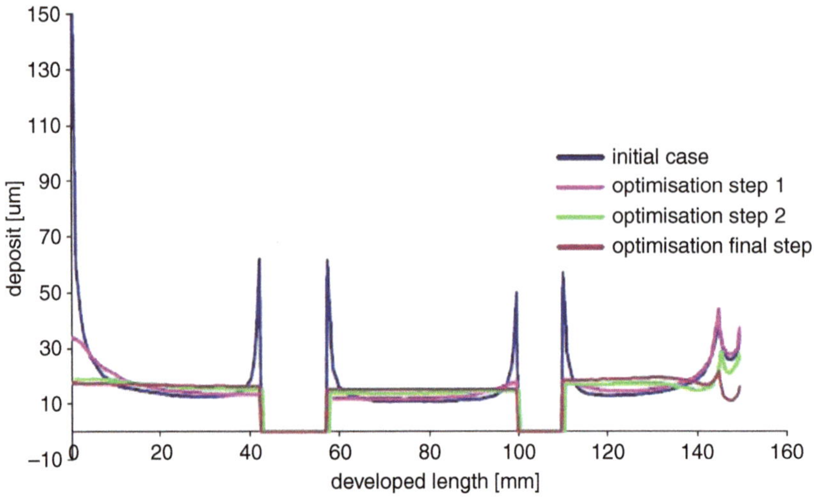

Fig. 7 Cr deposition over the plating surface for each optimization step

Table 2 Current densities and Cr thickness during the optimization process (case 2)

Case		Min value	Max value	Average value
Step 1	d[μm]	10.51	46.13	16.25
	J[A/m^2]	2,928	8,399	3,946
Step 2	d[μm]	11.9	41.47	16.08
	J[A/m^2]	3,205	7,760	3,946
Final step	d[μm]	10.64	24.49	16.08
	J[A/m^2]	2,954	5,320	3,946

The processed parameters are as following: the plating time 30 min, the average current density 4,000 A/m^2, the main current 110 A and the main voltage source 4.86 V. In this case, the weight gain obtained is 3.21 g.

The evolution of the Cr deposition on the plating surface, during the optimization process is given in the Fig. 7. The edge effects are once again significantly reduced and the thickness of the layer deposit is kept constant on the active zone (around 15 μm) as in the previous case.

In Table 2 are given the values of the obtained thickness of the Cr depositions and the current densities (min, max and average values) for each optimization step.

4 Conclusion

A comparison between the obtained results using the two practical optimization approaches is given in Fig. 8.

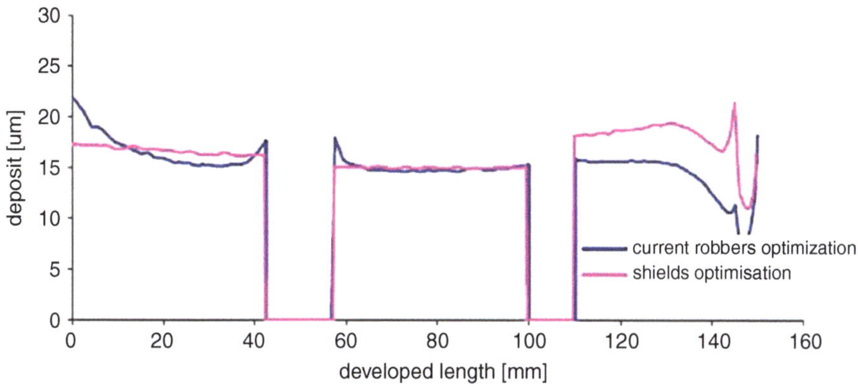

Fig. 8 Cr deposition over the plating surface: current robbers vs. shielding systems

The chosen current robbers system guarantee an uniform Cr layer deposition over the active plating zone with 1.32% in terms of standard deviation, while the shields system ensure the layer uniformity with 1.76% in terms of standard deviation. The current robbers system has better performances in term of the Cr thickness uniformity, but it is more expensive due to the fact that it consumes more material from the electrolyte, respectively 3.37 g instead of 3.21 g with the shielding system. In terms of currents, the first practical optimization approach needs a higher current, respectively 138 A instead of 110 A. In conclusion, the advantage of the robbers system in comparison with the shielding system is the higher uniformity of the Cr thickness depositions, but with a higher average value of the used current and a higher consumption of material from the electrolyte. The research will be extended with new practical optimization tools but using as analysis tool the extended finite element method (XFEM).

Acknowledgements This work was supported by CNCSIS-UEFISCSU under research IDEI program, grant number ID_2538/2008.

References

1. Druesne, F., Afzali, M., Mouton, R.: A New 3D Simulation and Design Tool of Electroplating. Plating and Surf. Finishing, Tech. J. AESF, June, (2002)
2. Prentice, G.A., Tobias, C.W.: ÒA Survey of Numerical Methods and Solutions for Current Distribution Problems. J. Electrochem. Soc. **129**, 72 (1982)
3. Alkire, R., Bergh, T., Sani, T.L.: Predicting Electrode Shape Change with Use of Finite Element Methods, J. Electrochem. Soc., **125**, 1981–1988 (1978)
4. Deconinck, J.: Current distribution and electrode shape change in electrochemical systems. A boundary element approach, Lecture Notes in Engineering no. 75, Springer, Berlin (1992)
5. Nernst, W.: Z. Phys. Chem. **47** (1904)

6. Bortels, L.: The Multi-Dimensional Upwinding Method as a Simulation Tool for the Analysis of Multi-Ion Electrolytes Controlled by Diffusion, Convection and Migration, PhD thesis, Fakulteit Toegepaste Wetenschappen, Vrije Universiteit Brussel (1996)
7. Purcar, M., Bortels, L., Van Bossche, B., Deconinck, J: Ò3D electrochemical machining computer simulationsÓ, J. Mater. Process. Tech. **149**, 472–478 (2004)
8. Elsyca, Belgium http://www.elsyca.be

Streamer Line Modeling

Thomas Christen, Helmut Böhme, Atle Pedersen, and Andreas Blaszczyk

Abstract After reviewing some basics of dielectric withstand of air insulation, we introduce two procedures for an improved prediction of streamer paths in complex geometries. Although based on the electric background field, we generalize conventional models that usually consider paths starting at a field maximum and traveling along field lines. The new approaches are able to explain both streamer inception points different from field maxima as well as deviations of the streamer path from field lines, and may help to further optimize dielectric withstand of high voltage devices.

1 Introduction

Design optimization of air-insulated electrical devices refers to maximization of the dielectric withstand with respect to detrimental gas discharges. For this, both the electric field distribution and the critical failure mechanism must be known [1–4]. In practice even the electric field calculation in a real 3-d geometry can be highly nontrivial because of geometrical complexity, e.g., due to large aspect ratios, and the presence of charged dielectric interfaces. The electric background field $\mathbf{E} = -\nabla\varphi$ is obtained from the Poisson equation for the electric potential φ. Often it reduces to

T. Christen (✉) · A. Blaszczyk
ABB Corporate Research, 5405 Baden, Switzerland
e-mail: thomas.christen@ch.abb.com, andreas.blaszczyk@ch.abb.com

H. Böhme
Dresden, Germany (previously with ABB Corporate Research)
e-mail: boehme.w75@t-online.de

A. Pedersen
SINTEF Energy Research, Norway (formerly with High Voltage Laboratory, Swiss Federal Institute of Technology, 8092 Zurich, Switzerland)
e-mail: Atle.Pedersen@sintef.no

B. Michielsen and J.-R. Poirier (eds.), *Scientific Computing in Electrical Engineering SCEE 2010*, Mathematics in Industry 16, DOI 10.1007/978-3-642-22453-9_19,

the Laplace equation as in most high voltage engineering applications space charges can be neglected. However, breakdown associated with gas discharges does involve space charges, which necessitates a self-consistent solution of the Poisson equation and the equations for the charge carriers. Except for special cases [5–8], such a complete simulation of gas discharges is infeasible today for real device geometries because of computational difficulties. Furthermore, besides discharge inception that can sometimes be evaluated from the Laplacian field, the judgement on withstand must include propagation and further development of the electric discharge. For instance, one must know whether a discharge initializing streamer stops after a short distance or develops further into a destructive electric arc connecting two counter-electrodes.

The main challenge of this paper concerns the prediction of streamer behavior, particularly its most likely path, from the knowledge of the Laplacian background field. We will focus on positive streamers as they are more critical than negative ones. Because the model behind our consideration is phenomenological, and because of the intrinsic erratic behavior of streamers, one can only expect approximate answers with statistical character. However, already those can be of high value for designing electrical devices.

2 Dielectric Breakdown of Air Insulation

Physical understanding of electrical discharge and spark phenomena in electrical engineering is based on a vast number of empirical facts, combined with various more or less sophisticated models [1–4]. Only recently the increase of computational power jumpstarted a reconsideration of the subject, leading to improved theories and simulation results (see, e.g., [9]). The basic steps of a dielectric breakdown in air insulated high voltage equipment are inception, streamer propagation, maybe streamer-leader transition for sufficiently large electrode separation, and arc formation after electrodes are connected by the conductive channel. A phenomenological understanding of withstand voltage U_w as a function of the electrode distance and for inhomogeneous field is sketched in Fig. 1 and covers the following steps. For the following, we will assume that $\mathbf{E}$ is known in the spatial region of interest, e.g., from computation.

2.1 Streamer Inception

After the appearance of an initial electron in a critical high-field volume, an electron avalanche starts to develop. If a sufficient number N_c of electron generations are produced due to impact ionization, a self-propagating streamer head forms. The *(streamer) inception criterion* is $\int_\gamma \alpha_{\mathrm{eff}}(E)dx = \ln(N_\mathrm{c})$ where α_{eff} is the E-dependent effective ionization coefficient including ionization, electron

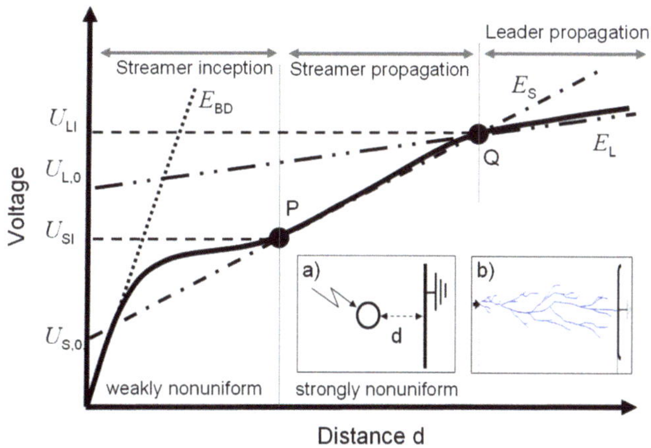

Fig. 1 Illustration of U_w (*thick curve*) in normal air as a function of the electrode distance for an inhomogeneous field configuration (e.g., sphere-plate electrodes (inset a)). Gas breakdown field E_{BD}; streamer inception voltage U_{SI} ($\approx$ voltage level at P); streamer head voltage $U_{S,0}$; streamer propagation field E_S (slope indicated by *dashed-dotted line*); leader inception voltage U_{LI} ($\approx$ voltage level at Q); leader head potential $U_{L,0}$; leader internal field E_L (slope indicated by *dashed-double-dotted line*). For weakly nonuniform fields, U_{SI} limits the withstand voltage (cross-over at P). The range relevant for streamer line modeling lies between P and Q. (Inset b) Sketch of a streamer bunch with many branches and connecting a tip electrode with a plate electrode

attachment, and detachment. Here, $E = |\mathbf{E}|$, and the integration path γ starts at the point with maximum field, follows the field line as long as $\alpha_{\text{eff}}(E) > 0$, and ends where the critical field value, E_{BD} given by $\alpha_{\text{eff}}(E_{BD}) = 0$, is reached. There exist a number of empirically determined fit functions for $\alpha_{\text{eff}}(E)$ (see for instance, [3,10]).

Typical values are [3, 4, 10]: $E_{BD} \approx 2.5$ kV/mm, a few mm for the critical streamer length γ, and $\ln(N_c) \approx 9$–21 (higher values correspond to lower fields, and $\ln(N_c) \approx 18.4$ should be used for strongly inhomogeneous fields in typical medium and high voltage devices). The inception voltage U_{SI} is based on the Laplacian background field and implicitly determined by the inception criterion. If the field is only weakly nonuniform, the streamer will short the electrodes immediately when the inception voltage is reached, which defines then withstand. Path selection is not so critical since its length is typically a few millimeters; it is thus sufficient to integrate along a field line.

In most literature, electrode inception is discussed because in simple arrangements the maximum field is located there. However, real devices contain usually additional solid insulation and the highest field values may occur away from electrodes, leading to *electrodeless inception* [11]. Then, a streamer dipole forms where at the same time positive and negative streamer heads are generated and separate [7]. The underlying physics differs from inception at an electrode. In particular, initiation is assumed to be related to electron detachment, which occurs in normal bulk air at about 3.5–4 kV/mm [12]. Detachment from shallow surface traps at solid dielectric surfaces may happen at lower fields; the inception criterion to be applied in this case is not fully clear yet.

2.2 Streamer Propagation

A streamer head will propagate towards the opposite electrode with typical velocities of 10^5–10^7 m/s. The exact value depends on a number of parameters, like the applied voltage [13], and is correlated to other streamer properties like the diameter [14]. The propagating streamer can be understood as a self-sustaining ionization wave, which is driven partially by the recombination-induced light-emission that leads to photo-electrons that initiate avalanches, and partially by the space-charge induced high field in front of the streamer head, which leads to the growth of these avalanches. The streamer can reach the counter electrode only if the applied voltage is large enough to maintain the propagation process. An estimate of the distance d_S until a streamer stops is given by an equal area rule based on the hypothesis of constant field ($E_S \approx 0.5$ kV/mm for air) in the streamer channel; a fact with reasonable empirical validation for electrode gaps from 5 cm to about 2 m. For shorter distances ($d < 5$ cm), the field is typically higher, while for larger gaps ($d > 2$ m) leader transition, as explained below, must be taken into account. The equal area rule (*streamer propagation criterion*) for the propagation length d_S is given by $\int_0^{d_S} E(\mathbf{x})dx \approx E_S d_S$ along a field line. The lowest voltage value U_w (the *withstand voltage*) that enables the streamer to connect the electrodes (i.e., $d_S = d$) in a strongly inhomogeneous field, is given by $U_w = U_S$, where approximately $U_S = U_{S,0} + E_S\, d$. Here d is the length of the streamer path (*clearance*) connecting the electrodes, and $U_{S,0} \approx 24$ kV characterizes the streamer head [15]. Sometimes, E_S is identified as the *streamer propagation field*, which is interpreted as the external field required for streamer propagation [13].

Application of these rules, between points P and Q in Fig. 1, requires knowledge of the streamer path. The fact that steamer propagation in air is driven by photo-ionization, which can be seen as a nonlocal effect, has two consequences. First, the finite absorption length and the finite avalanche size for photo-ionization imply that the seed for an avalanche is created somewhere on a semi-spherical surface in a finite distance in front of the streamer head. Hence, the streamer proceeds not solely straight-forward along a field line, but is with a certain probability deflected sideways. This *erratic behavior* might also lead to streamer branching via creation of two differently directed avalanches at the same time, which leads finally to streamer bundles (see Fig. 1 inset b). Secondly, in a situation where the field vector, $\mathbf{E}$, is not collinear with the gradient of the field strength, or with $\nabla \cdot \mathbf{E}^2$, the streamer path will be deflected systematically from the field-line direction, because the probability of the next streamer step is higher to be directed towards higher field strength. Even if steamer branching takes place deterministically via a Laplacian type of instability [16], this background-field induced effect should be expected to influence the path direction, as the streamer is not ideally conductive.

2.3 *Streamer-Leader Transition and Leader Propagation*

Streamer-leader transition (Q in Fig. 1) [1] occurs when the current in the stem of the streamer bunch is sufficiently high to heat up this channel to more than 5,000 K necessary for a thermal plasma with increased channel conductivity. Leader transition is strongly influenced by capacitive coupling with the surrounding electrodes and dielectric bodies, and can be predicted for simple geometries from empirical rules [17]. A mature leader channel can carry a current of about 1 A at a field of about 0.1 kV/mm.

One may write for distances larger than about 2 m [1], $U_{\mathrm{L}} = U_{\mathrm{L},0} + E_{\mathrm{L}}\, d$ with leader head potential $U_{\mathrm{L},0} \approx 0.6$–$0.8\,\mathrm{MV}$ and a leader channel field E_{L} of about 0.1–0.2 kV/mm. The leader can propagate over longer distances than a streamer, but with lower velocities of about 10^4–10^5 m/s. The leader dynamics is very erratic and the path almost unpredictable.

The relation $U_{\mathrm{w}} = U_{\mathrm{L}}$ is limited to very large voltages. As leaders must be avoided in practice, $U_{\mathrm{w}} = U_{\mathrm{LI}} \approx 1\,\mathrm{MV}$ is used for design issues. U_{LI} depends on the capacitive coupling between the propagating streamer and the environment, and is usually calculated from empirical rules (cf. [17]). At lower voltage levels and sizes below 2–3 m, leader transition is restricted to dielectric surfaces.

3 Modeling of Streamer Lines

Once the path of a streamer is known, its transit time can be estimated from its velocity along this line. Assuming 10^6 m/s, a typical distance of 10 cm requires a HV-pulse length of at least 100 ns; so the main issue is the determination of the line. In the following, we introduce two different and new improvements of streamer path prediction, which go beyond the usual simple approximation that identifies the path with the field line starting at the field maximum, from which the discharge paths observed in experiments often strongly deviate [11].

3.1 *Bundle of Field Lines*

Even if one assumes that the path can be associated with a field line, it is not obvious that the starting point, p, is associated necessarily with the maximum field value at the electrode. Indeed, p can be in a finite vicinity thereof. Relevant streamer lines (parameterized by p) must certainly fulfill both inception and propagation

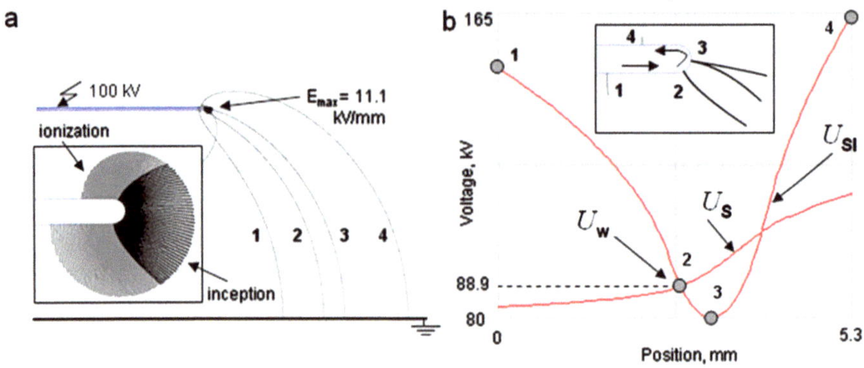

Fig. 2 (**a**) Streamer lines for parallel plate arrangement with voltage 100 kV and vertical distance 100 mm, (**b**) inception and streamer voltage distributions around the edge of the HV plate, and resulting withstand voltage

conditions, i.e., the voltage must be higher than both voltages for inception (U_{SI}) and for propagation (U_S, with d_S equal length of path connecting electrodes). It is then obvious that the most critical path among them has the starting point p associated with $U_w = \mathrm{Min_p\,Max}\,\{U_S, U_{SI}\}$. This is illustrated in Fig. 2 for an arrangement of a flat, 1 mm thick electrode disk placed horizontally a certain distance above a large grounded plate. Field lines start at the rounded edge with radius of curvature $r = 0.5$ mm. Figure 2 shows, as a function of p, the streamer inception voltage and the withstand voltage U_w. The latter is an increasing function of the length of the field line, while the former has a minimum at the rounded edge. Obviously, the minimum of the maximum of the two curves occurs at point 2. The critical streamer line has a length of about 120 mm (line 2). For a perpendicular plate configuration, where the edge is facing the grounded plate and the field line has the shortest possible length of 100 mm, the withstand voltage according to the streamer propagation criterion is slightly below 80 kV. The 10–15% withstand increase by changing the angle from perpendicular to parallel plates is in accordance with experimental observations [18], which support to usefulness of the minimax principle.

3.2 *Field Gradient based Streamer Deflection*

In reality the streamer path does not always follow an electric field line. Besides the erratic behavior, which may be deterministic or noise induced, as mentioned above one has to expect a deflection from the field line into regions of higher fields, if the vectors $\mathbf{E}$ and $\nabla \cdot \mathbf{E}^2$ are not collinear. For modeling, the deterministic part of the streamer path $\mathbf{x}(t)$ is assumed to be given by the solution of the ordinary differential equation, $d\mathbf{x}/dt = \mathbf{v}_S(\mathbf{x})$, where $\mathbf{x}$ may be interpreted as the streamer head location. We correct the main contribution $\mathbf{v}_S(\mathbf{x}) \propto \mathbf{E}(\mathbf{x})$ by an additional term

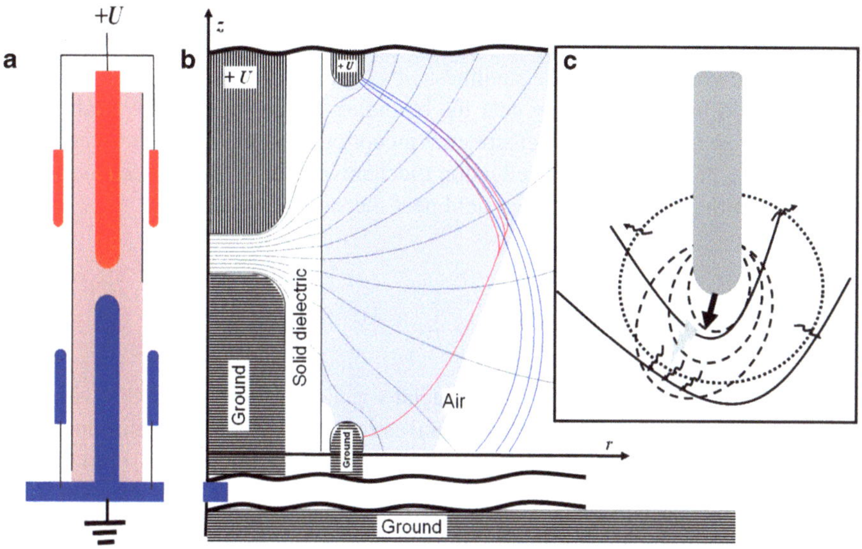

Fig. 3 (**a**) Sketch of an insulation rod with embedded and ring electrodes. (**b**) Simulation results: equipotential curves as *black solid lines*; region with $E > E_S$ in *blue*. Field lines (*blue*) starting in the critical high field region; streamer paths (*red*) according to $d\mathbf{x}/dt = \mathbf{v}_S$. Inset (**c**): deflection of a (positive) streamer path from the field-line direction of the external field. Recombination-radiation creates photo-electrons on a spherical surface within a certain distance from the streamer head. These act as first electrons of avalanches; this leads to a (weak) deflection from the field line direction into regions with higher field values. (*Solid*: equipotential curves; *dashed*: constant E; *dotted*: creation of photo-electrons (*arrows*))

$\propto \nabla \cdot \mathbf{E}^2$, such that $\mathbf{v}_S = k_1(E)\mathbf{E} + k_2\nabla \cdot \mathbf{E}^2$ is the streamer equation of motion. The factors $k_{1,2}$ have to be determined either empirically or from a more basic theory. In any case, they should satisfy two properties. First, k_1 must vanish when the external field drops below the streamer propagation field (cf. [3]), i.e., $k_1(E \leq E_S) \equiv 0$. Secondly, the term $k_2\nabla \cdot \mathbf{E}^2$ is typically a *small* correction (which could be derived from a perturbation approach). Indeed, if it predominates at large E, the streamer could propagate into high field regions against field lines, which is not reasonable in the framework of this deterministic consideration.[1] However, only if k_1 goes to zero, the second term will dominate, which keeps the streamer inside the streamer propagation region. The streamer path according to the streamer equation of motion is simulated for a special electrode arrangement shown in Fig. 3a. It consists of an insulating rod with two embedded electrodes and two additional electrode rings outside. In the following, bulk streamers that connect the rings through the air are considered. The grounded environment breaks the mirror symmetry at the center plane of the rod. Figure 3b shows the equipotential curves (black solid) and the

[1]Note that sparks can propagate partially against field lines of the external field [1], which is due to other reasons.

region with $E > E_S$ (blue). The result indicates that the streamer path remains in the region where the field is sufficiently high for propagation, and it is able to connect the two electrodes. The smallness of k_2 implies that a deviation of the path from the field lines occurs only near the boundary of this region. The final path is then neither sensitive to the exact streamer inception location, nor to the exact value of k_1. Figure 3c illustrates the deflection mechanism based on enhanced avalanche formation in the region with higher field near the streamer head.

4 Conclusion and Outlook

Knowledge on the discharge path is crucial for a determination of the withstand voltage of electrical devices. The traditional estimates based on the background field use the electric field lines starting from the point of maximum field. For complex geometries, this can be inappropriate as the critical path is (1) not the one that starts at maximum field and (2) often systematically deviates from field lines. We have proposed two phenomenological approaches that are able to improve the two items. They are still based on the background field and are thus only appropriate if the effect of space charges involved in the discharge is weak, which can be valid for streamers. It goes beyond the purpose of this article to combine the two items into a single algorithm for withstand prediction in arbitrary geometries. This task as well as elaboration of a deeper physical foundation and the experimental validation are postponed to future work.

References

1. Bazelyan, E.M., Raizer, Y.P.: Spark Discharge. CRC Press, New York (1998)
2. Kuffel, E., Zaengel, W., Kuffel, J.: High Voltage Engineering: Fundamentals. Butterworth-Heinemann, Oxford (2000)
3. Boeck, W., Pfeiffer, W.: Conduction and Breakdown in Gases. In: Webster, J.G. (ed.) Wiley encyclopedia of electrical and electronics engineering, Vol. 4, pp. 123–172. J. Wiley & Sons Inc., New York (1999)
4. Gallimberti, I.: The mechanism of the long spark formation. Colloque de physique **40**, 193–250 (1979)
5. Christen, T., Seeger, M.: Simulation of unipolar space-charge controlled electric fields J. Electrostat. **65**, 11 (2007)
6. Christen, T.: FEM Simulation of Space Charge, Interface and Surface Charge Formation in Insulating Media In: XVth International Symposium on High Voltage Engineering T84, Ljubljana, Slovenia, August 27–31 (2007)
7. Luque, A., Ratushnaya, V., Ebert, U.: Positive and negative streamers in ambient air: modeling evolution and velocities. J. Phys. D: Appl. Phys. **41**, 234005 (2008)
8. Franck, C., Seeger, M.: Application of High Current and Current Zero Simulations of High-Voltage Circuit Breakers. Contrib. Plasma Phys. **46**, 787–797 (2006)
9. Li, C., Ebert, U., Hundsdorfer, W.: 3D hybrid computations for streamer discharges and production of runaway electrons. J. Phys. D: Appl. Phys. **42**, 202003 (2009)

10. Petcharaks, K.: Applicability of the Streamer Breakdown Criterion to Inhomogenous Gas Gaps. Ph. D. Thesis No. 11192, ETH Zurich, 1995
11. Pedersen, A., Christen, T., Blaszczyk, A., Böhme, H.: Streamer inception and propagation models for designing air insulated power devices. CEIDP Conference Material, Virginia Beach, Oct. 2009
12. Pancheshnyi, S.: Role of electronegative gas admixtures in streamer start, propagation and branching phenomena. J. Phys. D: Appl. Phys. **40**, 645–653 (2005)
13. Allen, N.L., Mikropoulos, P.N.: Dynamics of streamer propagation in air. J. Phys. D: Appl. Phys. **32**, 213–219 (1999)
14. Briels, T.M.P., et al.: Positive and negative streamers in ambient air: measuring diameter, velocity and dissipated energy. J. Phys. D: Appl. Phys. **41**, 234004 (2008)
15. Nasser, E.: Der räumliche und zeitliche Entladungsaufbau im ungleichförmigen Feld bei positiver Spitze in atmosphärischer Luft. Arch. f. Elektrotech., XLIV (1959), pp. 157
16. Arrayas, M., Ebert, U., Hundsdorfer, W.: Spontaneous Branching of Anode-directed Streamers between Planar Electrodes. Phys. Rev. Lett. **88**, 174502 (2002)
17. Küchler, A.: Hochspannungstechnik. VDI Verlag. 2005
18. Böhme, H., Wolf, H.: Sharp edges in high voltage engineering; Die Durchschlagspannung von Entladestrecken mit Flachschienen bei Stoßspannung und Wechselspannung. Report GB 32/60 (unpublished) and diploma thesis TU Dresden 27/56

A Discontinuous Galerkin Formalism to Solve the Maxwell-Vlasov Equations. Application to High Power Microwave Sources

Laura Pebernet, Xavier Ferrieres, Vincent Mouysset, François Rogier, and Pierre Degond

Abstract In this paper, we present a *Particle-In-Cell* (PIC) method based on a Discontinuous Galerkin (DG) scheme to solve the Maxwell-Vlasov equations in time-domain. Comparisons with an other industrial software are given to validate the method.

1 Introduction

The main objective of this work is to propose an efficient solution for modelling High Power Microwave (HPM) sources by considering microwave/plasma interactions. For this kind of physical problem, we consider a collisionless and low density plasma. To describe the plasma dynamics in the presence of an electromagnetic field, represented by the Maxwell-Vlasovsystem, a Particle In Cell (PIC) method is adopted. The principle of a PIC method is to couple two solvers: one for the Maxwell part and another one to treat the displacement of macro-particles. Concerning this method, there already exist software based on the Finite Difference Time Domain (FDTD) scheme, which are very much used by the community (e.g. see [1]). However, the last ten years, a lot of studies on Discontinuous Galerkin (DG) methods [2] have shown more interest than the FDTD method for solving the Maxwell equations. In this paper, we are interested in applying a particular DG approach to solve Maxwell-Vlasov equations. As a first validation of the pertinence of our DG method, we will focus on the simulation of HPM sources and

L. Pebernet · X. Ferrieres (✉) · V. Mouysset · F. Rogier
ONERA, 2 avenue Edouard Belin 31055 Toulouse Cedex 4, France
e-mail: Laura.Pebernet@onera.fr; Xavier.Ferrieres@onera.fr; Vincent.Mouysset@onera.fr; Francois.Rogier@onera.fr

P. Degond
IMT, Université Paul Sabatier, 118 route de Narbonne 31062 Toulouse Cedex 9, France,
e-mail: pierre.degond@math.ups-tlse.fr

B. Michielsen and J.-R. Poirier (eds.), *Scientific Computing in Electrical Engineering SCEE 2010*, Mathematics in Industry 16, DOI 10.1007/978-3-642-22453-9_20,

in particular, on rendering the behaviour of a diode device. In Sect. 2, we describe the DG method proposed to solve the Maxwell-Vlasov system, and in Sect. 3, we present the simulation results for a diode configuration.

2 Mathematical Formalism

This section describes the mathematical formulation of the Maxwell-Vlasov problem and the chosen DG approximation to solve it.

2.1 Maxwell-Vlasov System

The kinetic model of a collisionless, weak density plasma is described by the evolution of a distribution function $\mathbf{f}_s = \mathbf{f}_s(\mathbf{v}, \mathbf{x}, t)$ for each particle species s. This function corresponds to the statistical average of the particles distribution in the phase space. In the non-relativistic case, it satisfies the Vlasov equation:

$$\frac{\partial \mathbf{f}_s}{\partial t} + \mathbf{v} \cdot \frac{\partial \mathbf{f}_s}{\partial \mathbf{x}} + \frac{q}{m}(\mathbf{E} + \mathbf{v} \times \mathbf{B}) \cdot \frac{\partial \mathbf{f}_s}{\partial \mathbf{v}} = 0, \tag{1}$$

where $\mathbf{x}$, $\mathbf{v}$, q and m are respectively, the position, the velocity, the charge and the mass of s. Equation (1) is coupled to the Maxwell equations which determine the electromagnetic fields $(\mathbf{E}, \mathbf{H})$, on a bounded computational domain Ω:

$$\begin{cases} \dfrac{\partial \mathbf{B}}{\partial t} + \nabla \times \mathbf{E} = 0, \\ \dfrac{\partial \mathbf{D}}{\partial t} - \nabla \times \mathbf{H} + \mathbf{J} = 0, \\ \nabla \cdot \mathbf{D} = \rho, \ \ \nabla \cdot \mathbf{B} = 0, \end{cases} \tag{2}$$

with $\mathbf{H} = \mu_0^{-1}\mathbf{B}$ and $\mathbf{D} = \varepsilon_0 \mathbf{E}$. The constants μ_0 and ε_0 are respectively the magnetic permeability and the electric permittivity of the medium. $\mathbf{J}$ and ρ are the electric current and charge densities representing the particle motion. They are defined by:

$$\rho(\mathbf{x}, t) = \sum_s q_s \int_{\mathbb{R}^3} f_s(\mathbf{x}, \mathbf{v}, t) d\mathbf{v}, \tag{3}$$

$$\mathbf{J}_s(\mathbf{x}, t) = \sum_s q_s \int_{\mathbb{R}^3} \mathbf{v} f_s(\mathbf{x}, \mathbf{v}, t) d\mathbf{v}. \tag{4}$$

On the boundary $\partial\Omega$ of the domain Ω, to simulate an infinite domain, we impose a Silver-Muller boundary conditions [3].

2.2 The Particle-In-Cell Method

One can show (see [4]) that the distribution function $\mathbf{f}_s$ is conserved along particles trajectories and that the positions $\mathbf{x}$ and velocities $\mathbf{v}$ of the particles are solutions of the characteristic equations (equations of motion):

$$\begin{cases} \dfrac{d\mathbf{x}}{dt} = \mathbf{v}, \\ \dfrac{d\mathbf{v}}{dt} = \dfrac{q}{m}\left(\mathbf{E} + \mathbf{v} \times \mathbf{B}\right). \end{cases} \tag{5}$$

In (5), $\mathbf{E}$ and $\mathbf{B}$, are solutions of the Maxwell equations (2), computed on a fixed mesh in the physical space which does not match with the positions of the particles. Hence, it is necessary to do interpolations between the positions of the particles and the fields in order to evaluate the coupling terms. This method of coupling (2) and (5), is called the *Particle-In-Cell* method. The main difficulty of this method is related to charge conservation. Indeed, the constraint on the discrete divergence of the electric field is not satisfied and to guarantee it, we use a hyperbolic correction [5]. The modified Maxwell equations are:

$$\begin{cases} \varepsilon_0 \dfrac{\partial \mathbf{E}_K}{\partial t} - \nabla \times \mathbf{H}_K + \chi \nabla \phi_K + \mathbf{J}_K = 0, \\ \mu_0 \dfrac{\partial \mathbf{H}_K}{\partial t} + \nabla \times \mathbf{E}_K = 0, \\ \mu_0 \dfrac{\partial \phi_K}{\partial t} - \chi \dfrac{\rho_K}{\varepsilon_0} + \chi \nabla \cdot \mathbf{E}_K = 0, \end{cases} \tag{6}$$

with $\chi \in \mathbb{R}^+$ and where ϕ_K is a scalar function. In the hyperbolic correction, χ is taken as a "penalisation term" to impose the electric divergence equation. In our simulations, χ is chosen around 5. This experimentally found value, allows us to guarantee the electric divergence relation and a stable numerical method without having a too small time step.

2.3 Approximation

For the spatial discretisation, we use a set of hexahedral elements K_i such that $\Omega = \bigcup_{i=1}^{N} K_i$. On each cell K, the Maxwell equations (6) are re-written as follows:

$$\begin{cases} \varepsilon_0 \dfrac{\partial \mathbf{E}_K}{\partial t} - \nabla \times \mathbf{H}_K + \chi \nabla \phi_K + \mathbf{J}_K + \beta^K_{\partial K} [\![\mathbf{H} \times \mathbf{n}_K]\!]_{\partial K} + \alpha^K_{\partial K} [\![\mathbf{n}_K \times (\mathbf{n}_K \times \mathbf{E})]\!]_{\partial K} \\ \qquad + \zeta^K_{\partial K} \chi [\![\mathbf{n}_K . \mathbf{E}]\!]_{\partial K} \mathbf{n}_K + \tau^K_{\partial K} \chi [\![\phi]\!]_{\partial K} \mathbf{n}_K = 0, \\ \mu_0 \dfrac{\partial \mathbf{H}_K}{\partial t} + \nabla \times \mathbf{E}_K + \gamma^K_{\partial K} [\![\mathrm{e} \times \mathbf{n}_K]\!]_{\partial K} + \delta^K_{\partial K} [\![\mathbf{n}_K \times (\mathbf{n}_K \times \mathbf{H})]\!]_{\partial K} = 0, \\ \mu_0 \dfrac{\partial \phi_K}{\partial t} - \chi \dfrac{\rho_K}{\varepsilon_0} + \chi \nabla \cdot \mathbf{E}_K + \eta^K_{\partial K} \chi [\![\mathbf{E} \cdot \mathbf{n}_K]\!]_{\partial K} + \theta^K_{\partial K} [\![\phi]\!]_{\partial K} = 0, \end{cases} \tag{7}$$

where $[\![\mathbf{u}]\!] = \mathbf{u}^{**} - \mathbf{u}^{*}$ defines the jump term on the boundary, ∂K, of K, and $\mathbf{u}^{*}$ and $\mathbf{u}^{**}$ are respectively the values of the traces from the inside and the outside of K at the considered surface.

The coefficients $\beta^K_{\partial K}, \alpha^K_{\partial K}, \tau^K_{\partial K}, \delta^K_{\partial K}, \gamma^K_{\partial K}, \eta^K_{\partial K}, \zeta^K_{\partial K}$ and $\theta^K_{\partial K}$ are chosen to ensure an equivalence between (6) and (7). Considering identical coefficients for all the cells, these values must satisfy:

- $1 + \beta - \gamma = 0, \alpha \geq 0$ and $\delta \geq 0$
- $1 - \tau - \eta = 0, \zeta \leq 0$ and $\theta \leq 0$

To approximate the system (7), for the electromagnetic fields **E** and **H**, we define the approximation U_h [2] such that:

$$U_h = \left\{ \mathbf{v}_h \in \left[\mathbf{L}^2(\Omega) \right]^3 \; : \; \forall K \in \mathscr{T}_h, \; DF^*_K \mathbf{v}_{h|K} o F_K \in [Q_r(\hat{K})]^3 \right\} \tag{8}$$

where $\forall r \in \mathbb{N}$, $Q_r(\hat{K})$ is the set of polynomials on $\hat{K} = [0,1]^3$ the orders of which are lower than or equal to r in each variable. Concerning the scalar corrector term ϕ, we define the following approximation space:

$$V_h = \{ \mathbf{v}_h \in \mathbf{L}^2(\Omega) \; : \; \forall K \in \mathscr{T}_h, \; \mathbf{v}_{h|K} o F_K \in Q_r(\hat{K}) \} \tag{9}$$

For each cell K, we denote the transformation between the cell and $\hat{K}$ by F_K. The basis functions φ_l for the electric and magnetic fields and the current density, are defined on K by a transformation $\varphi_l \circ F(\hat{\mathbf{x}}) = DF^{*-1}_K \hat{\varphi}_l(\hat{\mathbf{x}})$ of the basis function $\hat{\varphi}_l$ given on $\hat{K}$. DF_K is the Jacobian matrix of F and $\hat{\mathbf{x}}$ a point on $\hat{K}$. For a scalar functions, like the charge density and ϕ, the basis functions are defined by $\psi_l \circ F(\hat{\mathbf{x}}) = \hat{\psi}_l(\hat{\mathbf{x}})$. In $\hat{K}$, for a spatial approximation of order r, we introduce $(r+1)^3$ Gauss quadrature points, on each quadrature point we have 3 degrees of freedom for the electric, magnetic and current density terms and 1 degree of freedom for the charge term and ϕ. In $\hat{K}$, the basis functions $\hat{\varphi}_l$ are the tensor product of the Lagrangian polynomials $\hat{\psi}_l$ associated to the quadrature points $\hat{\mathbf{x}}_l$, $l = 1, (r+1)^3$.

By considering a Leap-Frog time approximation, the numerical scheme is stable under the condition:

$$\Delta t \leq \frac{2}{c_0} \mathcal{A}_1 + \frac{1}{c_0 \, \chi} \mathcal{A}_2 \tag{10}$$

where $\mathcal{A}_1$ and $\mathcal{A}_2$ are constants depending on the geometry [2,6]. We note that with this condition the greater the value of χ the smaller the time step must be.

2.4 Fields/Particles Interpolation

Consider one particle p with a position $\mathbf{x}_p = (x_p, y_p, z_p)$ and a velocity $\mathbf{v}_p = (v_{xp}, v_{yp}, v_{zp})$, located in the cell K. The electric and magnetic fields $(\mathbf{E}_p, \mathbf{H}_p)$ in $\mathbf{x}_{\mathbf{l}_0}$, the point associated to the degree of freedom $\mathbf{l}_0$ and closest to $\mathbf{x}_p$, are given by $\mathbf{E}_p = DF_K^{*-1}(\hat{\mathbf{x}}_{\mathbf{l}_0})\big(E^i_{K,\mathbf{l}_0}\big)_{i=1,\cdots,3}$ and $\mathbf{H}_p = DF_K^{*-1}(\hat{\mathbf{x}}_{\mathbf{l}_0})\big(H^i_{K,\mathbf{l}_0}\big)_{i=1,\cdots,3}$, where $\hat{\mathbf{x}}_{\mathbf{l}_0}$ is the corresponding point in the reference element.

To evaluate the coefficients of the equations relating the electric field and the corrector term ϕ, we need to know $\mathbf{J}_p$ and ρ_p, generated by the particle p, at the degrees of freedom of the fields. These terms are approximated in the spaces U_h and V_h, as follows:

$$\mathbf{J} \circ F_K = \sum_{i=1}^{3} \sum_{\mathbf{l} \in \{1,\cdots,r+1\}^3} J^i_{K,\mathbf{l}} DF_K^{*-1} \hat{\psi}_{l_1} \mathbf{e}^i, \rho \circ F_K = \sum_{\mathbf{l} \in \{1,\cdots,r+1\}^3} \rho_{K,\mathbf{l}} \hat{\psi}_{l_1} \quad (11)$$

To evaluate the terms $\mathbf{J}_{K,\mathbf{l}_0}$ and $\rho_{K,\mathbf{l}_0}$ in (11), we state that the particle p is associated to the nearest degree of freedom $\mathbf{l}_0$, in the element $K \in \mathscr{T}_h$ (see [7]).

3 Numerical Experiments

In this section, we show that our DG-PIC method is able to simulate a diode configuration. It corresponds to a particle device involving a field emission surface submitted to an large perpendicular electric field. Our numerical tests are done in two phases: first, we check the simulation of a coaxial line with the DG scheme (7), and then we introduce the particles to achieve the diode simulation. In all the simulations, we take a Q_2 approximation (order 2 in space) and an averaged spatial size of the cells equal to 0.01 m.

3.1 TEM Mode in a Coaxial Line

We apply our DG scheme to study a coaxial line where the 3D configuration is given by an inner radius r_0, an outer radius r_1 and a length L, respectively equal to 0.1 m, 0.2 m and 0.6 m. We impose a voltage generator $V(t)$ between the anode and the cathode and we put an absorbing boundary condition located at the extremities of the geometry to simulate an infinite line. The mathematical expression of the voltage

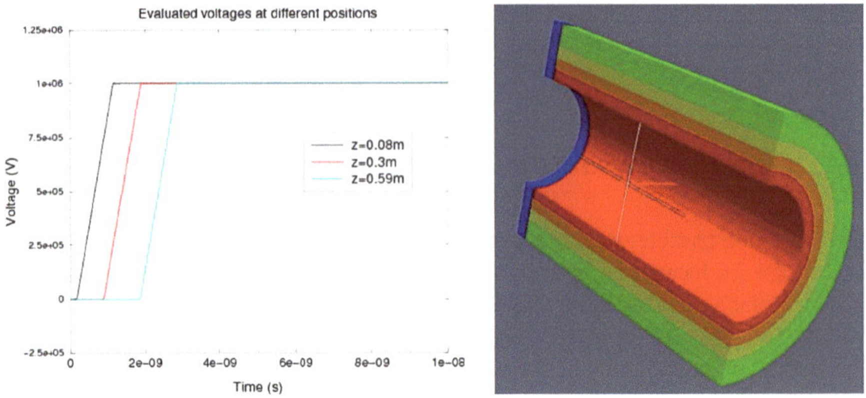

Fig. 1 Evaluated voltages at different positions on the coaxial line (*at left*) and evaluated electric field in the volume at $t = 10^{-8}$ s (*at right*). The colours represent the magnitude of the fields

generator is given by: $V(t) = \frac{V_0\,t}{10^{-9}}$ if $t < 10^{-9}$ s and $V(t) = V_0$ if $t > 10^{-9}$ s, with $V_0 = 4\,10^6$ V.

For this configuration, an analytical solution exists corresponding to a TEM mode propagating along the axis of the cylindre cavity (taken to be parallel to the z-coordinate). The transverse electromagnetic field distribution is given by $E(t,z,r,\theta) = (V(t-z/c))/(r\,Log(\frac{r_1}{r_0}))$ and $H(t,z,r,\theta) = (k \times E(t,z,r,\theta))/(Z_0)$, where t is the time, (z,r,θ) are the cylindre coordinates and $Z_0 = \sqrt{\frac{\mu_0}{\varepsilon_0}} \simeq 377\,\Omega$ is the vacuum impedance.

Figure 1 gives the voltages obtained at different locations on the coaxial line and the evaluated electric field in the volume at $t = 10^{-8}$ s. The results are in agreement with the analytic solution and allow us to validate our DG model on this configuration.

3.2 *Diode Configuration*

The diode is modelled by two finite cylinders with the same radii as in the previous configuration. The length of the inner cylinder is equal to 0.3 m whereas the outer one has radius 0.4 m. The source is given by the voltage generator given above. In the proposed configuration, we allow the inner cylinder (cathode) to emit particles on a surface, when the electric component field normal to this surface is higher than a breakdown value taken to be 2.5×10^7 V/m in this example. When a surface is authorised to emit, in its neighbouring volume, we introduce a random number of particles. The initial velocity and position of each particle is also taken randomly. From these positions, by using (5), we evaluate for each particle, its displacement through the different cells of the mesh to determine the final cell corresponding to its

Fig. 2 Positions of the particles obtained with the DG method at different times. The colour represents the energy of the particle

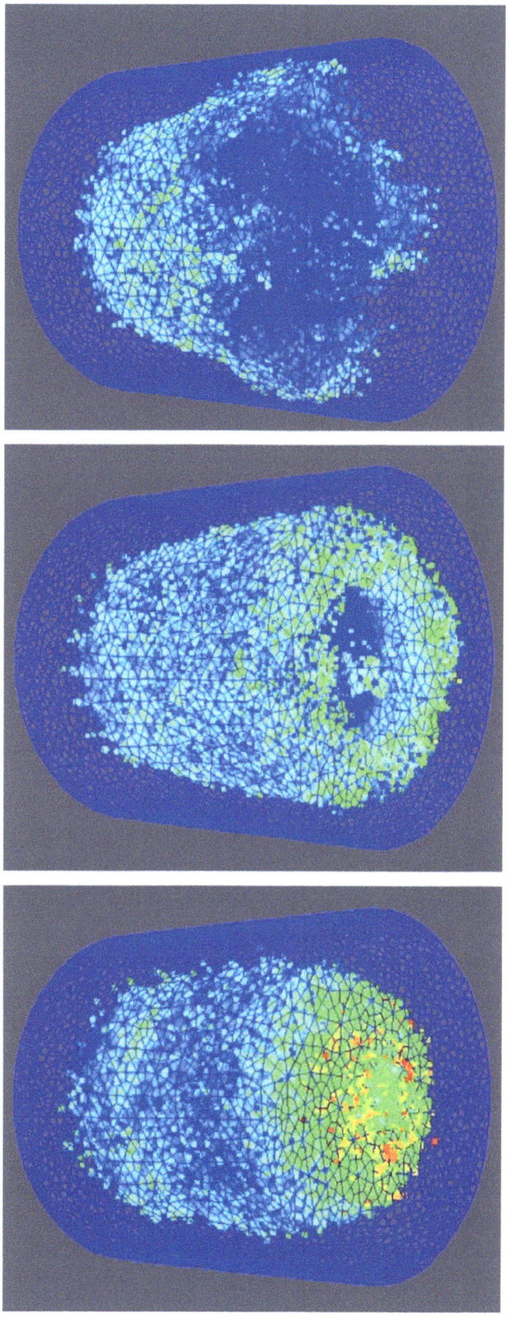

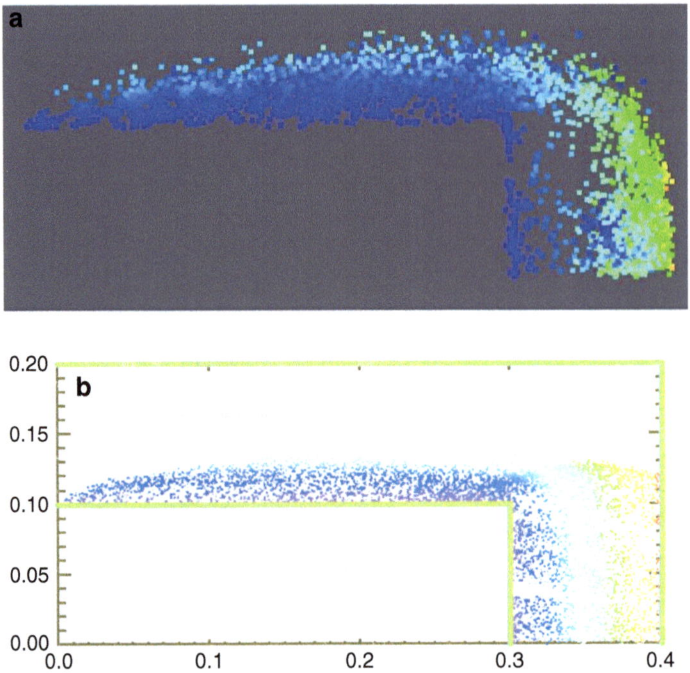

Fig. 3 Comparison of numerical results obtained for our 3D DG-PIC code (*at left*) and a 2D FDTD code (*at right*) at $t = 1.5 \times 10^{-8}$ s. The colour represents the energy of the particle

new position. More detail on the emission mechanism and the propagation process is given in [7].

Figure 2 shows the positions of the particles emitted from the cathode at the start, the middle and the end of the simulation. We observe on this figure that, first, the particles are attracted by the anode, then they are refocusing towards the cathode. This is the correct physical behaviour. Indeed, during the time, the movement of the particles generates current densities which decrease the electric field between the cathode and the anode. Finally, theses fields are not sufficient to allow the particles to reach the anode. In this 3D simulation, we have approximatively 1.6×10^6 particles at the end of the computation.

In Fig. 3, we compare, at $t = 1.5 \times 10^{-8}$ s, the positions of the particles emitted from the cathode obtained using our DG method and a 2D FDTD axisymmetric code, which can be considered as a good reference solution. We observe that, qualitatively, the shape of the two clouds of particles are similar, and thus that, our DG approach gives the correct physical behaviour. Now to compare quantitative results, we are interested in impedances values.

Figure 4 shows the values of input (at 0.03 m) and output (at 0.3 m) impedances evaluated with our 3D DG method (in red) and the 2D FDTD code (in black). The impedance values must be taken at a stable state, that is, when the currents and

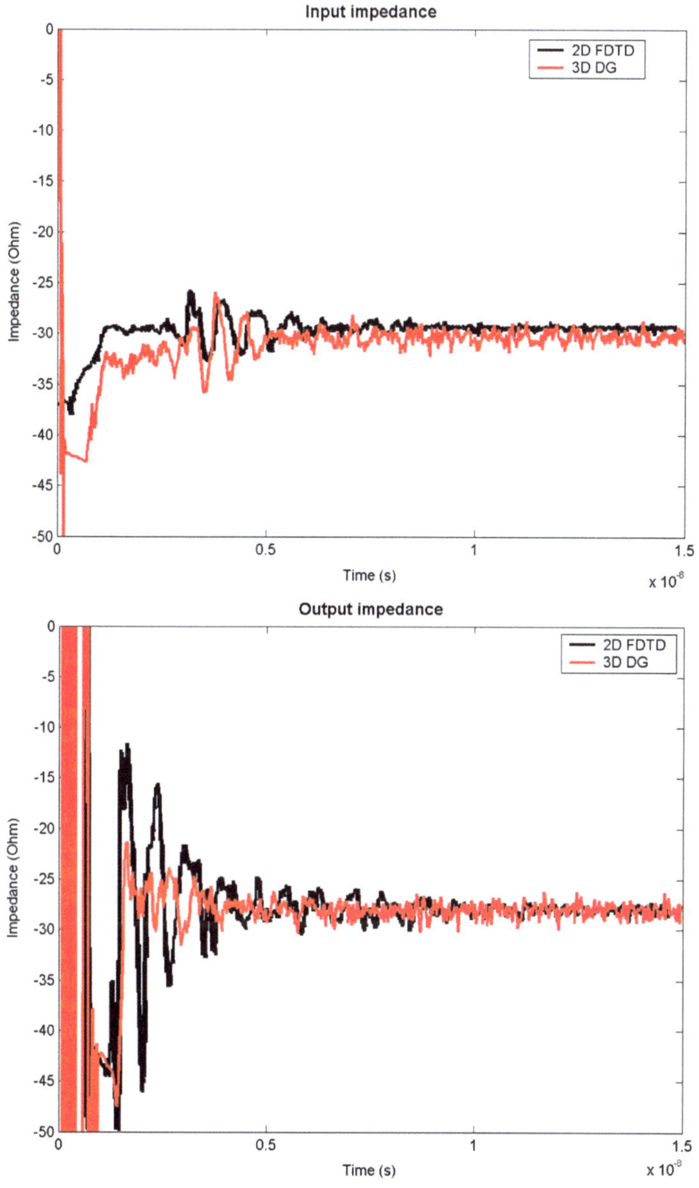

Fig. 4 Comparison of input (*at left*) and output (*at right*) impedances obtained with a 2D FDTD code (*in black*) and our 3D DG-PIC code (*in red*)

the voltages are constant in the time. In the figure, the correct impedance values correspond to the end of the curves. As we can see, the two solutions are very similar with an input impedance equal to 30 Ω and an output impedance equal to 27 Ω. Hence, we can consider that our 3D DG approach is satisfactory for this application.

4 Conclusion

In this paper, we have proposed a method for solving the Maxwell-Vlasov equations in the time domain using a DG method. The mathematical formulation and the approximation of the scheme have been described briefly and a numerical example illustrates the applicability. Today, the method has proved its efficiency for solving the Maxwell problem. Concerning the Maxwell-Vlasov system, we have shown that the method can be applied and gives satisfactory results. Considering its capacity to evaluate accurate fields with a low cost, we hope to obtain a fast and efficient method for solving the Maxwell-Vlasov equations, by taking few cells together with a high order spatial approximation. Currently, studies are in progress to evaluate the advantages of this method for solving the Maxwell-Vlasov problem in terms of CPU time, memory storage and accuracy.

Acknowledgement The main part of this study has been partly funded by CEA.

References

1. Birdsall, C.K., Langdon, A.B.: Plasma physics via computer simulation. Institute of Physics, Bristol (1991)
2. Cohen, G., Ferrieres, X., Pernet, S.: A spatial high-order hexahedral discontinuous Galerkin method to solve Maxwell's equations in time domain. J. Comput. Phys. **217**, 340–363 (2006)
3. Barucq, H., Delaurens, F., Hanouzet, B.: Method of absorbing boundary conditions: phenomena of error stabilization. SIAM J. Numer. Anal. **35**, 1113–1129 (1998)
4. Barthelmé, R.: Le problème de conservation de la charge dans le couplage des équations de Vlasov et de Maxwell. Ph.D. thesis, Louis Pasteur Université, Strasbourg (2005)
5. Munz, C. D., Schneider, R., Sonnendrücker, E., Voss, U.: Maxwell's equations when the charge conservation is not satisfied. Comptes Rendus de l'Académie des Sciences - Series I - Mathematics **328**, 431–436 (1999)
6. Rodriguez-Áros, Á., Rogier, F.: A Galerkin Discontinuous Method preserving The Gauss law in the numerical approximation of Maxwell's equations JSO MAHPSO, Dec., 7th 2009, Toulouse
7. Pebernet, L.:Etude d'un modèle PIC dans une approximation Galerkin Discontinue pour les équations de Maxwell-Vlasov. Recherche d'une solution hybride non conforme efficace. Ph.D. thesis, University Paul-Sabatier, Toulouse (2010)

Part III
Coupled Problems

Introduction

The present part is devoted to coupled problems appearing in electrical engineering. Coupled problems introduce new and complex problems which cannot always be solved by simply combining existing methods adapted to the subproblems. Most of the time, a comprehensive dedicated approach is needed to account for the interactions in the problem. In particular, the various subproblems can have their own time scale and, as a result, conflicting stability requirements to cope with. The papers in this part show recent progress in computational methods for solving multidisciplinary problems of industrial interest.

The first paper by F. Freschi and M. Repetto (an invited speaker at the conference) presents an overview of Tonti diagrams. These diagrams represent a fundamental duality between topological boundary relations on the one hand and differential relations on the other hand. For that reason, Tonti diagrams describe the basic nature of many different physical models. From a computational point of view, this underlying common structure makes it possible to establish a corresponding relation between finite dimensional algebraic topological operators and their duals, i.e., discretised differential operators. These concepts are illustrated by means of an induction heating problem including the nonlinear effects of temperature on the magnetic characteristics beyond the Curie point.

In their contribution, R. Appali et al. investigate soliton collision in biomembranes and nerves. Collision of solitons is an interesting phenomenon related to the stability of soliton solutions of nonlinear differential equations. The authors present simulations for pairs of solitons moving in opposite directions at the same velocity, demonstrating that these solitons collide elastically and produce small-amplitude noise travelling at higher velocity.

The paper by F. Denz, E. Gjonaj, and T. Weiland presents a combined experimental and numerical procedure for modelling zinc-oxide varistor based surge arresters. In a series of experiments, measurements on single-varistor disks exposed to two-millisecond current pulses are taken. Subsequently, the measured data is used to establish the nonlinear electro-thermal characteristics of the zinc-oxide under electrical stress. Using this information, an accurate finite element model with coupled thermal and electric fields can be constructed. This approach is applied to calculate the transient voltage and temperature distribution within a complete surge arrester unit.

The paper by M. Zubert et al. offers a new accurate behavioural static model of SiC Merged PiN Schottky (MPS) diodes. This model is dedicated to static and quasistatic electro-thermal simulations of MPS diodes for industrial applications. The model parameters were extracted using the Weighted Least Square (WLS) method for a few selected commercially available SiC MPS diodes. Additionally, the PSPICE Analogue Behavioural Model (ABM) implementation, the relevance of which has been statistically proven, is also presented. The thermal behaviour of the devices was taken into account using the lumped Cauer canonical networks extracted from electro-thermal measurements.

In the contribution by G. Alì et al. a dynamic iteration scheme is proposed for a coupled system of electric circuit and distributed semiconductor (pn-diode) model equations. The device is modelled using the drift-diffusion (DD) equations and the circuit by means of modified nodal analysis (MNA) equations. Analytic divergence and convergence results are verified numerically.

The contribution by S. Schöps, A. Bartel and H. De Gersem adresses multirate time integration of field/circuit coupled problems using Schur complements. When using distributed magnetoquasistatic field models as additional elements in electric circuit simulation, the field equations produce large symmetric linear systems that have to be solved. The naive coupling and solving (using direct solvers) is not always efficient, as the electric circuit is only coupled via coils, which are often only represented by a small subset of the unknowns. As such, the Schur complement approach is revisited. The method can be given a "physical" interpretation and it is shown that a heuristic for bypassing Newton iterations makes efficient multirate time integration for the field/circuit coupled model possible.

Tonti Diagrams and Algebraic Methods for the Solution of Coupled Problems

Fabio Freschi and Maurizio Repetto

Abstract Tonti diagrams highlight a common structure of several physical laws describing different phenomena. From a computational viewpoint, this underlying common structure allows to build topological operators (discrete counterpart of differential operators) only once, and they can be used to easily assemble the stiffness matrices and the coupling terms of the various problems. An application of this concept to the coupled electromagnetic-thermal problem of induction heating is presented in this work, taking into account the nonlinear effects of temperature on the magnetic characteristic beyond the Curie point.

1 Introduction

One of the most important ideas in the work of prof. Tonti and his algebraic formulation of physical theories [1], is the rigorous classification of physical variables and equations which characterize the mathematical description of a physical problem. As it will be detailed in the next section, this description is tied to a discretized space-time structure and this fact makes it suitable for its numerical implementation. In addition, this theoretical scheme is not specifically related to one particular physical phenomenon but, as a fundamental issue, highlights the common space-time structure underlying many physical theories. For this reason Tonti's work is a natural framework for developing a multiphysics numerical technique, where Tonti diagrams are a useful tool to display the functional relations between quantities which can be directly implemented in a numerical procedure which will be hereinafter called *Cell Method*, CM. After highlighting the form of Tonti diagrams and CM, the paper presents the application of the above mentioned

F. Freschi · M. Repetto (✉)
Dipartimento di Ingegneria Elettrica, Politecnico di Torino, Torino, Italy
e-mail: fabio.freschi@polito.it; maurizio.repetto@polito.it

B. Michielsen and J.-R. Poirier (eds.), *Scientific Computing in Electrical Engineering SCEE 2010*, Mathematics in Industry 16, DOI 10.1007/978-3-642-22453-9_21,

computational structure to couple, in a strong way, the electromagnetic and thermal fields in the case of induction heating of materials where coupling effects and nonlinearities present in the problem will be addressed by means of CM.

2 Tonti Diagrams and Cell Method

The theoretical work of prof. Tonti [2] is mainly directed to the classification of physical variables inside different physical theories. His attention has been devoted not only to the study of the structure, but also to its implementation in a numerical procedure for field analysis. As a complete description of Tonti's work, at least for electromagnetic theory, can be found in [1], here only the main concepts of the structure will be outlined. It must be remarked that many of the concepts that will be described are already present inside other numerical techniques for field solution, like finite volume method [3], finite difference time domain [4], finite integration technique [5], etc. but without a strict formal definition. As it will be clear in the next Sections, this rigorous definition of quantities will allow the set-up of a ready-to-implement numerical formulation of the problem.

2.1 Classification of Variables

A differential treatment of physical equations involves the use of variables which are defined in a point-wise fashion which enables the use of partial differential operators like gradient, curl and divergence. For example, the Laplace and Poisson equations are written in terms of these operators. If an analytical solution of a problem can not be found and its numerical analysis is needed, the solution equation must be discretized by some numerical technique, for instance finite element method, which translates the differential equation in a system of algebraic equations. In a different way, the use of domain related variables allows a direct expression of problem equations in algebraic form. Domain related variables, or *global variables*, GV, are instead associated to some elements of a space-time discretization. For example, in this way, there will be no use of point-wise magnetic flux density but of magnetic flux defined on an oriented surface. Domain variables are naturally associated to points, lines, surfaces and volumes of a space discretization. Crucial for the classification is the subdivision of variables in two classes: source variables, usually the causes of a physical field like heat production, electric charge, electric current, etc. and configuration variables which describe the state of a field distribution like temperature, electric potential, magnetic flux, etc.

2.2 Classification of Equations

Problem equations express constraints between the variables. Also these links can be subdivided in two categories: *topological* and *constitutive* equations. Topological equations translate in mathematical terms functional relations between variables, for instance a balance equation relating the production of some physical quantity inside a volume to what it is transferred to the exterior through the bounding surfaces. When expressed in terms of global variables on a discretized space-time structure, these equations do not depend on metric of space, for instance the balance equation above cited does not change if the volume is big or small, In addition, many physical theories share the same kind of topological equations.

It must be remarked that orientation of geometrical entities is also important: in a balance equation, a volume is related to its bounding surfaces, their orientation and the one of the volume possibly are not in agreement. In this case a sign must be attached, for instance, to some flux contributions on bounding surfaces. These orientation information are expressed as ± 1 coefficients. All these factors, called *incidence numbers* can be efficiently grouped together in matrices called *topological matrices.*

The second kind of equations is related to the specific physical problem and are linking variables by means of material properties and of metric information, they are called constitutive equations and link global variables of different kinds [6–8].

2.3 Space Discretization

The presence of two different kinds of GV is naturally coupled with the definition of two different sets of geometrical entities. The relation between these two sets is given by *duality*. In a space of D dimensions a duality links together a p dimensional entity of one mesh complex to a $D - p$ one of the other complex. One complex is called *primal* and the other *dual* and this last one is indicated by the $\sim$ sign. Topological equations are tied to one cell complex since they link one entity to its boundary, on the contrary constitutive equations express a connection between GV of different kinds and thus defined one on primal and the other on the dual complex. It must be remarked that *duality* relationship between geometrical objects does not imply their *orthogonality*, so that also duality concept on unstructured discretization can be used.

2.4 Tonti Diagrams

Tonti diagrams constitute an efficient classification tool that can arrange all the previous concepts: two columns are used for placing different geometrical entities, one for the primal and one for the dual complex and their associated variables. Vertical connections are of topological type linking entities to their boundaries, while horizontal connections link GV defined on primal to those on dual complex and are thus of constitutive type. While in Fig. 1 the Tonti diagram is making reference only to spatial entities, its structure can become more complex if also

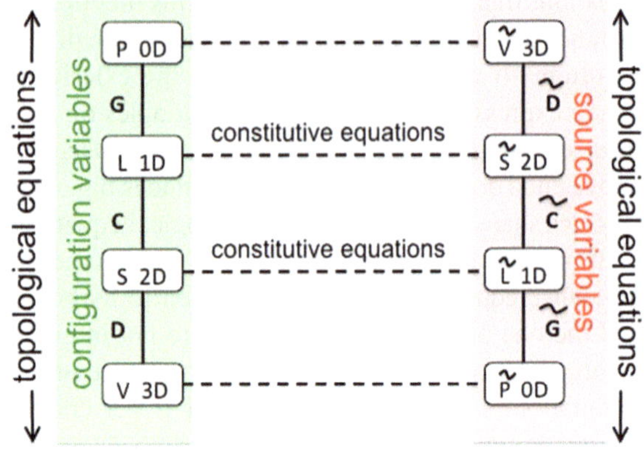

Fig. 1 Basic structure of Tonti diagrams with topological and constitutive constraints. G, C, D are topological matrices on primal complex as their dual counterparts $\tilde{G}$, $\tilde{C}$, $\tilde{D}$

time dimension is added. Also in this case discretization of time axis is applied so time entities become time instants t and time intervals τ. This aspect reflects in a doubling of each vertical column where one is referred to t and the other to τ. In addition, also duality in space discretization can be invoked with horizontal links connecting quantities defined on instants to ones defined on intervals.

3 Electromagnetic-Thermal Coupling

The problem of electromagnetic and thermal coupling arises in many industrial applications where induced eddy currents are used to heat up conductive workpieces. Different applications are possible from the melting of metals, usually under controlled conditions, to the local heating of material surface to obtain some particular surface characteristics like in the case of surface hardening. In this last case the problem is usually addressed by the analysis is the spatial distribution of the power transferred to and temperature of the workpiece which should be enough to cause martensitic transformation in the metallic structure. The analysis of the problem is difficult for the following reasons:

- There is a strong coupling between the electromagnetic and thermal problems because material characteristics, mainly electrical conductivity, is temperature dependent and so sharp changes in its value can be experienced during the heating up of the piece.
- Due to flux skin effect, the phenomenon is concentrated in a very thin layer of the material, usually less than some millimeters, and the thickness of this layer is temperature dependent on material characteristics.
- In case of ferromagnetic materials saturation effects are crucial to the magnetic flux density distribution, since a saturated material has an apparent magnetic permeability value lower than the one in the linear zone and this fact allows for a larger thickness of the flux penetration layer in the workpiece.

- In addition for ferromagnetic material, temperature is influencing also the material magnetic permeability value which decreases continuously with temperature but which experiences an abrupt change at the Curie point where magnetic behavior switches from the ferromagnetic to non-magnetic state [9].

Due to the mentioned difficulties, several approximated approaches have been proposed which neglect part of the phenomenon: one approximation can be introduced by the decoupling of the two phenomena: in electromagnetic solution material characteristics are considered to be independent on temperature and only magnetic nonlinear effects are considered. This approach can be considered valid to model the very first instants of the heating process [10].

Electromagnetic and thermal problems are coupled by material characteristics and sources: dependance of material characteristic, electrical and thermal conductivity and thermal capacity, on temperature; source of the thermal problem is the volume heat generation due to induced currents. The coupling is strict in the sense that, during the thermal transient, the variations of material characteristics cannot be neglected. The diagrams relevant to the two phenomena allows to highlight the topological and constitutive relations between variables. If the two phenomena are treated by the same spatial mesh, at least for the region subject to heating, most of the topological operators are common between them. For instance gradient matrix G is the same both for the equation relating the electric voltage to electric scalar potential $\mathrm{u} = \mathrm{G}\varphi$ and for the thermal equation $\Gamma = \mathrm{G}\theta$ relating thermal gradients to temperature.

Notwithstanding the similarities imposed in the space domain by the common structure of the equations, the peculiar characteristics of the two phenomena are different in terms of time-scale and in material behavior. Electromagnetic time-constants are at least two orders of magnitudes smaller than the thermal ones. Due to this fact, electromagnetic phenomena can be considered in steady state with respect to the variation of electrical conductivity versus temperature. At the same time nonlinear characteristic of the ferromagnetic material heavily influences the eddy current skin effect. Due to saturation of ferromagnetic material the skin-depth of eddy currents deviates from the one defined for linear materials and this aspect is crucial for a correct modeling of the phenomenon. As a matter of fact some approximations can be used to avoid a thorough nonlinear solution by means of equivalent material modeling. In this way ferromagnetic material is replaced by an equivalent linear non-homogeneous one allowing the use of time-harmonic formulation [10]. Time-decoupling allows thus the use of the most efficient and accurate formulation for each problem in time: approximated nonlinear time harmonic formulation for eddy currents [11] and time-stepping for the thermal conduction. Another physical law that increases difficulties in solving the nonlinearity, is the dependence of magnetic material properties on temperature: at the Curie temperature the ferromagnetic behavior of iron disappears and its magnetic permeability becomes equal to that of vacuum. This abrupt change is particularly important for the distribution of eddy currents inside iron due to the dependence of eddy current penetration on the permeability. The increase of skin

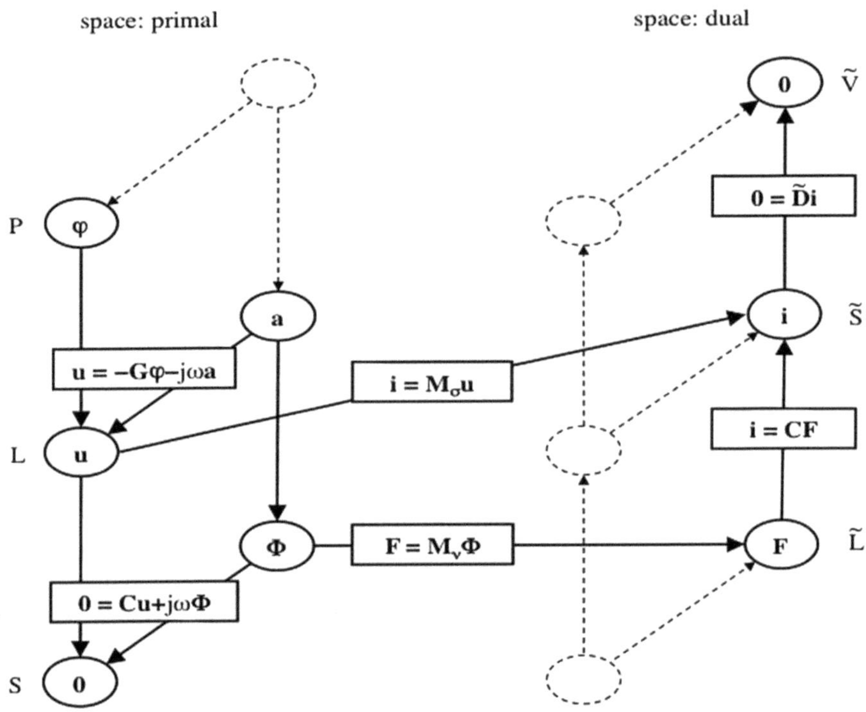

Fig. 2 The Tonti diagram for time-harmonic eddy current problem formulated in terms of magnetic vector and electric scalar potentials

depth at the Curie point influences also thermal generation because local values of current densities decrease and so the volume heat generation. This fact is responsible for a natural stabilization of temperature on the Curie value. With the previous assumptions, the source coupling term is related to the thermal production due to Joule losses. By making reference to quantities shown in Figs. 2 and 3, this term is computed starting by electrical currents i and voltages u evaluated respectively on dual and primal complex and transferring the resulting electrical energy W on the dual volume surrounding the primal node. A detailed analysis about this process can be found in [12].

Tonti diagrams for electromagnetic in time-harmonic formulation and for thermal flow in transient state are reported respectively in Figs. 2 and 3. The resultant equations are:

$$\begin{bmatrix} C^T M_\nu C + j\omega M_\sigma & M_\sigma G \\ j\omega G^T M_\sigma & G^T M_\sigma G \end{bmatrix} \begin{bmatrix} \underline{a} \\ \underline{\varphi} \end{bmatrix} \begin{bmatrix} i_S \\ 0 \end{bmatrix} \tag{1}$$

$$M_c \frac{d\theta}{dt} + G^T M_\lambda G\theta = W_{coupling} \tag{2}$$

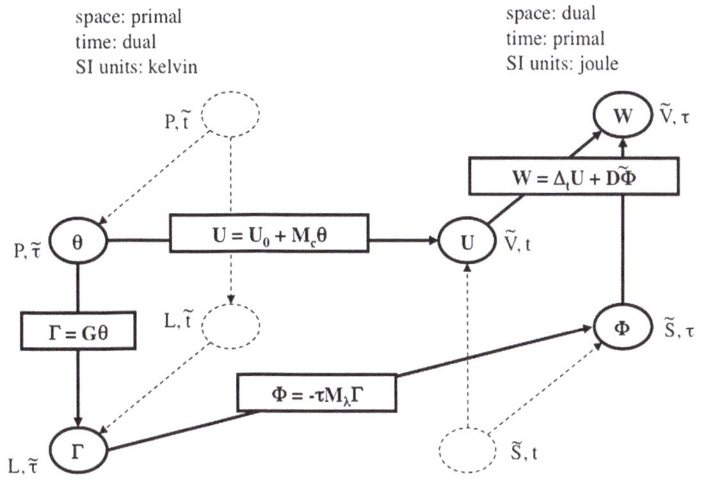

Fig. 3 The Tonti diagram for time-varying thermal conduction problem

Fig. 4 Time evolution of eddy currents along the thickness of a ferromagnetic material during transient evolution

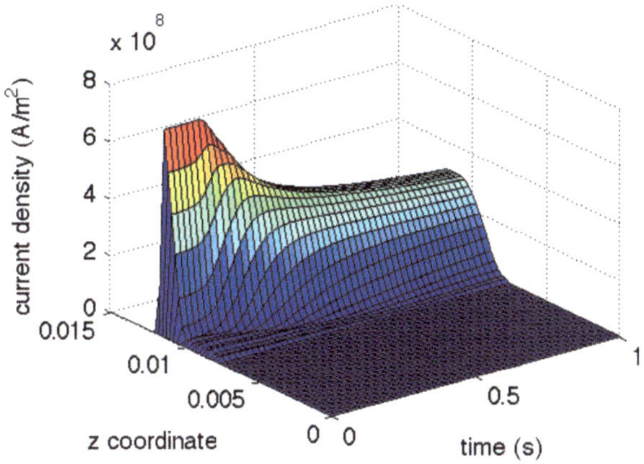

whereM_ν, M_σ are the constitutive magnetic and electric matrices for the eddy current formulation, whereas M_c, M_λ are the thermal capacity and thermal conductivity matrices. Unknowns of the problem are the magnetic vector and electric scalar potentials $\underline{a}$ and $\underline{\varphi}$ (in phasor domain) and the instantaneous temperature θ. The coupling term is the thermal energy generation $W_{coupling}$ inside dual volumes due to eddy currents.

The coupled approach is applied to a three dimensional problem of induction heating where a coil fed with medium frequency current (8 kHz) is heating up a slab of ferromagnetic material with Curie temperature value around 1,200 K. Details on geometry can be found in [10]. In Fig. 5, the variation of temperature on a line orthogonal to the material surface is reported. As it can be seen, the transition of material properties at the Curie temperature gives a upper limit to the temperature on the surface.

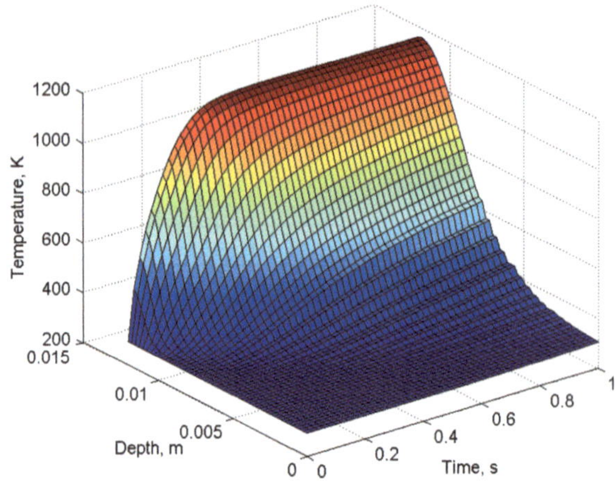

Fig. 5 Time evolution of temperature along the thickness of a ferromagnetic material during transient evolution

4 Conclusions

Multi-physics and coupled problems are nowadays of great interest due to the larger potentialities offered by computational power. Even if different kinds of coupling between numerical procedures are possible, the use of a common theoretical framework for all involved phenomena is of great help in a rigorous computation of coupling terms. In this viewpoint the work of Tonti has created a framework where different physical theories can live together on the same space-time discretization. Tonti diagrams, as a synthesis of the whole space-time discretization, are a valuable tool for a correct formulation tied to the physical nature of the problems. The translation of the abstract theoretical work in a computational procedure through the Cell Method seem thus to be a natural environment where coupled problems can be studied.

References

1. Tonti, E.: Finite formulation of electromagnetic field. IEEE Trans. Magn. **38**, 333–336 (2002)
2. Tonti, E.: Discrete physics. http://www.dic.units.it/perspage/discretephysics/
3. Eymard, R., Gallouët, T., Herbin, R.: Finite volume methods. In: Ciarlet, P., Lions, J. (eds.) Handbook of Numerical Analysis, vol. VII, pp. 713–1020. North-Holland/Elsevier, Amsterdam (2000)
4. Yee, K.: Numerical solution of boundary value problems involving Maxwell equations in isotropic media. IEEE Trans. Antenn. Propag. **AP-14**, 302–307 (1966)
5. Weiland, T.: A discretization method for the solution of Maxwell's equations for six-components fields. Electron. Comm. AEUE **31**, 116–120 (1977)
6. Repetto, M., Trevisan, F.: Global formulation for 3D magneto-static using flux and gauged potential approaches. Int. J. Numer. Meth. Eng. **60**, 755–772 (2004)
7. Specogna, R., Trevisan, F.: Discrete constitutive equations in $a - \chi$ geometric eddy-current formulation. IEEE Trans. Magn. **41**(4), 1259–1263 (2005)

8. Codecasa, L., Specogna, R., Trevisan, F.: Symmetric positive-definite constitutive matrices for discrete eddy-current problems. IEEE Trans. Magn. **43**(2), 510–515 (2007)
9. Aliferov, A., Dughiero, F., Forzan, M.: Coupled magneto-thermal FEM model of direct heating of ferromagnetic bended tubes. IEEE Trans. Magn. **46**(8), 3217–3220 (2010)
10. Canova, A., Dughiero, F., Fasolo, F., Forzan, M., Freschi, F., Giaccone, L., Repetto, M.: Simplified approach for 3d nonlinear induction heating problems. IEEE Trans. Magn. **45**(3), 1855–1858 (2009)
11. Du Terrail, Y., Sabonnadiere, J., Masse, P., Coulomb, J.: Nonlinear complex finite elements analysis of electromagnetic field in steady-state AC devices. IEEE Trans. Magn. **20**(4), 549–552 (1984)
12. Alotto, P., Freschi, F., Repetto, M.: Multiphysics problems via the cell method: The role of tonti diagrams. IEEE Trans. Magn. **46**(8), 2959–2962 (2010)

Soliton Collision in Biomembranes and Nerves- A Stability Study

Revathi Appali, Benny Lautrup, Thomas Heimburg, and Ursula van Rienen

Abstract Collision of moving solitons is an interesting phenomena which is closely related to the stability of solitons. We study the head-on collision of solitons in a recently introduced model for biomembranes and nerves. We conduct simulations for pairs of solitons moving in opposite directions with the same velocity. It is found that these stable solitons collide elastically and it results a small amplitude noise traveling with higher velocity. We have also examined the energy loss of the solitons after collision.

1 Introduction

The functional success of electrically stimulated brain implants eg. Deep Brain Stimulation (DBS) depends on the basic understanding of signal propagation in the nerve cells. Mathematical models of pulse propagation in these cells play a major role in further investigation of the interaction of these nerve cells with the electrodes. One such mathematical description of the nerve pulse propagation is "soliton model". Soliton model is based on the propagation of a localized density wave in the axon membrane [1, 3]. The important requirement of the model is the empirically known lipid phase transitions slightly below the physiological temperatures. Soliton models predict the exact pulse propagation velocities in myelinated nerves. The propagation velocities are closely related to the lateral sound velocities in the nerve membrane [1]. During compression, the appearance of a voltage pulse seems to be

R. Appali · U. van Rienen (✉)
Institute of General Electrical Engineering, University of Rostock, Justus-von-Liebig-Weg 2
18059 Rostock, Germany,
e-mail: revathi.appali@uni-rostock.de, ursula.van-rienen@uni-rostock.de

B. Lautrup · T. Heimburg
Niels Bohr Institute, Blegdamsvej 17, DK-2100, Copenhagen, Denmark
e-mail: lautrup@nbi.dk, theimbu@nbi.dk

B. Michielsen and J.-R. Poirier (eds.), *Scientific Computing in Electrical Engineering SCEE 2010*, Mathematics in Industry 16, DOI 10.1007/978-3-642-22453-9_22,

a consequence of the piezo-electric nature of partially charged and asymmetric cell membrane [2]. Moreover, the soliton model explains the reversible temperature and heat exchanges observed in connection with the nerve pulse. Another advantage of a soliton-based description of pulse propagation in nerves is its predictive power [1].

Lautrup et al. demonstrated that the soliton[1] solutions are stable with respect to small amplitude fluctuations and robust in the presence of dissipation. This shows that the solitons can propagate under realistic physiological conditions over the length scales of nerves (upto several meters eg., sciatic nerve in human) even in the presence of friction and lateral heterogeneities [3]. In this paper, we examined the stability of the solitons with the help of collision studies, which was not considered in reference [3]. In the following section, we will discuss the model from [3].

2 Soliton Model

The nerve pulse propagation in a myelinated nerve can be described by (3)

$$\frac{\partial^2}{\partial \tau^2}\Delta\rho^A = \frac{\partial}{\partial z}\left[\left(c_0^2 + p\Delta\rho^A + q\left(\Delta\rho^A\right)^2 + ...\right)\frac{\partial}{\partial z}\Delta\rho^A - h\frac{\partial^4}{\partial z^4}\Delta\rho^A\right] \quad (1)$$

Here,

- $\Delta\rho^A$ is the change in lateral density of the membrane $\Delta\rho^A = \rho^A - \rho_0^A$.
- ρ^A is the lateral density of the membrane.
- ρ_0^A is the equilibrium lateral density.
- c_0 is the velocity of small amplitude sound.
- p and q are the parameters determined from sound velocity and density dependence.
- h is the parameter to set the linear scale of the propagating pulse.

The empirical equilibrium value of ρ_0^A is 4.035 10^{-3} g/m^2 and the low frequency sound velocity c_0 is 176.6 m/s. The coefficients p and q were fitted to measured values of the sound velocity as a function of density.

We work with the dimensionless variables u, x and t defined in [3] as

$$u = \frac{\Delta\rho^A}{\rho_0^A} \qquad x = \frac{c_0}{h}z \qquad t = \frac{c_0^2}{\sqrt{h}}\tau \qquad B_1 = \frac{\rho_0}{c_0^2}p \qquad B_2 = \frac{\rho_0^2}{c_0^2}q \quad (2)$$

Equation (1) takes the following form with these variables

[1]We use the term "soliton" synonymous to "solitary wave". Since the localized solutions pass through each other and dissipate some energy, which is not the case for genuine solitons.

$$\frac{\partial^2 u}{\partial t^2} = \frac{\partial}{\partial x}\left(B(u)\right)\frac{\partial u}{\partial x} - \frac{\partial^4 u}{\partial x^4} \tag{3}$$

with

$$B(u) = 1 + B_1 u + B_2 u^2 \tag{4}$$

Here the parameter values are chosen as $B_1 = -16.6$, $B_2 = 79.5$ from [3]. We consider u as a function of $\xi = x - \beta t$ as in [2].

$$\beta^2 \frac{\partial^2 u}{\partial \xi^2} = \frac{\partial}{\partial \xi}\left(B(u)\frac{\partial u}{\partial \xi}\right) - \frac{\partial^4 u}{\partial \xi^4} \tag{5}$$

Equation (4) is known to have exponentially localized solitonic solutions which propagate without distortion for a finite range of sub-sonic velocities [3].

2.1 Analytical Solution

Localized solitonic solutions of (5) are given by (as in [3])

$$u(\xi) = \frac{2a_+a_-}{(a_+ + a_-) + (a_+ - a_-)cosh(\xi\sqrt{1-\beta^2})} \tag{6}$$

where $u = a_\pm$ are the real roots of the right hand side of the integrated equation, for the velocity range $\beta_0 <| \beta |< 1$ (Fig. 1).

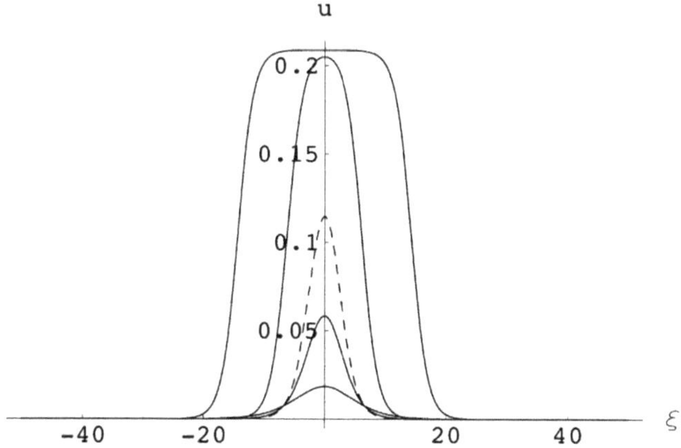

Fig. 1 Soliton profiles for velocities for $\beta = 0.95, 0.85, 0.734671, 0.65$ and $\beta_0 + 4 \times 10^{-9}$ respectively from bottom to up. Adapted from [3]

$$a_{\pm} = \frac{-B_1}{B_2}\left(1 \pm \sqrt{\frac{\beta^2 - \beta_0^2}{1 - \beta_0^2}}\right) \quad . \tag{7}$$

- The amplitude of the soliton decreases with the velocity β.
- The width of the soliton diverges for $\beta \to \beta_0$ and $\beta \to 1$.
- The soliton has a minimum width at $\beta = 0.734671$, shown in dashed line.

2.2 *Numerical Analysis*

To investigate the questions concerning the stability of the solitons of (6), B. Lautrup et.al have considered the model numerically in [3]. In this contribution, the stability of the solitonic solutions for infinitesimal perturbations was carried out along with the effect of dissipation on the soliton propagation. The model, as a system of two first order partial differential equations as mentioned in the reference [3] is used for our numerical consideration (see (8)).

$$\frac{\partial u}{\partial t} = \frac{\partial v}{\partial x} \qquad \frac{\partial v}{\partial t} = \frac{\partial f}{\partial x} \tag{8}$$

with

$$f = u + \frac{1}{2}B_1 u + \frac{1}{3}B_2 u^2 - \frac{\partial w}{\partial x} \quad and \;\; w = \frac{\partial u}{\partial x} \tag{9}$$

To realize the soliton propagation, the model in the above form was solved numerically with Finite Element Method (FEM) in COMSOL Multiphysics 3.5a®. The general form in the classical partial differential equation (PDE) mode of COMSOL Multiphysics® was employed with periodic boundary conditions. The analytical solution of $\beta = 0.734671$, "minimum width" was chosen as initial condition. Dispersion was found in the solution. The energy of the soliton was found to decrease during the propagation. The algorithm in COMSOL docs not yield full numerical stability (Fig. 2).

The stable numerical solution of (7) can be obtained by using a variant of the two-step Lax-Wendroff method as described in [3]. This was executed in C++ [4] and Mathematica® by the authors of [3] and the same is executed here for collision studies.

3 Collision Studies

We have investigated the head-on collision of two pulses with identical amplitudes and opposite velocities. It is known that pulses are blocked upon collision [5]. The FitzHugh-Nagumo model [6, 7], which is a simplified form of the Hodgkin-Huxley model [8], allows for both the cancellation and penetration of pulses depending

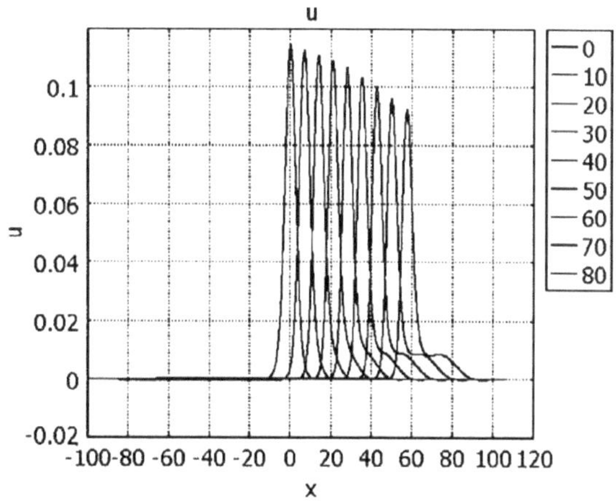

Fig. 2 Propagation of minimal width soliton. PDE solved with time stepping $\Delta t = 0.001$ and $\Delta x = 0.1$. The shape of the pulse is not conserved by the numerical algorithm in FEM based Comsol Multiphysics 3.5a®. Length of the periodic lattice has been increased here to depict the change of soliton shape during the propagation

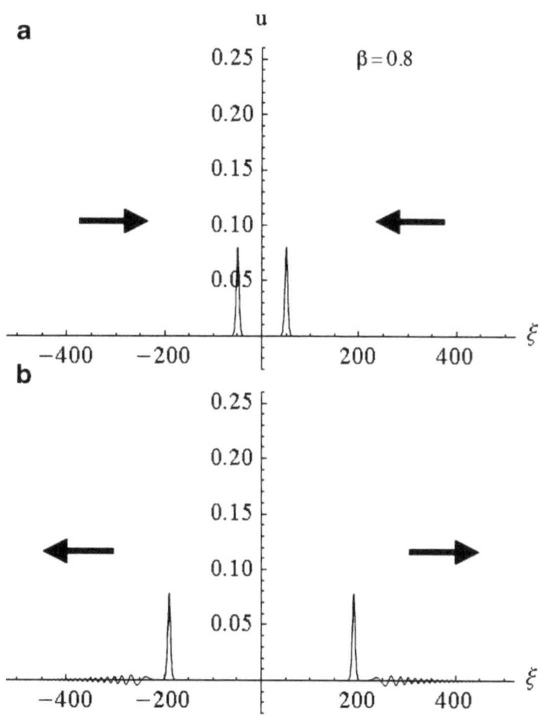

Fig. 3 Collision of two solitons before (**a**) and after collision (**b**) shown for $\beta = 0.8$. Small amplitude noise travelling ahead of the post-collision pulses for $\beta = 0.8$ that carries a very small fraction of the overall energy is obtained. The same was achieved for solitons of different velocity and amplitude

on parameters [9]. Since the soliton model is based on adiabatic and reversible physics without dissipation [10], here we have investigated collisions in the absence of friction. Figure 3 shows two identical solitons with $\beta = 0.8$ before and after collision. Small amplitude noise travelled ahead of the post-collision pulses with a very small energy in the order of $\ll 1\%$ compared to that of the solitons . The same was found for solitary pulses with different velocities and amplitudes.

The functional dependency of the sound velocity on density was given by (4). It represents a quadratic approximation to the experimental data and yields a satisfactory description in the density regime between solid and liquid membrane state [9]. However, when large amplitude pulses collide, (4) allows the density transiently to exceed the density of the solid phase ($u \approx 0.25$). Considering this as unphysical, a "soft barrier" at the density of the solid phase is introduced in (4):

$$B(u) = (1 + B_1(u) + B_2(u^2))(1 + e^{100(x-0.26)}) \tag{10}$$

This modification of (4) is only relevant at the moment for the collision of two large amplitude solitons. The result of such a collision of two solitons with β close to the minimum velocity β_0, given by (7), is shown in Fig.4. The soliton fell apart to a sequence of solitons and some additional low amplitude noise. This effect pronounces with the velocity closer to minimum velocity. Such decomposition into several pulses was not seen in the absence of the soft barrier. We compared the largest pulse energy after the collision to the energy before collision (Fig. 5).

The energy density of a soliton has both potential and kinetic energy contributions and can be calculated by using a Lagrangian formalism. (Adapted from [10])

$$e = \frac{c_0{}^2}{\rho_0{}^A}(\Delta\rho^A)^2 + \frac{p}{3\rho_0{}^A}(\Delta\rho^A)^3 + \frac{q}{6\rho_0{}^A}(\Delta\rho^A)^4 \tag{11}$$

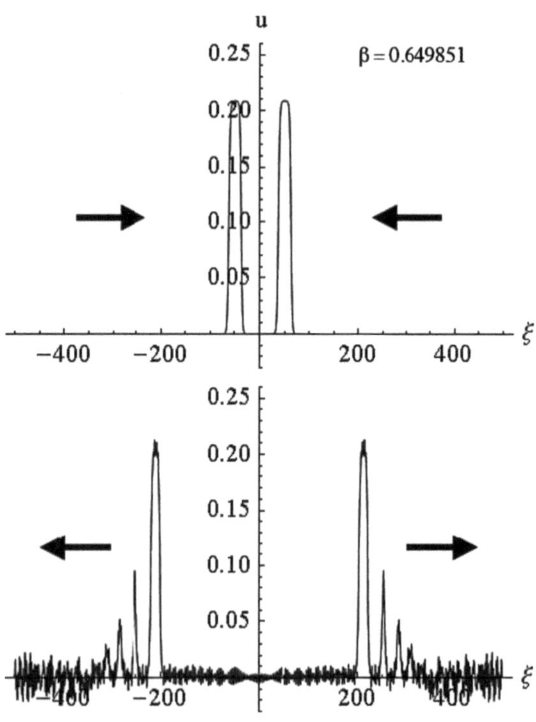

Fig. 4 Collision of two solitons before (**a**) and after collision (**b**) for $\beta = 0.649850822$ (close to maximum amplitude) and the additional condition of a maximum density change of $u = 0.25$. The pulse falls apart into several solitary pulses with different amplitude and velocity, and some small amplitude noise

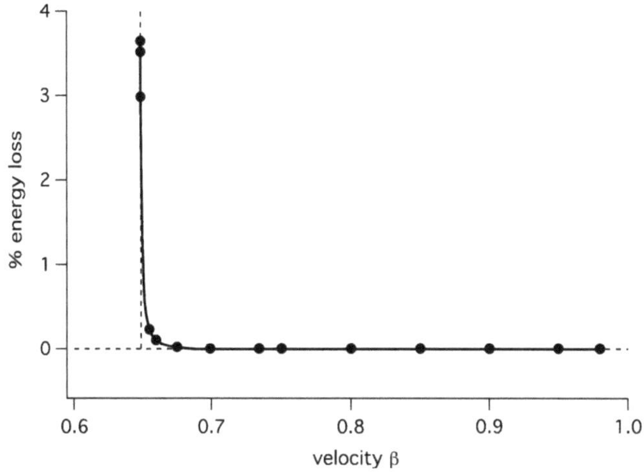

Fig. 5 Energy loss of soliton after collision in %. The energy content of the largest pulse after collision with the pulse before the collision are compared. Only when the pulses reach their maximum amplitude and minimum velocity, dissipation becomes significant

Even for the near-limiting case the fraction of energy lost into smaller amplitude solitons and small amplitude noise is $<4\%$ for the most extreme case studied. Thus, we observed most of the energy of the major soliton was conserved in collisions even after a maximum density was enforced.

4 Conclusion

The soliton model of nerve pulse propagation with the modified Good-Boussinesq equation [11] is explained. The analytic form of the solitons is given in Sect. 2.1. We moved on to numerical analysis of the model Sect. 2.2 to realize the solitary propagation (with periodic boundary conditions) using FEM based software Comsol Multiphysics®. Unexpectedly, the numerical solution of the PDE was discrepant from the analytical solution given in [3]. The pulse amplitude was found to be decreasing during the propagation. This can be attributed to an inherent problem of numerical dispersion in the software of Comsol Multiphysics®. Simulations were then carried out in C++ and Mathematica® to self-implement the numerical method and to obtain energy-loss less soliton propagation. Finally, the stability of the model is then tested with the aid of collision studies. In the context of our model, pulses pass through each other "almost undisturbed" with the generation of only small amounts of small amplitude noise. If a maximum density is introduced, as seems reasonable for the crystalline lipid matrix, large amplitude solitons can decay into a series of solitons. However, even under these extreme conditions, the bulk of the energy remains in the maximum amplitude soliton. Our model does not offer a description of the cancellation of pulses as suggested in other models but opens up a new possibility of passing through almost undisturbed and conserving the maximum energy even upon maximum density enforcement.

Acknowledgements We thank DFG for funding our project "welisa" under GRK 1505/1. Thanks to Andrew. D. Jackson for the meaningful discussions and to Kiran Kumar Sriperumbudur for his thoughtful suggestions.

References

1. Andersen, S., Jackson, A.D., Heimburg, T.: Towards a thermodynamic theory of nerve pulse propagation. Progr. Neurobiol. **88**, 104–113 (2009)
2. Heimburg, T.: Physical properties of biological membranes. In: Bohr, H. (ed.) Encyclopedia of Applied Biophysics, pp. 593–616. Wiley-VCH, Weinheim (2009)
3. Lautrup, B.,Jackson, A.D., Heimburg, T.: The stability of solitons in biomembranes and nerves. physics.bio-ph, arXiv: physics/0510106v1, 2008.
4. Press, W.H., Teukolsky, S.A., Vetterling, W.T., Flannery, B.P.: Numerical recipes in C, 2nd edn. Cambridge University Press, U.K. (1994)
5. Tasaki, I.: Collision of two nerve impulses in the nerve fiber. Biochem. Biophys. Acta. **3**, 494–497 (1949)
6. FitzHugh, R.: Impulses and physiological states in theoretical models of nerve membrane. Biophys. J. **1**, 445–466 (1961)
7. Nagumo, J., Arimoto, S., Yoshizawa, S.: An active pulse transmission line simulating nerve axon. Proc IRE. **50**, 2061–2070 (1962)
8. Hodgkin, A.L., Huxley, A.F.: A quantitative description of membrane current and its application to conduction and excitation in nerve. J. Physiol. **117**(4), 500–544 (1952)
9. Argentina, M., Coullet, P., Krinsky, V.: Head-on collision of waves in an excitable Fitzhugh-Nagumo system:a transition from wave annihilation to classical wave behavior. J. Theor. Biol. **205**, 47–52 (2000)
10. Heimburg, T., Jackson, A.D.: On soliton propagation in biomembranes and nerves. Proc. Natl. Acad. Sci. USA **102**, 9790–9795 (2005)
11. Manoranjan, V.S., Ortega, T., Sanz-Serna, J.M.: Soliton and antisoliton interactions in the good Boussinesq equation. J. Math. Phys. **29**(9), 1964–1968 (1988)

Nonlinear Characterization and Simulation of Zinc-Oxide Surge Arresters

Frank Denz, Erion Gjonaj, and Thomas Weiland

Abstract A combined experimental and numerical procedure to model zinc-oxide varistor based surge arresters is presented. In a series of experiments, measurements on single varistor disks exposed to two millisecond current pulses are taken. Thereafter, the measurement data are used to establish the nonlinear electro-thermal characteristics of the ZnO ceramics under electrical stress. Using this information, an accurate finite element model with coupled thermal and electric fields can be constructed. This approach is applied to calculate the transient voltage and temperature distribution within a complete surge arrester unit.

1 Introduction

Metal-oxide(MO) varistors are commonly used as active components in high-voltage surge arresters to protect power lines from lightning or switching over-voltages. The geometry of a high-voltage MO surge arrester is generally simple. Along a vertical axis a number of varistor disks, which are typically made of zinc-oxide ceramics, are stacked up, surrounded by a housing of insulating porcelain or polymer material. The overall behavior of the surge arrester depends mostly on the dynamic response of these varistors upon electrical stress as well as on the geometrical layout of the device, which determines the capacitive coupling of the single varistors to the environment.

Two important criteria are usually considered in the design of MO surge arresters. The first criterion is the limitation of the non-uniformity of the voltage distribution along the arrester column. Non-uniform voltage grading is a major concern, as it affects the physical degradation of the varistor disks inside a surge arrester

F. Denz (✉) · E. Gjonaj · T. Weiland
Technische Universität Darmstadt, Institut für Theorie Elektromagnetischer Felder, Schloßgartenstr. 8, 64285 Darmstadt, Germany
e-mail: denz@temf.tu-darmstadt.de

B. Michielsen and J.-R. Poirier (eds.), *Scientific Computing in Electrical Engineering SCEE 2010*, Mathematics in Industry 16, DOI 10.1007/978-3-642-22453-9_23,

negatively [1,2]. The second criterion is the thermal stability of the arrester. Thermal stability is largely determined by the heating behavior of the ZnO elements, when the electric energy of a surge is dissipated. The electrical and thermal phenomena are not independent from each other. Electric fields and temperature are mutually connected. The temperature distribution inside of an arrester influences the electric fields. Equally, the temperature distribution between the individual ZnO elements, both under normal operation as under transient overvoltages, influences the voltage profile along the arrester. Therefore, the design process for surge arresters requires a highly complex analysis, which should imperatively include the mutual coupling between electrical and thermal phenomena.

The analysis of surge arresters is traditionally based on circuit models with empirically determined parameters, e.g., [3]. A more accurate approach is the transient finite element analysis considering the geometry of the complete arrester [4]. This type of analysis, however, suffers in particular from deficient knowledge about the electro-thermal characteristics of the varistor material. The ZnO ceramics is characterized by an electrical conductivity which varies by many orders of magnitude with applied voltage. In addition, the electrical conductivity is very sensitive to temperature. Another material parameter which has some influence on the validity of the simulations is the temperature-dependent heat capacity of ZnO. These data are partially provided by the manufacturers, for example in the form of UI-characteristics of the varistor, but they are insufficient for reliable simulations. The numerically calculated voltage data for a single varistor disk differ largely from measurement data, when manufacturer-provided characteristics are used. This indicates that a more accurate electro-thermal characterization of the ZnO material is necessary before large-scale surge arrester simulations can be performed.

A combined experimental and numerical procedure for the analysis of surge arresters is proposed. The basic idea is to obtain the temperature-dependent varistor characteristics from few direct measurements of the residual voltage for single varistor disks exposed to appropriate current pulses. The approach used for extracting the nonlinear varistor parameters from measurement data is described in Sect. 2.2. In Sect. 3.1, the numerical procedure for coupled electro-thermal simulations incorporating these parameters is introduced. Numerical simulation results for a single varistor disk as well as for a complete surge arrester unit are presented in Sects. 3.2 and 3.3, respectively.

2 Characterization of the Varistor

This section is composed of two parts. In the first one, the procedure to obtain measurement data for recent zinc-oxide varistors is explained. In the second part, the treatment of the data to derive an estimation of electrical conductivity will be described.

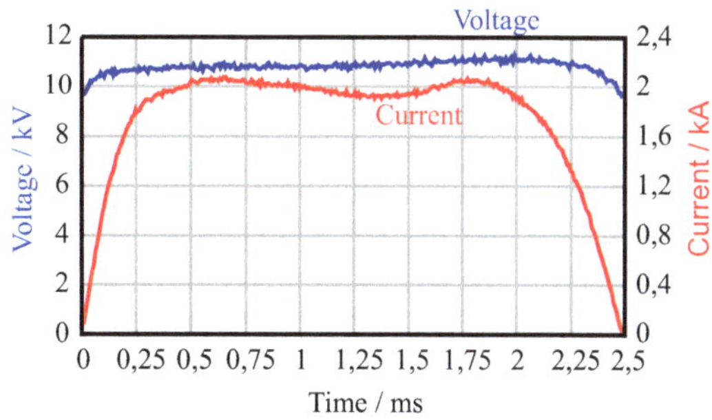

Fig. 1 Applied current impulse and measured voltage profile for the ZnO disks at 20°C

2.1 *Measurement Procedure*

A series of measurements was done in the impulse current lab of TU Darmstadt, to characterize the electro-thermal behavior of ZnO varistors. Six of the available varistors, which had the same type and dimensions (radius $\approx$ 3 cm, height $\approx$ 3.5 cm) and an almost identical rated voltage, were identified. They were heated in an oven until their temperature reached 300°C, before they were taken out. At different temperatures the individual varistors were subjected to a long-duration current impulse with a virtual peak time of 2 ms. The shape of this current impulse and the corresponding voltage for one of these measurements is shown in Fig. 1. Please note the temporary decrease of current, while voltage continues to rise, and the occurrence of different voltages for the same current, which suggests that the rise of temperature affects conductivity.

The test setup corresponds to the setups used for measurements according to the international standards by IEEE [5] and IEC [6]. The examined varistors were placed under pressure between two aluminum electrodes. The bottom electrode was connected with ground, while an impulse current was injected through the top electrode. The aluminum electrodes were replaced after each measurement to guarantee comparable measurement conditions. In this way, problems with the mechanical degradation of the electrodes were avoided and the temperature of the contact electrode is known, which would allow a repetition of the experiment. During the measurements voltage and the electric current flowing through the varistor were recorded for the estimation of electrical conductivity.

2.2 *Extraction of Material Parameters*

Four material parameters have an effect on the electro-thermally coupled simulations of surge arresters, which are: electrical conductivity, electrical permittivity, thermal conductivity and volumetric heat capacity.

Electrical conductivity is the most relevant parameter, since it determines the behavior of the zinc-oxide varistors. Unfortunately, the uncertainty about its value is very large, even though it is essential to know with relative accuracy the conductivity in the nonlinear region and how temperature affects it to simulate varistor devices successfully.

The existing and available material curves for electrical conductivity were insufficient for the planned numerical simulations. First, they provided no information about the dependence on temperature. However, temperature is critical for any explanation of the varistor behavior during current impulses, when their temperature rises significantly. Second, the curves were inaccurate in the nonlinear region. Such material curves are obtained by interpolation between data points, which are determined by a variety of methods. For low electric field strengths some data points are obtained from DC or AC measurements, while different impulse currents provide data points for high electric field strength. The strongly nonlinear region in between, which is critical for the numerical simulations, is generally of small or no interest. In consequence, there are few or no data points and the interpolation is poor, even though conductivity increases by several orders of magnitude. For these two reasons, it became necessary to extract the material parameters from our own measurements.

The other material parameters were set to values obtained from various sources. For the extraction of electrical conductivity, it is also necessary to know the volumetric heat capacity. Almost 30 years ago, Lat [7] had shown experimentally the existence of an approximately quadratic relationship between thermal energy and temperature for varistors from which heat capacity was derived.

Since no better values for electrical permittivity were available, a value obtained from low-frequency measurements was used ($\epsilon_r = 800$). Fortunately, the results for the two millisecond-impulse do not vary much with permittivity. For heat conductivity a value of 26 W/Km was assumed.

In the following, the method to establish a functional relationship between electrical conductivity, field strength and temperature is detailed. The measurements, which were described in the previous section, provided voltage-time and current-time curves for several different initial temperatures. Assuming that the field and current were homogeneous inside the varistor, current density and electric field strength can be obtained for every sampling point by dividing through the surface area or height of the varistor.

In the next step, it is assumed that the electric conductivity corresponds to the ratio of current density and field strength for any set of data i.

$$\sigma_i = \frac{J_i}{E_i} \tag{1}$$

This is only valid, if the second part in the following equation relating to the displacement current is small in comparison to the ohmic current.

$$J_i = \sigma_i E_i + \frac{\partial}{\partial t} (\varepsilon E_i) \tag{2}$$

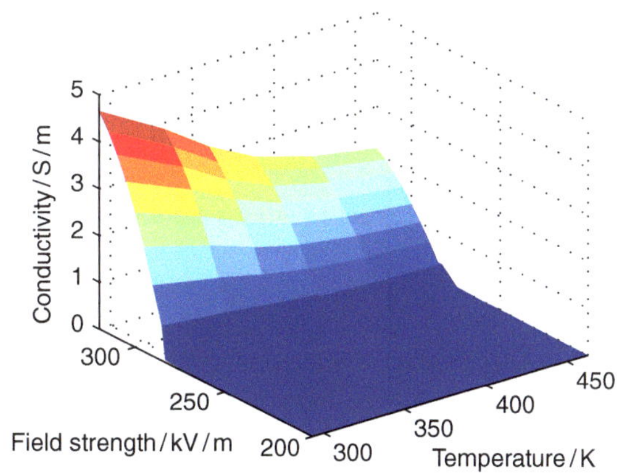

Fig. 2 Extracted electrical conductivity characteristics as a function of temperature and field strength of the ZnO varistor

Except at the very beginning and at the end, this is the case for the given current impulse.

Besides values for electrical conductivity and field strength, an approximate value of temperature is needed for every data point. Temperature could not be measured together with voltage and current, but it is possible to calculate the temperature at any moment from initial temperature of the individual measurement, ohmic losses and volumetric heat capacity. The following equation was solved numerically:

$$T(t) = T_0 + \int_{\tau=0}^{t} \frac{J(\tau)E(\tau)}{c_v(T(\tau))} \mathrm{d}\tau \tag{3}$$

Now, electrical conductivity, field strength and temperature are known approximately for each sampling point. By multivariate regression an adequate model for the relationship $\sigma = \sigma(E, T)$ was sought. Estimating the logarithm of conductivity instead or additionally is certainly preferable to a direct estimation, particularly to obtain satisfactory estimated values for lower field strengths. In Fig. 2 the selected nonlinear characteristic of conductivity in the range of interest is shown.

3 Simulations

3.1 Simulation Procedure

Numerical modeling of MO varistors requires the coupled solution of a heat conduction and of a electroquasistatics (EQS) problem. The relevant equations are:

$$c_v \frac{\partial T}{\partial t} - \nabla \cdot (\lambda \nabla T) = q_v \tag{4}$$

$$\nabla \cdot \left(\frac{\partial}{\partial t} (\varepsilon \nabla \Phi) \right) + \nabla \cdot (\sigma \nabla \Phi) = 0 \tag{5}$$

whereby Φ is a scalar potential, T temperature, ε permittivity, λ heat conductivity and q_v the heat loss density.

The two differential equations are coupled by the volumetric heat loss density

$$q_v = \mathbf{J} \cdot \mathbf{E} = \sigma E^2 , \tag{6}$$

and the dependence of electrical conductivity on temperature $\sigma = \sigma(E, T)$.

Because of the nonlinearity of the system it is necessary to operate in the time-domain. Furthermore, the solutions for temperature and electric potential are not obtained by solving one large system simultaneously, but separately and using the most recent solution of the other partial problem until convergence is achieved for any time-step. Thus, a consistent solution is guaranteed.

Electro-quasistatics solver (EQS) and heat flow solver are executed repeatedly until the solution for potential or temperature has converged below a user-defined threshold value. After convergence of both partial problems, convergence of temperature and electric potential combined is examined before advancing in time (Fig. 3).

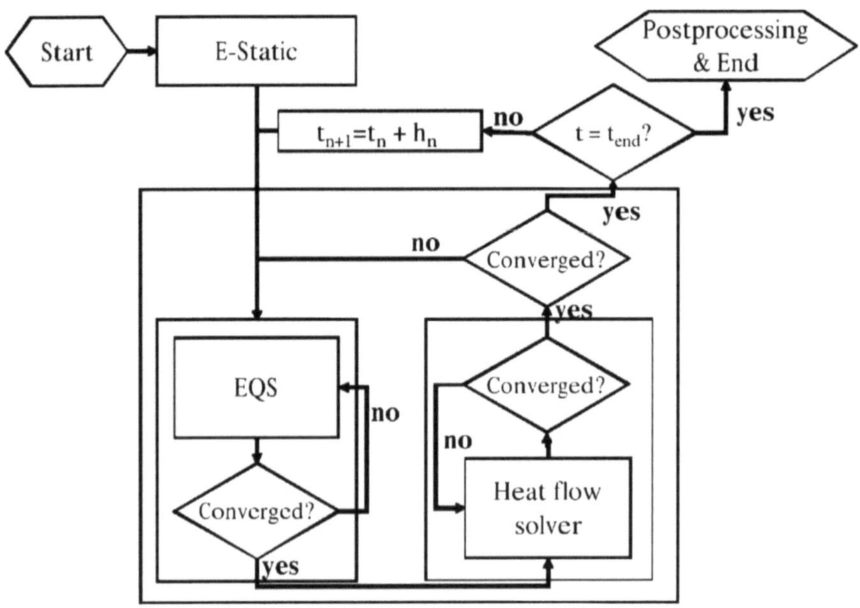

Fig. 3 Flowchart of the simulation process

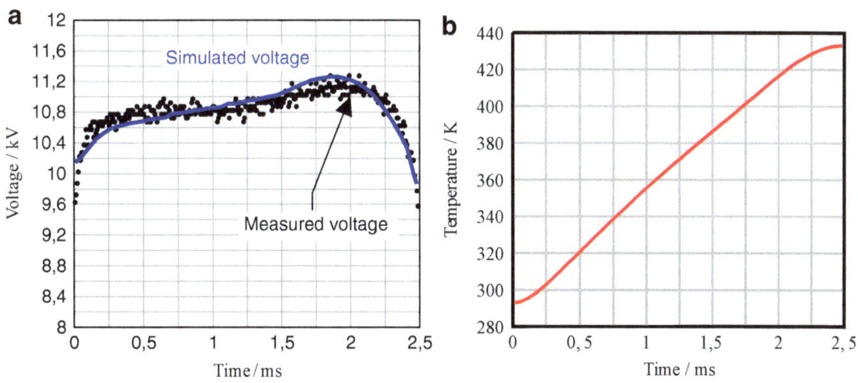

Fig. 4 Computational results for replication of measurements with finite elements (**a**) Comparison between measured and simulated voltage (*blue*: simulated; *black*: measured) (**b**) Simulated increase of temperature of the bulk of the zinc-oxide varistor during impulse

3.2 Validation of Electrical Conductivity Curve

The validity of the varistor model was investigated by numerical simulations of the single-disk measurement setup described in Sect. 2.1 using the same parameters as before. In Fig. 4 it can be seen that the simulated voltage resembles the measured voltage over the entire time of the impulse. Simultaneously, the considerable Joulean losses imply that the temperature of the varistor increases significantly. In slightly less than two milliseconds the temperature of the varistor increases by approximately 140 K.

3.3 AC Modeling of a Surge Arrester Unit

In this section the extracted material parameters are used to simulate a 50 Hz AC voltage signal, which constitutes the standard operating mode of a surge arrester. In that case, however, the peak voltage has to be low enough, so that the heat losses are not excessively large. This implies that the voltage is mostly below the lower limit for which the nonlinear conductivity was estimated. For this region the curve was extrapolated to provide a slowly decreasing conductivity.

The geometry is similar to Fig. 5 with a sequence of zinc oxide varistors and aluminum disks along the central axis. The dimensions were taken from a 3EP2 arrester by Siemens, which had been modified for measurements of the heat distribution. Some details had been eliminated, e.g., supporting rods, or simplified, notably the porcelain housing. Standard values were used for porcelain, air and metals. The differences in conductivity between copper and iron on the one hand and air and porcelain on the other were so large, that the metal parts were simulated

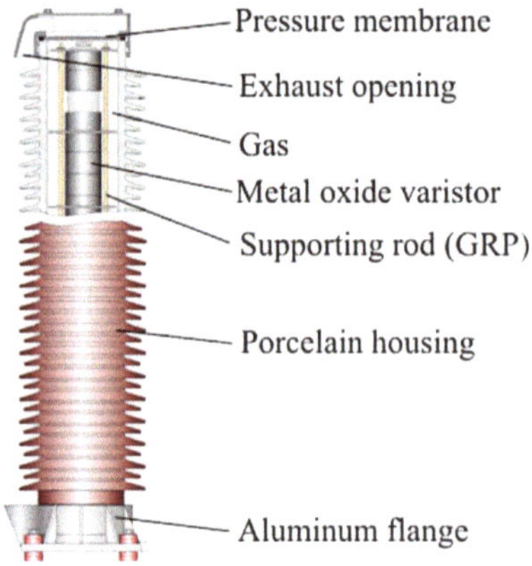

Fig. 5 Cross-section through a unit of a porcelain-housed surge arrester [8]

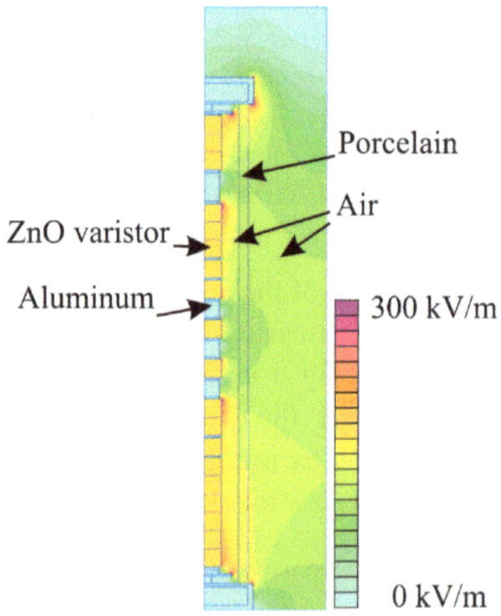

Fig. 6 Electric field strength at an arrester unit according to the distribution of effective electric potential

as so-called floating potentials. Peak voltage was set to a value, at which the corresponding electric field intensities inside the varistor reach temporarily the range of the estimated parameters for nonlinear conductivity without generating major thermal losses. By simulating several periods it is possible to calculate a distribution of effective voltage. In Fig. 6 the electric field strength, which corresponds to this voltage distribution is shown.

4 Conclusion

In this paper the parameters to describe the electrical conductivity of zinc-oxide varistors as function of field strength and temperature were extracted with sufficient accuracy to reproduce their reaction to current impulses. In principle the method can be applied to simulate arbitrary impulse shapes or geometries, though nonresistive effects become of greater importance for shorter impulses and the restrictions to the size of time-steps results in large computational costs, particularly for greater geometries or high current impulses. The simulation of the arrester unit has shown that it is feasible to simulate practical electro-thermal problems which involve the nonlinear conductivity of varistor materials.

References

1. Csendes, Z.J., Hamann, J.R.: Surge Arrester Voltage Distribution Analysis by the Finite Element Method. IEEE Trans. Power Apparatus Syst. **100**, 1806–1813 (1981)
2. Haddad, A.: ZnO surge arresters. In: Haddad, A., Warne, D.F. (ed.) Advances in High Voltage Engineering pp. 191–256. IEE, London (2004)
3. Bayadi, A., Harid, N., Zehar, K., Belkhiat, S.: Simulation of metal oxide surge arrester dynamic behavior under fast transients. International Conference on Power Systems Transients, IPST 2003, New Orleans, September 28 – October 2 (2003)
4. Zheng, Z., Boggs, S.A., Toshiya, I., Nishiwaki, S.: Computation of arrester thermal stability. IEEE Trans. Power Deliv. **25**, 1526–1529 (2010)
5. IEEE Power Engineering: IEEE Standard for Metal-Oxide Surge Arresters for AC Power Circuits (>1 kV) – IEEE Std C62.11-2005. IEEE, New York (2006)
6. IEC: International Standard 60099-4: Surge Arresters – Part 4: Metal-oxide surge arresters without gaps for a.c. systems. IEC, Geneva (2006)
7. Lat, M.V.: Thermal properties of metal oxide surge arresters. IEEE Trans. Power Apparatus Syst. **102**, 2194–2202 (1983)
8. Hinrichsen, V.: Metal-Oxide Surge Arresters – Fundamentals. Siemens AG, Berlin (2001)
9. Matsuoka, M.: Nonohmic properties of zinc oxide ceramics. Jpn. J. Appl. Phys. **10**, 736–746 (1971)
10. Bartkowiak, M., Comber, M.G., Mahan, G.D.: Energy handling capability of ZnO varistors. J. Appl. Phys. **79**, 8629–8633 (1996)
11. Einzinger, R.: Metal oxide varistors. Ann. Rev. Mat. Sci. **17**, 299–321 (1987)
12. Greuter, F., Blatter, G.: Electrical properties of grain boundaries in polycrystalline compound semiconductors. Semic. Sci. Technol. **5**, 111–137 (1990)
13. Clarke, D.: Varistor ceramics. J. Am. Ceram. Soc. **82**, 485–502 (1999)

Behavioural Electro-Thermal Modelling of SiC Merged PiN Schottky Diodes

M. Zubert, M. Janicki, M. Napieralska, G. Jablonski, L. Starzak, and A. Napieralski

Abstract This paper presents a new accurate behavioural static model of SiC Merged PiN Schottky (MPS) diode. This model is dedicated to static and quasi-static electro-thermal simulations of MPS diodes for industrial applications. The model parameters were extracted using the Weighted Least Square (WLS) method for a few selected commercially available SiC MPS diodes. Additionally, the PSPICE Analogue Behavioural Model (ABM) model implementation is also given. The relevance of the model has been statistically proven. The thermal behaviour of the devices was taken into account using the lumped Cauer canonical networks extracted from electro-thermal measurements.

1 Introduction

Silicon carbide devices are one of the most promising semi-conductor devices for power industrial applications. These devices offer, at least theoretically, excellent thermal properties and high operating frequencies as-well-as high power levels. Currently, the most frequently used SiC devices are the Merged PiN Schottky (MPS) diodes. Unfortunately, still there are no available models, which would be able to predict device characteristics, even the static ones, in a relatively wide range of operating temperatures. On the other hand, the correct prediction of the device behaviour is required for the robust design of the state-of-the-art power equipment. Obviously, there are models provided by manufacturers and the classical SPICE embedded diode models as-well-as physical models [3], but none of them is able to produce accurate temperature dependent device characteristics for SiC MPS.

M. Zubert (✉) · M. Janicki · M. Napieralska · G. Jablonski · L. Starzak · A. Napieralski
Department of Microelectronics and Computer Science, Technical University of Lodz, Lodz, Poland
e-mail: mariuszz@dmcs.pl; janicki@dmcs.pl; mnapier@dmcs.pl; gwj@dmcs.pl; starzak@dmcs.pl; napier@dmcs.pl

B. Michielsen and J.-R. Poirier (eds.), *Scientific Computing in Electrical Engineering SCEE 2010*, Mathematics in Industry 16, DOI 10.1007/978-3-642-22453-9_24,

Here, we propose a new static model for SiC MPS diodes, whose parameters can be identified from measurements. The reverse and forward diode model as-well-as its SPICE Analogue Behavioural Modelling (ABM) implementation are presented in the following sections.

2 Behavioural Model of MPS Diode

The proposed behavioural static model of MPS SiC diodes was created based on a series of measurements carried out on various commercially available SiC MPS diodes provided by different manufacturers and rated for voltages ranging from 300 V to 600 V and currents from 20 A to 2 A respectively. The measurements in the thermal static conditions were taken with the Tektronix 576 Curve-Tracer System. Originally, the model was developed for the SDP04S60 diode (2-nd generation MPS SiC diodes). The measurements were taken for the following case temperature values: −2 °C, 2.5 °C, 15 °C, 25 °C and 35 °C ÷ 120 °C with the step of 5 °C. Then, the model was verified for the SDP10S30, CSD04060, CSD10030 (all 2-nd gen.) and C3D04060 (3-rd gen.) diodes for the case temperatures ranging from 25 °C to 150 °C with the step of 25 °C. The detailed measurement procedure was described in [4]. The proposed electro-thermal behavioural static model of MPS diodes, pictured in Fig. 1, in the electrical domain can be summarized using the following equations for the reverse and the forward bias respectively:

$$I_{\text{rev}}\left(V_{\text{rev}}, T\right) = \beta\left(T\right) \cdot \exp\left(V_{\text{rev}} \cdot \alpha\left(T\right)\right) \tag{1}$$

$$I_{\text{fwd}}\left(V_{\text{fwd}}, T\right) = \exp \frac{V_{\text{fwd}} - V_{\text{intrsc}}\left(T\right) - R_{\text{s}} I_{\text{fwd}}\left(V_{\text{fwd}}, T\right)}{V_{\text{r}}\left(T\right)} \tag{2}$$

where: I_{rev}, I_{fwd}, V_{rev}, V_{fwd} - currents [A] and voltages [V] for reverse and forward bias ($V_{rev} \geq 0$, $V_{fwd} \geq 0$), see Fig. 5 R_s - internal parasitic resistance [˙]; V_{intrsc} –

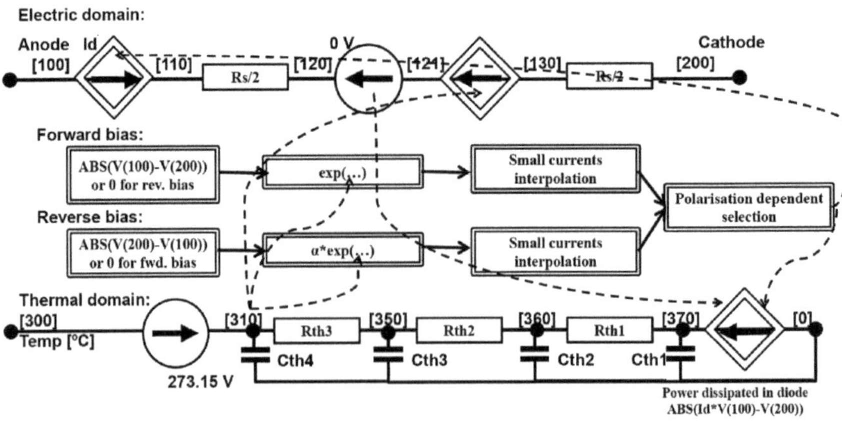

Fig. 1 Simplified analogue behavioural model of MPS SiC diode

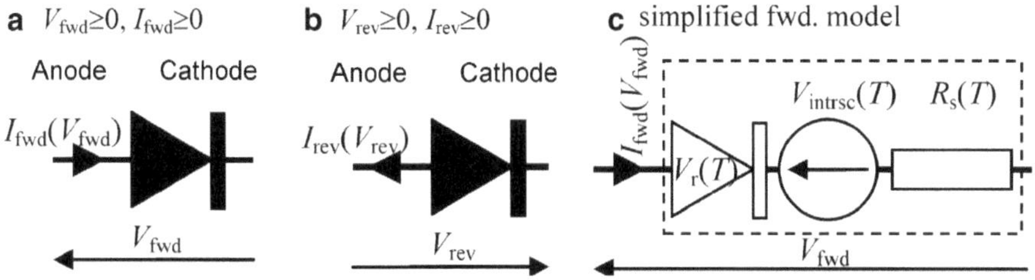

Fig. 2 Comparison of the measured SiC diode rev. characteristics with the proposed model (1)

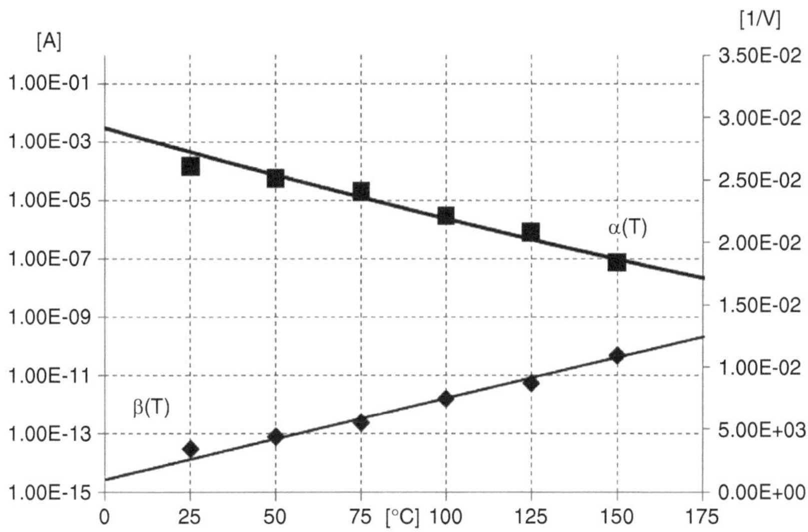

Fig. 3 Estimation of $\alpha\,(T)$ and $\beta\,(T)$ for SDP04S60 diode, see (4), (3) and (1)

internal voltage drop [V]; V_r, α, β - empiric coefficients [V, V^{-1}, A]; T - case temperature [°C].

The parameters in the above expressions can be accurately approximated using the following formulas (see Fig. 3):

$$\beta\,(T) = \beta_1 \cdot \exp\,(\beta_2 \cdot T) \tag{3}$$

$$\alpha\,(T) = \alpha_0 + \alpha_1 \cdot T + \alpha_2 \cdot T^2 \tag{4}$$

$$V_{\text{intrsc}}\,(T) = V_{\text{intrsc1}} + V_{\text{intrsc2}} \cdot (T + 273.15) \tag{5}$$

$$V_r\,(T) = V_{\text{ref}} \cdot (1 + \alpha_{\text{ref},1} \cdot \Delta T + \alpha_{\text{ref},2} \cdot \Delta T^2) \tag{6}$$

$$\Delta T = T - T_{nom} \tag{7}$$

The static model parameters for each device were extracted based on the measurements. The nominal case temperature T_{nom} assumed for the extraction of

Table 1 Estimated parameters for proposed MPS diode model

	2-nd generation of SiC MPS diodes				3-rd generation
	The model design	The model verification			
Diode	**SDP04S60**	**SDP10S30**	**CSD04060**	**CSD10030**	**C3D04060**
R_s	0.02533977	0.02656187	0.02464326	0.02529571	0.0253904
Reverse bias:					
α_2	5.41723813568926E-08	2.10144959397976E-07	Equation (8) is used instead of (1)	2.09920550874107E-07	0
α_1	-7.80429621880507E-5	-1.72626905326908E-4		-1.88345913175174E-4	-4.127549E-05
α_0	0.0291607049966412	0.0633278592970076		0.0322715235840343	0.01778526
β_1	2.57980343601669E-15	1.80333526117792E-14		6.29080916898752E-11	1.521115E-12
β_2	0.064468696717636	0.0564502636696925		0.08215764156411	0.03894504
Comment	$25 \div 150\,°C$; $CL0.99$	$25 \div 150\,°C$; $CL0.99$	$25 \div 150\,°C$	$25 \div 150\,°C$; $CL0.99$	$25 \div 150\,°C$
Forward bias:					
$V_{intrsc1}$	1.02334	1.069501	0.9699216	0.6585694	0.948265
$V_{intrsc2}$	-1.287796E-3	-1.264594E-3	-1.118851E-3	-1.109617E-3	-1.042935E-3
V_{ref}	0.2525095	0.1293532	0.2130102	0.1399051	0.0273944
$\alpha_{ref,1}$	2.572823E-3	1.154090E-3	2.688235E-3	2.842473E-3	2.70945E-3
$\alpha_{ref,2}$	1.849108E-5	-7.411367E-7	4.285593E-5	-4.556804E-6	0
Comment	Measurements 5098; $-2 \div 125\,°C$; $CL0.99$	Measurements 420; $25 \div 150\,°C$; $CL0.95$	Measurements 235; $25 \div 150\,°C$; $CL0.99$	Measurements 366; $25 \div 150\,°C$;$CL0.99$	Measurements 1825; $25 \div 150\,°C$;$CL0.99$
Thermal domain:					
R_{Th1}, C_{Th1}	1.11, 8.20E-04		0.271, 7.07E-04		
R_{Th2}, C_{Th2}	1.77, 7.53E-04		1.69, 1.64E-03		
R_{Th3}, C_{Th3}	0.924, 0.229		0.794, 0.220		
R_{Th4}, C_{Th4}	0.353, 13.3		0.351, 13.4		

The break down voltage (BV) were taken from data sheets

parameters given in Table 1 equalled 27 °C. The proposed model (1–7) is consistent with all tested MPS diodes, except the reverse characteristic of the CSD04060 diode (see Fig. 4). For the correct simulation of its behaviour, the following special fitted model could be applied:

$$I_{\text{rev}}\left(V_{\text{rev}}, T\right) = \exp\left(\frac{a + b \cdot V_{75} + c \cdot {V_{75}}^{2}}{1 + d \cdot V_{75} + e \cdot {V_{75}}^{2}}\right) \tag{8}$$

$$V_{75} = \frac{V_{rev} + f + g \cdot T}{1 + h \cdot T} \tag{9}$$

where V_{75} – reverse voltage drop at 75 °C. This equation takes into account the break down behaviour of MPS diode.

The extracted parameter values for the case temperatures in the range of $25 \div 150\,^{\circ}\mathrm{C}$ are: $a = -23.7661$, $b = -0.0231843$, $c = 0.0000549752$, $d = 0.00603505$, $e = -7.75449E-6$, $f = -12.438$, $g = 0.17909$, $h = 0.000104971$. The second break-down behaviour has not been experimentally observed for the other tested diodes (see Fig. 2a).

The thermal behaviour has been modelled with the thermal impedance synthesised in the canonical Cauer network, where the MPS diode die, the die attach, the heat slug and the interface to the heat-sink are represented with R_{th1}, $C_{th1} \div R_{th3}$, C_{th4} respectively (see Listing 1). The external heat-sink and environment has been modelled using R_{th4}, R_{th5} and C_{th5}. For the details on the thermal impedance measurements, refer to [1].

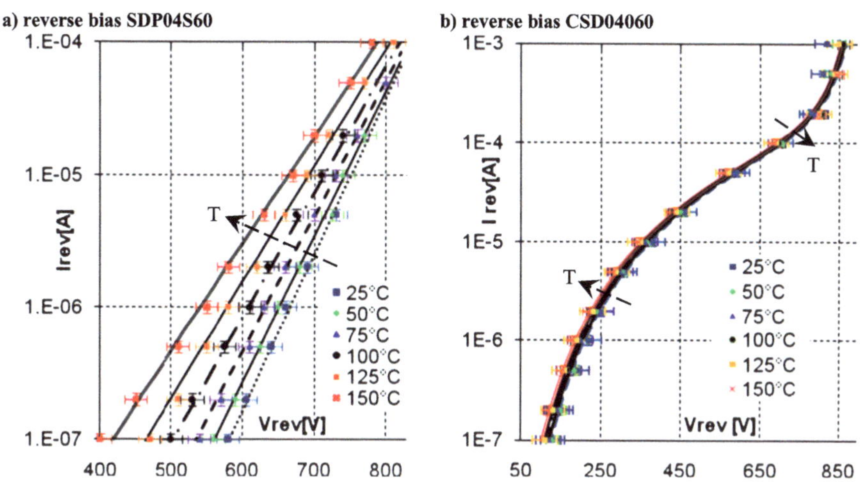

Fig. 4 (**a**,**b**) The currents and voltages direction for fwd. and rev. bias. (**c**) internal parameters for reverse and forward bias

3 Parameter Estimation Procedure

All parameters in this paper were estimated employing the Weighted Least Square (WLS) method, using the following objective function: $J = (\Delta z)^T R^{-1} (\Delta z)$, where $\Delta z = z - h(x)$ are the estimated measurement residuals. The measurement vector (z) is constructed as follows:

$$z = [\Delta z_\bullet]_{n\times 1} = \left[V_{fwd,1}, I_{fwd,1}, \ldots, V_{fwd,i}, I_{fwd,i}, \ldots, V_{fwd,n}, I_{fwd,n}\right]^T \quad (10)$$

where $\left(V_{fwd,i}, I_{fwd,i}, T_i\right)$ is the i-th measurement triple (voltage, current and temperature). The $h(x) = [h_{\bullet,\bullet}]_{(2n)\times(n+7)}$ is the nonlinear function relating measurements to the system state vector (x), containing the information on the estimated parameters ($p_\bullet$) and the residuals ($\Delta z_\bullet$)

$$x = [x_\bullet]_{(n+7)\times 1} = \left[\, \Delta z_1 \ldots \Delta z_n \; p_1 \ldots p_7 \right]^T \quad (11)$$

The estimated parameters are associated with (1) written here in a more convenient form without the explicit exponent function:

$$\begin{aligned} V_{fwd,i} &= [(p_2 \cdot T_i + p_1) + p_3] \cdot \log I_{fwd,i} + p_4 \cdot T_i \\ &\quad + I_{fwd,i} \cdot p_5 \cdot \left[1 + p_6 \cdot (T_i - T_{nom}) + p_7 \cdot (T_i - T_{nom})^2\right] \end{aligned} \quad (12)$$

The nonlinearity of (12) is overcome by the Gauss-Newton method applied to the Taylor series expansion, which leads to the iterative solution of the so-called Normal Equation (NE):

$$\left(H^T R^{-1} H\right) \cdot \Delta x = H^T R^{-1} \Delta z \quad (13)$$

$$x^{NEW} := x + \Delta x \quad (14)$$

where the non-vanishing elements of jacobian $h(x)$

$$H = [H_{\bullet,\bullet}]_{2n\times(n+7)} = \left[\frac{\partial h(x)}{\partial x}\right] \quad (15)$$

are calculated using the following rules

$$\begin{aligned} H_{i,n+1} &= \frac{\partial V_{fwd,i}}{\partial p_1}, \ldots, \; H_{i,n+7} = \frac{\partial V_{fwd,i}}{\partial p_7}, \; H_{2i,i} = 1, \; H_{2i-1,1} \\ [5pt] &= V_{fwd,i}, \; H_{2i,1} = x_i \end{aligned} \quad (16)$$

for $i = 1, \ldots, n$. The weights (R^{-1}) take into account the accuracy of each measurement and are computed as the reciprocals of the measurement standard deviation ($\delta_{Vfwd,i}{}^2, \delta_{Ifwd,i}{}^2$):

$$R = diag\left\{\delta_{Vfwd,1}{}^2, \delta_{Ifwd,1}{}^2, \ldots, \delta_{Vfwd,i}{}^2, \delta_{Ifwd,i}{}^2, \ldots, \delta_{Vfwd,n}{}^2, \delta_{Ifwd,n}{}^2\right\} \quad (17)$$

It should be underlined that the applied formulation of the WLS method assures fast convergence in 6 ÷ 7 iteration steps. The other classical formulations of WLS method (e.g. Hybrid QR, Hatchel) lead to the ill-conditioning and singularity. The χ^2-test shows that all the models are consistent with the measurements at the confidence level (CL) of 0.99 except for one of the devices as shown in Table 1. The very high consistency of the model with the behaviour of the MPS diodes is visible also in Figs. 4 and 5 representing the reverse and forward characteristics respectively.

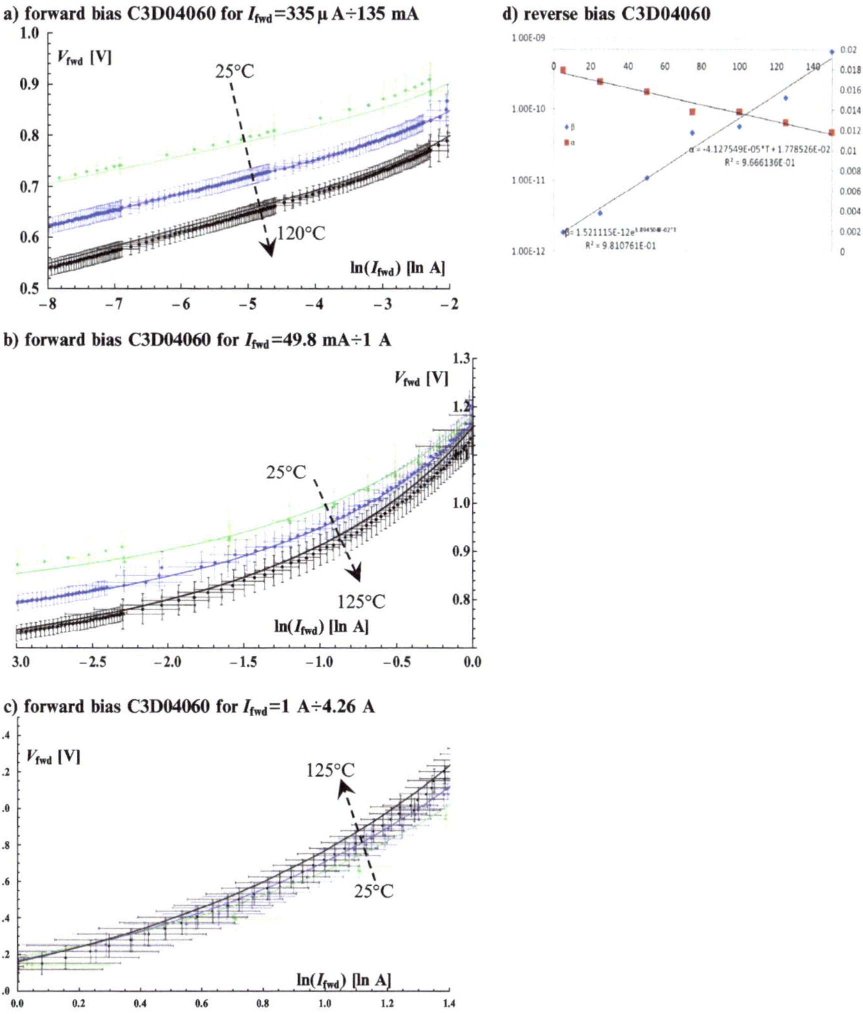

Fig. 5 Comparision of the measured SiC diode forward characteristics with the proposed model. Temperatures: 25 °C (*green*), 75 °C (*blue*), 125 °C for C3D04060 (*black*). Y-axis: V_{fwd} [V]; X-axis: $ln\left(I_{fwd}\right)$; The measurement deviation is presented using error bars (**a**) forward bias C3D04060 for $I_{fwd} = 335\mu\text{A} \div 135\,\text{mA}$ (**b**) forward bias C3D04060 for $I_{fwd} = 49.8\text{mA} \div 1\,\text{A}$ (**c**) forward bias C3D04060 for $I_{fwd} = 1\text{A} \div 4.26\,\text{A}$ (**d**) reverse bias C3D04060

4 Model Implementation

The PSPICE MPS diode model implementation is shown in Listing 1 at the end of the paper. This listing contains also an additional interpolation block for small voltages $V_{dLeft} \leq V_{rev} \leq 0$ and $0 \leq V_{fwd} \leq V_{dRight}$ forcing zero currents for unbiased diodes. The proposed model takes into account the full electro-thermal coupling. The external environment (e.g. heat sinks) should be connected to the model between the nodes 300 and 0. For the presented simulations, the typical heat sink and environment thermal impedance has been used. The model has been proposed for the 2-nd generation of MPS SiC diodes but is also consistent with 3-rd MPS diode (C3D04060). The simulation results obtained for the new model are compared with the measurements. The diode current I_d and he power P_d dissipation obtained from the manufacturer diode models are not consistent with the real MPS diode.

5 Conclusions

The proposed behavioural static model of SiC MPS diodes showed a very good agreement of the simulated forward and reverse characteristics with the measurements of real devices and definitely produced much better results than the models provided by the device manufacturers. The parameter estimation procedure based on Weighted Least Square (WLS) method was successfully applied to the reformulated diode model given in (12). The main advantage of the proposed model is that it is given in a closed form, which allows its straightforward implementation in the behavioural extensions of modern simulators, such as Berkeley SPICE and PSPICE, or as an embedded model (eg. [2]). The main drawback of the proposed approach is the necessity to force zero current for unbiased devices and the lack of a direct physical interpretation of model parameters.

Acknowledgements This work was supported partly by the grant of Polish Ministry of Science and Higher Education 0312/R/T02/2008/04 and the Internal University Grant K25/DZST/1/2010.

References

1. Banaszczyk, J., Janicki, M., Vermeersch, B., De Mey, G., Napieralski, A.: Application of advanced thermal analysis method for investigation of internal package structure. In: 14-th International Conference on Mixed Design of Integrated Circuits and Systems - MIXDES2007, pp. 553–558 (2007)
2. Napieralski, A., Starzak, L., Swiercz, B., Zubert, M.: chap. Web-based modelling tools. In: Power HVMOS Devices Compact Modeling. Springer, Berlin (2010)
3. Werber, D., Wachutka, G.: Physical modeling of sic devices based on the optical characterization of their internal electrothermal behaviour. In: International Semiconductor Device Research Symposium ISDRS09, pp. 1–2 (2009)

4. Zubert, M., Napieralska, M., Jablonski, G., Starzak, L., Janicki, M., Napieralski, A.: Static electro-thermal model of sic merged pin schottky diodes. In: 10-th International Seminar on Power Semiconductors - ISPS10, pp. 227–232 (2010)

Listing 1

```
 DMCS model presentation: diode C3D04060SS, steady-state
* (c) DMCS 2010
.OPTIONS EXPAND WIDTH=132
* MODEL DEFINITION:     A    K  Thermal
.SUBCKT DMCSC3D04060SS 100 200 300 + PARAMS: Rs=0.0253904 RsThC=7.788199E-5 + RTh1=0.271 CTh1=7.07E-04 +
RTh2=1.69  CTh2=1.64E-03 + RTh3=0.794 CTh3=0.220 +            CTh4=13.4 + Alpha2=0.0
Alpha1=-4.127549E-05 Alpha0=0.01778526 + Beta1=1.521115E-12 Beta2=0.03894504 + Vintrsc1=0.948265
Vintrsc2=-1.042935E-3 + Vref=0.0273944 AlphaRef1=2.70945E-3 AlphaRef2=0.0 + VdLeft=100.0 VdRight=0.01
***Forward:
EVDPlus 500 0 VALUE={ IF( V(100,110)>=0, V(100,110), 0.0 ) } EIFwd   600 0 VALUE={
exp((V(500)-(Vintrsc1+ Vintrsc2*V(300)))/ +
(Vref*(1+AlphaRef1*(V(300)-300.15)+AlphaRef2*(V(300)-300.15)*(V(300)-300.15)))) }
* Interpolation: [0,VdRight]
EIFwdInterp 601 0 VALUE={ + V(500)*exp((VdRight-(Vintrsc1+ Vintrsc2*V(300)))/ +
(Vref*(1+AlphaRef1*(V(300)-300.15)+AlphaRef2*(V(300)-300.15)*(V(300)-300.15)))) }
***Reverse:
EVDMinus 510 0 VALUE={ IF( V(100,110)<0, V(110,100), 0.0 )     } EBetaT   810 0
VALUE={Beta1*exp(Beta2*V(310)) } EAlphaT  820 0 VALUE={ Alpha2*V(310)*V(310)+Alpha1*V(310)+Alpha0  }
EIRev    800 0 VALUE={ V(810)*exp(V(820)*V(510))     }
* Interpolation: [VdLeft,0]
EIRevInterp 801 0 VALUE={ (Exp(V(820)*VdLeft)*V(510)*V(810)* + (VdLeft*(2-V(820)*VdLeft) +
V(510)*(V(820)*VdLeft-1)))/VdLeft*VdLeft}
***All:
*Simplest: GIFwdRev 100 110 VALUE={ IF( V(100,110)>=0,  V(600), -V(800))  }
GIFwdRev 100 110 VALUE={ IF( V(100,110)>=0, + IF(V(500)>VdRight,V(600),V(601)), +
IF(V(100,110)<VdLeft,-V(800),-V(801)) ) } RRs1 120 110  {0.5*Rs} VIProbe 120 121 0 ERsT 121 130 VALUE={
RsThC*V(310)*(V(120,130))    } RRs2 130 200  {0.5*Rs}
***Thermal domain:
VK2C 300 310 273.15 RTh3 350 310 {RTh3} RTh2 360 350 {RTh2} RTh1 370 360 {RTh1} CTh4 310 0 {CTh4} IC=9.9
CTh3 350 0 {CTh3} IC={0.9+9.0*(Rth2+Rth1)/(Rth3+Rth2+Rth1)} CTh2 360 0 {CTh2}
IC={0.9+9.0*(Rth1)/(Rth3+Rth2+Rth1)} CTh1 370 0 {CTh1} IC={0.9}
* Can be used for steady-state and dynamic:
GTh1 0 370 VALUE={ABS(V(100,200)*I(VIProbe))} .ENDS
* APPLICATION:
VIN 10 0 2.5 VthTemp 20 0 300.0 Rth5 21 20 0.588780 ; air tunell Cth5 21 0 106.849460 Rth4 22 21 0.353
; heat sink XD1 10 0 22 DMCSC3D04060SS
*.DC VIN -850 0 -5 VthTemp LIST 300
.DC VIN 0 2.0 0.01 VthTemp LIST 300 .PROBE .END
```

A Convergent Iteration Scheme for Semiconductor/Circuit Coupled Problems

Giuseppe Alì, Andreas Bartel, Markus Brunk, and Sebastian Schöps

Abstract A dynamic iteration scheme is proposed for a coupled system of electric circuit and distributed semiconductor (pn-diode) model equations. The device is modelled by the drift-diffusion (DD) equations and the circuit by MNA-equations. The dynamic iteration scheme is investigated on the basis of discrete models and coupling via sources and compact models. The analytic divergence and analytic convergence results are verified numerically.

1 Introduction

Distributed semiconductor models typically result in partial differential equations (PDEs). The trend of miniaturization in electronics industry leads to devices of growing complexity, where – due to smaller signals – parasitic effects become more and more important. Description of semiconductor devices by compact models enforces time-consuming parameter fitting and might result in sub-circuits of several hundred parameters for the description of a single device [5]. Thus the coupling of PDE-models and circuit simulation becomes desirable.

Efficient techniques to couple the different subsystems are needed. We propose a simple Gauss-Seidel iteration, where the displacement current is explicitly modeled by an extracted capacitance.

G. Alì (✉)
Dipartimento di Matematica, Universitá della Calabria and INFN-Gruppo c. Cosenza, I-87036 Arcavacata di Rende (CS), Italy
e-mail: g.ali@mat.unical.it

A. Bartel · M. Brunk · S. Schöps
Institut für Numerische Analysis, FB C, Bergische Universität Wuppertal, Gaussstrasse 20, 42119 Wuppertal, Germany
e-mail: bartel@math.uni-wuppertal.de; brunk@math.uni-wuppertal.de; schoeps@math.uni-wuppertal.de

B. Michielsen and J.-R. Poirier (eds.), *Scientific Computing in Electrical Engineering SCEE 2010*, Mathematics in Industry 16, DOI 10.1007/978-3-642-22453-9_25,

In this section the basic models are introduced: for the semiconductor device the drift-diffusion (DD) equations and for the surrounding circuit the modified nodal analysis (MNA) equations. In the second section different coupling approaches are discussed. In the third section a dynamic iteration scheme is developed that guarantees unconditional convergence. This is confirmed by a simple numerical example. In the last section we draw some conclusions.

1.1 Device Model

Our np-diode shall be modelled on the domain $\Omega \subset R^d$ for $d = 1,2,3$ with $\partial\Omega = \Gamma = \Gamma_D \cup \Gamma_N$. The DD-model equations consist of conservation laws for the electron and hole densities n, p coupled to the Poisson equation for the electric potential V,

$$\partial_t n - q^{-1}\text{div}\, J_n = -R, \qquad J_n = \mu_n(U_T \nabla n - n\nabla V), \tag{1a}$$

$$\partial_t p + q^{-1}\text{div}\, J_p = -R, \qquad J_p = -\mu_p(U_T \nabla p + p\nabla V), \tag{1b}$$

$$\varepsilon_s \Delta V = q(n - p - C(x)), \quad J_{\text{tot}} = \int_{\Gamma_k} \{\varepsilon_s \partial_t \nabla V - (J_n + J_p)\}\, ds. \tag{1c}$$

Here $R = R(n,p)$ denotes the generation-recombination term, μ_n, μ_p are the mobility parameters and q is the elementary charge. The permittivity is given by ε_s, $C(x)$ is the doping concentration and U_T the thermal voltage. The total current J_{tot} leaving the device at terminal k given by $\Gamma_k \subset \Gamma_D$ incorporates the displacement current $\varepsilon_s \partial_t \nabla V$. Of course, $k = 1,2$, and due to charge conservation in the diode, the current is the same for both terminals Γ_k.

The model equations are supplemented with initial conditions for n, p and boundary conditions for V, n, p on the Dirichlet boundary Γ_D and for $J_n, J_p, \nabla V$ on the Neumann boundary Γ_N. For the simulations presented below (see Sect. 3) we discretized the DD-equations by use of exponentially fitted mixed finite elements as described in [4,9], since this allows for positivity preservation. Thus on a small time window the discretized equations can be written in the form

$$A_n(\mathbf{V})\, d_t \mathbf{n} + B_n(\mathbf{p},\mathbf{V})\, \mathbf{n} = f_n(\mathbf{p},\mathbf{V}), \qquad L\mathbf{V} = \mathbf{n} - \mathbf{p} - \mathbf{C} + f_V(\mathbf{v}_\text{D}), \tag{2a}$$

$$A_p(\mathbf{V})\, d_t \mathbf{p} + B_p(\mathbf{p},\mathbf{V})\, \mathbf{p} = f_p(\mathbf{n},\mathbf{V}), \qquad \mathbf{i}_\text{D} = j_D(\mathbf{n},\mathbf{p},\mathbf{V}), \tag{2b}$$

with regular matrices A_n, A_p, B_n, B_p, L, where the recombination term typically is semi-linearized by use of old values of n (in (2a)) or p (in (2b)), see [4] for details. Here the bold symbols represent the vectors containing the discrete approximations of the corresponding values. $\mathbf{i}_\text{D}$ is the discrete approximation of J_{tot} and $\mathbf{v}_\text{D}$ denotes the applied voltage drop, which is determined by the surrounding circuit. The boundary conditions are incorporated in the functions f_n, f_p and f_V, respectively.

We note that standard finite element or finite difference discretization allow for the same representation.

Alternatively the displacement current can be expressed equivalently in terms of the time derivative of the applied voltage drop, [1]; this yields

$$\mathbf{i}_{\mathrm{D}} = C_{\mathrm{D}} \frac{\mathrm{d}}{\mathrm{d}t} \mathbf{v}_{\mathrm{D}} - \mathbf{i}_{\mathrm{SD}} \quad \text{with} \quad \mathbf{i}_{\mathrm{SD}} := j_{SD}(\mathbf{n}, \mathbf{p}, \mathbf{V}). \tag{2c}$$

For a cubic diode with length l and cross-section A, where a 1-d model is sufficient, it holds: $C_{\mathrm{D}} = \frac{\varepsilon_s A}{l}$.

1.2 Circuit Model

The extended circuit reads in the flux/charge oriented form of the MNA [8]:

$$\mathbf{A}_{\mathrm{C}} \frac{\mathrm{d}}{\mathrm{d}t} \mathbf{q} + \mathbf{A}_{\mathrm{R}} \mathbf{g}_{\mathrm{R}}(\mathbf{A}_{\mathrm{R}}^{\top} \mathbf{u}, t) + \mathbf{A}_{\mathrm{L}} \mathbf{i}_{\mathrm{L}} + \mathbf{A}_{\mathrm{V}} \mathbf{i}_{\mathrm{V}} + \mathbf{A}_{\mathrm{I}} \mathbf{i}(t) + \mathbf{A}_{\mathrm{D}} \mathbf{i}_{\mathrm{D}} = \mathbf{0}, \tag{3a}$$

$$\frac{\mathrm{d}}{\mathrm{d}t} \Phi - \mathbf{A}_{\mathrm{L}}^{\top} \mathbf{u} = \mathbf{0}, \qquad \mathbf{A}_{\mathrm{V}}^{\top} \mathbf{u} - \mathbf{v}(t) = \mathbf{0}, \tag{3b}$$

$$\mathbf{q} - \mathbf{q}_{\mathrm{C}}(\mathbf{A}_{\mathrm{C}}^{\top} \mathbf{u}, t) = \mathbf{0}, \qquad \Phi - \Phi_{\mathrm{L}}(\mathbf{i}_{\mathrm{L}}, t) = \mathbf{0}, \tag{3c}$$

with given functions $\mathbf{q}_{\mathrm{C}}(\mathbf{v}, t)$, $\mathbf{g}_{\mathrm{R}}(\mathbf{v}, t)$, $\Phi(\mathbf{i}, t)$, $\mathbf{v}_{\mathrm{S}}(t)$ and $\mathbf{i}_{\mathrm{S}}(t)$ denoting the constitutive relations for charges, resistances, fluxes, voltage and current sources, respectively. Matrices $\mathbf{A}_{\star}$ denote network incidences and $\mathbf{i}_{\mathrm{D}}$ is the current through the diode. The unknowns of the network are charges $\mathbf{q}(t)$, fluxes $\Phi(t)$, node potentials $\mathbf{u}(t)$, except ground, and currents $\mathbf{i}_{\mathrm{L}}(t)$, $\mathbf{i}_{\mathrm{V}}(t)$ through inductors and voltage sources.

1.3 Coupling

The structure of the equations allows two representations of the circuit-device coupling: (a) coupling by plain sources (*source coupling*), in which the distributed device model takes the displacement current into account, i.e., $\mathbf{i}_{\mathrm{s}}$ is defined by (2b), or, (b) coupling, where the displacement current is described in terms of circuit variables, i.e., (2c) is treated as an additional circuit equation (*coupling with capacitance*), see Fig. 1. In both settings the voltage drop $\mathbf{v}_{\mathrm{D}}$ in the circuit is supplied as a boundary condition to the device model. In the case of a monolithic coupling, where all equations are solved in one system, those two representations are equivalent, but in the case of a weak coupling by a dynamic iteration scheme, they exhibit different behavior. – Setting (b) reflects standard compact model design.

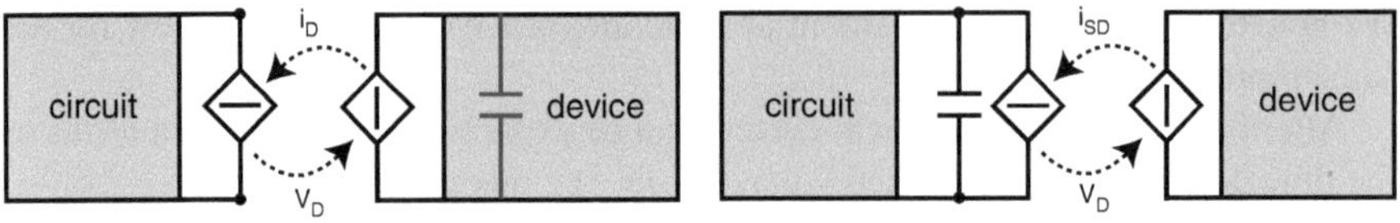

(a) Source coupling. (b) Coupling with Capacitance.

Fig. 1 Coupling with displacement current in device (a) and in circuit (b). (**a**) Source coupling (**b**) Coupling with capacitance

1.3.1 Source Coupling

Spatial discretization of the DD-equations (2) yields an index-1 DAE for any given $\mathbf{v}_\mathrm{D}(t)$ [4]. Assuming the standard loop and cutset conditions, [7], the circuit equations (3) are index-1 as well (for given $\mathbf{i}_\mathrm{D}(t)$). Assuming the overall system to be of index-1 [10], it can be written in semi-explicit form:

$$\begin{aligned} \dot{\mathbf{y}}_d &= \mathbf{f}_d(\mathbf{y}_d, \mathbf{z}_d), & \dot{\mathbf{y}}_c &= \mathbf{f}_c(\mathbf{y}_c, \mathbf{z}_c, \mathbf{z}_d), \\ \mathbf{0} &= \mathbf{g}_d(\mathbf{y}_d, \mathbf{z}_d, \mathbf{z}_c), & \mathbf{0} &= \mathbf{g}_c(\mathbf{y}_c, \mathbf{z}_c, \mathbf{z}_d), \end{aligned} \tag{4}$$

with $\partial \mathbf{g}_d / \partial \mathbf{z}_d$ and $\partial \mathbf{g}_c / \partial \mathbf{z}_c$ regular. The differential variables of the diode and circuit are $\mathbf{y}_d := (\mathbf{n}, \mathbf{p})$ and $\mathbf{y}_c := (\mathbf{q}, \Phi)$, respectively, while the algebraic unknowns are $\mathbf{z}_d = (\mathbf{V}, \mathbf{i}_\mathrm{D})$ and $\mathbf{z}_c := (\mathbf{u}, \mathbf{i}_\mathrm{L}, \mathbf{i}_\mathrm{V})$. The vector $\mathbf{V}$ describes the space discrete electric potential and $\mathbf{i}_\mathrm{D}$ the device current, defined in (1c). Due to spatial discretization the values of $\mathbf{n}, \mathbf{p}$ in some discretization nodes may turn into algebraic variables. This does not pose any problem as long as the index-1 assumptions hold. Thus this case is not considered in the following.

In this setting all node potentials $\mathbf{u}$ are algebraic variables of the circuit and so is $\mathbf{v}_\mathrm{D} = \mathbf{A}_\mathrm{D}^\top \mathbf{u}$. Hence only $\mathbf{z}_c$ enters the algebraic equations of the device $\mathbf{g}_d$. The diode current is also algebraic, but appears, depending on the circuit's topology, in the differential $\mathbf{f}_c$ and in the algebraic equation $\mathbf{g}_c$ of system (4).

1.3.2 Coupling with Capacitance

Now substituting $\mathbf{i}_\mathrm{D}$ by (2c) in the current balance equation (3a), we end up with a slightly different system of equations

$$\begin{aligned} \dot{\mathbf{y}}_d &= \mathbf{f}_d(\mathbf{y}_d, \mathbf{z}_d), & \dot{\mathbf{y}}_c &= \mathbf{f}_c(\mathbf{y}_c, \mathbf{z}_c, \mathbf{z}_d), \\ \mathbf{0} &= \mathbf{g}_d(\mathbf{y}_d, \mathbf{z}_d, \mathbf{y}_c), & \mathbf{0} &= \mathbf{g}_c(\mathbf{y}_c, \mathbf{z}_c), \end{aligned} \tag{5}$$

with differential unknowns $\mathbf{y}_d := (\mathbf{n}, \mathbf{p})$ and $\mathbf{y}_c := (\mathbf{q}, \Phi, \mathbf{P}_D \mathbf{u})$ and algebraic unknowns $\mathbf{z}_d = (\mathbf{V}, \mathbf{i}_\mathrm{SD})$ and $\mathbf{z}_c := (\mathbf{Q}_D \mathbf{u}, \mathbf{i}_\mathrm{L}, \mathbf{i}_\mathrm{V})$, where $\mathbf{Q}_D$ is a projector onto the kernel of $\mathbf{A}_\mathrm{D}^\top$ and $\mathbf{P}_D$ its complement, as typically defined in circuit index analysis, [7]. In this notation the node potentials are split because the capacitance C_D is not written in charge oriented form. The advantages of the charge/flux oriented MNA, i.e., charge conservation, are still respected because of the linearity of C_D.

Due to the capacitive path between the coupling nodes the voltage drop $\mathbf{v}_\mathrm{D}$ belongs to the differential variables $\mathbf{y}_c$ and only this enters the device subsystem $\mathbf{g}_d$ and, in turn, the device current $\mathbf{i}_\mathrm{SD}$ enters only the circuit's differential equation $\mathbf{f}_c$.

2 Dynamic Iteration Schemes

To simulate the coupled system of circuit and discretized device equations efficiently and reliably we propose a dynamic DAE-DAE iteration scheme that ensures stability and speeds up convergence. The key is the coupling via capacitive branches, [3], which is ensured here by fitting the capacitance $C_\mathrm{D} = \frac{\varepsilon_s A}{l}$ for our 1-d model. For higher dimensional models see [1].

In a dynamic iteration scheme, the time interval of interest is split into time windows that are treated sequentially. On these windows, each subsystem is solved independently by a problem-specific time-integrator. The mutual input from the subsystems is considered as a known functions, only dependent on time. The exchange of data between the subsystems is organized in a Gauss-Seidel-like iteration scheme.

The stability and convergence of the iteration depends on the order in which the problems are computed, [2]. For both coupling approaches we may start by computing the diode model or the circuit model first. We will see, that for the coupling with parallel capacitance the iteration will converge, independent of the particular order.

2.1 Source Coupling

Let us start with the source coupling for the case, that the device is simulated first. The circuit solution $\mathbf{y}_c^{(0)}(t), \mathbf{z}_c^{(0)}(t)$ is considered as known and from that the new device solution $\mathbf{y}_d^{(1)}(t), \mathbf{z}_d^{(1)}(t)$ is obtained. Afterwards the circuit solution $\mathbf{y}_c^{(1)}(t), \mathbf{z}_c^{(1)}(t)$ is updated using the new solution of the device.

$$\begin{aligned} \dot{\mathbf{y}}_d^{(1)} &= \mathbf{f}_d(\mathbf{y}_d^{(1)}, \mathbf{z}_d^{(1)}), & \dot{\mathbf{y}}_c^{(1)} &= \mathbf{f}_c(\mathbf{y}_c^{(1)}, \mathbf{z}_c^{(1)}, \mathbf{z}_d^{(1)}), \\ \mathbf{0} &= \mathbf{g}_d(\mathbf{y}_d^{(1)}, \mathbf{z}_d^{(1)}, \mathbf{z}_c^{(0)}), & \mathbf{0} &= \mathbf{g}_c(\mathbf{y}_c^{(1)}, \mathbf{z}_c^{(1)}, \mathbf{z}_d^{(1)}). \end{aligned} \tag{6}$$

According to [2] we can find a maximum time step size H_0 such that the iteration scheme is convergent for any time step size $H \leq H_0$ if

$$\alpha := \left\| \left(\frac{\partial \mathbf{g}}{\partial \mathbf{z}^{(1)}} \right)^{-1} \left(\frac{\partial \mathbf{g}}{\partial \mathbf{z}^{(0)}} \right) \right\| < 1, \quad \text{where } \mathbf{g}(\cdot) := (\mathbf{g}_d(\cdot), \mathbf{g}_c(\cdot)), \ \mathbf{z} := (\mathbf{z}_d, \mathbf{z}_c). \tag{7}$$

Thus, in (6) the dependence of $\mathbf{g}_d$ on the old iterate $\mathbf{z}_c^{(0)}$ can cause divergence, cf. [2] and see example below. A similar contractivity condition occurs for the coupling in reversed order of the subsystems. Clearly the condition vanishes in both orders if the dependence of $\mathbf{g}_d$ on $\mathbf{z}_c$ turns into a differential dependency (for the device-first approach), or if $\mathbf{g}_c$ does not depend on the algebraic variable $\mathbf{z}_d$ (for the circuit-first approach), [3]. This happens for capacitive paths between the device pins and this will be exploited next.

2.2 *Coupling with Capacitance*

In the case of the coupling with parallel capacitances, the contraction factor vanishes regardless of the order of the subsystems. When starting with the device, i.e,

$$\begin{aligned} \dot{\mathbf{y}}_d^{(1)} &= \mathbf{f}_d(\mathbf{y}_d^{(1)}, \mathbf{z}_d^{(1)}), & \dot{\mathbf{y}}_c &= \mathbf{f}_c(\mathbf{y}_c^{(1)}, \mathbf{z}_c^{(1)}, \mathbf{z}_d^{(1)}), \\ \mathbf{0} &= \mathbf{g}_d(\mathbf{y}_d^{(1)}, \mathbf{z}_d^{(1)}, \mathbf{y}_c^{(0)}), & \mathbf{0} &= \mathbf{g}_c(\mathbf{y}_c^{(1)}, \mathbf{z}_c^{(1)}), \end{aligned} \tag{8}$$

the only occurrence of an old iterate in an algebraic constraint is the differential variable $\mathbf{y}_c^{(0)}$ in $\mathbf{g}_d$. Hence the contraction factor vanishes. On the other hand, when starting with the circuit subproblem

$$\begin{aligned} \dot{\mathbf{y}}_c^{(1)} &= \mathbf{f}_c(\mathbf{y}_c^{(1)}, \mathbf{z}_c^{(1)}, \mathbf{z}_d^{(0)}), & \dot{\mathbf{y}}_d^{(1)} &= \mathbf{f}_d(\mathbf{y}_d^{(1)}, \mathbf{z}_d^{(1)}), \\ \mathbf{0} &= \mathbf{g}_c(\mathbf{y}_c^{(1)}, \mathbf{z}_c^{(1)}), & \mathbf{0} &= \mathbf{g}_d(\mathbf{y}_d^{(1)}, \mathbf{z}_d^{(1)}, \mathbf{y}_c^{(1)}), \end{aligned} \tag{9}$$

then there is no dependence on old iterates in algebraic constraints at all; thus again the convergence is unconditional, according to [2].

3 Numerical Results

Next, we visualize the above results by the simulation of a simple series connection of a voltage source, a resistor and an amplified silicon pn-diode (1d). The resistance is given as $R = 1\Omega$, the voltage source is given by $v(t) = \sin(\omega t)$V with a frequency $\omega = 2\pi 10^{11}$ Hz. The diode consists of a 50 nm n-region doped with C_0 and a 50 nm p-region doped with $-C_0$. The output current of the diode is amplified by the factor 1,500, since this causes α in (7) to be greater than one. This makes our example on the one hand rather academic, but on the other hand it illustrates that even in simple setups divergence occurs. Further parameters of the diode are given in Fig. 2b.

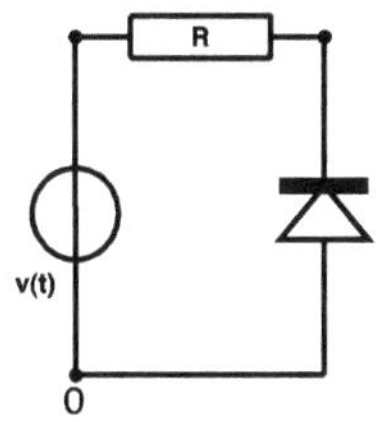

(a) Example circuit

Parameter	Physical meaning	Numerical value
q	elementary charge	$1.6 \cdot 10^{-19}$ As
ε_s	permittivity constant	10^{-10} As/Vm
U_T	thermal voltage at $T_L = 300K$	0.026 V
μ_n/μ_p	low-field carrier mobilities	$0.15/0.045\,\mathrm{m^2/Vs}$
C_0	maximum doping concentration	$10^{23}\,\mathrm{m^{-3}}$
l	length	10^{-7} m
A	cross-section	$10^{-14}\,\mathrm{m^2}$

(b) Physical parameters for a silicon *pn*-junction diode.

Fig. 2 Circuit and device parameters (**a**) Example circuit (**b**) Physical parameters for a silicon pn-junction diode

For the following simulations we applied a constant time step size of $\Delta t = 0.1$ ps and simulate our circuit und until $T = 10$ ps. On each time window, we accomplish ten iterations and compare the network variables computed with our dynamic iteration scheme below, to a monolithic reference solution. – The reference solution is made to verify the convergence of the dynamic iteration scheme to the solution of the monolithic systems. Therefore it is computed with the same step size.

In the case of circuit first and a parallel capacitance, the algorithm reads:

0) **Initialization.** Set first time window to T_n with $n := 0$.
1) **Guess.** Get a circuit solution $(\mathbf{y}_c^{(0)}, \mathbf{z}_c^{(0)})$ on T_n.
2) **Solve the DAE initial value problems.**

   a) Time-integration of the network on T_n

$$\dot{\mathbf{y}}_c^{(1)} = \mathbf{f}_c(\mathbf{y}_c^{(1)}, \mathbf{z}_c^{(1)}, \mathbf{z}_d^{(0)}), \qquad \text{with } \mathbf{y}_c^{(1)}(t_n) = \mathbf{y}_{c,n}$$
$$\mathbf{0} = \mathbf{g}_c(\mathbf{y}_c^{(1)}, \mathbf{z}_c^{(1)})$$

   b) Time-integration of the circuit on T_n

$$\dot{\mathbf{y}}_d^{(1)} = \mathbf{f}_d(\mathbf{y}_d^{(1)}, \mathbf{z}_d^{(1)}), \qquad \text{with } \mathbf{y}_d^{(1)}(t_n) = \mathbf{y}_{d,n}$$
$$0 = \mathbf{g}_d(\mathbf{y}_d^{(1)}, \mathbf{z}_d^{(1)}, \mathbf{y}_c^{(1)}),$$

3) **Sweep Control.** If e.g. $||\mathbf{y}^{(1)} - \mathbf{y}^{(0)}|| >$ tol, then repeat the step, i.e., set $(\mathbf{y}_c^{(0)}, \mathbf{y}_d^{(0)}) := (\mathbf{y}_c^{(1)}, \mathbf{y}_d^{(1)})$ and go to Step 2), otherwise Step 4)
4) **Next window.** If $t_{n+1} < T$, then set new initial values $\mathbf{y}_{\star,n+1} := \mathbf{y}_\star^{(1)}(t_{n+1})$ and proceed to the next time window $n := n + 1$, go to Step 1).

In the other cases, the algorithms read analogously. For the numerical comparisons presented below, we replaced the sweep control in the above algorithm by a fix number of ten iteration per time step. In case of reversed order (diode first) or the source coupling approach, the algorithm has been adjusted accordingly.

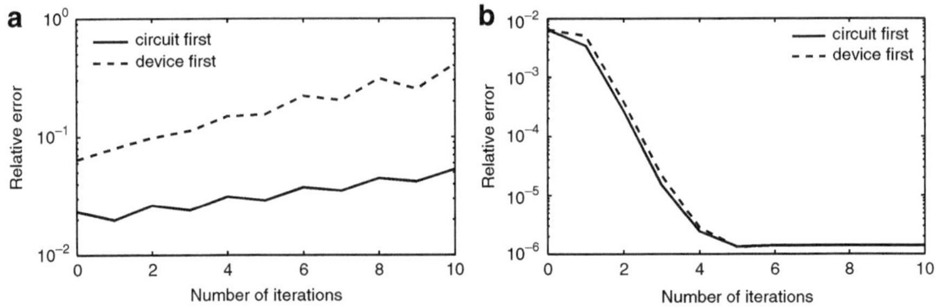

Fig. 3 Relative error of network components between 2.2–2.3 ps (**a**) Source coupling (**b**) Coupling with capacitance

3.1 Source Coupling

First we use the source coupling approach and simulate the simple circuit accordingly. The dynamic iteration scheme does not converge. In Fig. 3a we depict the relative error of the network components, i.e. the deviation from the reference solution, against the number of iterations in the time interval 2.2–2.3 ps. We choose this interval, since there simulation breaks down. We clearly see, that for both orders – device or circuit first – the iteration scheme clearly diverges. The different starting values for the two orders is due to bad convergence in the previous time windows and is the result of error propagation.

3.2 Coupling with Capacitance

In the second approach we extract the capacitive behavior of the diode and model this by a parallel linear capacitance $C_D = 1{,}500\frac{\varepsilon_s A}{l} = 1.5 \times 10^{-14}$ F. In turn, we compute the diode current $\mathbf{i}_{\mathrm{SD}}$ without consideration of the displacement current. In contrast to the source coupling approach, we observe a convergent algorithm. In Fig. 3b we depict the relative error (i.e. the relative deviation from the monolithic reference solution) of the network components against the number of iterations in the interval 2.2–2.3 ps, where the source coupling algorithm broke down. We clearly see, that we get convergence with the capacitance in parallel. Moreover, we observed significantly better convergence on the previous time windows, what we easily can deduct from the coinciding starting values for both orders.

Circuit first performs slightly better, which can be physically motivated, since the circuit drives the diode. – Thus the suggested algorithm is coupling with capacitance and circuit first.

4 Conclusion

For a PDAE-system consisting of coupled subsystems of circuits and devices, dynamic iteration schemes are often based on the source coupling approach. We have shown that for a simple examples this might lead to divergent iteration schemes. We tackled this problem by a dynamic iteration scheme based on the modeling of capacitive effects in the semiconductor devices. The capacitive effects are extracted from the device model and modeled by an additional capacitive path in the circuit. We have shown that this approach leads – in accordance with the theory in [2] – to a convergent dynamic iteration scheme independent of the order of computation of the different subsystems. We note that we do not simply add a capacitance to the circuit in order to aid convergence, but extract the capacitance from the device model. Thus, it is ensured that the modification does not change our coupled system.

The extracted capacitance can also be regarded as a predictor for the capacitive behavior of the semiconductor device and thus it also can be regarded as a compact model for this effect. Thus, our approach also fits into the framework of [6].

With the suggested coupling via the extracted capacitance we get convergence for both orders – circuit first or device first. However, we observe slightly better convergence for the circuit first approach. We shortly note that this is due to the dependence of g_d on y_c. A deeper analysis of this effect is subject to ongoing research.

Acknowledgements The authors acknowledge partial support from the EU within the ICESTARS project, grant number FP7/2008/ICT/214911, from the German Academic Exchange Service "DAAD Jahresprogramm für Doktoranden" and the post-doc program of the "FG Mathematik and Informatik" of the Bergische Universität Wuppertal.

References

1. Alì, G., Bartel, A., Günther, M.: Parabolic differential-algebraic models in electrical network design. SIAM J. Mult. Model. Sim. **4**(3), 813–838 (2005)
2. Arnold, M., Günther, M.: Preconditioned dynamic iteration for coupled differential-algebraic systems. BIT **41**(1), 1–25 (2001). DOI 10.1023/A:1021909032551
3. Bartel, A.: Partial differential-algebraic models in chip design - thermal and semiconductor problems. Ph.D. thesis, Technische Universität Karlsruhe, Düsseldorf (2004). VDI Verlag
4. Brunk, M., Kværnø, A.: Positivity preserving discretization of time dependent semiconductor drift-diffusion equations. to apper in APNUM (2010)
5. Denk, G.: Circuit simulation for nanoelectronics. In: Anile, A., Alì, G., Mascali, G. (eds.) Scientific Computing in Electrical Engineering, pp. 13–20. Springer, Berlin (2006)
6. Ebert, F.: On partitioned simulation of electrical circuits using dynamic iteration methods. Ph.D. thesis, Technische Universität, Berlin (2008)
7. Estévez Schwarz, D., Tischendorf, C.: Structural analysis of electric circuits and consequences for MNA. Int. J. Circ. Theor. Appl. **28**(2), 131–162 (2000)

8. Feldmann, U., Günther, M.: CAD-based electric-circuit modeling in industry I: mathematical structure and index of network equations. Surv. Math. Ind. **8**(2), 97–129 (1999)
9. Marini, D., Pietra, P.: New mixed finite element schemes for current continuity equations. COMPEL **9**, 257–268 (1990)
10. Selva Soto, M., Tischendorf, C.: Numerical analysis of DAEs from coupled circuit and semiconductor simulation. Appl. Numer. Math. **53**(2-4), 471–488 (2005)

Multirate Time Integration of Field/Circuit Coupled Problems by Schur Complements

Sebastian Schöps, Andreas Bartel, and Herbert De Gersem

Abstract When using distributed magnetoquasistatic field models as additional elements in electric circuit simulation, the field equations contribute with large symmetric linear systems that have to be solved. The naive coupling and solving (using direct solvers) is not always efficient, since the electric circuit is coupled only via coils, which are often represented only by a small subset of the unknowns. We revisit the Schur complement approach, give a physical interpretation and show that a heuristics for bypassing Newton iterations allow for efficient multirate time-integration for the field/circuit coupled model.

1 Introduction

Circuit simulators assemble the underlying equations element-wise, usually by modified nodal analysis (MNA). Each element contributes with an element stamp that describes the current/voltage relation and possibly internal equations. This results in a system of Differential Algebraic Equations (DAEs). In our case of the field/circuit problem, parts of this system stem from Maxwell's equations.

In the next section, we summarize the mathematical model for coupled electric circuits with maqnetoquasistatic (MQS) field devices. Our point of view stresses the usual assembly via stamping during time discretization. In the following section, we introduce a Schur complement approach for the MQS stamp, cf. [1, 2]. The

S. Schöps (✉) · A. Bartel
Institut für Numerische Analysis, Bergische Universität Wuppertal, Gaussstrasse 20, 42119 Wuppertal, Germany
e-mail: bartel,schoeps@math.uni-wuppertal.de

S. Schöps · H. De Gersem
Wave Propagation and Signal Processing Research Group, Katholieke Universiteit Leuven - Campus Kortrijk, Etienne Sabbelaan 53, 8500 Kortrijk, Belgium
e-mail: Herbert.DeGersem@kuleuven-kortrijk.be

B. Michielsen and J.-R. Poirier (eds.), *Scientific Computing in Electrical Engineering SCEE 2010*, Mathematics in Industry 16, DOI 10.1007/978-3-642-22453-9_26,

fourth section deals with the corresponding computational cost. Then, in section five, a bypassing technique of the Jacobian, similar to simplified Newton, and the bypassing of the right-hand-side are interpreted and employed as a multirate time-integration scheme. Also bypassing is a common technique in classical circuit simulation, but here the energy balances of field and circuit are taken into account. The paper is completed by numerical results and conclusions.

2 Mathematical Model Description and Time Discretization

Common circuit simulators use MNA to assemble the circuit equations element-wise. Each element contributes with differential and algebraic relations to the underlying DAE, [3]. To the list of basic elements, the magnetoquasistatic (MQS) device is added (with subscript M), which allows the coupling to field effects while still using MNA. For each element we have a model $\mathbf{f}_e$ consisting of current balances for the network nodes (except ground) and additional constitutive relations for non-current defining elements, which gives

$$\mathbf{f}(\dot{\mathbf{y}}, \mathbf{y}, t) := \sum_e \mathbf{Q}_e \mathbf{f}_e(\dot{\mathbf{y}}_e, \mathbf{y}_e, t) + \mathbf{Q}_M \mathbf{f}_M(\dot{\mathbf{y}}_M, \mathbf{y}_M) = 0 \tag{1}$$

using for element e: local variables $\mathbf{y}_e$ and generalized topology matrices $\mathbf{Q}_e$ that map local to global variables, such that holds $\mathbf{y}_e = \mathbf{Q}_e^\top \mathbf{y}$. The global unknowns $\mathbf{y}$ consist of the node potentials, whose differences define the respective voltage drop $\mathbf{v}_e$ at each element, and of several currents in particular the currents through MQS devices $\mathbf{i}_M$, [4]. All currents contribute to the balances required by Kirchhoff's Current Law (KCL), which is included in (1).

The field distribution of the MQS device is described in terms of the degrees of freedom of the discretized magnetic vector potential (MVP) $\mathbf{a} = \mathbf{a}(t)$, e.g. by the finite integration technique (FIT) or the finite element method (FEM):

$$\mathbf{M}\dot{\mathbf{a}} + \mathbf{K}(\mathbf{a})\mathbf{a} = \mathbf{X}\mathbf{i}_M, \tag{2a}$$

$$\mathbf{X}^\top \dot{\mathbf{a}} = \mathbf{v}_M - \mathbf{R}\mathbf{i}_M. \tag{2b}$$

Equation (2a) stems from the continuous curl-curl equation, where $\mathbf{M}$ and $\mathbf{K}(\mathbf{a})$ denote the singular conductivity matrix and the curl-curl matrix with the nonlinear reluctivity $(\mathbf{a})$ employed. $\mathbf{K}(\mathbf{a})$ includes gauging (e.g. grad-div) and boundary conditions, [5], such that a positive definite matrix pencil $\left(\frac{1}{h}\mathbf{M} + \mathbf{K}\right)$ is obtained. The problem is completed with a initial value $\mathbf{a}_0$. The columns of the coupling matrix $\mathbf{X} = [\mathbf{X}_1, \ldots, \mathbf{X}_p]$ distribute the branch currents $\mathbf{i}_M$ on the spatial grid, [6].

The second equation is the coupling equation: it relates the branch voltage $\mathbf{v}_M$ to the MVP and to the branch current using linear DC resistance $\mathbf{R}$. All common conductor types (solid, stranded and foil conductors) can be realized in (2) by the structure of the conductivity matrix, [6]. Summing up, the field model $\mathbf{f}_M$ reads:

$$\mathbf{f}_M\left(\dot{\mathbf{y}}_M, \mathbf{y}_M\right) := \begin{bmatrix} \mathbf{i}_M \\ \mathbf{X}^\top \dot{\mathbf{a}} - \mathbf{v}_M + \mathbf{R}\mathbf{i}_M \\ \mathbf{M}\dot{\mathbf{a}} + \mathbf{K}(\mathbf{a})\mathbf{a} - \mathbf{X}\mathbf{i}_M \end{bmatrix}, \quad \text{where } \mathbf{y}_M = \begin{bmatrix} \mathbf{i}_M \\ \mathbf{v}_M \\ \mathbf{a} \end{bmatrix}, \tag{3}$$

where the first row contains the contribution to the KCL and the last row represent the curl-curl equation (2a).

Typically, circuit simulators use BDF schemes for time discretization. This gives for constant step size h a nonlinear system at each discrete time t_n for $\mathbf{y}_n \approx \mathbf{y}(t_n)$, [7]:

$$\mathbf{f}\left(\frac{1}{h}\rho\mathbf{y}_n, \mathbf{y}_n, t_n\right) = 0 \quad \text{with} \quad \frac{1}{h}\rho\mathbf{y}_n := \frac{1}{h}\sum_{i=0}^{k} \alpha_i \mathbf{y}_{n-i} \approx \dot{\mathbf{y}}_n$$

using coefficients α_i (k-th order BDF). As usually Newton-Raphson is applied:

$$\mathbf{J}_n^{(i)}\mathbf{y}_n^{(i+1)} = -\mathbf{f}_n^{(i)} + \mathbf{J}_n^{(i)}\mathbf{y}_n^{(i)} \quad \text{with} \quad \mathbf{J}_n^{(i)} := \frac{\partial \mathbf{f}}{\partial \mathbf{y}_n}\left(\frac{1}{h}\rho\mathbf{y}_n^{(\mathrm{i})}, \mathbf{y}_n^{(\mathrm{i})}, t_n\right) \tag{4}$$

$$\mathbf{f}_n^{(i)} := \mathbf{f}\left(\frac{1}{h}\rho\mathbf{y}_n^{(\mathrm{i})}, \mathbf{y}_n^{(\mathrm{i})}, t_n\right).$$

Due to the structure of (1), the assembly of the Newton system (4) is performed by a cycle over all circuit elements (which can be organized in parallel), such that

$$\mathbf{J}_n^{(i)} := \sum_{e,M} \mathbf{Q}_e \mathbf{J}_e^{(i)} \mathbf{Q}_e^\top \text{ with } \mathbf{J}_e^{(i)} := \frac{\alpha_0}{h}\frac{\partial \mathbf{f}_e^{(i)}}{\partial \dot{\mathbf{y}}_e} + \frac{\partial \mathbf{f}_e^{(i)}}{\partial \mathbf{y}_e}, \; \mathbf{f}_e^{(i)} := \mathbf{f}_e\left(\frac{1}{h}\rho\mathbf{y}_e^{(i)}, \mathbf{y}_e^{(i)}, t_n\right) \tag{5}$$

suppressing the time index n. This resembles the element-wise assembly in FEM. Each contribution ("stamp"), $\mathbf{J}_e^{(i)}$, $\mathbf{f}_e^{(i)}$, consists of inner and external variables, i.e., variables used only inside the particular element and variables related to other elements by the simulator, [2].

In the following we want to speed up solving the Newton system by elimination of the MVP $\mathbf{a}$. Therefore we work out the MQS stamp and revisit the Schur complement next.

3 MQS Stamp and Schur Complement

For the MQS model (3) in terms of $\mathbf{y}_M^\top = (\mathbf{i}_M^\top, \mathbf{v}_M^\top, \mathbf{a}^\top)$, we obtain the following Jacobian stamp (for BDF time discretization):

$$\mathbf{J}_M^{(i)} := \begin{bmatrix} \mathbf{I} & 0 & 0 \\ \mathbf{R} & -\mathbf{I} & \frac{\alpha_0}{h}\mathbf{X}^\top \\ -\mathbf{X} & 0 & \mathbf{K}_h^{(i)} \end{bmatrix} \qquad \text{with } \mathbf{K}_h^{(i)} := \underbrace{\left.\frac{\mathrm{d}\mathbf{K}(\mathbf{a})}{\mathrm{d}\mathbf{a}}\right|_{\mathbf{a}=\mathbf{a}^{(i)}}}_{=:\mathbf{k}_a(\mathbf{a}^{(i)})} + \frac{\alpha_0}{h}\mathbf{M} \tag{6}$$

and differential reluctivity matrix $\mathbf{k}_a(\mathbf{a}^{(i)})$. The function-evaluation stamp reads:

$$\mathbf{f}_M^{(i)} := \begin{bmatrix} \mathbf{I} & 0 & 0 \\ \mathbf{R} & -\mathbf{I} & 0 \\ -\mathbf{X} & 0 & \mathbf{K}(\mathbf{a}^{(i)}) \end{bmatrix} \mathbf{y}_M^{(i)} + \begin{bmatrix} 0 \\ \mathbf{X}^\top \\ \mathbf{M} \end{bmatrix} \frac{1}{h}\rho\mathbf{a}^{(i)}.$$

and the right-hand side contribution is given by:

$$\mathbf{r}_M^{(i)} := -\mathbf{f}_M^{(i)} + \mathbf{J}_M^{(i)}\mathbf{y}_M^{(i)} = \frac{1}{h}\begin{bmatrix} 0 \\ \mathbf{X}^\top \\ \mathbf{M} \end{bmatrix} \left(\alpha_0\mathbf{a}^{(i)} - \rho\mathbf{a}^{(i)}\right) + \begin{bmatrix} 0 \\ 0 \\ \mathbf{k}_a(\mathbf{a}^{(i)}) - \mathbf{K}(\mathbf{a}^{(i)}) \end{bmatrix} \mathbf{a}^{(i)}. \tag{7}$$

For the MQS devices, only the current/voltage relation of the series connection of a (nonlinear) inductor and a resistor needs to be unveiled to the host circuit simulator. But the inner variables $\mathbf{a}$ is not used outside the MQS stamp, it can be eliminated from the Newton system by the well-known Schur complement, that is, to indeed reduce the element stamp. – This is especially beneficial for all kinds elements with rather large stamps, e.g. semiconductors, [2], or MQS device, [1], since more compact stamps are obtained, which fit better into the overall framework. – Removing $\mathbf{a}$ yields a reduced stamp in terms of $\widetilde{\mathbf{y}}_M^\top = (\mathbf{i}_M^\top, \mathbf{v}_M^\top)$. The reduced Jacobian reads

$$\widetilde{\mathbf{J}}_M^{(i)} := \begin{bmatrix} \mathbf{I} & 0 \\ \mathbf{R} + \frac{\alpha_0}{h}\mathbf{L}_h^{(i)} & -\mathbf{I} \end{bmatrix} = \begin{bmatrix} \mathbf{I} & 0 & 0 \\ 0 & \mathbf{I} & -\frac{\alpha_0}{h}\mathbf{X}^\top \left(\mathbf{K}_h^{(i)}\right)^{-1} \end{bmatrix} \mathbf{J}_M^{(i)} \begin{bmatrix} \mathbf{I} & 0 \\ 0 & \mathbf{I} \\ 0 & 0 \end{bmatrix} \tag{8}$$

with generalized inductance matrix

$$\mathbf{L}_h^{(i)} := \mathbf{X}^\top \left(\mathbf{K}_h^{(i)}\right)^{-1} \mathbf{X} \tag{9}$$

using $\mathbf{K}_h^{(i)}$ from (6) and corresponding reduced right-hand side contribution:

$$\widetilde{\mathbf{r}}_M^{(i)} = \begin{bmatrix} 0 \\ \widetilde{\mathbf{r}}_{M,v}^{(i)} \end{bmatrix} \quad \text{where} \quad \widetilde{\mathbf{r}}_{M,v}^{(i)} = \frac{1}{h}\mathbf{X}^\top \left(\mathbf{I} - \frac{\alpha_0}{h}\left(\mathbf{K}_h^{(i)}\right)^{-1}\mathbf{M}\right)\left(\alpha_0\mathbf{a}^{(i)} - \rho\mathbf{a}^{(i)}\right)$$
$$- \frac{\alpha_0}{h}\mathbf{X}^\top \left(\mathbf{K}_h^{(i)}\right)^{-1} \left(\mathbf{k}_a(\mathbf{a}^{(i)}) - \mathbf{K}(\mathbf{a}^{(i)})\right)\mathbf{a}^{(i)}.$$

We notice that the MVP needs still to be computed to evaluate the reduced right-hand side (and the nonlinear material curve). Equation (9) corresponds the common inductance extraction approach, [8], but in addition the Schur complement takes eddy current effects into account (due to the conductance matrix). – Moreover, the dimension of the reduced stamp is independent of the space discretization of the field problem. Thus the spatial mesh can be refined and coarsened during the time-integration without restarting the host-simulator. Nevertheless, the reduction comes with additional cost.

4 Computational Cost for Schur Complement (Direct Solver)

For the Schur complement in the Newton iteration $i + 1$, we need to compute $\mathbf{L}_h^{(i)}$. Applying a direct solver, the matrix $\mathbf{K}_h^{(i)}$ has to be factorized (one LU decomposition) and forward/backward substitutions for the vector potentials in each branch:

$$\mathbf{K}_h^{(i)}\mathbf{a}_{M,j}^{(i)} = \mathbf{X}_j \qquad (\text{for } j = 1, \ldots, p), \quad \text{s.t.} \quad \mathbf{L}_h^{(i)} = \mathbf{X}^\top \mathbf{a}_M^{(i)} \tag{10}$$

by sparse inner products. Also, the MVP for the right-hand-side voltage must be computed. To this end, we project onto the MVP defining equation inside the Newton iteration (derived from Jacobian (6) and right-hand side (7)):

$$\mathbf{K}_h^{(i)}\mathbf{a}^{(i+1)} = \mathbf{r}_{M,a}^{(i)} + \mathbf{X}\mathbf{i}_M^{(i+1)}, \quad \mathbf{r}_{M,a}^{(i)} := \frac{1}{h}\mathbf{M}\big(\alpha_0\mathbf{a}^{(i)} - \rho\mathbf{a}^{(i)}\big) + \big(\mathbf{k}_a(\mathbf{a}^{(i)}) - \mathbf{K}(\mathbf{a}^{(i)})\big)\mathbf{a}^{(i)}.$$

Thus we compute the remaining term $\mathbf{a}_V^{(i)}$ by forward/backward substitutions from:

$$\mathbf{K}_h^{(i)}\mathbf{a}_V^{(i)} = \mathbf{r}_{M,a}^{(i)}, \tag{11}$$

and obtain for the MVP

$$\mathbf{a}^{(i+1)} = \mathbf{a}_V^{(i)} + \mathbf{a}_M^{(i)}\mathbf{i}_M^{(i+1)}.$$

Moreover, we obtain for the reduced right-hand side the simplification:

$$\widetilde{\mathbf{r}}_{M,v}^{(i)} = \frac{1}{h}\mathbf{X}^\top \left(\alpha_0\mathbf{a}^{(i)} - \rho\mathbf{a}^{(i)} - \mathbf{a}_V^{(i)}\right).$$

Thus one LU-decomposition and $p + 1$ forward/backward substitutions are necessary for the Schur complement. The choice of solver for the Schur complement is independent of the solver used in circuit host simulator. So, for example an iterative method such as block-PCG could be used. Such a procedure should support multiple right-hand-sides, as [9], for efficiency. A further advantage of iterative methods applied to 3D problems is the *weak gauging* introduced by the iterative solver, [10], such that further gauging, as employed here, becomes unnecessary.

5 Bypassing as Multirate Time Integration

The generalized inductance matrix depends on the saturation (BH-curve), but this effect is rather slow compared to other time rates of the electric circuit, e.g. the switching frequency of transistors. Saturation depends on the supplied energy

$$\mathbf{E}(t_n) = \mathbf{E}_0 + \int_0^{t_n} \mathbf{i}_M(s)\mathbf{v}_M(s)\mathrm{d}s \tag{12}$$

with the initial energy level of the device $\mathbf{E}_0$. The relevant time scale of the nonlinearity is given by the integral above, even if the applied voltage is a much faster signal. We approximate (12) by

$$\mathbf{E}_n \approx \mathbf{E}_0 + h\left(\sum_{j=0}^{n-1} \mathbf{i}_M(t_j)\mathbf{v}_M(t_j) + \mathbf{i}_M^{(i)}\mathbf{v}_M^{(i)}\right)$$

and compare this to the initial energy $\mathbf{E}_0$. Consequently, updates of the material are often superfluous: whenever it behaves (nearly) linearly, only one forward/backward substitution for the right-hand-side per iteration is necessary. This allows an interpretation as a simplified Newton algorithm, where the Jacobian (8) is frozen across several iterations and possibly several time steps if the (relative) change of energy does not exceed a threshold and if the reluctivity is (nearly) constant.

Furthermore, if the material is rather linear the right-hand-side evaluation can be bypassed as well: then the vector potential needs no update and the distributed field problem is decoupled from the circuit, where it is represented by an inductance matrix, similarly to the co-simulation approach, [8]. The algorithm reads

0) compute $\mathbf{a}^{(i)}$
1) approximate the energy $\mathbf{E}_n^{(i)}$
2) **if** norm$(\mathbf{E}_n^{(i)} - \mathbf{E}_0) > tol$

then evaluate material curve ${}^{(i)} := (\mathbf{a}^{(i)})$

2a) **if** $||\nu^{(i)} - \nu^{(i-1)}|| > tol$
then compute $\mathbf{L}_h^{(i)}$ and $\mathbf{v}_M^{(i)}$
else bypass matrix update $\mathbf{L}_h^{(i)} := \mathbf{L}_h^{(i-1)}$ and $\mathbf{v}_M^{(i)} := \mathbf{v}_M^{(i-1)}$

else bypass material update $\nu^{(i)} := \nu^{(i-1)}$ and $\mathbf{L}_h^{(i)} := \mathbf{L}_h^{(i-1)}$, $\mathbf{v}_M^{(i)} := \mathbf{v}_M^{(i-1)}$
3) return to host simulator.

where the change in the energy level is by measured by a relative norm. This algorithm unburdens the host simulator from solving unnecessarily large system of equations, while still having a suitable Jacobian information at hand. The drawback are additional Newton iterations due to the inferior convergence of simplified Newton, but solving a sequence of reduced system. If eddy currents included into the model, the inductance matrix $\mathbf{L}_h = \mathbf{L}(\mathbf{a}, h)$ depends on the time step size h and therefore the matrix must be recomputed or interpolated for any change of h.

6 Computational Results

All presented methods, the standard Newton without Schur complement, and the ones using the Schur approach, i.e., simplified and bypassed Newton, have been implemented in the Framework of the CoMSON DP. In Fig. 1a a simple example circuit is shown, where a pulse width modulated (PWM) voltage source is connected to the primary side of a transformer. The PWM is switching at 20 kHz, Fig. 1b. The secondary side is connected by a resistor. This transformer has a highly nonlinear behavior that is simulated until its saturation phase is reached, Fig. 1c.

In the beginning of the start-up phase, $t \leq 0.03$ s, both, the simplified and bypassed Newton methods detect the linearity in the material and skip the superfluous evaluations and the LU decompositions, Fig. 2b, although the applied voltages and currents are fast switching due to the PWM. The standard Newton method employed here follows the rather naive procedure to evaluate the material in every iteration, see Table 1. In the highly nonlinear saturation phase, $0.03\,\text{s} < t \leq 0.06$, all methods require more Newton iterations per step and update the Jacobian almost at every time step. Bypassing element evaluations implies a linearity assumption and as a consequence the Newton iteration will require less Jacobian updates, Fig. 2a, but with the drawback of a larger error. After the saturation level is reached, $t > 0.06$, the field problem behaves again rather linearly and the updates of the simplified and bypassing Newton are clearly reduced.

The performance of this approach depends on the choice of the error norms, tolerances and device characteristics. We found the heuristic to be insensitive to changes in the parameters, especially for rather linear or fully saturated models, because the change in the saturation cause the high computational costs. For example in an induction machine, where the saturation rotates, we are forced to recompute the Schur complement more often, but the rotation is still determined by the energy and not by the frequency of a pulsed input. Especially in those

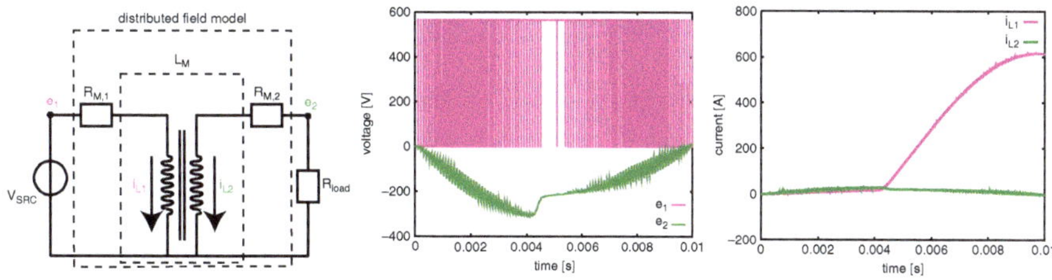

Fig. 1 Transformer: setup and reference solution (**a**) Transformer circuit (**b**) Applied nodal potentials (**c**) Currents through coils

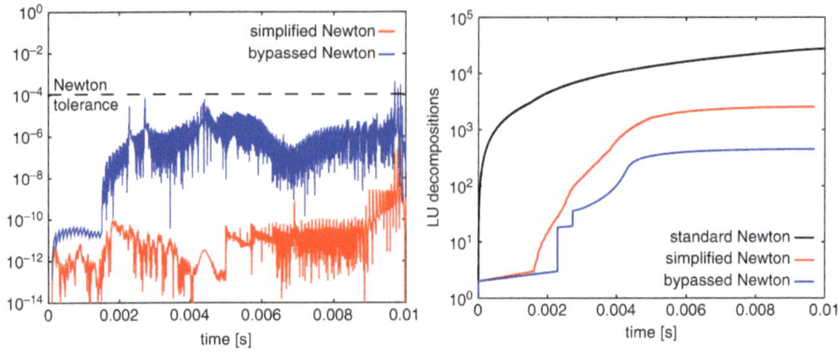

Fig. 2 Transformer: reference solution, errors and decompositions. (**a**) Relative errors with respect to the reference solution obtained by full Newton-Raphson (**b**) Number of LU decompositions, full Newton requires one decomposition per iteration

Table 1 Transformer: computational costs

	Decompositions	Forward/ backward substitutions	Stamp evaluations	Time
Full Newton	23,371	27,936	27,936	20 h
Simplified Newton	2,531	36,460	31,398	1 h
Bypassed Newton	450	3,171	20,449	25 min

applications one can further optimize the method and interpolate from previous Schur complements in dependence of the rotor angle and reuse them in the stationary phase.

When using an adaptive step-size control it should reflect that the recomputation of the inductance matrix $\mathbf{L}_h$ should be avoided if the step size h changes only insignificantly. On the other hand the application of an adaptive step size control to problems with pulsed inputs as described here is not recommended due to high amounts of rejected steps. Thus a fixed step size is here typically no additional constraint.

7 Conclusion

Applying the Schur complement approach to MQS devices yields small element stamps that are equivalent to the constitutive relation of the series connection of inductors and resistors. The additional costs of the complement computation can be neglected if solvers with multiple right-hand side techniques are available. The presented heuristics to bypass Newton iterations reduce the computational costs clearly and they automatically detect when full Newton iterations are necessary. Due to bypassing, both problems are quasi decoupled and the time-integration of the circuit is cheapened because only basic elements are evaluated. This decoupling exploits the multirate time behavior of the coupled system if present.

Acknowledgements This work was partially supported by the EU within the ICESTARS project, grant number FP7/2008/ICT/214911, by the German Academic Exchange Service "DAAD Jahresprogramm für Doktoranden" and the post-doc program of the "FG Mathematik und Informatik" of the Bergische Universität Wuppertal.

References

1. Vaananen, J.: Circuit theoretical approach to couple two-dimensional finite element with external circuit equations. IEEE Trans. Magn. **32**(2), 400–410 (1996)
2. Feldmann, U., Miyake, M., Kajiwara, T., Miura-Mattausch, M.: On local handling of inner equations in compact models. In: Roos and Costa: Scientific Computing in Electrical Engineering SCEE 2008, pp. 143–150. Springer, Berlin (2010)
3. Feldmann, U., Günther, M.: CAD-based electric-circuit modeling in industry I: mathematical structure and index of network equations. Surv. Math. Ind. **8**(2), 97–129 (1999)
4. Bartel, A., Baumanns, S., Schöps, S.: Structural analysis of electrical circuits including magnetoquasistatic devices. Submitted.
5. Schöps, S., Bartel, A., De Gersem, H., Günther, M.: DAE-index and convergence analysis of lumped electric circuits refined by 3-D MQS conductor models. In: Roos and Costa: Scientific Computing in Electrical Engineering SCEE 2008, pp. 341–350. Springer, Berlin (2010)
6. De Gersem, H., Weiland, T.: Field-circuit coupling for time-harmonic models discretized by the finite integration technique. IEEE Trans. Magn. **40**(2), 1334–1337 (2004)
7. Hairer, E., Nørsett, S.P., Wanner, G.: Solving Ordinary Differential Equations II: Stiff and Differential-Algebraic Problems, 2nd edn. Springer Series in Computational Mathematics. Springer, Berlin (1996)
8. Schöps, S., De Gersem, H., Bartel, A.: A cosimulation framework for multirate time-integration of field/circuit coupled problems. IEEE Trans. Magn. **46**(8), 3233–3236 (2010)
9. Wimmer, G., Steinmetz, T., Clemens, M.: Reuse, recycle, reduce (3r) – strategies for the calculation of transient magnetic fields. Appl. Numer. Math. **59**(3-4), 830–844 (2008)
10. Clemens, M., Weiland, T.: Iterative methods for the solution of very large complex-symmetric linear systems in electrodynamics. In: CMCMM, vol. 2, pp. 1–7 (1996)

Part IV
Circuit and Device Modelling and Simulation

Introduction

As stated in the introduction to this book, with regard to the modelling of configurations for which propagation delay is negligible, macroscopic electromagnetic interactions can be modelled using the "quasistatic" asymptotic approximation to the Maxwell equations. The specific properties of the equations of quasistatic electromagnetic interactions produce the vast domain of circuit theory. The geometrical size of configurations that can be handled by circuit theory is limited by the rate of change in the states. The smaller the configuration we have to study, the faster the changes we can accurately handle. On the other hand, on the scale of the interior of an elementary electronic device like a transistor, the Maxwell macroscopic equations themselves are not appropriate and one has to resort to quantum physical models and statistical physics. This part presents advances in both types of approaches.

In the analysis of electronic circuits, one has to deal with the nonlinearities of electronic devices. Therefore, many solution techniques rely on evolution equations in which the time derivative of the electronic state is computed from the state itself and its past history. The uniqueness problems and the stability of the algorithms to advance the (necessarily discretised) states in time represent major challenges in numerical analysis. Alternative techniques rely on "entire-domain" expansions, in which the evolution of a system's state over a given time interval is represented as a sum of global basis functions over the interval. Such methods bring with them different mathematical problems of solvability and numerical accuracy. This part puts forward papers on both techniques.

The first paper in this part, written by H. Thornquist (an invited speaker at the conference) and E. Keiter, presents recent advances in parallel computation techniques in "full-chip" circuit simulations, including new preconditioning strategies for the pertaining system matrices.

The following two papers are related to the general problem of "modelling uncertainty," which complicates any physical modelling. Even when we solve our model equations with great accuracy there is no guarantee that we have a very precise correspondence with the pertaining physical observations. Computations are made based on the assumption that the coefficients of the model equations are representative of the materials and their geometries. However, any simulation result should be accompanied by an indication of the sensitivity of the essential aspects of the result to changes in the most pertinent model coefficients.

The paper by F. Veersé, J. Besnard and H. Filiol proposes a method for the analysis of the sensitivity of steady-state mismatch deviations in a circuit due to device model parameter deviations. Their method avoids the costly multiple computations required by, for example, a Monte Carlo-type approach to the problem. The domain of validity of the method is discussed and illustrated in numerical examples. The paper by R. Pulch addresses a modelling uncertainty problem by considering it as a stochastic process problem. The author uses a polynomial chaos method to compute the propagation of the uncertainty in a model parameter, the period in a forced oscillation, to the uncertainty in the resulting electronic state. The method is illustrated in a numerical example.

The following three papers deal with steady-state problems with oscillators. The paper by M. Hulkkonen et al. proposes improvements in harmonic balance methods both with regard to the initial estimates of the oscillator frequency and amplitude and to the convergence. Numerical experiments validate the proposed methods. The paper by R. Mirzavand et al. presents a new gauge technique for the Newton Raphson method for finding the periodic steady state (PSS) of free-running oscillators in the time domain. The method is tested on a benchmark problem and is shown to perform better than more conventional phase-shift condition methods. The next paper by M. Gourary et al. puts forward a new general method that uses frequency-dependent admittance matrices to model the parasitic coupling of oscillators. The error estimates given for the explicitly computed locking frequency are confirmed by a revealing SPICE simulation.

The following two papers address problems arising when different time scales are significant in a single computation. The paper by V. Savcenco et al. proposes a method for solving power system problems showing different time scales in the solution of the evolution equations. The partitioning of the system components according to the maximum allowed time step is done automatically on the basis of the system's topology. The paper by K. Bittner and E. Dautbegovic presents a method in which the time evolution of the state of an electronic circuit is discretised using a Galerkin method. Their method includes a wavelet-based adaptive refinement that results in a highly accurate solution with a performance comparable to that of time-stepping methods.

The remaining four papers in this part treat various aspects of device modelling. The paper by C. de Falco et al. presents a mathematical study of a model for polymer solar cells in the form of a system of nonlinear diffusion-reaction partial differential equations. In addition to an existence proof, an algorithm for the solution is proposed and a numerical example for a photovoltaic cell is discussed. The paper by G. Mascali and V. Romano presents a detailed analysis of a model for semiconductors. The authors use the Schrödinger-Poisson-Boltzmann equations and show how to obtain appropriate closure relations accounting for various scattering processes. A numerical solution scheme is illustrated on the example of a nanoscale silicon diode. The paper by Y. Li and Y.-C. Chen presents a geometric programming method for the optimisation of doping profiles in MOSFET devices. The paper by V. Romano and A. Rusakov explores a new way of modelling the heating of a semiconductor. The model equations are obtained by means of a maximum entropy principle. An iterative solution method is used to obtain stationary solutions and numerical simulations are shown, illustrating the differences to simpler approaches to the heating problem on a two-dimensional configuration.

Advances in Parallel Transistor-Level Circuit Simulation

Heidi K. Thornquist and Eric R. Keiter

Abstract Parallel transistor-level circuit simulation has the potential to significantly impact the need for reliably determining parasitic effects for modern feature sizes. Incorporating parallelism into a simulator at both coarse and fine-grained levels, through the use of message-passing and threading paradigms, is supported by the advent of inexpensive clusters, as well as multi-core technology. However, its effectiveness is reliant upon the development of efficient parallel algorithms for traditional "true SPICE" circuit simulation. In this paper, we will discuss recent advances in fully parallel transistor-level full-chip circuit simulation, concluding with scaling results from a newer strategy for the parallel preconditioned iterative solution of circuit matrices.

1 Introduction

At modern technology nodes (45 nm and below) analog style, SPICE-accurate simulation can be a significant (and prohibitive) development bottleneck. Traditional circuit simulation, originally made popular by the Berkeley SPICE program[17], does not scale well beyond tens of thousands of unknowns, due to reliance on direct matrix solver methods.

Over the years a number of algorithms for so-called "fastSPICE" tools have been developed to enable faster, larger-scale circuit simulation. Often based on

H.K. Thornquist (✉) · E.R. Keiter
Electrical Systems Modeling Department, Sandia National Laboratories
P.O. Box 5800, Albuquerque, NM 87185-0316
e-mail: hkthorn@sandia.gov; erkeite@sandia.gov

Sandia National Laboratories is a multi-program laboratory managed and operated by Sandia Corporation, a wholly owned subsidiary of Lockheed Martin Corporation, for the U.S. Department of Energy's National Nuclear Security Administration under contract DE-AC04-94AL85000.

B. Michielsen and J.-R. Poirier (eds.), *Scientific Computing in Electrical Engineering SCEE 2010*, Mathematics in Industry 16, DOI 10.1007/978-3-642-22453-9_27,

circuit-level partitioning algorithms [1,3,5], such tools can be applied to much larger problems, but the approximations inherent to such algorithms can break down for modern feature sizes. In particular, for state-of-the-art modern VLSI design, high levels of integration between functional modules and interconnects are subject to prohibitive parasitic effects, rendering such tools unreliable. As a result, the ability to perform transient simulation of the full-system is desirable for modern feature sizes.

With the advent of inexpensive computer clusters, as well as multi-core technology, the potential exists for performing large-scale, SPICE-accurate simulation. In fact, parallelism can be incorporated into a simulator at both coarse and fine-grained levels, through the use of message-passing and threading paradigms. The development of efficient parallel algorithms for circuit simulation is an active area of research, which has seen some success, but has not reached its full potential.

2 Background

Traditional circuit simulation, such as SPICE [17], is based on the set of nonlinear differential algebraic equation (DAEs)

$$f(x(t)) + \frac{d}{dt}q(x(t)) - b(t) = 0 \tag{1}$$

where $x(t) \in \mathbb{R}^N$ is the vector of nodal voltages and branch currents, q and f are functions representing the dynamic and static circuit elements (respectively), and $b(t) \in \mathbb{R}^M$ is the input vector. This set of equations, which are more generally expressed as $F(x, x') = 0$, is solved by numerical integration methods, resulting in the nested solver loop in Fig. 1.

Transient analysis of these circuit equations (1) employs an implicit time discretization scheme, such as Backward Euler or the trapezoid rule, that requires the solution to a sequence of nonlinear equations, $F(x) = 0$. Typically, Newton's method is used to solve these nonlinear equations, resulting in a sequence of linear systems

$$Ax = b$$

that involve the conductance, $G(t) = \frac{df}{dx}(x(t))$, and capacitance, $C(t) = \frac{dq}{dx}(x(t))$, matrices. During the DC operating point (DCOP) calculation, the q terms are not present in (1), so the linear system only involves the conductance matrix.

For transient and DC analysis, the computational expense is in repeatedly solving linear systems of equations, which are at the center of the nested solver loop (Fig. 1). Solving these linear systems requires their assembly, which depends upon device evaluations for the whole circuit. So, normally, the computational expense is dominated by either the device evaluations or the numerical method used to solve the linear systems. Techniques for improving the efficiency of device evaluations are relatively straightforward when compared with the numerical issues encountered during the linear system solve.

Fig. 1 Nested solver loop

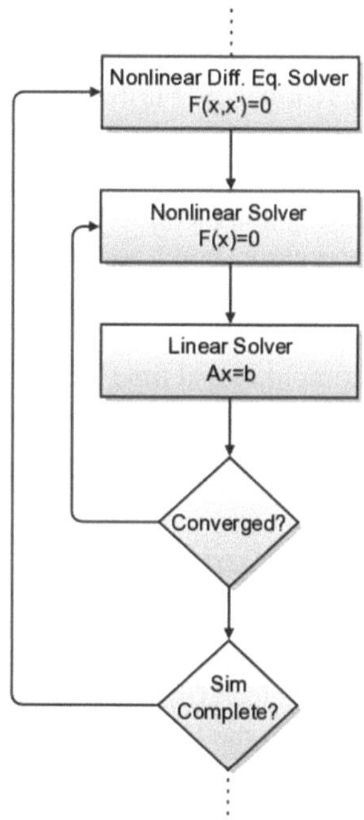

The linear systems solved during transient and DC analysis are typically sparse, have heterogeneous non-symmetric structure, and are often ill-conditioned. As such, iterative matrix solvers have historically not been the first choice for circuit simulation, and direct sparse solvers [9, 15] have been the industry standard approach. Direct solvers have the advantage of reliability and ease of use, and for smaller problems direct methods are usually faster than iterative methods. However, direct solvers typically scale poorly with problem size and become impractical when the linear system has hundreds of thousands of unknowns or more.

3 Parallel Transistor-Level Circuit Simulation

Recent development of inexpensive computer clusters, as well as multi-core technology, has resulted in significant interest for efficient parallel circuit simulation. Many commercial tools, like HSPICE [4], have integrated multithreading approaches to obtain reasonable speedups on a small number of cores. Several numerical techniques for parallel circuit simulation have been investigated, including Fröhlich [11], who used a multi-level Newton approach in the TITAN simulator; Peng et al. [19] used a domain decomposition approach that relied on a combination of direct and iterative solvers; and Dong et al. [10] emulated hardware pipelining for time integration by computing circuit solutions at multiple adjacent time points in the WavePipe simulator. MAPS [24] provides a framework that runs multiple simulation strategies in parallel, with synchronization, to ensure more robust performance. Furthermore, interest has developed recently around parallel SPICE acceleration using graphical processing units (GPUs) [2, 13].

Parallelism can be integrated into every step of the nested solver loop shown in Fig. 1. Furthermore, parallelism can be achieved through both coarse-scale (multi-processor) and fine-scale (multi-threaded) approaches. A composition of these two approaches will provide circuit simulation with the best performance impact on the widest variety of parallel platforms. As discussed before, the majority of the computational time is spent in device evaluations and linear solvers, so this section will focus on parallelism pertaining to those specific tasks.

3.1 Parallel Load Balance

Circuit problems tend to be heterogeneous, so the optimal parallel load balance for device load (matrix and residual vector assembly) functions will likely be different than for the linear solution phase. Device loads and linear solves each happen once per Newton iteration, so over the course of a long run the combined cost of both will comprise the bulk of the wall clock simulation time.

For smaller problems, the load phase should dominate run time. As the problem size increases, the linear solve phase will dominate, and it should scale super-linearly, while the loads should scale linearly. This is because linear solution methods are generally communication intensive, while the communication volume required during the loads is relatively small.

As a result, the device load phase can be balanced by taking into account only the computational work required, while balancing the matrix structure must minimize communication volume. How this communication volume is measured or optimized is an active area of research for many types of numerical simulation problems. Since the load and solve phases have different load balance requirements, it makes sense to have completely different parallel load balance for each. The relative amount of time spent in each phase is problem-dependent.

Figure 2 is a simple illustration of a coarse-scale load balancing approach for device evaluation and matrix structure that can be used in circuit simulation. For many circuits of interest, a naive load balance, in which the total number of devices is evenly divided among the available processors will demonstrate very good parallel scaling. For circuits that are very heterogeneous, weights can be applied to different device types to achieve a better balance. Once the appropriate coarse-scale balancing of the devices across processors is achieved, fine-scale balancing that uses multi-threading to accelerate each processors computations can be used.

The middle box in Fig. 2 represents the communication necessary to accommodate both load balances. This is dependent upon the partitioning of the linear system, which is a much more difficult and complex issue. Unlike the device evaluation, a naive partitioning will generally not suffice. In practice, this partitioning of the

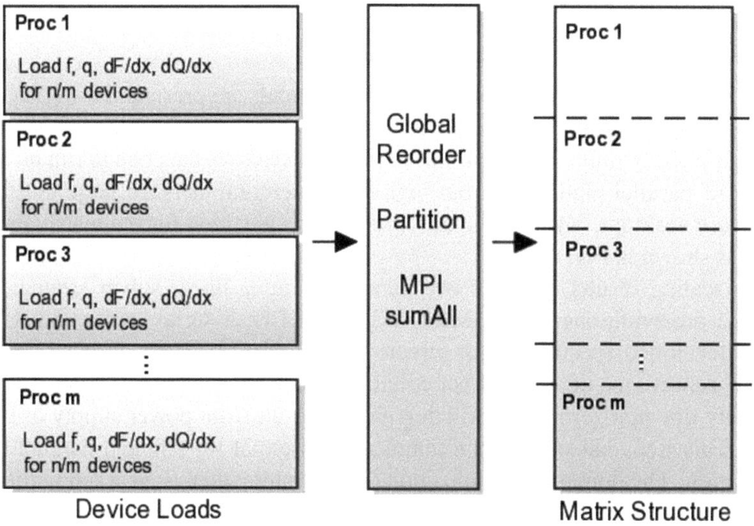

Fig. 2 Parallel load balance for device evaluation and matrix structure

matrix structure is chosen to accelerate the computation of the linear system solution or make the solution method more robust.

3.2 *Parallel Linear Solvers*

Iterative matrix solution methods (e.g. GMRES [20]) have been made to scale well to much larger problem sizes for other types of physical problems. However, the conventional wisdom has generally held that such methods are not effective for traditional circuit simulation. There has been some progress on the use of iterative methods for circuit simulation, notably Basermann [7] and Bomhof [8], both of whom relied on distributed Schur-complement based preconditioners. Harmonic Balance simulators commonly use iterative methods to remain matrix-free. Multigrid methods have successfully been applied to power-grid simulation by several authors [18, 23, 25]. Recently, a preconditioning strategy has been developed to generate an efficient parallel load balance for the matrix structure of Fig. 2 [14].

Direct sparse linear solvers are the industry standard for solving the linear systems generated during DC or transient analysis, so it would make sense to consider using parallel direct linear solvers like SuperLU_DIST [16] or PARDISO [21]. However, most of these solvers are designed to be general purpose, which does not allow them to be efficient on circuit-specific matrix structure. Furthermore, there is still an issue of scaling with parallel direct solvers that will limit the size of linear system (and thus size of circuit) for which they will be effective.

4 Scaling Study Using Xyce

In this section, results from some scaling experiments are presented using Xyce [6]. Xyce is designed from the "ground up" to be a parallel simulator, and is based on a message-passing implementation (MPI) [12]. This allows the code to run on a wide variety of parallel platforms, from high-end supercomputers, to large clusters, to multi-core desktops. Xyce uses a different parallel partition for the matrix load and solve, as shown in Fig. 2.

The scaling results compare two parallel iterative linear solver strategies that generate preconditioners for GMRES [20] to KLU, a serial sparse direct linear solver developed specifically for circuit simulation [22]. Both iterative strategies initially remove the dense rows (or columns) that correspond to columns (or rows) with only one non-zero entry, which typically result from power supply or ground nodes. This step, called singleton removal, is essential for efficient parallel matrix distribution. The domain decomposition (DD) strategy then uses graph partitioning on the resulting symmetrized graph to reduce communication and a local fill-reducing ordering on the block diagonal before performing an incomplete LU (ILU) factorization.

The block triangular form (BTF) strategy first uses singleton removal, then performs a block triangular form reordering of the resulting matrix. Hypergraph partitioning is used on the block graph to reduce communication and a direct factorization (KLU) is used on the block diagonal, resulting in a block Jacobi preconditioner. More details regarding the DD and BTF strategies can be found in [14].

4.1 Experimental Setup

The results presented are obtained from the transient simulation of a large application-specific integrated circuit (ASIC), with around a half-million devices, using Xyce. All computations are performed on a cluster with 2.2 GHz AMD four-socket, quad-core processors with 32 GB DDR2 RAM and an Infiniband interconnect using the OFED software stack. Each node of the machine has a total of sixteen cores, and the user can request anywhere from one to sixteen cores per node. If less than sixteen cores per node are used, the memory is evenly divided between the cores, and more memory is available for each core.

This comparison also examines the performance of loading values into the Jacobian matrix and the residual vector, which includes the device evaluation. The scaling is done relative to the serial simulation performance, where KLU is used as the linear solver. The simulations were run on 8, 16, 32, and 64 processors (cores) where four processors per node (ppn) were used. Thus, 2, 4, 8, and 16 nodes were used to perform this study.

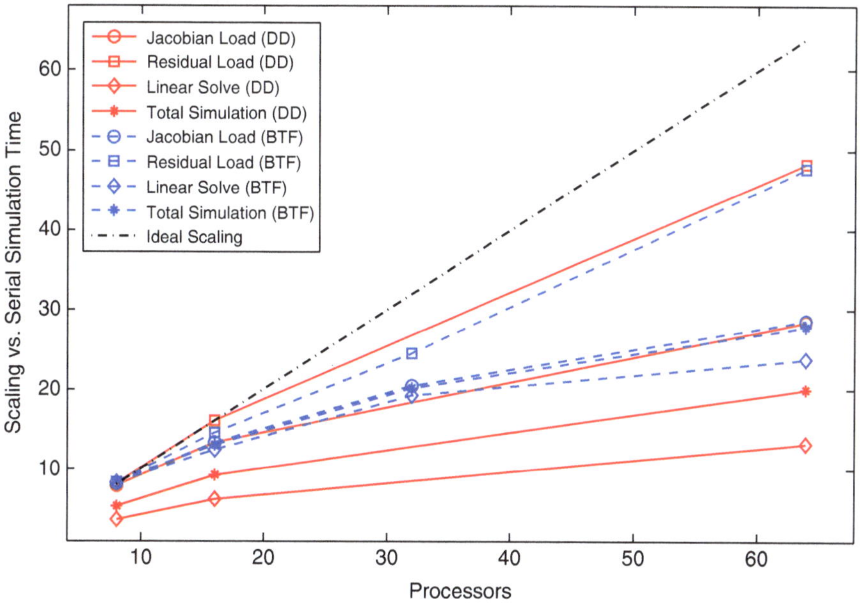

Fig. 3 Xyce scaling study on large ASIC using four processors per node

4.2 Numerical Results

Figure 3 illustrates the scalability of the DD and BTF linear solver strategies over an increasing number of processors. The Jacobian and residual load scale about the same for the two strategies, as one would expect. It should be noticed that the residual load has the best performance, which supports the conjecture that a naive distribution of devices to processors is sufficient. However, with respect to the linear solver strategies, BTF is almost twice as fast as the DD strategy.

For this ASIC, the BTF strategy is a better choice than DD with respect to overall parallel scaling and robustness. When the DD strategy is used, the total simulation scaling falls midway between the Jacobian load and linear solve scalings. This indicates that the linear solve is a bottleneck to overall parallel performance, representing a larger fraction of the total runtime. However, with the BTF linear solver strategy, the total scaling is consistent with the Jacobian load scaling, demonstrating that the BTF-based linear solve does not impact the overall simulation scaling. On 32 processors, using the DD strategy resulted in a simulation failure, illustrating that this solver strategy may not be as robust as the BTF strategy for this ASIC.

Finally, Fig. 3 illustrates that past 32 processors, only a moderate speedup in the simulation is observed. For the BTF strategy, increasing the processors from 32 to 64 only yields an additional speedup of 1.4x. For any fixed problem size, this roll-off is to be expected beyond a certain number of processors. However, it should be noted that the overall runtime on 32 processors is approximately twenty times faster than the serial case, which is still a substantial improvement.

These numerical results highlight the parallel scalability of the DD and BTF iterative linear solver strategies on a single large ASIC. These strategies have been tested previously [14] on a more diverse suite of circuits to illustrate their limitations. Both are ineffective on circuits with feedback structure or a large number of parasitics. A different domain decomposition approach [19] that divides up the circuit into linear and nonlinear domains, will be more effective when a large number of parasitics are present. The results illustrate that, if an effective preconditioner is available, parallel transistor-level circuit simulation can be scalable.

5 Conclusion

In this paper, advances in parallel techniques for transistor-level circuit simulation have been discussed. It was argued that efficient parallel circuit simulation requires integration of large and small scale parallelism into every step of the nested solver loop. Specific attention was given to parallelism issues in the device evaluation and linear solver.

While not as robust as direct solvers, iterative solver strategies have the potential to enable scalable parallel simulation. Scaling results were presented to show that such strategies can reduce the total simulation time by up to a factor of twenty compared to the serial solver KLU on 32 processors. Ultimately, a robust and scalable parallel transistor-level circuit simulator will require a comprehensive strategy for device evaluation and linear solvers to obtain good performance.

Acknowledgements The authors would like to thank David Day, Erik Boman, Mike Heroux, Ray Tuminaro, and Scott Hutchinson for many helpful discussions. We would also like to thank the Xyce development team, including Tom Russo and Rich Schiek.

References

1. Cadence UltraSim. http://www.cadence.com/products/custom_ic/ultrasim/
2. Nascentric OmegaSim. http://www.nascentric.com/omegasim_gx.html
3. Synopsys HSIM. http://www.synopsys.com/Tools/Verification/AMSVerification/CircuitSimulation/HSIM
4. Synopsys HSPICE. http://www.synopsys.com/Tools/Verification/AMSVerification/Circuit-Simlation/HSPICE
5. Synopsys NanoSim. http://www.synopsys.com/Tools/Verification/AMSVerification/Circuit-Simulation/Pages/NanoSim.aspx
6. Xyce Parallel Circuit Simulator. http://xyce.sandia.gov
7. Basermann, A., Jaekel, U., Nordhausen, M.: Parallel iterative solvers for sparse linear systems in circuit simulation. Fut. Gen. Comput. Sys. **21**(8), 1275–1284 (2005)

8. Bomhof, C., vanderVorst, H.: A parallel linear system solver for circuit simulation problems. Num. Lin. Alg. Appl. **7**(7–8), 649–665 (2000)
9. Davis, T.A.: Direct Methods for Sparse Linear System. SIAM, Philadelphia (2006)
10. Dong, W., Li, P., Ye, X.: Wavepipe: Parallel transient simulation of analog and digital circuits on multi-core shared-memory machines. In: IEEE/ACM Design Automation Conference, pp. 238–243 (2008)
11. Fröhlich, N., Riess, B., Wever, U., Zheng, Q.: A new approach for parallel simulation of VLSI-circuits on a transistor level. IEEE Trans. Circ. Sys. Part I **45**(6), 601–613 (1998)
12. Gropp, W., Lusk, E., Doss, N., Skjellum, A.: A high-performance, portable implementation of the MPI message passing interface standard. Parallel Computing **22**(6), 789–828 (1996)
13. Gulati, K., Croix, J.F., Khatr, S.P., Shastry, R.: Fast circuit simulation on graphics processing units. In: Proceedings of the 2009 Conference on Asia and South Pacific Design Automation, pp. 403–408. IEEE Press (2009)
14. Heidi K. Thornquist et al.: A parallel preconditioning strategy for efficient transistor–level circuit simulation. In: IEEE/ACM International Conference on Computer-Aided Design (ICCAD), pp. 410–417 (2009)
15. Kundert, K.: Sparse matrix techniques. In: A. Ruehli (ed.) Circuit Analysis, Simulation and Design. North-Holland, New York (1987)
16. Li, X.S., Demmel, J.W.: SuperLU_DIST: A scalable distributed-memory sparse direct solver for unsymmetric linear systems. ACM Trans. Math. Softw. **29**(2), 110–140 (2003)
17. Nagel, L.W.: Spice 2, a computer program to simulate semiconductor circuits. Tech. Rep. Memorandum ERL-M250 (1975)
18. Nassif, J.N.K.S.R., Najm, F.N.: A multigrid-like technique for power grid analysis. IEEE Trans. Computer-Aided Des. **21**, 1148–1160 (2002)
19. Peng, H., Cheng, C.K.: Parallel transistor level circuit simulation using domain decomposition methods. In: Proceedings of the 2009 Conference on Asia and South Pacific Design Automation, pp. 397–402. IEEE Press (2009)
20. Saad, Y., Schultz, M.H.: GMRES: a generalized minimal residual algorithm for solving nonsymmetric linear systems. SIAM J. Sci. Comput. **7**(3), 856–869 (1986)
21. Schenk, O., Gärtner, K.: Solving unsymmetric sparse systems of linear equations with PARDISO. Fut. Gen. Comput. Sys. **20**(3), 475–487 (2005)
22. Stanley, K., Davis, T.: KLU: a 'Clark Kent' sparse LU factorization algorithm for circuit matrices. In: SIAM Conference on Parallel Processing for Scientfic Computing (PP04) (2004)
23. Sun, K., Zhou, Q., Mohanram, K., Sorensen, D.C.: Parallel domain decomposition for simulation of large-scale power grids. In: IEEE/ACM International Conference on Computer-Aided Design (ICCAD), pp. 54–59 (2007)
24. Ye, X., Dong, W., Li, P., Nassif, S.: MAPS: Multi-algorithm parallel circuit simulation. In: IEEE/ACM International Conference on Computer-Aided Design, pp. 73–38 (2008)
25. Zhuo, C., Hu, J., Zhao, M., Chen, K.: Power grid analysis and optimization using algebraic multigrid. IEEE Trans. Computer-Aided Des. **27**, 738–751 (2008)

Sensitivity-Based Steady-State Mismatch Analysis for RF Circuits

Fabrice Veersé, Joël Besnard, and Hubert Filiol

Abstract Based on the assumption of small parameter variations a sensitivity–based analysis method is proposed for the computation of the steady–state mismatch deviations of periodic and quasi-periodic circuits. Unlike classical Monte-Carlo or worst-case analyses this approach does not require several (time) periodic steady-state analyses to be performed. In addition it provides detailed information on the relative contributions of the different device model parameters. Some numerical examples illustrate the potential and limitations of the approach.

1 Introduction

With the size of process nodes reaching a few (tens of) nanometers, it has become mandatory to account for the effect induced by device variations (in the range of 10–30%) while verifying the circuit design. This may be done using worst-case or Monte-Carlo transistor-level simulations, analyzing repeatedly the circuit for different sets of device model parameters. But the computational resources needed for simulating the periodic (or quasi–periodic) steady–state of RF circuits and the number of simulations required to achieve a given accuracy level often make the Monte-Carlo approach unaffordable in practice. Likewise the worst-case analysis requires a number of steady–state simulations largely dependent on the number of device parameters, which according to [1] can exceed 800 for the BSIM4 model from Berkeley [2].

F. Veersé (✉) · J. Besnard · H. Filiol
Mentor Graphics (Ireland) Ltd. French Branch, Immeuble Le Viséo Bât. B, 110 rue Blaise Pascal, Inovallée Montbonnot, 38334 Saint Ismier Cedex, France
e-mail: Fabrice_Veerse@mentor.com; Joel_Besnard@mentor.com; Hubert_Filiol@mentor.com

B. Michielsen and J.-R. Poirier (eds.), *Scientific Computing in Electrical Engineering SCEE 2010*, Mathematics in Industry 16, DOI 10.1007/978-3-642-22453-9_28,

On the contrary, the method proposed in this paper uses sensitivity information to avoid such repeated steady-state analyses, making it a method of choice for predicting the mismatch of RF circuits.

A similar approach was used by Oehm and Schumacher [3] to derive a DC-voltage mismatch deviation, under the hypothesis of small-magnitude Gaussian random mismatch. The authors scaled the deviation of each mismatch parameter by its sensitivity and computed the mismatch deviation as the square root of their added squared values. The extension of this approach to transient simulations, using the adjoint method to compute sensitivities [4–6], was studied e.g. in [7–10].

A different approach is proposed in [11] for the analysis of mismatch effects on transient performance of periodic circuits at steady state: a periodic noise simulation is performed using some auxiliary noise sources modeling the DC mismatch in device parameters, and the resulting noise power spectral density is interpreted in terms of performance variation.

In the spirit of [7–10] the method proposed in this paper extends the approach of [3] to the steady-state simulations of periodic or quasi-periodic circuits, using the adjoint method. Unlike the approach in [11] no modification of the circuit (auxiliary noise source) is needed and the interpretation of the results is as straightforward as for DC simulations.

In the following section, the method used to compute harmonic–balance steady–state sensitivities is summarized. It is used in the next section for the derivation of a method for steady–state mismatch analysis. Finally, some numerical experiments illustrate the effectiveness and some limitations of the approach.

2 Harmonic–Balance Steady–State Sensitivity Analysis

The computation of the circuit (quasi-)periodic steady-state using the harmonic balance (HB) method [12] amounts to solving the nonlinear system of equations

$$\begin{aligned}\mathbf{F}(\mathbf{X}_{\mathrm{HB}}(\mathbf{p}),\mathbf{p}) &= \Omega\Gamma\mathbf{Q}(\Gamma^{-1}\mathbf{X}_{\mathrm{HB}}(\mathbf{p}),\mathbf{p}) + \Gamma\mathbf{I}(\Gamma^{-1}\mathbf{X}_{\mathrm{HB}}(\mathbf{p}),\mathbf{p}) \\ &\quad + Y(\mathbf{p})\mathbf{X}_{\mathrm{HB}}(\mathbf{p}) + \mathbf{B} = 0,\end{aligned} \tag{1}$$

where Γ and Γ^{-1} are the direct and inverse Fourier transforms, Ω is a diagonal matrix expressing the equivalent of time-differentiation in frequency domain, $\mathbf{Q}$ and $\mathbf{I}$ are vectors gathering the instantaneous charges $\mathbf{q}(\mathbf{x}_{\mathrm{HB}}(t_i,\mathbf{p}),\mathbf{p})$ and currents $\mathbf{i}(\mathbf{x}_{\mathrm{HB}}(t_i,\mathbf{p}),\mathbf{p})$, with $\mathbf{x}_{\mathrm{HB}}(\mathbf{p}) = \mathit{\Gamma}^{-1}\mathbf{X}_{\mathrm{HB}}(\mathbf{p})$ being the steady–state unknowns in time domain. The $Y(\mathbf{p})$ matrix accounts for the frequency-dependent elements and $\mathbf{B}$ is the vector of harmonic components of independent sources. The vector $\mathbf{p}$ is made of parameters such as transistor widths and lengths.

Partial differentiation of this system with respect to $\mathbf{X}_{\mathrm{HB}}$ leads to the following expression of the HB Jacobian matrix:

$$J(\mathbf{X}_{\mathrm{HB}}(\mathbf{p}),\mathbf{p}) = \Omega\Gamma C(\mathbf{x}_{\mathrm{HB}}(\mathbf{p}),\mathbf{p})\Gamma^{-1} + \Gamma G(\mathbf{x}_{\mathrm{HB}}(\mathbf{p}),\mathbf{p})\Gamma^{-1} + Y(\mathbf{p}) \tag{2}$$

where $C(\mathbf{x}_{\mathrm{HB}}(\mathbf{p}),\mathbf{p})$ and $G(\mathbf{x}_{\mathrm{HB}}(\mathbf{p}),\mathbf{p})$ are block-diagonal matrices with entries $\partial\mathbf{q}(\mathbf{x}_{\mathrm{HB}}(t_i,\mathbf{p}),\mathbf{p})/\partial\mathbf{x}$ and $\partial\mathbf{i}(\mathbf{x}_{\mathrm{HB}}(t_i,\mathbf{p}),\mathbf{p})/\partial\mathbf{x}$ respectively.

Differentiating the HB residual (1) with respect to the parameters $\mathbf{p}$, one obtains:

$$J(\mathbf{X}_{\mathrm{HB}}(\mathbf{p}),\mathbf{p})\frac{\mathrm{d}\mathbf{X}_{\mathrm{HB}}(\mathbf{p})}{\mathrm{d}\mathbf{p}} + \frac{\partial\mathbf{F}(\mathbf{X}_{\mathrm{HB}}(\mathbf{p}),\mathbf{p})}{\partial\mathbf{p}} = 0. \tag{3}$$

The sensitivity of an output expression $g(\mathbf{X}_{\mathrm{HB}}(\mathbf{p}),\mathbf{p})$ with respect to the parameters $\mathbf{p}$ is given by

$$\frac{\mathrm{d}g(\mathbf{X}_{\mathrm{HB}}(\mathbf{p}),\mathbf{p})}{\mathrm{d}\mathbf{p}} = \frac{\partial g(\mathbf{X}_{\mathrm{HB}}(\mathbf{p}),\mathbf{p})}{\partial\mathbf{X}_{\mathrm{HB}}}\frac{\mathrm{d}\mathbf{X}_{\mathrm{HB}}(\mathbf{p})}{\mathrm{d}\mathbf{p}} + \frac{\partial g(\mathbf{X}_{\mathrm{HB}}(\mathbf{p}),\mathbf{p})}{\partial\mathbf{p}} \tag{4}$$

Using (3) left-multiplied by $\mathbf{w}^*$ where the * superscript denotes conjugate transposition and $\mathbf{w}$ is the solution of the adjoint system

$$[J(\mathbf{X}_{\mathrm{HB}},\mathbf{p})]^*\,\mathbf{w} = \left(\frac{\partial g(\mathbf{X}_{\mathrm{HB}},\mathbf{p})}{\partial\mathbf{X}_{\mathrm{HB}}}\right)^*, \tag{5}$$

to express the first term in the right-hand side of (4), shows that the sensitivity of the output expression $g(\mathbf{X}_{\mathrm{HB}}(\mathbf{p}),\mathbf{p})$ with respect to the parameters $\mathbf{p}$ can be computed from

$$\frac{\mathrm{d}g(\mathbf{X}_{\mathrm{HB}}(\mathbf{p}),\mathbf{p})}{\mathrm{d}\mathbf{p}} = -\mathbf{w}^*\frac{\partial F(\mathbf{X}_{\mathrm{HB}}(\mathbf{p}),\mathbf{p})}{\partial\mathbf{p}} + \frac{\partial g(\mathbf{X}_{\mathrm{HB}}(\mathbf{p}),\mathbf{p})}{\partial\mathbf{p}} \tag{6}$$

3 Steady-State Mismatch Analysis

The proposed mismatch analysis is a special application of the above sensitivity analysis, based on statistical deviations of device model parameters. Considering that the output quantity of interest $y = g(\mathbf{X}_{\mathrm{HB}}(\mathbf{p}),\mathbf{p})$ usually depends on the device model parameters $\mathbf{p}$ only implicitly through the dependence of the circuit steady–state vector $\mathbf{X}_{\mathrm{HB}}(\mathbf{p})$, we let $\partial g/\partial\mathbf{p} = 0$ in the sequel.

Assuming that the parameters $\mathbf{p}$ are unbiased ($\bar{\mathbf{p}} = E(\mathbf{p}) = \mathbf{0}$) and given their covariance matrix $\Sigma_{\mathbf{p}} = E((\mathbf{p}-\bar{\mathbf{p}})(\mathbf{p}-\bar{\mathbf{p}})^T)$, one obtains the first-order approximation

$$\Sigma_y = E\left((y-\bar{y})(y-\bar{y})^T\right) \approx \left(\frac{\mathrm{d}g}{\mathrm{d}\mathbf{p}}\right)\Sigma_{\mathbf{p}}\left(\frac{\mathrm{d}g}{\mathrm{d}\mathbf{p}}\right)^T. \tag{7}$$

That is, the covariance matrix of the output quantity $y = g(\mathbf{X}_{\mathrm{HB}}(\mathbf{p}))$ *may be estimated from the covariance matrix of the device model parameters* $\mathbf{p}$ *and the sensitivity of this output with respect to these parameters.*

If the parameters stored in the vector $\mathbf{p}$ are uncorrelated, the above equation reduces to

$$\sigma_y = \sqrt{\sum_i \left(\frac{\mathrm{d}g}{\mathrm{d}p_i}\right)^2 \sigma_{p_i}^2} = \sqrt{\sum_i \left(\frac{\mathrm{d}g}{\mathrm{d}p_i}\sigma_{p_i}\right)^2} \tag{8}$$

where σ_z indicates the standard deviation of the quantity z and the summation extends over the number of individual parameters p_i.

Let p_i be a transistor model parameter (*e.g.* its width or length), then (6) for steady–state sensitivity specializes to

$$\frac{\mathrm{d}g}{\mathrm{d}p_i}\sigma_{p_i} = -\mathbf{w}^* \frac{\partial F}{\partial p_i}\sigma_{p_i} = -\mathbf{w}^* \left(\Omega\Gamma \frac{\partial \mathbf{Q}(\mathbf{x}_{\mathrm{HB}}, \mathbf{p})}{\partial p_i}\sigma_{p_i} + \Gamma \frac{\partial \mathbf{I}(\mathbf{x}_{\mathrm{HB}}, \mathbf{p})}{\partial p_i}\sigma_{p_i}\right) \tag{9}$$

where $\mathbf{w}$ is the solution of the adjoint system (5).

The term $(\partial \mathbf{Q}/\partial p_i)\,\sigma_{p_i}$ in the above equation may be obtained by gathering the instantaneous deviations of charges

$$\frac{\partial \mathbf{q}(\mathbf{x}_{\mathrm{HB}}(t_j), \mathbf{p})}{\partial p_i}\sigma_{p_i} \approx \left[\mathbf{q}(\mathbf{x}_{\mathrm{HB}}(t_j), \mathbf{p} + \sigma_{p_i}\mathbf{e}_{\mathrm{i}}) - \mathbf{q}(\mathbf{x}_{\mathrm{HB}}(t_j), \mathbf{p})\right] \tag{10}$$

for all the FFT sampling instants t_j. The current deviation term $(\partial \mathbf{I}/\partial p_i)\,\sigma_{p_i}$ is treated similarly. *This amounts to perturbing the model parameters for each instance of the corresponding device model by a standard deviation and evaluating the resulting current and charges variations, for all instants* t_j.

Besides this evaluation of the device models, the main computational cost of the proposed approach to compute the mismatch deviation σ_y consists in solving the *linear* adjoint system (5). This is a far smaller effort than that required by either a Monte–Carlo analysis or a worst–case one, and even less than that required by the steady–state analysis (*i.e.* solving the *nonlinear* system (1)).

But whereas the Monte-Carlo and worst-case analyses properly deal with deformed output distributions, our perturbation approach is more suited to linear and weakly nonlinear problems with small Gaussian parameter deviations producing Gaussian or almost Gaussian output deviations (see also the discussion in Sect. 5).

4 Numerical Experiments

In this section some steady-state mismatch analysis results are reported for a simple series-RC filter and a nonlinear active-load amplifier, evidencing the usefulness and some limitations of the approach.

4.1 An RC Filter

The circuit contains a 1 kΩ resistor connected to a periodic voltage source and a 1 nF grounded capacitor. The voltage source has a 1V DC component surimposed with a 10 mV sinusoidal component oscillating with a 1 MHz frequency. Gaussian deviations for the resistor and capacitor values are specified as a percentage of their respective nominal values, and the 4-σ performance variation of the first-harmonic component of the voltage at the output node connecting the resistor and the capacitor is computed using a Monte-Carlo analysis with 1000 runs, a worst-case analysis and the proposed steady-state mismatch analysis.

Results are provided in Table 1 where the first column indicates the values of the resistor and capacitor standard deviations as a percentage of their nominal values. The third column reports the standard deviations computed by the Monte-Carlo method and our harmonic-balance steady-state mismatch analysis (this data is not available from the worst-case analysis). The next two columns give the maximum and minimum values of the magnitude of the first-harmonic of the output voltage computed by the different methods; for the steady-state mismatch analysis these values are set equal to the nominal value plus-or-minus four times the corresponding standard deviation. The remaining three columns provide similar information for the phase of the first-harmonic of the output voltage.

For the 0.3% and 3% Gaussian deviations the steady-state mismatch analysis computes performance variations in close agreement with the results of the Monte-Carlo analysis, and closer than those of the worst-case analysis.

A limitation of the steady-state mismatch analysis is evidenced by the 30%-deviation results: with such large parameter deviations the approximations made in (7) and (10) are not justified anymore. And although Gaussian distributions are used for the varying parameters and the circuit is linear, second-order effects arise from the nonlinearity of the computation of the magnitude and the phase of the first-harmonic of the output voltage. These second-order effects are clearly noticeable

Table 1 Steady-state mismatch results for first-harmonic component of RC filter

σ_R, σ_C	Methods	First-harmonic *magnitude* variation Nominal value: 1.572E−03			First-harmonic *phase* variation Nominal value: −1.7010E+02		
		Standard deviation	Minimum value	Maximum value	Standard deviation	Minimum value	Maximum value
0.3%	Monte Carlo	6.557E−06	1.551E−03	1.591E−03	3.804E−02	−1.711E+02	−1.709E+02
	Worst Case		1.536E−03	1.609E−03		−1.712E+02	−1.707E+02
	Mismatch	5.865E−06	1.548E−03	1.595E−03	3.776E−02	−1.711E+02	−1.708E+02
3%	Monte Carlo	6.575E−05	1.380E−03	1.784E−03	3.815E−01	−1.721E+02	−1.697E+02
	Worst Case		1.259E−03	2.013E−03		−1.728E+02	−1.684E+02
	Mismatch	5.897E−05	1.336E−03	1.808E−03	3.796E−01	−1.725E+02	−1.694E+02
30%	Monte Carlo	1.065E−03	5.692E−04	9.805E−03	7.072E+00	−1.767E+02	−7.866E+01
	Worst Case		3.287E−04	9.698E−03		−1.781E+02	−1.041E+02
	Mismatch	6.274E−04	−9.376E−04	4.081E−03	4.039E+00	−1.871E+02	−1.548E+02

from the unsymmetry of the Monte-Carlo minimum and maximum values about the nominal one, that unravels a non-Gaussian distribution of the output magnitude and phase. Owing to the linearity of the circuit this limitation and such second-order effects would not be present if mismatch deviations of the real and imaginary parts of the output were computed instead of those of its magnitude and phase.

The steady-state mismatch results are consistent with the indication in [9] that the mismatch analysis based on sensitivities computed via the adjoint method are generally in good agreement with those of the Monte-Carlo analysis when the coefficient of variance of the elements is less than 20%.

4.2 An Active-Load Amplifier

To determine whether the steady-state mismatch analysis could be of any value in practice for a nonlinear circuit, an active-load amplifier using 0.25-micron SPICE level 3 models is considered. Gaussian mismatch deviations are specified on the zero-bias threshold voltage and the surface mobility following the approach in [13]. Results are provided in Table 2 for the magnitude of the DC-component of the output voltage, and in Table 3 for the magnitude and phase of its first-harmonic component.

The steady-state mismatch results are in good agreement with those of the Monte-Carlo analysis, and in better agreement than those of the worst-case analysis.

The ratio of the computed standard deviation to the corresponding nominal value are found to be less than 20% and could be used as an indication of the validity of

Table 2 Steady-state mismatch results for DC-component of active-load amplifier

Methods	DC-component magnitude variation Nominal value: 1.126E+00		
	Standard deviation	Minimum value	Maximum value
Monte Carlo	6.106E−02	9.394E−01	1.344E+00
Worst Case		7.624E−01	1.472E+00
Mismatch	6.140E−02	8.803E−01	1.371E+00

Table 3 Steady-state mismatch results for first-harmonic component of active-load amplifier

Methods	First-harmonic *magnitude* variation Nominal value: 4.692E−02			First-harmonic *phase* variation Nominal value: 8.865E+01		
	Standard deviation	Minimum value	Maximum value	Standard deviation	Minimum value	Maximum value
Monte Carlo	7.158E−04	4.457E−02	4.903E−02	5.438E−02	8.843E+01	8.880E+01
Worst Case		4.129E−02	5.142E−02		8.832E+01	8.888E+01
Mismatch	7.169E−04	4.406E−02	4.979E−02	5.311E−02	8.843E+01	8.886E+01

the results provided by the steady-state mismatch analysis whenever a Monte-Carlo analysis is not affordable.

5 Discussion

In Sect. 3 the independence of device mismatch parameters p_i was assumed. This hypothesis seems natural since mismatch variations are by definition local to each device and thus uncorrelated between devices. Unfortunately not all SPICE device-model formulations are based on physical and independent parameters that could be used for the described steady-state mismatch analysis. At the same time, disregarding existing correlations between model parameters is not an option, since the computed performance variations due to the mismatch effects would likely be under-estimated or over-estimated, depending on the signs of the sensitivities (see (7)). One possibility for accounting for correlations is to perform a principal component analysis (PCA) [14] to identify uncorrelated linear combinations of mismatch device parameters, and to perform the steady-state mismatch analysis using these combinations as independent parameters. The global performance variation computed by the analysis will then account correctly for correlations between the original mismatch device parameters and it will identify the uncorrelated linear combinations of parameters that most contribute to this variation. Due to the presence of correlations between parameters, isolating the relative contribution of each mismatch device parameter may not be feasible anymore.

In [3], it is assumed also that the random variations of the device-model parameters are Gaussian and small in amplitude. This is not strictly necessary as long as there exist a mapping between the original distribution and a Gaussian one (see e.g. [8] where a similar approach for transient simulations is applied with log-normal parameter distributions), and the current and charge deviations (10) remain small enough for the first-order approximation in (7) to be valid. Whenever this linear pertubation approach is not valid due to nonlinearities, second-order effects cannot be neglected anymore. One possibility to compute them is to use second-order adjoint sensitivities [15]. However the absence of second-order derivatives in device models leads the authors in [15] to compute second-order sensitivities via perturbations of first-order derivatives (computed by the adjoint model), with a cost proportional to the number of parameters. With such a cost scaling with the number of parameters, a more general approach based on interpolation of central moments [16] constitutes a worthy alternative.

As a cost proportional to the number of parameters is likely to be prohibitive for the mismatch analysis of RF circuits, the steady-state mismatch analysis proposed in this paper may prove a valuable tool for a cheap estimation of the performance variation together with some indication of the parameter variations that are likely to be responsible for them; as long as its limitations are well-understood and not overlooked. The coefficient of variance of the elements together with the ratio between the computed standard deviation and the nominal output value could be useful indicators of the validity of the results.

Acknowledgements The authors would like to thank the members of the Analog and RF simulation teams at Mentor Graphics for their support and the stimulating working atmosphere. The anonymous reviewers are acknowledged for pointing many typos and for the interest shown through additional questions.

References

1. Denk, G.: Circuit simulation for nanoelectronics. In Anile, A., Alì, G., Mascali, G. (eds.) Scientific Computing in Electrical Engineering SCEE 2004, Mathematics in Industry, vol. 9, pp. 13–20. Springer, Berlin (2006)
2. BSIM4 manual. Departement of Electrical and Computer Science, University of California, Berkeley (2000) http://www-device.EECS.Berkeley.EDU/~\bsim
3. Oehm, J., Schumacher, K.: Quality assurance and upgrade of analog characteristics by fast mismatch analysis option in network analysis environment. IEEE J. Solid-State Circ. **28**(7), 865–871 (1997)
4. Director, S.W., Rohrer, R.A.: Generalized adjoint network and network sensitivities. IEEE Trans. Circ. Theory **16**(8), 318–323 (1969)
5. Ilievski, Z., Xu, H., Verhoeven, A., Schilders, W.H.A., Mattheij, R.M.M.: Adjoint transient sensitivity analysis in circuit simulation. In: Ciuprina, G., Ioan, D. (eds.) Scientific Computing in Electrical Engineering SCEE 2006, Mathematics in Industry, vol. 11, pp. 183–189. Springer, Berlin (2007)
6. Ilievski, Z.: Model order reduction and sensitivity analysis. Ph.D. Thesis, Eindhoven University of Technology (2010). URL http://alexandria.tue.nl/extra2/201010770.pdf
7. Graupner, A., Schwarz, W., Schüffny R.: Statistical analysis of analog structures through variance calculation. IEEE Trans. Circ. Syst. I: Fund. Theory Appl. **49**(8), 1071–1078 (2002)
8. Haüsler, R., Kinzelbach, H.: Sensitivity-based Stochastic Analysis Method for Power Variations. In: Proceedings of the 9. ITG/GMM-Fachtagung Entwicklung von Analogschaltung mit CAE-Methoden (ANALOG'06), Dresden, Germany, September 27–29, ITG-FB report, 196, VDE Verlag, Berlin Offenbach (2006)
9. Yuan, F.: Analysis of stochastic behaviour of linear circuits using first-order second-moment and adjoint network techniques. Electr. Lett. **33**(9), 766–768 (1997)
10. Gu, B., Gullapalli, K., Zhang, Y., Sundareswaran, S.: Faster Cell Characterization using Adjoint Sensitivity Analysis. In: Proceedings of the IEEE 2008 Custom Integrated Circuits Conference (CICC), pp. 229–232. San Jose, CA, September 21–24 (2008)
11. Kim, J., Jones, K.D., Horowitz, M.A.: Fast, non-Monte Carlo estimation of transient performance variation due to device mismatch. In: Proceedings of the 44th ACM/IEEE Design Automation Conference (DAC'07), pp. 440–443, San Diego, CA, June 4–8 (2007)
12. Nastov, O.J., Telichevsky, R., Kundert, K., White, J.: Fundamentals of Fast Simulation Algorithms for RF Circuits. Proc. IEEE **95**(3), 600–621 (2007)
13. Pelgrom, M., Duinmaijer, A., Webers, A.: Matching properties of MOS transistors. IEEE J. Solid-State Circ. **24**(5), 1433–1439 (1989)
14. Jollife, I.T.: Principal Component Analyis. 2nd ed., Springer, New York (2002)
15. Ye, X., Li, P., Liu, F.Y.: Exact time-domain second-order adjoint-sensitivity computation for linear circuit analysis and optimization. IEEE Trans. Circ. Sys.–I: Regular Papers **57**(1), 236–248 (2010)
16. Zhang, M., Olbrich, M. Seider, D., Frerichs, M., Kinzelbach, H., Barke, E.: CMCal: An accurate analytical approach for the analysis of process variations with non-Gaussian parameters and nonlinear functions. In: Proceedings of the Design, Automation & Test in Europe Conference 2007 (DATE'07), pp. 1–6, April 16–20, Nice, France (2007)

Modelling and Simulation of Forced Oscillators with Random Periods

Roland Pulch

Abstract In nanoelectronics, the miniaturisation of circuits causes uncertainties in the components. An uncertainty quantification is achieved by the introduction of random parameters in corresponding mathematical models. We consider forced oscillators described by time-dependent differential algebraic equations, where a random period appears. A corresponding uncertainty quantification results from a modelling based on a transformation to a unit time interval. We apply the technique of the generalised polynomial chaos to resolve the stochastic model. Thereby, a Galerkin approach yields a larger coupled system of differential algebraic equations satisfied by an approximation of the random process. We present numerical simulations of an illustrative example.

1 Introduction

Uncertainty quantification becomes important in nanoelectronics, since the down-scaling of circuits produces undesired but inevitable variations in the components. In the mathematical models, corresponding physical parameters are substituted by random variables to describe the uncertainties. We consider the traditional modelling of electric circuits by differential algebraic equations (DAEs), where the time-dependent solution becomes a random process now.

On the one hand, forced oscillators with random parameters, where the period of the input signals is constant and deterministic, have been investigated in [5–8]. On the other hand, autonomous oscillators with random parameters have been considered in [10]. Thereby, the period depends on the random parameters, since

R. Pulch (✉)
Lehrstuhl für Angewandte Mathematik und Numerische Mathematik, Bergische Universität Wuppertal, Gaußstr. 20, D-42119 Wuppertal, Germany
e-mail: pulch@math.uni-wuppertal.de

B. Michielsen and J.-R. Poirier (eds.), *Scientific Computing in Electrical Engineering SCEE 2010*, Mathematics in Industry 16, DOI 10.1007/978-3-642-22453-9_29,

no input signals appear. Now we analyse the case of forced oscillators, where the period of the input signals is assumed to be a random variable modelling an own uncertainty. This setting represents a mixture of the two previous cases.

Since the period is given by a random variable, the domain of dependence differs for a single oscillation. We achieve a model for uncertainty quantification by a transformation to a unit time interval as for autonomous oscillators. The stochastic model can be resolved by a quasi Monte-Carlo simulation, for example. We use the technique of the generalised polynomial chaos (gPC), see [1,2,12], in the numerical simulation to investigate a more sophisticated approach. A Galerkin method results in a larger coupled system of DAEs, which yields an approximation of the periodic random process. To illustrate the modelling and the simulation, we apply a transistor amplifier supplied by an input with random period as test example.

2 Modelling of Uncertainties in Period

The mathematical modelling of electric circuits is based on approaches, which typically yield systems of DAEs, see [3]. We consider general systems of the form

$$A(\mathbf{p})\mathbf{x}'(t,\mathbf{p}) = \mathbf{f}(t,\mathbf{x}(t,\mathbf{p}),\mathbf{p}), \tag{1}$$

where $\mathbf{x} : [t_0, t_1] \to \mathbb{R}^n$ represents unknown node voltages, branch currents and possibly other quantities. The singular matrix $A \in \mathbb{R}^{n\times n}$ and the right-hand side $\mathbf{f}$ include physical parameters $\mathbf{p} = (p_1,\ldots,p_q)^\top$ from some relevant set $Q \subseteq \mathbb{R}^q$. Hence the solution $\mathbf{x}$ of (1) depends on time as well as the parameters. If the matrix A is regular, then the system (1) consists of implicit ordinary differential equations (ODEs).

We assume that the chosen parameters exhibit some uncertainties. Consequently, we replace the parameters by independent random variables $\mathbf{p} : \Omega \to Q$ according to some probability space $(\Omega, \mathscr{A}, \mu)$, i.e., a sample space Ω, a sigma-algebra $\mathscr{A}$ over Ω and a probability measure μ. We use a classical random distribution for each parameter like Gaussian, uniform, beta, etc. Given a function $f \in L^1(\Omega)$ depending on the random parameters, we denote the expected value by

$$\langle f(\mathbf{p})\rangle := \int_\Omega f(\mathbf{p}(\omega))\,\mathrm{d}\mu(\omega) = \int_Q f(\mathbf{p})\rho(\mathbf{p})\,\mathrm{d}\mathbf{p} \tag{2}$$

with the probability density function $\rho : Q \to \mathbb{R}$. For two functions $f, g \in L^2(\Omega)$ depending on the random parameters, the expected value $\langle f(\mathbf{p})g(\mathbf{p})\rangle$ represents an inner product according to the Hilbert space $L^2(\Omega)$. We also apply the expected value (2) to vector-valued or matrix-valued functions by components.

We investigate forced oscillators, i.e., the right-hand side of (1) includes periodic input signals with the period T. Now let the period T also be a random variable. Two cases imply the same model:

(i) Although the period corresponds to the input signals, it is chosen in dependence on the selected parameters, i.e., $T = T(\mathbf{p})$. In contrast to the case of autonomous oscillators, see [10], we assume that the period can be evaluated directly for a given tuple of parameters. Hence the period inherits some uncertainties from the random parameters.
(ii) The period is considered as an additional independent random parameter due to an own uncertainty. Thus we introduce the period to the set of parameters. Without loss of generality, we can write $T = T(\mathbf{p})$ as in the scenario (i), where the special case $T(\mathbf{p}) = p_1$ is given, for example.

Consequently, let $\mathbf{f}(t + T(\mathbf{p}), \cdot, \mathbf{p}) = \mathbf{f}(t, \cdot, \mathbf{p})$ for all $t \in \mathbb{R}$ and each $\mathbf{p} \in Q$ in (1). We assume that the solution of the dynamical system (1) inherits the periodicity, i.e,

$$\mathbf{x}(t + T(\mathbf{p}), \mathbf{p}) = \mathbf{x}(t, \mathbf{p}) \qquad \text{for all } t \text{ and each } \mathbf{p} \in Q. \tag{3}$$

We restrict our attention to a single cycle of each periodic solution. Since a single cycle is given in a time interval $[0, T(\mathbf{p})]$, the domain of dependence differs in case of random parameters. We want to compare the periodic solutions, which represent realisations for different random parameters. In particular, an expected value and a corresponding variance describe a kind of comparison of the realisations with respect to the underlying random distribution. However, a direct definition of an expected value or a variance corresponding to the single cycles of the random process is not feasible, because the domains of dependence differ.

As for autonomous oscillators, see [10], we transform the given time intervals $t \in [0, T(\mathbf{p})]$ into the unit interval $\tau \in [0, 1]$. The transformed solution reads

$$\tilde{\mathbf{x}}(\tau, \mathbf{p}) := \mathbf{x}(\tau T(\mathbf{p}), \mathbf{p}) \qquad \text{for each } \mathbf{p} \in Q \tag{4}$$

with the independent variable τ. It follows the periodicity

$$\tilde{\mathbf{x}}(\tau + 1, \mathbf{p}) = \tilde{\mathbf{x}}(\tau, \mathbf{p}) \qquad \text{for all } \tau \text{ and each } \mathbf{p} \in Q. \tag{5}$$

The same relations are given for the input signals in the right-hand side of (1).

The transformation (4) changes the DAEs (1) into the equivalent system

$$A(\mathbf{p})\tilde{\mathbf{x}}'(\tau, \mathbf{p}) = T(\mathbf{p})\, \mathbf{f}(\tau T(\mathbf{p}), \tilde{\mathbf{x}}(\tau, \mathbf{p}), \mathbf{p}). \tag{6}$$

Due to (5), the corresponding periodic boundary conditions read

$$\tilde{\mathbf{x}}(0, \mathbf{p}) = \tilde{\mathbf{x}}(1, \mathbf{p}) \qquad \text{for each } \mathbf{p} \in Q. \tag{7}$$

We apply the stochastic model (6), (7) in case of random periods, where the solution is the periodic random process $\tilde{\mathbf{x}}$. The original random process $\mathbf{x}$ satisfying (1) is obtained via the transformation (4). The expected value of $\tilde{\mathbf{x}}$ can be seen as a reference shape of the oscillations in the standardised time interval $[0, 1]$, where the locations are relative to the input signals. The variance of $\tilde{\mathbf{x}}$ characterises the discrepancies with respect to the reference shape.

3 Numerical Simulation

The stochastic model (6), (7) can be resolved by a quasi Monte-Carlo simulation, for example. Common numerical techniques yield the solutions of the resulting periodic boundary value problems like multiple shooting methods, finite difference methods or harmonic balance. Typically, a large number of samples is required to achieve sufficiently accurate approximations.

Alternatively, we derive a technique based on the gPC, see [1,2,12]. The gPC has already been applied to forced oscillators with constant periods in [5–8]. Assuming finite second moments, the random process satisfying (6) can be represented via

$$\tilde{\mathbf{x}}(\tau, \mathbf{p}(\omega)) = \sum_{i=0}^{\infty} \mathbf{v}_i(\tau) \Phi_i(\mathbf{p}(\omega)). \tag{8}$$

A complete set of basis polynomials $\Phi_i : Q \to \mathbb{R}$ is involved, where we consider an orthonormal system, i.e., $\langle \Phi_i \Phi_j \rangle = \delta_{ij}$ with the Kronecker delta. Each random distribution implies a corresponding polynomial basis. The multivariate polynomials are just products of the orthogonal univariate polynomials. Hence the basis polynomials are known explicitly. The time-dependent coefficient functions satisfy the equation

$$\mathbf{v}_i(\tau) = \langle \tilde{\mathbf{x}}(\tau, \mathbf{p}) \Phi_i(\mathbf{p}) \rangle. \tag{9}$$

The series (8) converges point-wise for each τ in $L^2(\Omega)$. The coefficient functions (9) inherit the smoothness of the random process under certain assumptions.

The unknown coefficient functions can be determined by either a stochastic collocation or the stochastic Galerkin approach, see [11, 12]. In a stochastic collocation method, approximations of the probabilistic integrals (9) are computed. We apply the stochastic Galerkin method in the following. A truncation of the series (8) at the mth term yields an approximation of the random process. Inserting this finite approximation in the DAEs (6) causes the residual

$$\mathbf{r}(\tau, \mathbf{p}) := A(\mathbf{p}) \left(\sum_{i=0}^{m} \mathbf{v}_i'(\tau) \Phi_i(\mathbf{p}) \right) - T(\mathbf{p})\, \mathbf{f} \left(\tau T(\mathbf{p}), \sum_{i=0}^{m} \mathbf{v}_i(\tau) \Phi_i(\mathbf{p}), \mathbf{p} \right).$$

The Galerkin approach demands that the residual is orthogonal with respect to the space spanned by the applied basis functions, i.e.,

$$\langle \mathbf{r}(\tau, \mathbf{p}) \Phi_l(\mathbf{p}) \rangle = \mathbf{0} \qquad \text{for each } \tau \text{ and } l = 0, 1, \ldots, m.$$

It follows the larger coupled system of DAEs

$$\sum_{i=0}^{m} \langle \Phi_l(\mathbf{p}) \Phi_i(\mathbf{p}) A(\mathbf{p}) \rangle \mathbf{v}_i'(\tau) = \left\langle \Phi_l(\mathbf{p})\, T(\mathbf{p})\, \mathbf{f} \left(\tau T(\mathbf{p}), \sum_{i=0}^{m} \mathbf{v}_i(\tau) \Phi_i(\mathbf{p}), \mathbf{p} \right) \right\rangle \tag{10}$$

for $l = 0, 1, \ldots, m$. A constant matrix appears in the left-hand side of the complete system. The coefficient functions inherit the periodicity of the random process due to (9). Hence we arrange the boundary conditions

$$\mathbf{v}_l(0) = \mathbf{v}_l(1) \qquad \text{for } l = 0, 1, \ldots, m. \tag{11}$$

The periodic boundary value problem (10), (11) can be solved by common numerical techniques again. Often the probabilistic integral in the right-hand side of (10) cannot be calculated explicitly. Gaussian quadrature yields an approximation of the right-hand side evaluations.

A special case appears in case of a constant matrix, i.e., $A(\mathbf{p}) = A_0$. Due to the orthogonality of the basis polynomials, the coupled system (10) simplifies to

$$A_0 \mathbf{v}_l'(\tau) = \left\langle \Phi_l(\mathbf{p})\, T(\mathbf{p})\, \mathbf{f}\left(\tau T(\mathbf{p}), \sum_{i=0}^{m} \mathbf{v}_i(\tau)\Phi_i(\mathbf{p}), \mathbf{p}\right)\right\rangle \tag{12}$$

for $l = 0, 1, \ldots, m$. Hence the constant matrix corresponding to the left-hand side of the complete system becomes block-diagonal.

In contrast to a Monte-Carlo simulation, the gPC problem (10), (11) has to be solved just once. A simulation based on the larger coupled system from the stochastic Galerkin method is often more efficient than a quasi Monte-Carlo simulation in case of linear systems of differential equations (ODEs, DAEs or PDEs). The above approach is also feasible for linear systems (1) with time-dependent inputs. However, autonomous oscillators are described by nonlinear systems of ODEs or DAEs in most instances. In the nonlinear case, the efficiency of the stochastic Galerkin approach requires further investigations.

The solution of boundary value problems of dynamical systems with random parameters via the gPC using either a stochastic collocation or the stochastic Galerkin approach is analysed more detailed in [9].

4 Illustrative Example

We apply a transistor amplifier shown in Fig. 1(left). A mathematical modelling yields a nonlinear system of DAEs for the unknown five node voltages, see [4]. The differential index of the DAEs is one. We arrange the input signal

$$U_{\text{in}}(t) = 0.4 \sin\left(\tfrac{2\pi}{T} t\right)$$

with period T. A corresponding periodic solution for $T = 0.01$ is depicted in Fig. 1(right). The voltages U_1, U_2, U_3 exhibit the form of sine waves. In contrast, the voltage U_4 and the output voltage U_5 behave nonlinearly due to the transistor.

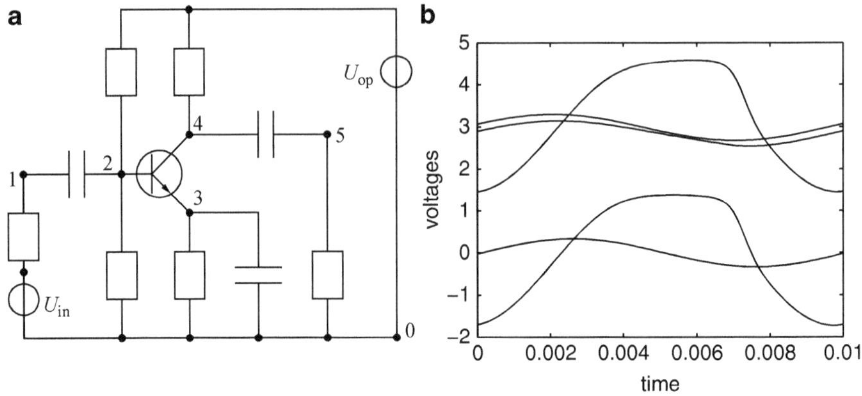

Fig. 1 Circuit of transistor amplifier (left) and deterministic periodic solution for $T = 0.01$ (right)

Now we consider a random period

$$T(p) = 0.01(1 + 0.1p),$$

where p represents a standardised random variable with a beta distribution according to the probability density function

$$\rho(p) = C(\alpha, \beta)(1 - p)^{\alpha}(1 + p)^{\beta} \qquad \text{for } -1 \leq p \leq 1$$

with a constant $C(\alpha, \beta)$. Hence the random period itself is distributed of beta type. We choose $\alpha = \beta = 2$. It follows the expected value $\langle T(p) \rangle = 0.01$, the standard deviation $\sigma(T(p)) \doteq 3.8 \cdot 10^{-4}$ and the range of the random period includes variations up to 10%. Although the modelling as well as the numerical simulations can include more random parameters, the other physical parameters are chosen deterministic for simplicity in this example.

The gPC expansion (8) includes the Jacobi polynomials. We apply polynomials up to degree $m = 3$. The larger coupled system exhibits the structure (12). The periodic boundary value problem (11), (12) is solved by a finite difference method using asymmetric formulas of second order (BDF2). Figure 2 illustrates the resulting approximations of the expected values (degree 0) and the standard deviations corresponding to the five node voltages. The expected values are similar to the deterministic solution shown in Fig. 1(right). The standard deviation of the voltages U_1, U_2, U_3 is relatively low. In contrast, the voltages U_4 and U_5 feature a subdomain in time (near $\tau = 0.7$), where a relatively high standard deviation appears. The standard deviation of U_4 and U_5 is nearly the same, since the shape of the oscillations agrees. Furthermore, Fig. 3 illustrates the coefficient functions (9) of the output voltage U_5. The magnitude of the coefficient functions decreases rapidly for increasing degree, which reflects the convergence of the gPC representation (8).

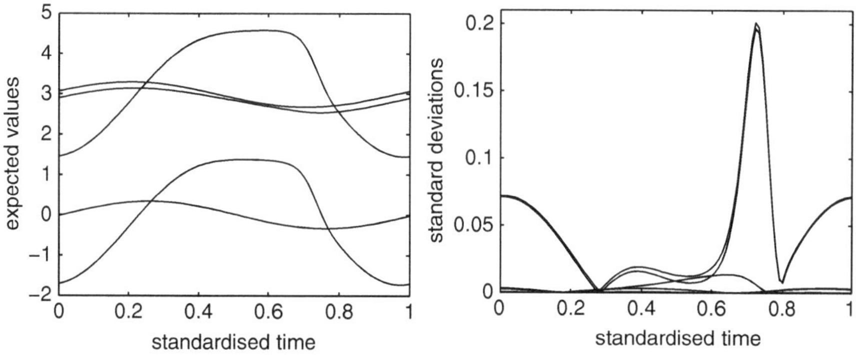

Fig. 2 Expected values (left) and standard deviations (right) for node voltages with random period

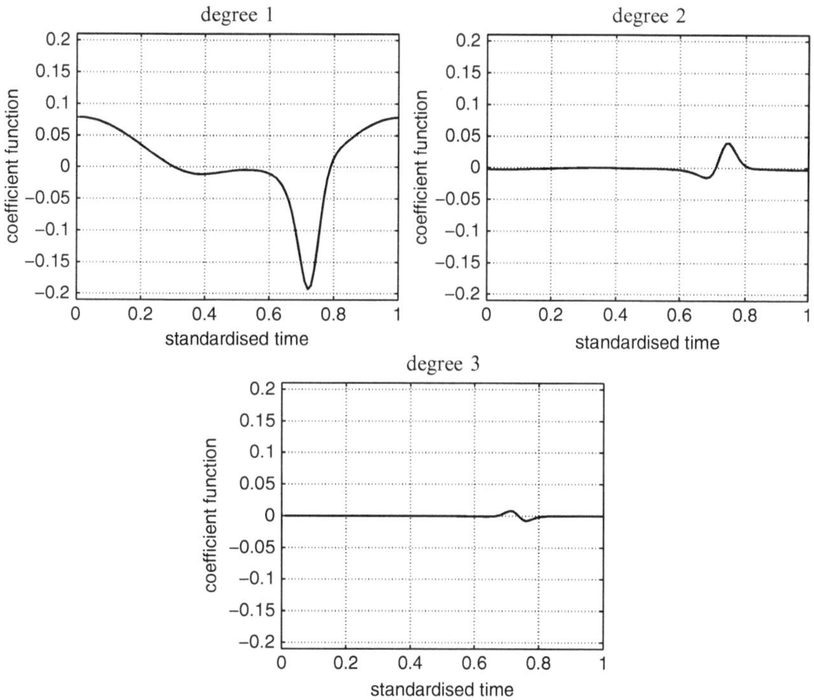

Fig. 3 Coefficient functions of output voltage in gPC

For comparison, we compute a reference solution via a quasi Monte-Carlo simulation using $K = 1000$ samples. The periodic boundary value problems (6), (7) are resolved by a finite difference method of second order again. Alternatively, we solve the periodic boundary value problems (11), (12) for different orders m now.

Table 1 Maximum differences between approximations from gPC for different order m and approximations from quasi Monte-Carlo simulation using $K = 1000$ samples corresponding to output voltage U_5

	$m = 1$	$m = 2$	$m = 3$	$m = 4$	$m = 5$
expected value	$2.6 \cdot 10^{-3}$	$8.4 \cdot 10^{-5}$	$5.1 \cdot 10^{-6}$	$4.3 \cdot 10^{-6}$	$4.5 \cdot 10^{-6}$
variance	$4.5 \cdot 10^{-3}$	$2.2 \cdot 10^{-4}$	$5.3 \cdot 10^{-5}$	$3.6 \cdot 10^{-5}$	$3.7 \cdot 10^{-5}$

The maximum absolute differences between the approximations of the expected values and the variances corresponding to the output voltage are shown in Table 1. The differences corresponding to the other node voltages have the same or a smaller magnitude. As hoped for, the accuracy of the gPC approximations improves for increasing order m. In particular, a linear approximation ($m = 1$) is not sufficiently accurate, whereas nonlinear polynomials of a low order yield an adequate numerical solution. The differences do not decrease for $m \geq 4$ any more. To achieve a better agreement for large orders m, the number K of samples has to be increased in the Monte-Carlo simulation and a higher accuracy has to be demanded in all involved finite difference methods.

5 Conclusions

A modelling of forced oscillators with random periods has been introduced, which defines a corresponding random process. We have constructed a numerical technique based on the generalised polynomial chaos for solving the stochastic model. A Galerkin approach changes the underlying system of differential algebraic equations into a larger coupled system of differential algebraic equations. We presented numerical simulations of a test example, which confirm that the stochastic Galerkin approach is feasible in this application. Further investigations are required for statements on the efficiency of the technique in comparison to stochastic collocation methods or quasi Monte-Carlo simulations. In particular, more test examples have to be considered for a discussion of the efficiency.

References

1. Augustin, F., Gilg, A., Paffrath, M., Rentrop, P., Wever, U.: Polynomial chaos for the approximation of uncertainties: chances and limits. Euro. J. Appl. Math. 19, 149–190 (2008)
2. Ghanem, R.G., Spanos, P.D.: Stochastic Finite Elements: A Spectral Approach. (rev. ed.) Dover, New York, (2003)
3. Günther, M., Feldmann, U., ter Maten, E.J.W.: Modelling and discretization of circuit problems. In: Schilders, W.H.A., ter Maten, E.J.W. (eds.) Handbook of Numerical Analysis, vol. XIII, Numerical Methods in Electromagnetics, pp. 523–659, Elsevier, North-Holland (2005)

4. Hairer, E., Wanner, G.: Solving Ordinary Differential Equations II: Stiff and Differential-Algebraic Problems (2nd ed.) Springer, Berlin (2002)
5. Lucor, D., Karniadakis, G.E.: Adaptive generalized polynomial chaos for nonlinear random oscillators. SIAM J. Sci. Comput. **26** (2), 720–735 (2004).
6. Lucor, D., Su, C.H., Karniadakis, G.E.: Generalized polynomial chaos and random oscillators. Int. J. Numer. Meth. Eng. **60**, 571–596 (2004).
7. Pulch, R.: Polynomial chaos for analysing periodic processes of differential algebraic equations with random parameters. Proc. Appl. Math. Mech. **8**, 10069–10072 (2008).
8. Pulch, R.: Polynomial chaos for the computation of failure probabilities in periodic problems. In: Roos, J., Costa, L. (eds.), Scientific Computing in Electrical Engineering SCEE 2008, Mathematics in Industry, vol. 14, pp. 191–198, Springer, Berlin (2010)
9. Pulch, R.: Polynomial chaos for boundary value problems of dynamical systems. to appear in: Appl. Numer. Math.
10. Pulch, R.: Modelling and simulation of autonomous oscillators with random parameters. to appear in: Math. Computers in Simulation, **81**(6), 1128–1143 (2011)
11. Xiu, D., Hesthaven, J.S.: High-order collocation methods for differential equations with random inputs. SIAM J. Sci. Comput. **27** (3), 1118–1139 (2005)
12. Xiu, D.: Fast numerical methods for stochastic computations: A review. Comm. Comput. Phys. **5** (2-4), 242–272 (2009)

Initialization of HB Oscillator Analysis from Transient Data

Mikko Hulkkonen, Mikko Honkala, Jarmo Virtanen, and Martti Valtonen

Abstract Oscillation frequency and amplitude of a free-running oscillator are commonly solved with harmonic balance (HB) method using an oscillator probe. This usually requires optimization. Poor initial values of oscillation may lead to unsuccessful optimization or will at least require a great number of optimization cycles. Therefore, two methods to initialize HB oscillator analysis from transient data are presented. These methods improve the initial estimates of oscillator frequency and amplitude. In addition, techniques to improve convergence of the analysis by initializing HB voltages from transient data and using an oscillator probe pulse are discussed. The efficiency of the methods is examined and verified through numerical experiments.

1 Introduction

Applying the harmonic balance (HB) method to free-running oscillators is difficult since the oscillation frequency is not known beforehand. Different approaches to oscillator analysis have been studied in [1] and [2]. One common way to solve free-running oscillator problems with the HB method is to use single, multiple, or multi-harmonic probes [3].

In the APLAC circuit simulator [4], the HB oscillator problem is solved by optimization. The optimization variables are the probe voltage V_{osc} and the oscillation frequency f_{osc}, which are optimized to make the probe voltage V_{osc} equal to the HB voltage across the probe element at the fundamental oscillation frequency f_{osc}.

M. Hulkkonen (✉) · M. Honkala · J. Virtanen · M. Valtonen
Aalto University School of Science and Technology, Department of Radio Science and Engineering, P.O. Box 13000, FI-00076 AALTO, Finland
e-mail: mikko.hulkkonen@tkk.fi; mikko.a.honkala@tkk.fi; jarmo.virtanen@tkk.fi; martti.valtonen@tkk.fi

B. Michielsen and J.-R. Poirier (eds.), *Scientific Computing in Electrical Engineering SCEE 2010*, Mathematics in Industry 16, DOI 10.1007/978-3-642-22453-9_30,

The user is required to set initial values for V_{osc} and f_{osc}. A poor initial frequency or amplitude value can lead to unsuccessful optimization or will at least require a great number of optimization cycles to succeed [5]. Therefore, algorithms have been developed to improve the user-defined initial values of oscillation frequency and amplitude before optimization. Also, initialization of HB voltages for optimization are presented.

2 Initialization of HB Analysis

Two separate methods were developed to enhance the initialization of HB analysis, namely, an FFT-based method and a zero crossing method. Each method can be used to initialize both the oscillation frequency and the oscillation amplitude of the probe element [3]. In this section, the functionality and the algorithms used by these methods are presented. Also, the use of initial oscillation probe pulse and initialization of HB voltages to speed up the analysis and improve the convergence are introduced.

2.1 *Transient Analysis Set-Up*

Both methods use data obtained from the transient analysis. The length of the transient analysis is determined by two parameters: the initial oscillation frequency f_{osc} set by the user and its minimum value f_{oscmin}. The transient analysis is run by default up to

$$t_{end} = \frac{\text{coeff}}{f_{oscmin}} + t_0, \tag{1}$$

where the default values of coeff and t_0 are 1 and $5/f_{osc}$, respectively. These values are the same that are used by default in the APLAC simulator when user requests transient assisted HB simulation, that is a valid request also for other circuits than oscillators. Both the zero crossing and the FFT methods use the data of transient analysis when $t_0 \leq t \leq t_{end}$. The collected data is the transient voltage over the probe element.

2.2 *FFT-Based Oscillation Frequency Detection*

After the transient analysis is run for the oscillator circuit to obtain the oscillation waveform, a Fourier transform is utilized to get the frequency-domain spectrum of the oscillator. A spectral line having the largest magnitude gives an approximation of the oscillation frequency. Depending on the sampling rate, the accurate oscillation frequency may be situated between the sampled frequency points. Therefore, parabolic interpolation with three frequency points (at the location of the maximum

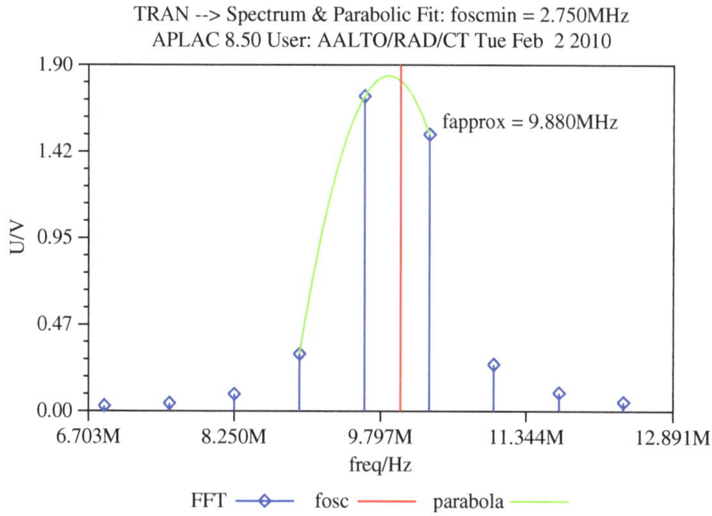

Fig. 1 Example of parabolic interpolation of frequency. Fosc indicates the correct oscillation frequency. Fapprox is the frequency estimated by the interpolation

and at two neighboring points) is used to determine a more accurate estimate for the oscillation frequency

$$f_{osc} = \frac{i_{max} + d}{\Delta t \cdot N_{FFT}}, \tag{2}$$

where Δt is the time between analyzed points, N_{FFT} is the number of FFT points, i_{max} is the index of the spectral line having the largest magnitude, and d specifies how far the accurate maximum is from i_{max}. The value of d is interpolated as follows

$$d = \frac{1}{2} \cdot \frac{V_{max-1} - V_{max+1}}{V_{max-1} - 2 \cdot V_{max} + V_{max+1}}, \tag{3}$$

where V_{max} and $V_{max\pm1}$ are the spectrum values nearest to the detected maximum.

Figure 1 presents an example of the interpolation. The vertical line is the correct oscillation frequency (10.01MHz in Fig. 1), and the parabolic interpolation estimates 9.88MHz.

2.3 *Oscillation Frequency Detection From Transient Zero Level Crossings*

Similar to the FFT-based frequency detection, this method uses data obtained from the transient simulation. The DC voltage level is found from the time-domain response as the average of analyzed points starting from t_0. Next, the DC value is

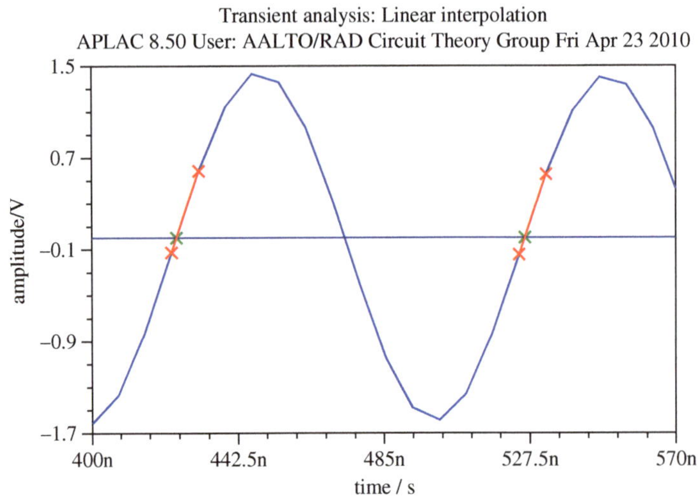

Fig. 2 Example of linear interpolation. The points used for interpolation are indicated by the x-markers

removed from the response and the period is determined from the zero crossings in the waveform. Accuracy of the zero crossings is improved by using linear interpolation.

An example of the linear interpolation is in Fig. 2. The horizontal line is the DC level, the x-markers connected with lines show the points used for interpolation.

2.4 Estimation of the Voltage Probe Amplitude

The amplitude of the voltage probe V_{osc} can be estimated from the transient analysis.

When the FFT-based method is used, the new value for the probe voltage V_{osc} is the spectrum peak value. From the transient waveform, the voltage can be computed as follows: $V_{osc} = (V_{max} - V_{min})/2$, where V_{max} and V_{min} are the minimum and maximum values, respectively, of the transient waveform.

For some oscillators, the time interval of the transient analysis may not be long enough to start the oscillator properly. Therefore, the oscillation amplitude is typically only a fraction of the correct one. Because of this, both algorithms can optionally use an adaptive V_{osc} estimation. In this case, the amplitude from transient analysis V_{amp} is compared to $V_{osc,init}$, the initial value of V_{osc} given by user. If the amplitude obtained from the transient analysis is too small, the analysis is continued until it is large enough. The condition $V_{amp} > \epsilon \cdot V_{osc,init}$, where ϵ is a predefined limit, determines this. If the required amplitude is not reached in a reasonable amount of transient analysis time, the user-specified $V_{osc,init}$ is used.

2.5 Initialization of HB Voltages

In order to improve the convergence of the HB analysis at the first point of optimization, the spectral voltages of every node can be initialized from the transient data to correspond to the estimated oscillation frequency f_{osc}. In this case, the transient analysis is continued by one period of f_{osc} and the voltage waveforms for each node are stored. From these voltages, FFT is used to calculate the spectral voltages which are then written to a separate file called guess file. If this optional guess file is omitted, the HB voltages are initialized to DC voltages.

2.6 Oscillator Probe Element Pulse

As stated before, some oscillators may start oscillating rather slowly. In order to accelerate the start-up of the oscillator, the oscillator probe element can inject a smooth and short sinusoidal pulse to the oscillator in the beginning of the transient analysis. The pulse was selected so that its value at $t = 0$ and $t = t_2$ (end of the pulse) is zero. The value of this voltage pulse is

$$v = \begin{cases} V_{osc} \cdot (t/t_1)\cos(2\pi f_{osc} t), & \text{when } t \leq t_1 \\ V_{osc}\cos(2\pi f_{osc} t), & \text{when } t_1 < t \leq t_2 \\ 0, & \text{when } t > t_2, \end{cases}$$

where $t_1 = 1/(4f_{osc})$ and $t_2 = 3/(4f_{osc})$. The variables f_{osc} and V_{osc} have the user-defined initial values. As the shape of the pulse is determined by the user-specified value of f_{osc}, the transient simulation response would be too deterministic if the shape of the pulse would have been pure sinusoidal. The waveform of the pulse is presented in Fig. 3 when $f_{osc} = 1\text{GHz}$ and $V_{osc} = 1\text{V}$.

3 Implementation

The algorithms have been implemented in the APLAC simulator. Two different frequency estimation methods, FFT and zero crossing (ZeroC), can be chosen, and four different initialization modes presented in Table 1 have been implemented.

The user has to insert the voltage probe element into the circuit and has to specify the initial value for its amplitude. Also, the oscillation frequency and optionally its lower and upper boundaries have to be defined. The length of the transient analysis is determined from the boundaries. The optimization method used can be chosen from APLAC's existing optimization methods [4], the most common being MinMax, NelderMead and Gradient methods.

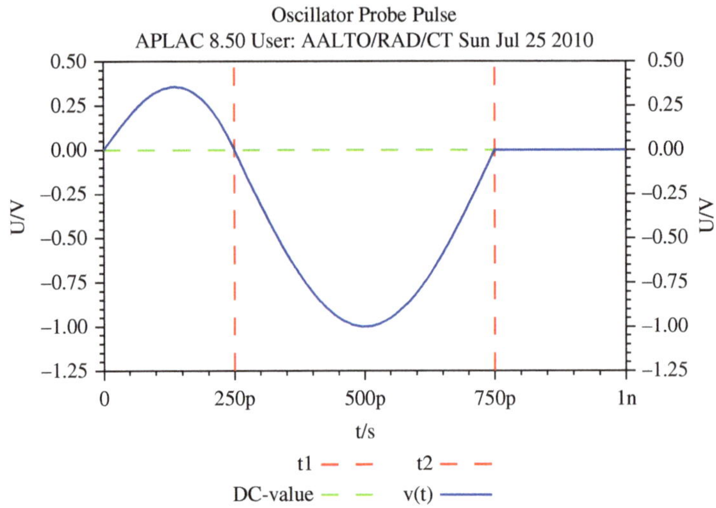

Fig. 3 Oscillator probe element pulse waveform

4 Results

Both methods have been tested with four oscillator circuits, namely, colpitts, VCO, pierce, and VHF oscillators. Several test setups were used in the testing. Characteristic results for the number of HB iterations and speedups compared to the default oscillator analyzing method are presented in Figs. 4 and 5. These tests were done with initialization mode 2 of Table 1. The corresponding test setups are in Table 2.

5 Conclusion

In general, both methods, the FFT method and the zero-crossing method, provide estimated initial values that improve the performance of the oscillator optimization analysis. Optimization methods are, however, quite sensitive to the initial values, and depending on the values of analysis and/or optimization parameters, the optimization algorithm does not always benefit for the improved initial values. There is also large variation in the behaviour between optimization methods MinMax and NelderMead, partly due to the internal differences of these methods. In some cases, regardless of good initial values, the optimization may end up in some local minimum or, for some algorithm-specific reason, get driven in completely wrong direction. This results in smaller speedups and a greater occurrence of poor convergence when selecting the unfit method.

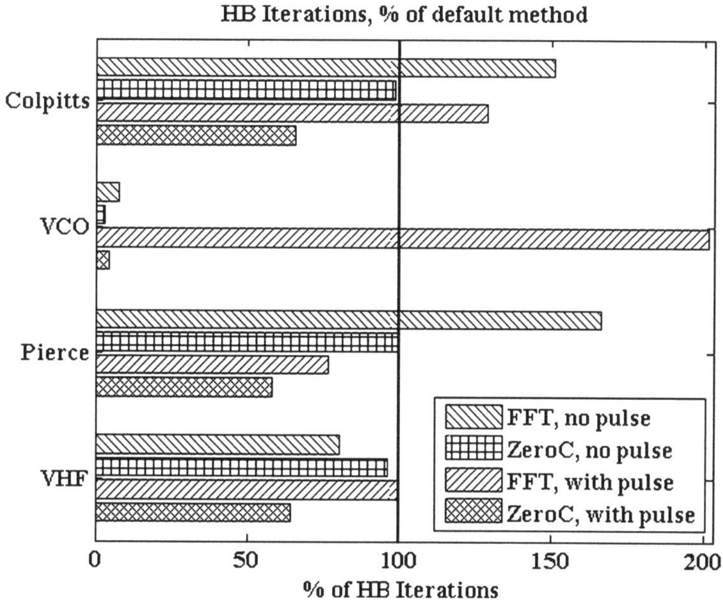

Fig. 4 HB iterations compared to the default oscillator method. Pulse indicates the use of the initial pulse

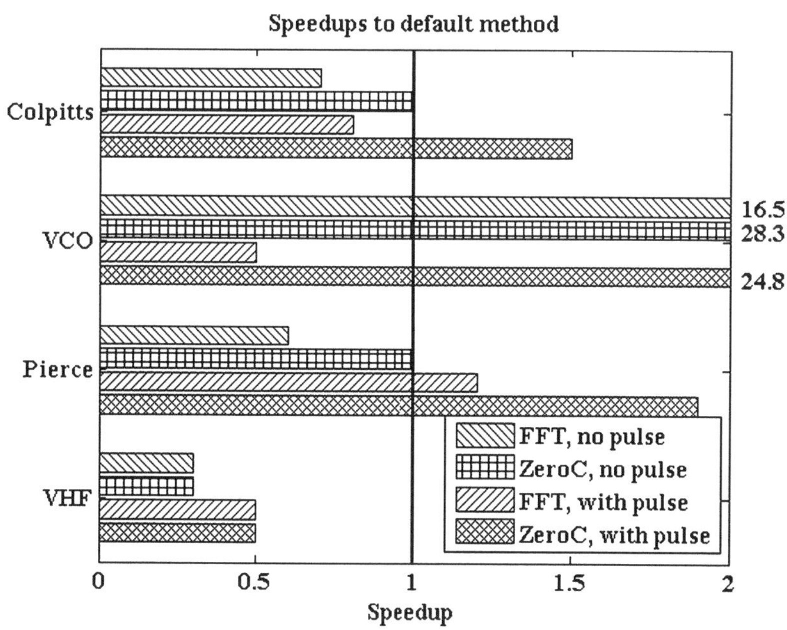

Fig. 5 Speedups compared to the default oscillator method. Pulse indicates the use of the initial pulse

Table 1 Initialization modes. 'gfile' indicates the usage of the guess file

HBOSCGUESS	Variables initalized
0	f_{osc}
1	f_{osc}, gfile
2	f_{osc}, V_{osc}, gfile
3	f_{osc}, V_{osc}

Table 2 Test setups for Figs. 4 and 5

Circuit	f_{osc} Initial	V_{osc} Initial	V_{osc} Limits	Optimization method	t_0	FFT points
Colpitts	16 MHz	1.0	[0.01, 100]	MinMax	$5/f_{\mathrm{osc}}$	128
VCO	1 GHz	1.0	[0.01, 100]	MinMax	$5/f_{\mathrm{osc}}$	128
Pierce	300 kHz	1.0	[0.01, 100]	MinMax	$5/f_{\mathrm{osc}}$	128
VHF	100 MHz	0.1	[0.01, 100]	MinMax	$5/f_{\mathrm{osc}}$	128

As a result of the several simulations used for testing these methods, it can be stated that the methods do not necessarily reduce the total simulation time. The more important outcome is that they improve convergence and simulator robustness, which greatly improves the quality of the HB based oscillator analysis and makes it possible for the designer to obtain good results with less accurate initial values.

In conclusion, two methods to initialize the HB oscillator analysis, i.e., the estimation of the oscillation frequency and voltage, were developed and implemented into the APLAC simulator. The methods were tested on four real-life oscillator circuits and the results were good. Further research and development of the methods is also possible after more feedback based on real design problems is received.

Acknowledgements This work is a part of the collaborative research project ICESTARS within the ICT research program of EU FP7/2008/ICT/214911, and was funded by the European Union.

References

1. Hoube, S.H.M.J., Maubach, J.M.: Periodic steady-state analysis of free-running oscillators. In: van Rienen, U., Günther, M., Hecht, D. (eds.) Scientific Computing in Electrical Engineering, Lecture Notes in Engineering, vol. 18, pp. 217–224. Springer, Heidelberg (2001)
2. Lampe, S., Brachtendorf, H.G., ter Maten, E.J.W., Onneweer, S.: Robust limit cycle calculations of oscillators. In: van Rienen, U., Günther, M., Hecht, D. (eds.) Scientific Computing in Electrical Engineering, Lecture Notes in Engineering, vol. 18, pp. 233–240. Springer, Heidelberg (2001)
3. Brambilla, A., Gruosso, G., Storti Gajani, G.: Robust harmonic-probe method for the simulation of oscillators. IEEE Trans. CAS I **57**(9), pp. 2531–2541 (2010)
4. APLAC, http://www.awrcorp.com/USA/Products/APLAC
5. Virtanen, J., ter Maten, J., Beelen, T., Honkala, M., Hulkkonen, M.: Initial conditions and robust Newton-Raphson for Harmonic Balance oscillator analysis of free-running oscillators. European Conference on Mathematics for Industry (2010) (Submitted)

Robust Periodic Steady State Analysis of Autonomous Oscillators Based on Generalized Eigenvalues

R. Mirzavand Boroujeni, E.J.W. ter Maten, T.G.J Beelen, W.H.A. Schilders and A. Abdipour

Abstract In this paper, we present a new gauge technique for the Newton Raphson method to solve the periodic steady state (PSS) analysis of free-running oscillators in the time domain. To find the frequency a new equation is added to the system of equations. Our equation combines a generalized eigenvector with the time derivative of the solution. It is dynamically updated within each Newton–Raphson iteration. The method is applied to an analytic benchmark problem and to an LC oscillator. It provides better convergence properties than when using the popular phase-shift condition. It also does not need additional information about the solution. The method can also easily be implemented within the Harmonic Balance framework.

R.M. Boroujeni (✉) · E.J.W. ter Maten · W.H.A. Schilders
CASA group, Department of Mathematics and Computer Science, Eindhoven University of Technology, 5600 MB Eindhoven, The Netherlands
e-mail: R.Mirzavand.Boroujeni,E.J.W.ter.Maten,W.H.A.schilders@tue.nl

R.M. Boroujeni · A. Abdipour
Electrical Engineering Department, Amirkabir University of Technology, 15875-4413 Tehran, Iran
e-mail: RMirzavand,Abdipour@aut.ac.ir

E.J.W. ter Maten · T.G.J. Beelen
NXP Semiconductors, High Tech Campus 46, 5656 AE Eindhoven, The Netherlands
e-mail: Jan.ter.Maten,Theo.G.J.Beelen@nxp.com

E.J.W. ter Maten
Bergische Universität Wuppertal, Fachbereich C, Wicküler Park, Bendahler Str. 29, D-42285 Wuppertal, Germany
e-mail: Jan.ter.Maten@math.uni-wuppertal.de

B. Michielsen and J.-R. Poirier (eds.), *Scientific Computing in Electrical Engineering SCEE 2010*, Mathematics in Industry 16, DOI 10.1007/978-3-642-22453-9_31,

1 Introduction

Designing an oscillator requires a Periodic-Steady State (PSS) analysis. The PSS solution can be found by long time integration, starting from perturbing the DC-solution. It is needed for phase noise analysis [2]. Time integration is robust (it always works: the DC-solution is an unstable PSS solution), but the convergence can be very slow. Therefore dedicated solution methods have been presented based in time-domain, frequency-domain or by hybrid circuit-state representations [1, 7, 11]. In these methods the period T or the frequency f is an additional unknown. To make the solution unique an additional equation, like a phase-shift condition, is added [3, 5, 8]. The overall system of equations is solved by a Newton–Raphson method that needs initial estimates for the solution as well as for T (or f).
This paper presents a Newton–Raphson based method with a dynamic additional condition to find the PSS solution of a free-running oscillator in the time domain. Here generalized eigenvectors of the linearized circuit equations and the time derivative at each time step provide a new robust gauge equation for the Newton–Raphson equations. The method is applied to an analytic benchmark problem and to an LC oscillator. The efficiency of the method is verified through numerical experiments. It provides better convergence properties than when using the popular phase-shift condition. It also does not need additional information about the solution.

2 The Autonomous Oscillator Problem

The PSS problem for autonomous circuits on one overall period T is defined as a system of Differential-Algebraic Equations (DAEs) in the following form,

$$\frac{\mathrm{d}\mathbf{q}(\mathbf{x})}{\mathrm{d}t} + \mathbf{j}(\mathbf{x}) = \mathbf{0} \in \mathbb{R}^n, \tag{1}$$

$$\mathbf{x}(0) = \mathbf{x}(T), \tag{2}$$

where $\mathbf{x} = \mathbf{x}(t) \in \mathbb{R}^n$ and T are unknown; $\mathbf{q}$ and $\mathbf{j}$ are known functions of $\mathbf{x}$. In the above autonomous circuit, there is a non-trivial PSS solution in the absence of sources. Here the period T (or the frequency $f = 1/T$) is unknown and is determined by the system. By transforming the simulation time interval $[0, T]$ to the standard interval $[0, 1]$, f enters the above equations as a parameter

$$f\,\frac{\mathrm{d}\mathbf{q}(\mathbf{x})}{\mathrm{d}t} + \mathbf{j}(\mathbf{x}) = 0. \tag{3}$$

Taking f as extra unknown, we need an extra equation to complete the system. Usually one requires the additional constraint condition

$$\mathbf{c}^T\mathbf{x}(t_c) - c = 0, \tag{4}$$

to provide a non-zero value for some vector $\mathbf{c}$ which makes the phase-shift unique. For instance, one provides the value of a particular coordinate of $\mathbf{x}$ at some time t_c.

3 Newton Procedure

We discretize $[0, 1]$ using equidistant time points $t_i = i\,\Delta t$ for $i = 0, \ldots, N$ with $N\,\Delta t = 1$. Thus, $t_0 = 0$, $t_N = 1$. Let $\mathbf{x}_i$ approximate $\mathbf{x}(t_i)$ and $\mathbf{X} = \begin{bmatrix} \mathbf{x}_0 \cdots \mathbf{x}_{N-1} \end{bmatrix}^T$. We discretize (3) by applying Simpson's Rule on the (overlapping) sub-intervals $[t_{i-1}, t_{i+1}]$, for $i = 1, \ldots, N$, yielding

$$\mathbf{F}_i(\mathbf{X}, f) = f\,\frac{\mathbf{q}(\mathbf{x}_{i+1}) - \mathbf{q}(\mathbf{x}_{i-1})}{2\Delta t} + \frac{\mathbf{j}(\mathbf{x}_{i-1}) + 4\,\mathbf{j}(\mathbf{x}_i) + \mathbf{j}(\mathbf{x}_{i+1})}{6}, \quad i = 1, \ldots, N. \tag{5}$$

For $i = N - 1$ and $i = N$ we apply the periodicity constraint $\mathbf{x}_N = \mathbf{x}_0$ and $\mathbf{x}_{N+1} = \mathbf{x}_1$. Let $t_c = t_{k'}$ for some k' and redefine $\mathbf{c}$ to apply to $\mathbf{X}$. We write $\mathbf{q}_i = \mathbf{q}(\mathbf{x}_i)$ and similarly for $\mathbf{j}_i$. The Newton–Raphson method to solve the discrete systems becomes

$$\mathbf{M}^k \begin{bmatrix} \mathbf{X}^{k+1} - \mathbf{X}^k \\ f^{k+1} - f^k \end{bmatrix} = - \begin{bmatrix} \mathbf{F}(\mathbf{X}^k, f^k) \\ \mathbf{c}^T \mathbf{X}^k - c \end{bmatrix}, \tag{6}$$

in which $\mathbf{X}^k = \begin{bmatrix} \mathbf{x}_0^k \cdots \mathbf{x}_{N-1}^k \end{bmatrix}^T$ and

$$\mathbf{F}(\mathbf{X}, f) = \begin{bmatrix} f\,\frac{\mathbf{q}_2 - \mathbf{q}_0}{2\Delta t} + \frac{\mathbf{j}_0 + 4\mathbf{j}_1 + \mathbf{j}_2}{6} \\ \vdots \\ f\,\frac{\mathbf{q}_0 - \mathbf{q}_{N-2}}{2\Delta t} + \frac{\mathbf{j}_{N-2} + 4\mathbf{j}_{N-1} + \mathbf{j}_0}{6} \\ f\,\frac{\mathbf{q}_1 - \mathbf{q}_{N-1}}{2\Delta t} + \frac{\mathbf{j}_{N-1} + 4\mathbf{j}_N + \mathbf{j}_1}{6} \end{bmatrix}, \quad \mathbf{M}^k = \begin{bmatrix} \mathbf{A}^k & \mathbf{b}^k \\ \mathbf{c}^T & \delta \end{bmatrix}. \tag{7}$$

Here

$$\mathbf{A}^k = \left.\frac{\partial \mathbf{F}}{\partial \mathbf{x}}\right|_{\mathbf{X}^k, f^k} = f^k \cdot \mathbf{C}^k + \mathbf{G}^k, \quad \mathbf{b}^k = \left.\frac{\partial \mathbf{F}}{\partial f}\right|_{\mathbf{X}^k, f^k}, \tag{8}$$

for suitable matrices $\mathbf{C}$ and $\mathbf{G}$, that are composed by the local Jacobians $\mathbf{C}_i = \left.\frac{\partial \mathbf{q}}{\partial \mathbf{x}}\right|_{\mathbf{x}=\mathbf{x}_i}$ and $\mathbf{G}_i = \left.\frac{\partial \mathbf{j}}{\partial \mathbf{x}}\right|_{\mathbf{x}=\mathbf{x}_i}$ and the discretization step size,

$$\mathbf{C} = \frac{1}{2\Delta t} \begin{bmatrix} -\mathbf{C}_0 & \mathbf{0} & \mathbf{C}_2 & & \cdots & \mathbf{0} \\ \mathbf{0} & -\mathbf{C}_1 & \mathbf{0} & \mathbf{C}_3 & \cdots & \mathbf{0} \\ \vdots & \ddots & \ddots & \ddots & & \vdots \\ \mathbf{C}_0 & & \cdots & \mathbf{0} & -\mathbf{C}_{N-2} & \mathbf{0} \\ \mathbf{0} & \mathbf{C}_1 & \cdots & \mathbf{0} & \mathbf{0} & -\mathbf{C}_{N-1} \end{bmatrix}, \quad \mathbf{G} = \frac{1}{6} \begin{bmatrix} \mathbf{G}_0 & 4\mathbf{G}_1 & \mathbf{G}_2 & \mathbf{0} & \cdots & \mathbf{0} \\ \mathbf{0} & \mathbf{G}_1 & 4\mathbf{G}_2 & \mathbf{G}_3 & \cdots & \mathbf{0} \\ \vdots & \ddots & \ddots & \ddots & & \vdots \\ \mathbf{G}_0 & & \cdots & \mathbf{0} & \mathbf{G}_{N-2} & 4\mathbf{G}_{N-1} \\ 4\mathbf{G}_0 & \mathbf{G}_1 & \cdots & \mathbf{0} & \mathbf{0} & \mathbf{G}_{N-1} \end{bmatrix}. \tag{9}$$

Usually $\delta = 0$ in (7). The matrix $\mathbf{A}$ becomes badly conditioned when the Newton iterands converge. This is due to the fact that the time derivative of the PSS solution solves the linearized homogeneous circuit equations when linearized at the PSS solution. Hence when the discretization is exact this time derivative of the ultimate PSS is in the kernel of $\mathbf{A}$. Due to this conditioning problem the vectors $\mathbf{b}$ and $\mathbf{c}$ and (scalar) value δ are really needed to make the matrix $\mathbf{M}$ non-singular (otherwise one could use a Schur complement approach). $\mathbf{b}$ must have non-trivial components in $\text{Ker}(\mathbf{A})$ and in $\text{Ker}(\mathbf{A}^T)$, both. A similar statement holds for $\mathbf{c}$. Hence $\text{Ker}(\mathbf{A}) \not\perp \text{Ker}(\mathbf{A}^T)$.

4 Bordered Matrices

Theorem 1. *Let* $\mathbf{A}^+$ *be the Moore–Penrose inverse of* $\mathbf{A}$ *[6]. Define* $\mathbf{g}$, $\mathbf{h}$, $\mathbf{u}$, $\mathbf{v}$, α *by*

$$\begin{aligned}
\mathbf{g} = \mathbf{A}^+\mathbf{b}, \quad \mathbf{h} &= \mathbf{c}^*\mathbf{A}^+ \quad \text{(least squares approximations)},\\
\mathbf{u} = (\mathbf{I} - \mathbf{A}\,\mathbf{A}^+)\mathbf{b}, \quad \mathbf{v} &= \mathbf{c}^*(\mathbf{I} - \mathbf{A}^+\mathbf{A}) \quad \text{(projection errors)},\\
\alpha = \delta - \mathbf{c}^*\mathbf{A}^+\mathbf{b}.
\end{aligned}$$

Then $\mathbf{g}$, $\mathbf{h}$, $\mathbf{u}$ *and* $\mathbf{v}$ *satisfy*

$$\begin{aligned}
\mathbf{A}^+\mathbf{u} = 0, \quad \mathbf{v}\,\mathbf{A}^+ &= 0,\\
\mathbf{u}^+\mathbf{A} = 0, \quad \mathbf{A}\,\mathbf{v}^+ &= 0, \quad [(\mathbf{u}^+)^T \in \text{Ker}(\mathbf{A}^T), \quad \mathbf{v}^+ \in \text{Ker}(\mathbf{A})],\\
\mathbf{h}\,\mathbf{A} + \mathbf{v} = \mathbf{c}^*, \quad \mathbf{A}\,\mathbf{g} + \mathbf{u} &= \mathbf{b},\\
\mathbf{v}\,\mathbf{g} = 0, \quad \mathbf{h}\,\mathbf{u} &= 0,\\
\mathbf{h}\,\mathbf{A}\,\mathbf{A}^+ = \mathbf{h}, \quad \mathbf{A}^+\mathbf{A}\,\mathbf{g} &= \mathbf{g}, \quad \mathbf{h}\,\mathbf{A}\,\mathbf{g} = \delta - \alpha.
\end{aligned}$$

We are now able to derive more detailed expressions for the generalized inverse of a bordered matrix. See also [2,3] and [4,9] for cases where $\mathbf{u} = 0$ or $\mathbf{v} = 0$.

Theorem 2. *Let*

$$\mathbf{M} = \begin{bmatrix} \mathbf{A} & \mathbf{b} \\ \mathbf{c}^* & \delta \end{bmatrix}, \quad \tilde{\mathbf{M}} = \begin{bmatrix} \mathbf{A} & \mathbf{u} \\ \mathbf{v} & \alpha \end{bmatrix}. \tag{10}$$

Assume $\mathbf{u} \neq 0$ *and* $\mathbf{v} \neq 0$*, then*

$$\mathbf{M}^+ = \begin{bmatrix} \mathbf{A}^+ - \mathbf{g}\,\mathbf{u}^+ - \mathbf{v}^+\mathbf{h} - \delta\,\mathbf{v}^+\mathbf{u}^+ & \mathbf{v}^+ \\ \mathbf{u}^+ & 0 \end{bmatrix}, \quad \tilde{\mathbf{M}}^+ = \begin{bmatrix} \mathbf{A}^+ - \alpha\,\mathbf{v}^+\mathbf{u}^+ & \mathbf{v}^+ \\ \mathbf{u}^+ & 0 \end{bmatrix}. \tag{11}$$

The expression for $\tilde{\mathbf{M}}^+$ follows by checking the Moore-Penrose conditions [6]. For $\mathbf{M}^+$ we note that, when $\delta = \alpha + \mathbf{c}^*\mathbf{g}$,

$$\mathbf{M} = \begin{pmatrix} \mathbf{I} & 0 \\ \mathbf{h} & 1 \end{pmatrix} \tilde{\mathbf{M}} \begin{pmatrix} \mathbf{I} & \mathbf{g} \\ 0 & 1 \end{pmatrix}.$$

Hence

$$\mathbf{M}^{+} = \begin{pmatrix} \mathbf{I} & -\mathbf{g} \\ 0 & 1 \end{pmatrix} \tilde{\mathbf{M}}^{+} \begin{pmatrix} \mathbf{I} & 0 \\ -\mathbf{h} & 1 \end{pmatrix}.$$

Let $\text{Ker}(\mathbf{A}) = \langle \mathbf{a} \rangle$, $\text{Ker}(\mathbf{A}^T) = \langle \mathbf{a}_T \rangle$ ($\mathbf{a}$ and $\mathbf{a}_T$ unit vectors) and let $\mathbf{b} \in \langle \mathbf{a}_T \rangle$ and $\mathbf{c} \in \langle \mathbf{a} \rangle$. Then the most simple expressions appear because $\mathbf{g} = \mathbf{0}$, $\mathbf{h} = \mathbf{0}$, $\mathbf{u} = \mathbf{b}$, $\mathbf{v} = \mathbf{c}^*$. Furthermore, there also is robustness in the sense that if we have other choices then the bordered matrix may still be non-singular. Note that the lower right entries in $\mathbf{M}^+$ and $\tilde{\mathbf{M}}^+$ are zero (which may not happen for $\mathbf{M}$ or $\tilde{\mathbf{M}}$).
For the bordered matrix $\mathbf{M}^k$ in (7) the choice of $\mathbf{b}^k$ comes from the partial differentiation with respect to the chosen additional unknown f. The choice of $\mathbf{c}$ depends on the "gauge" equation that we add to the system. The matrix $\mathbf{A}$ is a matrix pencil, hence a choice for a generalized (kernel) eigenvector is best here. As equation we prefer the bi-orthogonality equation. This prevents all problems with determining the location of the oscillation and the range of values of the PSS solution.

5 Using Generalized Eigenvalue Methods

A proper dynamic expression within the loops for the vector c can increase the convergence rate of the Newton method. Generalized eigenvalue methods for matrix pencils are good candidates for obtaining a dynamic vector $\mathbf{c}$ to make $\mathbf{M}$ non-singular. Applying these methods in each Newton iteration gives the eigentriples $(\mathbf{v}, \mathbf{w}, \lambda)$ such that $[\lambda f\ \mathbf{C}+\mathbf{G}]\mathbf{v} = \mathbf{0}$ and $\mathbf{w}^T[\lambda f\ \mathbf{C}+\mathbf{G}] = \mathbf{0}$. Generalized eigenvalue methods are provided by the DPA (Dominant Pole Algorithm) and RQI (Raleigh Quotient Iteration) [10]. Here a combination of these methods (SARQI) is used to obtain a good accuracy and convergence rate.
The $\mathbf{v}$ and $\mathbf{w}$ have a bi-orthogonality relation with the matrix $\mathbf{C}$, $\mathbf{w}^T\mathbf{C}\mathbf{v} = 1$. In Sect. 3 we observed that in the limit when the Newton approximations are close to the exact solution, the right-hand side eigenvector $\mathbf{v}$ for the λ closest to 1 is close to $\mathrm{d}\mathbf{X}/\mathrm{d}t$ (up to a normalisation factor). Hence by approximating the bi-orthogonality relation by

$$\mathbf{w}^T \cdot \mathbf{C} \cdot \left. \frac{\mathrm{d}\mathbf{X}}{\mathrm{d}t} \right|_{\mathbf{X}=\mathbf{X}^k} - 1 = 0. \quad (12)$$

we obtain a good choice for a dynamic gauge equation within each iteration of the Newton method. To write (12) even in the form $\mathbf{c}^T\mathbf{X} - c = 0$, we express $\mathrm{d}\mathbf{X}/\mathrm{d}t$

into $\mathbf{X}$. Spectral differentiation [12] provides $\mathrm{d}\mathbf{X}/\mathrm{d}t = \mathbf{D} \cdot \mathbf{X}$ with good accuracy using some matrix $\mathbf{D}$. This results in a choice $\mathbf{c}^T = \mathbf{w}^T \cdot \mathbf{C} \cdot \mathbf{D}$ and $c = 1$.
We observe that we always can compare $\mathbf{v}$ with $\mathrm{d}\mathbf{X}/\mathrm{d}t$ for convergence. We may even consider $\mathbf{c}^T = \mathbf{v}^T$. Additionally we can compare λf_{old} with f. We finally note that spectral differentiation easily fits Harmonic Balance implementations.

6 Analytic Benchmark Oscillator

As an example, consider the analytic benchmark problem [7],

$$\frac{\mathrm{d}y}{\mathrm{d}t} = z + \varepsilon\left(1 - \sqrt{y^2 + z^2}\right) y,$$
$$\frac{\mathrm{d}z}{\mathrm{d}t} = -y + \varepsilon\left(1 - \sqrt{y^2 + z^2}\right) z. \tag{13}$$

The fact that we can tune convergence speed with ε makes this particular problem a suitable benchmark problem. For all ε the exact PSS solution of this problem is $y(t) = \sin(t - t_c)$, $z(t) = \cos(t - t_c)$, where t_c is some constant phase shift. The period $T = 2\pi$. By defining $r^2 = y^2 + z^2$, the system of equations (13) can be written in the form of (1),

$$\mathbf{x} = \begin{bmatrix} y(t) \\ z(t) \end{bmatrix}, \quad \mathbf{q} = \begin{bmatrix} y(t) \\ z(t) \end{bmatrix}, \quad \text{and} \quad \mathbf{j} = \begin{bmatrix} -\varepsilon\,(1 - r)\,y - z \\ -\varepsilon\,(1 - r)\,z + y \end{bmatrix}.$$

Starting with initial conditions $T_0 = 2.2\pi$, $y_0(t) = 1.5\sin(t + \pi/4)$, $z_0(t) = \cos(t)$, and $N = 101$ (100 actual time grid points), the PSS solution is obtained using the old phase-shift condition method and the new eigenvector condition method. Figure 1 shows the initial guess and the PSS solution of $y(t)$ for both methods when $\varepsilon = 0.1$. For both methods we determine the maximum of the normalized correction of the solution and the normalized frequency correction

$$\Delta\mathbf{X}^k\big|_{\text{Normalized}} = ||\mathbf{X}^{k+1} - \mathbf{X}^k||_\infty / ||\mathbf{X}^k||_\infty, \qquad \Delta f^k\big|_{\text{Normalized}} = |f^{k+1} - f^k| / |f^k|$$

during each k-th Newton–Raphson iteration; the results are presented in Fig. 2. The better convergence behaviour of the new method is clearly observed. Although the simulation time and memory usage of the old method with a good phase-shift condition are smaller than that of the new method, the former method does not converge without enough information about $\mathbf{x}$ (see the curves with a $\times$ mark). Because of the observed robustness on the non-singularity of $\mathbf{M}^k$ (Sect. 4), one may stop the dynamic update of the gauge equation when the process starts converging.

7 LC Oscillator

For many applications the free running oscillator can be modeled as an LC tank with a nonlinear resistor that is governed by the following differential equations for the unknowns v as the nodal voltage and i as the inductor current.

$$\begin{bmatrix} C & 0 \\ 0 & L \end{bmatrix} \begin{bmatrix} \dot{v}(t) \\ \dot{i}(t) \end{bmatrix} + \begin{bmatrix} \frac{1}{R} & 1 \\ 0 & -1 \end{bmatrix} \begin{bmatrix} v(t) \\ i(t) \end{bmatrix} + \begin{bmatrix} S \tanh(\frac{G_n v(t)}{S}) \\ 0 \end{bmatrix} = \mathbf{0} \tag{14}$$

$$v(0) = v_0, \quad i(0) = i_0. \tag{15}$$

where C, L and R are the capacitance, inductance and resistance, respectively. The voltage controlled nonlinear resistor is defined by the S and G_n parameters. For example, consider an oscillator designed for a frequency of 6 GHz with $L = 0.53\, nH, C = 1.33\, pF, R = 250\, \Omega, S = 1/R$, and $G_n = -1.1/R$. Starting with initial conditions $T_0 = 2.2\pi$, $v_0(t) = \sin(t)$, $i_0(t) = 0.2\sin(t)$, and $N = 101$ (100 actual grid points), the PSS solutions are obtained using the old phase-shift condition method and with the new eigenvector gauge method. The comparisons of the methods using the maximum of the normalized correction and the normalized frequency correction with respect to the iteration number k are presented in Fig. 3 showing similar improvement as in the previous example.

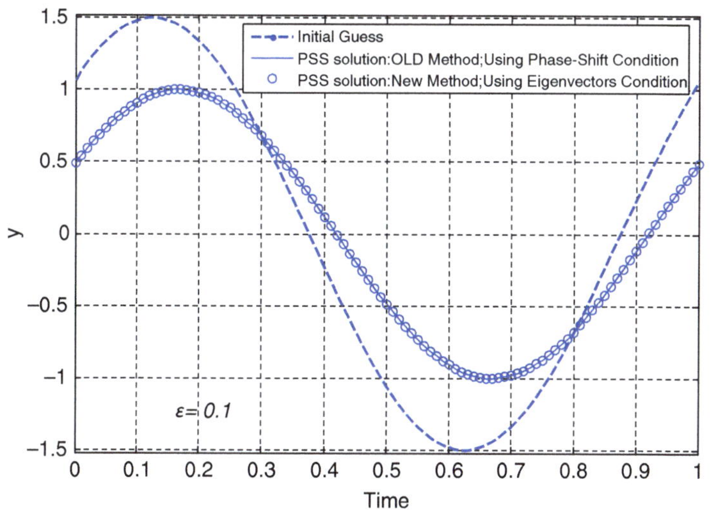

Fig. 1 Initial guess and PSS solution of $y(t)$ for different methods when $\varepsilon = 0.1$

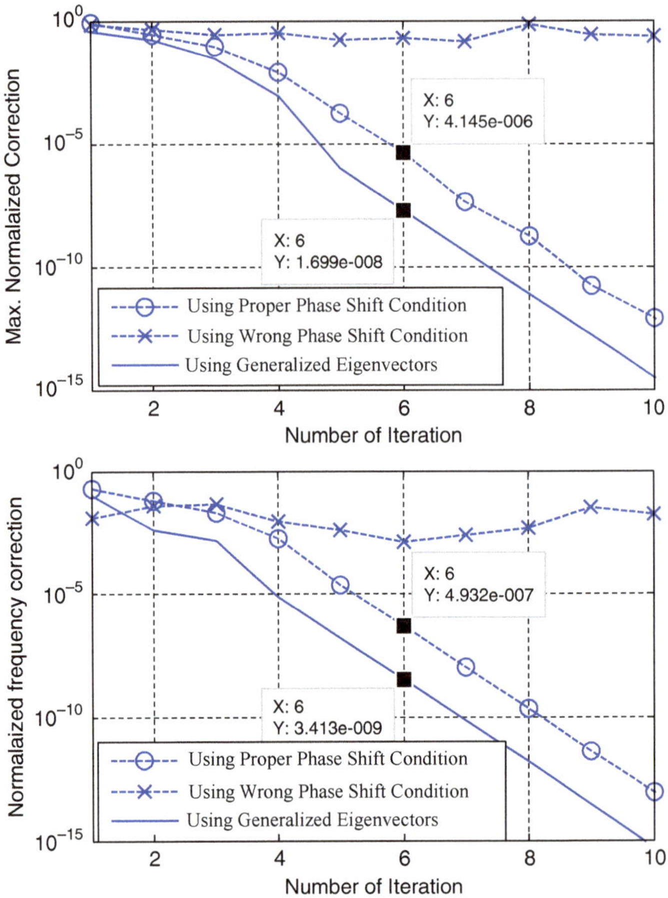

Fig. 2 Maximum of the normalized correction and normalized frequency correction for each iteration when $\varepsilon = 0.1$ for different methods

8 Conclusion

A new time-domain technique for the Newton–Raphson simulation of a free-running oscillator was presented. The generalized eigenvectors for the eigenvalue closest to 1 and the time derivative of the solution provide a robust gauge equation that is dynamically updated within each Newton–Raphson iteration. It was verified that the new method has better convergence properties compared to the popular phase-shift condition method and does not need additional information about the solution. The gauge equation also easily fits a Harmonic Balance environment.

Acknowledgements This work was supported in part by the Education and Research Institute for ICT and by the EU FP7/2008/ICT/214911 project ICESTARS.

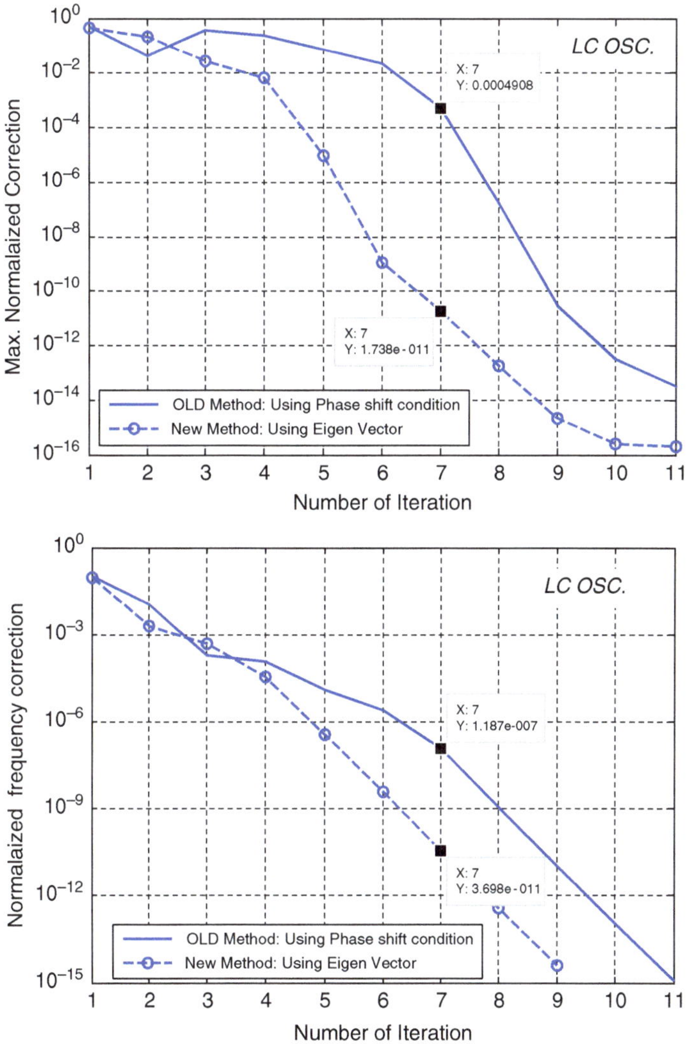

Fig. 3 Maximum of the normalized correction and normalized frequency correction for each iteration for different methods

References

1. Aprille, T.J., Trick, T.: Steady state analysis of nonlinear circuits with periodic inputs. In: Proceedings of IEEE 1972, pp. 108–114 (1972)
2. Brambilla, A., Maffezzoni, P., Storti Gajani, G.: Computation of period sensitivity functions for the simulation of phase noise in oscillators. IEEE Trans. Circ. Sys. – I: Regular Papers **52**(4), 681–694 (2005)

3. Brambilla, A., Gruosso, G., Storti Gajani, G.: Robust harmonic-probe method for the solution of oscillators. IEEE Trans. Circ. Sys. – I: Regular Papers **57**(9), 2531–2541 (2010)
4. Campbell, S., Meyer, C.D.: Generalized inverses of linear transformations. CL-56, SIAM – Society for Industrial and Applied Mathematics (2008) (orig. Pitman Publishing Limited, 1979) Philadelphia, USA
5. Duan, X., Mayaram, K.: Frequency-domain simulation of ring oscillators with a multiple probe method. IEEE Trans. Computer-Aided Des. Integr. Circ. Sys. (TCAD) **25–12**, 2833–2842 (2006)
6. Golub, G.H., Van Loan, C.F.: Matrix computations, third edition. The John Hopkins University Press, Baltimore, MD (1996) (orig. 1983)
7. Houben, S.H.M.J.: Circuits in motion - the simulation of electrical oscillators. PhD-Thesis, Eindhoven University of Technology (2003)
8. Lampe, S., Brachtendorf, H.G., ter Maten, E.J.W., Onneweer, S.P.: Robust limit cycle calculations of oscillators. In: van Rienen, U., Günther, M., Hecht, D. (eds.) Scientific Computing in Electrical Engineering. Lecture Notes in Engineering 18, Springer, Berlin, pp. 233–240 (2001)
9. Meyer Jr, C.D.: The moore-penrose invere of a bordered matrix. Linear Algebra Its Appl. **5**(4), 375–382 (1972) (not electronically available)
10. Rommes, J.: Methods for eigenvalue problems with applications in model order reduction. Ph.D. Thesis, Utrecht University, http://sites.google.com/site/rommes/software/ (2007)
11. Telichevesky, R., Kundert, K., Elfadel, I., White, J.: Fast simulation algorithms for RF circuits. In: Proceedings of the IEEE 1996 Custom Integrated Circuits Conference, pp. 437–444 (1996)
12. Trefethen, L.N.: Spectral Methods in MATLAB. SIAM, Philadelphia, http://www.comlab.ox.ac.uk/oucl/work/nick.trefethen/spectral/ (2000)

Mutual Injection Locking of Oscillators under Parasitic Couplings

M.M. Gourary, S.G. Rusakov, S.L. Ulyanov, and M.M. Zharov

Abstract The method to analyze the mutual injection locking of weakly coupled arbitrary oscillators is proposed. The couplings are defined by frequency-dependent admittance matrices. The algebraic system with respect to phases and common locking frequency is derived. For two oscillators the system is transformed to the single phase equation and explicit expression for the locking frequency. The accuracy comparison with SPICE simulation is presented.

1 Introduction

The analysis of coupled oscillators by SPICE simulation requires too high computational efforts, so some approaches based on the phase macromodels were proposed. In particular, the time-domain simulation with nonlinear phase macromodel [1, 2] provides the evaluation of locking, pulling and transient effects in coupled oscillators and PLLs. The more effective analysis of pure locking effects can be performed using steady-state phase equations of locked oscillators. These methods were applied both to the case when locked oscillators provide required functional properties of the design [3,4] and to the analysis of the undesirable mutual injection locking due to parasitic couplings in integrated circuits [5, 6]. However, proposed methods suffer from the lack of generality that is especially important for the analysis of parasitic locking. The following shortcoming can be pointed out:

- The published methods are directed to sinusoidal and/or weakly nonlinear oscillators but parasitic locking can occur between any types of oscillators.
- The interactions are usually defined by constant transfer factors from the output voltage of one oscillator to the injected current of another one. This allows

M.M. Gourary (✉) · S.G. Rusakov · S.L. Ulyanov · M.M. Zharov
IPPM RAS, 3, Sovetskaya, Moscow, Russia
e-mail: gourary@ippm.ru; rusakov@ippm.ru; ulyas@ippm.ru; zarov@ippm.ru

B. Michielsen and J.-R. Poirier (eds.), *Scientific Computing in Electrical Engineering SCEE 2010*, Mathematics in Industry 16, DOI 10.1007/978-3-642-22453-9_32,

to describe resistive couplings [6] but cannot capture couplings represented by linear networks with frequency-dependent transfer functions (e.g., inductances/capacitances or ground/supply couplings).

In this paper we propose a new method that eliminates the above mentioned limitations.

2 Injection Locking of the Oscillator under External Excitation

The analysis is based on the phase equation derived in [7,8] for an arbitrary oscillator under small excitation. The oscillator fundamental ω_0 is assumed to be close to the excitation frequency $\omega = \omega_0 + \Delta\omega$. The equation is represented in the form [7]

$$(\omega - \omega_0)/\omega_0 = W(\mathbf{B}, \phi) . \tag{1}$$

where ϕ is the locking phase, $\mathbf{B}$ is the harmonic balance (HB) excitation vector,

$$W(\mathbf{B}, \phi) \equiv \frac{1}{2} \sum_{k=-K}^{K} \sum_{l=1}^{L} \mathbf{B}_{kl} \mathbf{V}_{kl} \exp(-jk\phi) , \tag{2}$$

$\mathbf{V}$ is the perturbation projection vector (PPV) [1, 2] of the oscillator, l is a nodal index, k is a harmonic index. Because double-sided Fourier series is used the factor 1/2 appears in (2).

$W(\mathbf{B}, \phi)$ is a 2π periodic function and its maximal (W_{max}) and minimal (W_{min}) values for $0 \le \phi < 2\pi$ define the oscillator locking range

$$W_{min} \le (\omega - \omega_0)/\omega_0 \le W_{max} . \tag{3}$$

For excitation frequencies within the locking range (3) there are at least two solutions (Fig. 1a). The stable solution corresponds to the phase with $\frac{d}{d\phi} W(\mathbf{B}, \phi) < 0$ [9]. The locking region in the plane injection magnitude ($|\mathbf{B}|$) vs. frequency offset ($\Delta\omega$) is known as Arnold's tongue (Fig. 1b).

In integrated circuits injection currents $\mathbf{B}$ can be induced by periodic nodal voltages of some circuit device if there exists a parasitic interconnection between the oscillator nodes and the excitation nodes. We assume that the interconnection can be represented by a linear network with admittance matrix $\mathbf{Y}(\omega)$. The entries of the matrix are sufficiently small to provide the small-signal assumption in the derivation of (1) and to neglect input impedances of the oscillator. Hence the excitation vector $\mathbf{B}$ presented as a function of ω is defined as

$$\mathbf{B}_{kl}(\omega) = \textstyle\sum_{m=1}^{M} \mathbf{Y}_{lm}(k\omega)\mathbf{X}_{km} \text{ or } \mathbf{B}(\omega) = \tilde{\mathbf{Y}}(\omega)\mathbf{X} , \tag{4}$$

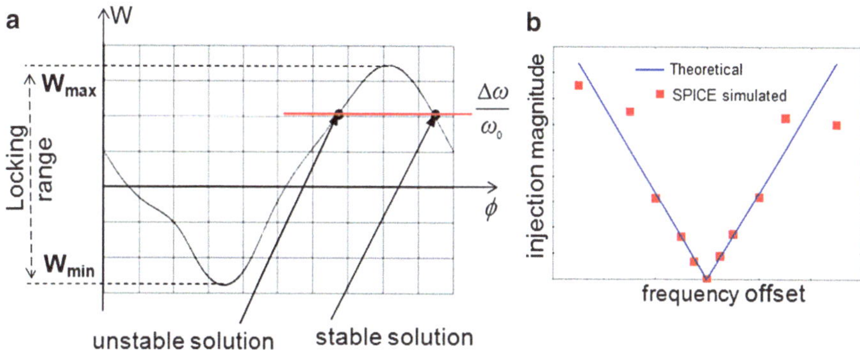

Fig. 1 Locking range and solutions of phase equation (**a**); Arnold's tongue (**b**), computed by (3) (theoretical) and by SPICE simulation (simulated)

where $\mathbf{X}$ is the HB vector of the nodal voltages, m are indexes of the excitation nodes, $\tilde{\mathbf{Y}}(\omega)$ is the HB block-diagonal matrix with blocks $\mathbf{Y}(k\omega)$.

Equation (1) is obtained by the linearization of full oscillator equations, and its error is evaluated as the second order value:

$$\Delta\omega/\omega_0 - W(\mathbf{B}, \phi) = O(||\mathbf{B}||^2) = O(\Delta\omega^2)\,. \tag{5}$$

Therefore we can perform transformations of (1) within the accuracy order (5). Particularly the denominator ω_0 in the left hand side of (1) can be replaced by $\omega + O(\Delta\omega)$ because

$$\Delta\omega/(\omega_0 + O(\Delta\omega)) = \Delta\omega/\omega_0 + O(\Delta\omega^2)\,. \tag{6}$$

Similarly injection currents (4) can be evaluated under frequency $\omega_0 + O(\Delta\omega)$ instead of ω due to

$$\begin{aligned} \mathbf{B}(\omega + O(\Delta\omega)) - \mathbf{B}(\omega) &= (\tilde{\mathbf{Y}}(\omega + O(\Delta\omega)) - \tilde{\mathbf{Y}}(\omega))\mathbf{X} \approx \frac{d}{d\omega}\tilde{\mathbf{Y}}(\omega)\cdot\mathbf{X}\cdot\Delta\omega \\ &= O(\tilde{\mathbf{Y}}(\omega)\cdot\mathbf{X}\cdot\Delta\omega)) = O(||\mathbf{B}(\omega)\Delta\omega||) = O(\Delta\omega^2)\,, \end{aligned} \tag{7}$$

and the linearity of $W(\mathbf{B}, \phi)$ (2) with respect to $\mathbf{B}$. The derivation of (7) is based on the following estimations: $O(||\frac{d}{d\omega}\tilde{\mathbf{Y}}(\omega)||) = O(||\tilde{\mathbf{Y}}(\omega)||)$ is obtained by assuming $\mathbf{Y}(\omega)$ to satisfy Lipschitz condition; $O(||\mathbf{B}(\omega)||) = O(\Delta\omega)$ is resulted from (1).

Known methods based on steady-state phase equations [3–5] use the Adler equation representing a special case of (1) for weakly-nonlinear LC oscillators [9]. Methods [1, 2, 6] use the nonlinear phase equation [10] which is based on Floquet theory for the linearized ODE system and also provides the accuracy order similar to (5). Thus using (1) allows us to develop the approach that provides for arbitrary oscillators the same accuracy as the methods [3–6] meant for sinusoidal oscillators only.

One can apply (4), (7) to obtain a new value ω_0^{self} of the fundamental after connecting small admittances $\mathbf{Y}^{\text{self}}$ to the oscillator. Such a circuit can also be considered as the initial oscillator excited by its own periodic steady-state (PSS) harmonics $\mathbf{X^0}$. The ambiguous phase of the self-locking oscillator can be fixed by setting $\phi = 0$. Then the substitution (4) into (1) results in the nonlinear equation for ω_0^{self}

$$(\omega_0^{\text{self}} - \omega_0)/\omega_0 = W(\tilde{\mathbf{Y}}^{\text{self}}(\omega_0^{\text{self}})\mathbf{X^0}, 0)\,. \tag{8}$$

Due to (7) ω_0^{self} in the right hand side of (8) can be replaced by ω_0 within the error order (5), and the obtained linear equation can be explicitly solved. The self-locking deviation of the fundamental ($\Delta\omega^{\text{self}} = \omega_0^{\text{self}} - \omega_0$) is represented by the expression

$$\Delta\omega^{\text{self}} = \omega_0 W(\tilde{\mathbf{Y}}^{\text{self}}(\omega_0)\mathbf{X}^0, 0) = \omega_0 \mathbf{V}^T \tilde{\mathbf{Y}}^{\text{self}}(\omega_0)\mathbf{X}^0\,, \tag{9}$$

where the superscript T denotes vector transpose.

3 Phase Equations for Mutually Locked Oscillators

Here we consider n oscillators with close fundamentals ω_i and known PPV $\mathbf{V}^i$. The effect of the waveforms $\mathbf{X}^j$ of jth oscillator on the excitation currents of the ith one is defined by the admittance matrix $\mathbf{Y}^{ij}(\omega)$. If all oscillators are locked with the common frequency ω and phases ϕ_i then we write (1) for ith oscillator as follows

$$\frac{\omega - \omega_i}{\omega_i} = \sum_{j=1}^{n} W^i(\tilde{\mathbf{Y}}^{ij}(\omega)\mathbf{X}^j, \phi_i - \phi_j)\,. \tag{10}$$

Here W^i is the function (2) with PPV $\mathbf{V}^i$. The set of locked oscillators produces free-running oscillations with arbitrary phase that can be fixed by setting $\phi_n = 0$. Thus (10) defines the system of nth order with n variables: ω, $\phi_1,\ldots,\phi_{n-1}$.

Diagonal terms ($j = i$) in (10) $W^i(\tilde{\mathbf{Y}}^{ii}(\omega)\mathbf{X}^i, 0)$ define the self-locking deviation of the fundamental and can be excluded from (10) by substituting $\omega_i = \bar{\omega}_i - \Delta\omega_i^{\text{self}}$ into the numerator of the left hand side. Here $\Delta\omega_i^{\text{self}}$ is defined by (9) with $\omega_0 = \omega_i$, $\tilde{\mathbf{Y}}^{\text{self}} = \tilde{\mathbf{Y}}^{\text{ii}}$. Then (10) is transformed to

$$(\omega - \bar{\omega}_i)/\omega_i = \textstyle\sum_{j\neq i} W^i(\tilde{\mathbf{Y}}^{ij}(\omega)\mathbf{X}^j, \phi_i - \phi_j)\,. \tag{11}$$

Taking into account accuracy considerations (5), (6) all denominators of the left hand sides in (11) can be replaced by a common value assumed to be ω_1 for all i. Due to (7) the same value can replace ω as the argument of admittance matrix. Then denoting $\mathbf{B}^{ij} = \tilde{\mathbf{Y}}^{ij}(\omega_1)\mathbf{X}^j$ we obtain the system

$$(\omega - \bar{\omega}_i)/\omega_1 = \textstyle\sum_{j\neq i} W^i(\mathbf{B}^{ij}, \phi_i - \phi_j)\,. \tag{12}$$

Equation (12) is linear with respect to the locking frequency. Hence we can eliminate ω and derive the system with respect to $n - 1$ phases ϕ_i.

Specifically for two oscillators the system (12) contains equations

$$(\omega - \bar{\omega}_1)/\omega_1 = W^1(\mathbf{B}^{1,2}, \phi_1)\,, \ (\omega - \bar{\omega}_2)/\omega_1 = W^2(\mathbf{B}^{2,1}, -\phi_1)\,, \qquad (13)$$

which yield the single phase equation (for $\phi = \phi_1$)

$$(\bar{\omega}_2 - \bar{\omega}_1)/\omega_1 = W^1(\mathbf{B}^{1,2}, \phi) - W^2(\mathbf{B}^{2,1}, -\phi)\,. \qquad (14)$$

Equation (14) can be numerically solved by sweeping $0 \leq \phi < 2\pi$. After that the locking frequency is obtained from (13) as

$$\omega = \bar{\omega}_1 + \omega_1 W^1(\mathbf{B}^{1,2}, \phi)\,. \qquad (15)$$

4 Dependencies on Coupling Factor

The interaction intensity of the oscillators can be characterized by the coupling factor c which is defined as admittance multiplier: $\mathbf{Y}(\omega) = c \cdot \mathbf{y}(\omega)$, where $\mathbf{y}(\omega)$ is the admittance matrix under unit value of the coupling factor. Then functions W^i in (13) are also linearly dependent on the factor:

$$W^i(\mathbf{B}^{ij}, \phi) = c \cdot w^{ij}(\phi)\,, \qquad (16)$$

where $w^{ij}(\phi) = W(\tilde{\mathbf{y}}^{ij}(\omega_1) \cdot \mathbf{X}, \phi)$.

Thus we can obtain the linear dependence of the self-coupling frequency deviation (9) on the coupling factor

$$\Delta\omega_0^{\text{self}} = c \cdot \delta\omega_0^{\text{self}}, \text{ where } \delta\omega_0^{\text{self}} = \omega_0 \mathbf{V}^T \tilde{\mathbf{y}}^{\text{self}}(\omega_0)\mathbf{X}^0\,. \qquad (17)$$

For two oscillators phase (14) can be transformed to

$$(\omega_2 - \omega_1)/\omega_1 = c \cdot w^{\text{diff}}(\phi)\,, \qquad (18)$$

where $w^{\text{diff}}(\phi) = w^{1,1}(0) + w^{1,2}(\phi) - w^{2,1}(\phi) - w^{2,2}(0)$.

From (18) we can obtain the expression for the locking range with linear dependence on the coupling factor

$$c \cdot w^{\text{diff}}_{min}(\phi) \leq (\omega_2 - \omega_1)/\omega_1 \leq c \cdot w^{\text{diff}}_{max}(\phi)\,. \qquad (19)$$

The condition (19) is similar to (3), and the locking region in the plane coupling factor vs. frequency offset ($\Delta\omega$) has the form of Arnold's tongue (Fig. 1b).

If (19) is satisfied then the solution of (18) at $\frac{d}{d\phi} w^{\text{diff}}(\phi) < 0$ defines an implicit phase dependence on the coupling factor $\phi(\Delta\omega/c)$, where $\Delta\omega = \omega_2 - \omega_1$. This dependence can be easily numerically evaluated. Then the dependence of locking frequency on the coupling factor c and the discrepancy of fundamentals $\Delta\omega$ is defined by the expression derived from (15)

$$\Delta\omega^{\mathrm{lock}} = c \cdot (\omega_1 \cdot w^{1,2}(\phi(\Delta\omega/c)) + \delta\omega_1^{\mathrm{self}}), \tag{20}$$

where $\Delta\omega^{\mathrm{lock}} = \omega - \omega_1$ is the deviation of locking frequency from the fundamental for the first oscillator.

Expressions (17), (19), (20) can be applied to analyze arbitrary oscillators after PSS and PPV harmonics of each oscillator are obtained.

5 Experimental Results

We performed experiments for 3- and 5-stages CMOS ring oscillators (RO-3, RO-5). Ring oscillators were used to illustrate the applicability of the proposed method to nonsinusoidal oscillators. The oscillators PSSs were obtained by HB simulations, and PPVs were determined by the method [10].

Locking frequencies evaluated by (17), (20) were compared with the results of SPICE simulations. We suppose SPICE simulations results to be accurate ones because we performed a number of simulations with decreasing tolerances until the varying of results ceased. Thus the difference between computed and simulated results is the error of (17), (20) due to neglecting of high order terms in (5)–(7).

Dependence (17) for self-coupled oscillator was verified by connecting grounded capacitance C to the output node of RO-3. In this case the scalar admittance is defined as $Y^{\mathrm{self}}(\omega) = -j\omega \cdot C$, where the minus sign is resulted from the inverse direction of excitation current assumed in the derivation of (1) in [7]. Thus from (17) we obtain the linear dependence of the self-coupling fundamental deviation on the capacitance value. The comparison with simulated results is presented in Fig. 2

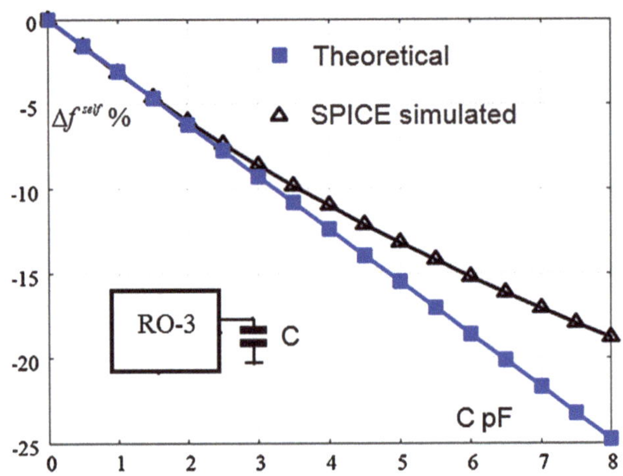

Fig. 2 Relative fundamental deviation ($\Delta f^{\mathrm{self}} = \frac{f_0^{\mathrm{self}} - f_0}{f_0} \cdot 100\%$) of 3-stages ring oscillator due to attached capacitance. Theoretical dependence is obtained by (17)

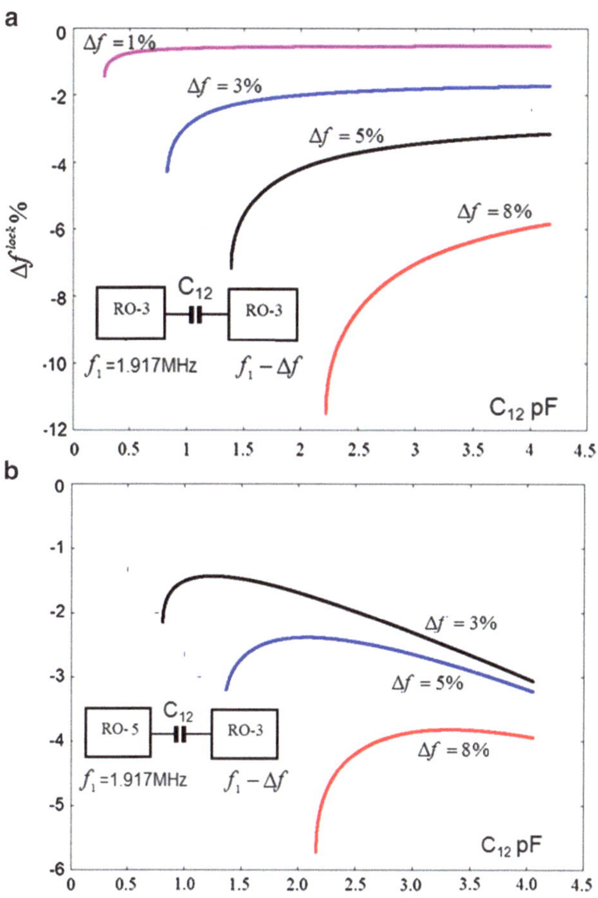

Fig. 3 Mutually locked ring oscillators: RO-3 and RO-3 (**a**), RO-5 and RO-3 (**b**). Theoretical dependencies of locking frequencies deviations ($\Delta f^{lock} = \frac{f^{lock} - f_1}{f_1} \cdot 100\%$) on the coupling capacitance (C_{12}) for various discrepancies of fundamentals ($\Delta f = \frac{f_2 - f_1}{f_1} \cdot 100\%$). The dependencies are evaluated by (20)

where one can see the approximately squared relationship between the error and the fundamental deviation in accordance with (5).

Dependencies (20) of locking frequencies were examined for two 3-stages CMOS ring oscillators coupled by the capacitance C_{12} between output nodes (Fig. 3a). Mutual admittances are: $Y^{1,1}(\omega) = Y^{2,2}(\omega) = -Y^{1,2}(\omega) = -Y^{2,1}(\omega) = -j\omega C_{12}$. Similar dependencies for coupled 3- and 5-stages CMOS ring oscillators are presented in Fig. 3b.

The comparisons with SPICE simulations for both cases are shown in Figs. 4, 5. It is seen that the errors of computed curves increase under the growth of the fundamentals deviation.

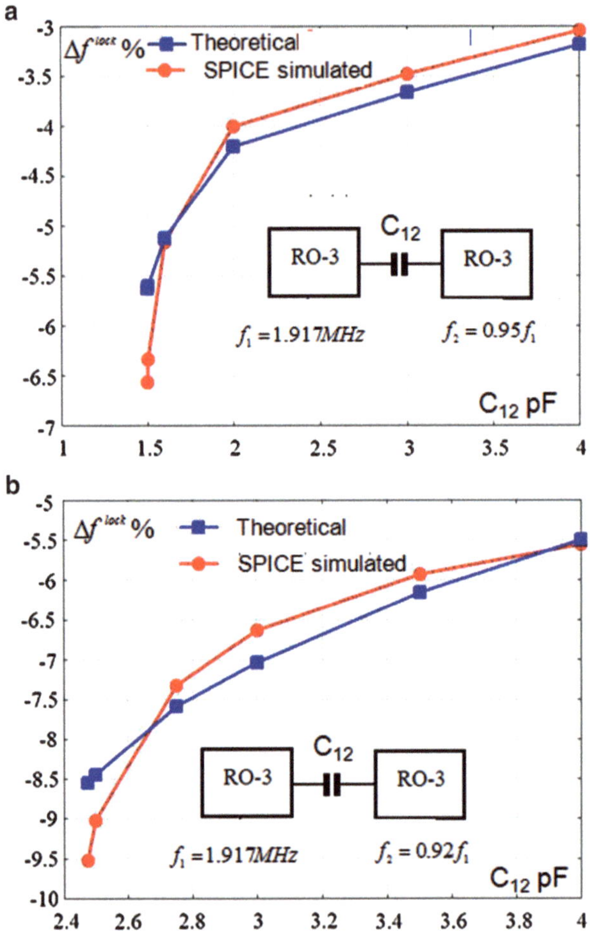

Fig. 4 Two 3-stages CMOS ring oscillators. Theoretical and simulated dependencies of locking frequency on the coupling capacitance (C_{12}) for $\Delta f = 5\%$ (**a**) and $\Delta f = 8\%$ (**b**)

Our experiments also showed the decrease of the computational efforts (CPU time) due to the proposed approach. SPICE simulation (one point in Figs. 4, 5) requires about 20 s. The proposed approach requires two HB simulations to obtain PSS and PPV for each oscillator (4 s). After that the evaluation of 20 parameterized curves like in Fig. 3 (2000 points per curve) is performed within 1 s.

6 Conclusion

This paper has presented a method to evaluate the locking frequencies for arbitrary oscillators weakly coupled by any linear interconnect networks. The method provides the same accuracy order as known methods meant for sinusoidal oscillators

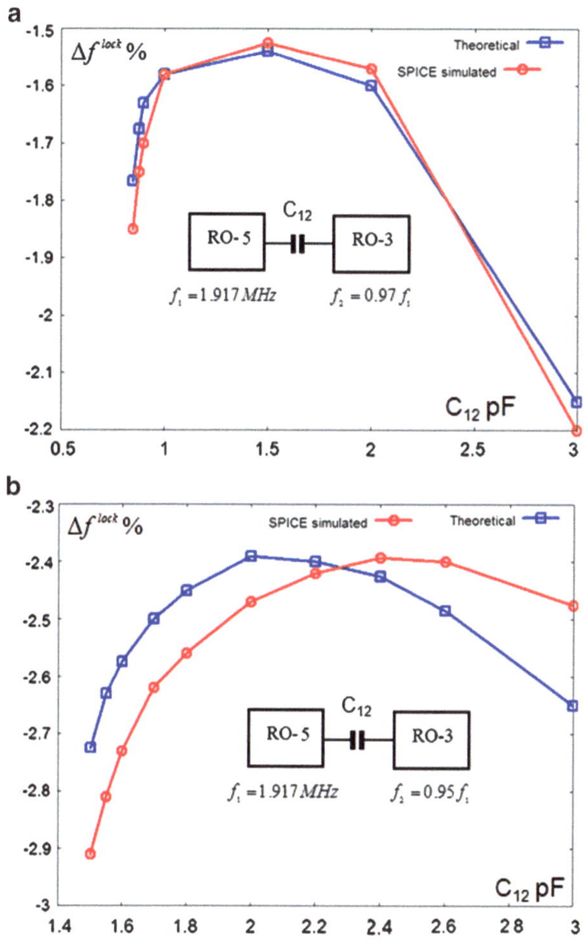

Fig. 5 3- and 5- stages CMOS ring oscillators. Theoretical and simulated dependencies of locking frequency on the coupling capacitance (C_{12}) for $\Delta f = 3\%$ (**a**) and $\Delta f = 5\%$ (**b**)

only. For two oscillators the evaluation algorithm includes the solution of the first order phase equation and the computation of explicit analytical expressions.

Acknowledgements The work was supported the RFBR grant no. 09-07-00029.

References

1. Lai, X., Roychowdhury, J.: Fast and accurate simulation of coupled oscillators using nonlinear phase macromodels. MTT-S Int. Microw. Symp. Dig. **2**, 871–874 (2005)
2. Harutyunyan, D., Rommes, J., ter Maten, J., Schilders, W.: Simulation of mutually coupled oscillators using nonlinear phase macromodels. IEEE Trans. Computer-Aided Des. **28**(10), 1456–1466 (2009)

3. York, R.A.: Injection- and Phase-Locking Techniques for Beam Control. IEEE Trans. Microw. Tech. **46**(11), 1920–1929 (1998)
4. Bonnin, M., Corinto, F.: Periodic oscillations in weakly connected cellular nonlinear networks. IEEE Trans. Circ. Syst.-I. **55**(6), 1671–1684 (2008)
5. Razavi, B.: Mutual injection pulling between oscillators. Proc. IEEE Custom Integr. Circ. Conf. 675–678 (2006)
6. Maffezzoni, P.: Synchronization analysis of two weakly coupled oscillators through a PPV macromodels. IEEE Trans. Circ. Syst.-I. **57**(3), 654–663 (2010)
7. Gourary, M.M., Rusakov, S.G., Ulyanov, S.L., Zharov, M.M., Mulvaney, B.J., Gullapalli, K.K.: Injection locking conditions under small periodic excitations. Proc. IEEE Int. Symp. Circ. Syst. 544–547 (2008)
8. Maffezzoni, P.: Analysis of oscillator injection locking through phase-domain impulse response. IEEE Trans. Circ. Syst.-I. **55**(5), 1297–1305 (2008)
9. Gourary, M.M., Rusakov, S.G., Ulyanov, S.L., Zharov, M.M., Mulvaney, B.J., Gullapalli, K.K.: Smoothed form of nonlinear phase macromodel for oscillators. Proc. IEEE/ACM Int. Conf. Computer-Aided Des. 807–814 (2008)
10. Demir, A., Long, D., Roychowdhury, J.: Computing phase noise eigenfunctions directly from steady-state Jacobian matrices. Proc. IEEE/ACM Int. Conf. Computer-Aided Des. 283–288 (2000)

Time Domain Simulation of Power Systems with Different Time Scales

Valeriu Savcenco, Bertrand Haut, E. Jan W. ter Maten, and Robert M.M. Mattheij

Abstract The time evolution of power systems is modeled by a system of differential and algebraic equations. The variables involved in the system may exhibit different time scales. In standard numerical time integration methods the most active variables impose the time step for the whole system. We present a strategy, which allows the use of different, local time steps over the variables. The partitioning of the components of the system in different classes of activity is performed automatically and is based on the topology of the power system.

1 Introduction

Modeling of power systems results in large differential-algebraic systems. These systems are built from the equations describing the network, the generators, the voltage regulators, the speed governors and the dynamic shunt loads. All together they form a non-linear system in semi-explicit form

$$\begin{aligned} y' &= f(t, y, z), \\ 0 &= g(t, y, z), \end{aligned} \tag{1}$$

V. Savcenco (✉) · E.J.W. ter Maten · R.M.M. Mattheij
Eindhoven University of Technology, P.O. Box 513, 5600 MB Eindhoven, The Netherlands
e-mail: v.savcenco@tue.nl; e.j.w.ter.maten@tue.nl; r.m.m.mattheij@tue.nl

B. Haut
Tractebel Engineering S.A, Avenue Ariane 7, 1200 Brussels, Belgium
e-mail: bertrand.haut@gdfsuez.com

E. Jan W. ter Maten
NXP Semiconductors B.V., High Tech Campus 46, 5656 AE Eindhoven, The Netherlands
e-mail: jan.ter.maten@nxp.com

B. Michielsen and J.-R. Poirier (eds.), *Scientific Computing in Electrical Engineering SCEE 2010*, Mathematics in Industry 16, DOI 10.1007/978-3-642-22453-9_33,

with initial values $y(0) = y_0$ and $z(0) = z_0$, such that $g(t_0, y_0, z_0) = 0$. It is assumed that the matrix $\frac{\partial g}{\partial z}$ is non singular and therefore system (1) has index one.

Time domain simulation is an important application for the dynamic security assessment of power systems [6]. The components involved in the systems are known to exhibit a wide range of time scales. A voltage wave propagation due to lightning lasts a few microseconds to milliseconds but a secondary frequency control may have a time duration of several minutes. A particular situation requiring numerical simulation is that a damaging event which occurs in one of the European countries should not affect other countries. For such problems, the multirate time stepping strategies can automatically detect strong local temporal activity and lead to large speed-ups in the simulation time [2, 5, 7]. With such methods different solution components can be integrated with different time steps.

2 Multirate Time Stepping

In this paper it will be assumed that the variables of the system (1) can be partitioned into fast and slow

$$y = [y_{fast}, y_{slow}] \quad \text{and} \quad z = [z_{fast}, z_{slow}]. \tag{2}$$

Our multirate time stepping strategy can be described as follows. For a given global time step $\tau = t_n - t_{n-1}$, we first compute a tentative approximation at the time level t_n for the both fast and slow variables. We accept the computed numerical solution for the slow components, while for the fast components the computation is redone with smaller time steps. During this refinement computation the subsystem

$$\begin{aligned} y'_{fast} &= f_{fast}(t, y_{fast}, z_{fast}, \omega), \\ 0 &= g_{fast}(t, y_{fast}, z_{fast}, \omega) \end{aligned} \tag{3}$$

is solved, where ω denotes the already computed values of the slow variables. During the refinement stage, values at the intermediate time levels of the slow components might be needed. These values can be obtained by interpolation.

The intervals $[t_{n-1}, t_n]$ are called time slabs. After each completed time slab the solutions are synchronized. In our approach, these time slabs are automatically generated, similar as in the single-rate approach, but without imposing temporal accuracy constraints on all components.

An important issue in our strategy is to determine the size of the time slabs. These could be taken large with a large multirate factor, or small with a lower multirate factor. A decision can be made based on an estimate of the number of components at which the solution needs to be calculated, including the overhead due to coupling.

In this paper we consider two levels of activity: slow variables and fast variables. One can also allow for more levels of activity. In this case, the desired accuracy does not necessary have to be achieved during the first refinement. The refinement can be continued until the error estimator is below a prescribed tolerance for all components.

3 Mixed Adams-BDF Time Integration Method

As the basic time integration method we use the mixed Adams-BDF method presented in [1]. The second-order Adams method is applied to the differential state variables and provides a reliable detection of unstable situations. It is symmetrically A-stable (the domain of stability coincides with the left complex half-plane) and thus does not suffer from the hyper stability. The second-order BDF method is used for the algebraic state variables, since it less sensitive to the variations in the algebraic equations than the Adams method. Detailed description and the coefficients for both methods can be found in [3].

3.1 Interpolation

For given approximations $w_{n-1} \approx w(t_{n-1})$ and $w_n \approx w(t_n)$ for the solution vector $w = [y, z]$, the multirate schemes can require an intermediate value $w_I(t_{n-1+\theta}) \approx w(t_{n-1} + \theta\tau)$ for $0 < \theta < 1$. This can be calculated by using linear or quadratic interpolation.

For the linear interpolation we use the values of w_{n-1} and w_n

$$w_I(t_{n-1+\theta}) = (1-\theta)w_{n-1} + \theta w_n \,. \tag{4}$$

For the quadratic interpolation we use the values of w_{n-1}, w_n, w'_{n-1} and w'_n

$$w_I(t_{n-1+\theta}) = \alpha_1 w_{n-1} + \alpha_2 w_n + \tau(\beta_1 w'_{n-1} + \beta_2 w'_n) \tag{5}$$

with

$$\alpha_1 = 1 - \theta^2 + 2\rho, \quad \alpha_2 = \theta^2 - 2\rho, \quad \beta_1 = \theta - \theta^2 + \rho, \quad \beta_2 = \rho\,, \tag{6}$$

where ρ is a free parameter satisfying the condition $-\frac{1}{2} \le \rho - \frac{1}{2}\theta^2 \le 0$. Particular cases are forward quadratic interpolation ($\rho = 0$) and backward quadratic interpolation ($\rho = \theta^2 - \theta$).

4 Partitioning Strategy

Partitioning of the variables in slow and fast can be fixed and given in advance, or it can vary in time and should be performed automatically during the time integration process.

In this section we present a strategy for automatic partitioning of the differential and algebraic variables. This strategy is based on the local time variation of the numerical solution of the system and on the topology of the power system.

A power system can be usually decomposed in two parts:

- A large network which consists of a set of nodes (each node introducing two variables) connected by a set of branches (lines, cables and transformers)
- A set of components (synchronous machines, motors, loads...) which are usually connected to a particular node.

This particular structure can be used to derive a dedicated partitioning strategy.

We first perform a single step with step size τ and using an error estimator we determine the variables which do not satisfy the criterion

$$e_i < Tol, \tag{7}$$

where e_i is the estimated local error for the variable i and Tol is a given tolerance. These variables will be called fast. The local error vector $\boldsymbol{e}$ is computed as the difference between the corrected solution and the predicted solution.

To allow for accurate computation of the fast variables, during the refinement stage, we also recompute the slow variables which are strongly coupled to the fast ones. The propagation of the fast status is performed as follows:

1. All the components which contain at least one fast variable are classified as fast.
2. All the nodes which contain at least one fast variable are classified as fast.
3. The connection node of a fast component is classified as fast.
4. The fast status of the nodes is then propagated through the network:
   (a) The graph G is defined as follows:
      - A node in G is defined for each electrical node;
      - An edge is defined between two nodes of G if there exists at least one branch linking the two corresponding electrical nodes;
      - A weight representing an "electrical distance" will be associated to each edge of G. Let us denote by C_1 and C_2 the two 2×2 sub-matrices of the admittance matrix coupling the pairs of variables associated nodes 1 and 2. The weight between node 1 and 2 is defined as

$$l_{12} = \min\left(\frac{1}{\|C_1\|_\infty}, \frac{1}{\|C_2\|_\infty}\right) \tag{8}$$

where

$$\|C\|_\infty = \max |C_{ij}|.$$

(b) Each node at a distance less than a given parameter tol_G from a fast node is classified as fast.

5. All the variables belonging to a fast node or a fast component are classified as fast and will therefore be updated during the refining phase.

The creation of a table containing, for each node, the list of strongly connected nodes can be efficiently (through a modified Dijkstra algorithm and a parallel implementation) performed off-line before the start of the simulation. With this off-line preparation, the cost of the above partitioning is almost negligible during the simulation.

5 Numerical Experiments

In this section we present numerical results for two test problems. For the results reported here we used quadratic interpolation to obtain missing component values. Linear interpolation was also tried and the results were nearly identical; this simply indicates that the interpolation errors are not significant in these tests.

The computational costs are presented in terms of number of function evaluations, number of Jacobian evaluations and number of Newton iterations. We estimate the total computation cost by means of formula

$$\begin{aligned} C = {} & 1.2 \cdot 10^{-7} N_{\text{FuncEval}} + 7.2 \cdot 10^{-7} N_{\text{JacEval}} \\ & + 5 \cdot 10^{-7} N_{\text{LUFactor}} + 5 \cdot 10^{-8} N_{\text{Newton}} \,. \end{aligned}$$

Here the coefficients represent the reference costs and are based on the benchmarks in a particular software package.

5.1 A Chain Test Problem

For our first test problem we consider a power system composed of a chain of 100 small subsystems connected by very long lines. Each subsystem comprises a generator and the corresponding controllers modeled by 30 equations, a step-up transformer and an impedant load. A schematic illustration of the chain is presented in Fig. 1. The resulting system contains 4970 variables, 3089 of which are algebraic.

A short-circuit of 100 ms is performed at the first high voltage busbar. During the very first second, this event strongly affects the beginning of the chain while the rest of the system remains more or less constant. The impact of the short-circuit propagates to the neighboring subsystems while being progressively damped.

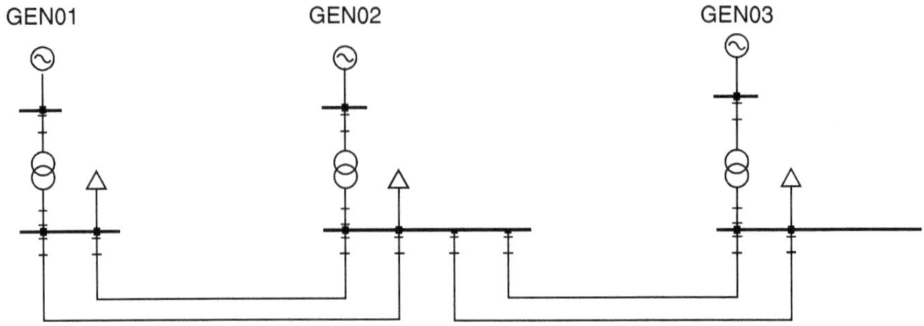

Fig. 1 Chain of 100 subsystems

Table 1 Errors and computational costs for the chain problem

	Single-rate	Multirate
$\|\|error\|\|_\infty$	$7.64 \cdot 10^{-2}$	$5.28 \cdot 10^{-2}$
$\|\|error\|\|_2$	$4.22 \cdot 10^{-5}$	$4.22 \cdot 10^{-5}$
N_{FuncEval}	184326	47102
N_{JacEval}	11892	15355
N_{Newton}	184326	47102
C	0.045	0.026

Table 1 shows the number of function evaluations, number of Jacobian evaluations, number of Newton iterations, estimated costs and the weighted L^2- and infinity-norm errors for the single-rate and multirate methods. From these results it is seen that a substantial improvement in number of function evaluations is obtained. For the single-rate method, the number of function evaluations is four times larger. Moreover, the error behavior of the multirate scheme is very good. The speed up in terms of estimated costs is smaller than the one based on the number of function evaluations. This reduction in speed up is due to large number of Jacobian evaluations. This is again visible from the results presented in the table. An improvement of the local Jacobian evaluation within multirate time stepping is needed.

Figure 2 shows the time points in which the solution for two variables, one fast and one slow, were computed. It is seen that the time steps used for the fast variable are much smaller than the ones used for the slow variable. The solution of the fast variable on this interval is computed by 26 time steps, whereas only 5 time steps are needed for the slow variable. In this simulation 70 fast variables were observed.

5.2 PEGASE Problem

As the second test we consider the PEGASE problem. This problem is a dedicated test case constructed by the PEGASE consortium [4]. The system modeled is loosely

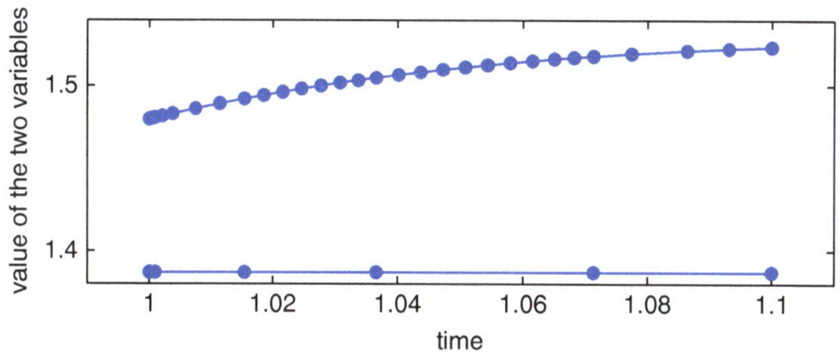

Fig. 2 Solution for two components

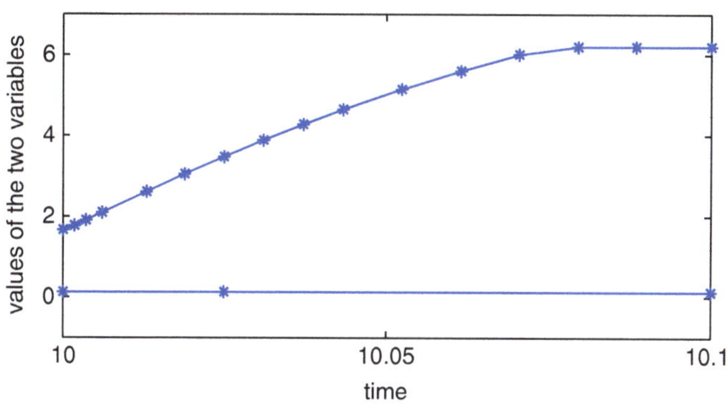

Fig. 3 Time evolution of one fast and one slow variable

inspired from the European transmission grid in terms of size (number of branches, nodes, generators, loads), topology and type of units (nuclear, hydro, TGV). The problem is modeled by a DAE system with 123463 variables, of which 50235 are algebraic.

We solve this problem on the time interval $0 < t < T = 10.1$. A short-circuit is performed in the southern Italy during the last 0.1 s of simulation time. We expect that this event will only have a local impact and hence, multirate method will be able to exploit this difference in the time scales.

Figure 3 shows the time points in which the solution for two variables, one fast located in Italy and one slow located in Luxembourg, were computed during the time interval when the short-circuit occurred. It is seen that the time steps used for the fast variable are much smaller than the ones used for the slow variable. The solution for the fast variable on this interval is computed by 15 time steps, whereas only 2 time steps are needed for the slow variable.

Table 2 shows the number of function evaluations, number of Jacobian evaluations, number of Newton iterations, estimated costs and the weighted L^2- and infinity-norm errors (measured with respect to an accurate reference solution) for the single-rate and multirate methods. From these results it is seen that a substantial

Table 2 Errors and computational costs for the PEGASE problem

	Single-rate	Multirate
$\|\|\text{error}\|\|_\infty$	$4.6 \cdot 10^{-2}$	$5.3 \cdot 10^{-2}$
$\|\|\text{error}\|\|_2$	$1.3 \cdot 10^{-4}$	$1.3 \cdot 10^{-4}$
N_{FuncEval}	4938600	1363740
N_{JacEval}	2592765	585950
N_{Newton}	4938600	1363740
C	4.00	0.94

improvement in cost is obtained. For the single-rate method the estimated costs are four times larger. Moreover, the error behavior of the multirate scheme is very good.

6 Conclusions

In this paper we presented a multirate time stepping strategy for systems of differential and algebraic equations resulting from modeling of power systems. The algorithm for dynamic partitioning of the components into slow and fast was described. Numerical experiments confirmed that the efficiency of time integration methods can be significantly improved by using large time steps for inactive components, without sacrificing accuracy.

Acknowledgements This work was performed in the context of the PEGASE project funded by European Community's 7th Framework Programme (http://www.fp7-pegase.eu/).

References

1. Astic, J.Y., Bihain, A., Jerosolimski, M.: The mixed Adams-BDF variable step size algorithm to simulate transient and long term phenomena in power systems. IEEE Trans. Power Syst. **9**, 929–935 (1994)
2. Chen, J., Crow, M.L.: A variable partitioning strategy for the multirate method in power systems. IEEE Trans. Power Syst. **23**, 259–266 (2008)
3. Hairer, E., Wanner, G.: Solving Ordinary Differential Equations II – Stiff and Differential-Algebraic Problems, 2nd edn., Springer Series in Comp. Math., 14, Springer, Berlin (1996)
4. Website of the PEGASE project, *http://www.fp7-pegase.eu/*
5. Savcenco, V., Hundsdorfer, W., Verwer, J.G.: A multirate time stepping strategy for stiff ODEs. BIT **47**, 137–155 (2007)
6. Stubbe, M., Bihain, A., Deuse, J., Baader, J.C.: Simulation of the dynamic behaviour of electrical power systems in the short and long terms. CIGRE 38-03 (1998)
7. Verhoeven, A., Tasic, B., Beelen, T., ter Maten, E.J.W., Mattheij, R.M.M.: Automatic partitioning for multirate methods. In: Ciuprina G., Ioan D. (eds.) Scientific Computing in Electrical Engineering. Springer, Berlin, 229–236 (2007)

Adaptive Wavelet-Based Method for Simulation of Electronic Circuits

Kai Bittner and Emira Dautbegovic

Abstract In this paper we present an algorithm for analog simulation of electronic circuits involving a spline Galerkin method with wavelet-based adaptive refinement. Numerical tests show that a first algorithm prototype, build within a productively used in-house circuit simulator, is completely able to meet and even surpass the accuracy requirements and has a performance close to classical time-domain simulation methods, with high potential for further improvement.

1 Introduction

Wavelet theory emerged during the twentieth century from the study of Calderon-Zygmund operators in mathematics, the study of the theory of subband coding in engineering and the study of renormalization group theory in physics. The common foundation for the wavelet theory was laid down at the end of the 1980s and beginning of the 1990s by work of Daubechies [1, 2], Morlet and Grossman [3], Donoho [4], Coifman [5], Meyer [6], Mallat [7] and others. Today wavelet-based algorithms are already in productive use in a broad range of applications [6–13], such as image and signal compression (JPEG2000 standard, FBI fingerprints database), speech recognition, numerical analysis (solving operator equations, boundary value problems), stochastics, smoothing/denoising data, physics (molecular dynamics, geophysics, turbulence), medicine (heart-rate and ECG analysis, DNA analysis) to name just a few. Recent approaches [14–18] to the problem of multirate envelope

K. Bittner (✉)
Bergische Universität Wuppertal, Gauß-Str. 20, 42097 Wuppertal, Germany
e-mail: bittner@math.uni-wuppertal.de

E. Dautbegovic
Infineon Technologies AG, 81726 Munich, Germany
e-mail: Emira.Dautbegovic@infineon.com

B. Michielsen and J.-R. Poirier (eds.), *Scientific Computing in Electrical Engineering SCEE 2010*, Mathematics in Industry 16, DOI 10.1007/978-3-642-22453-9_34,

simulation indicate that wavelets could also be used to address the qualitative challenge by a development of novel wavelet-based circuit simulation techniques capable of an efficient simulation of mixed analog-digital circuits [19].

The wavelet expansion of a function f is given as

$$f = \sum_{k \in \mathscr{I}} c_k \, \phi_k + \sum_{j=0}^{\infty} \sum_{k \in \Lambda_j} d_{jk} \, \psi_{jk}. \tag{1}$$

Here, j refers to a level of resolution, while k describes the localization in time or space, i.e., ψ_{jk} is essentially supported in the neighborhood of a point x_{jk}. The wavelet expansion can be seen as coarse scale approximation $\sum_{k \in \mathscr{I}} c_k \, \phi_k$ by the scaling functions ϕ_k complemented by detail information of increasing resolution j in terms of the wavelets ψ_{jk}.

In the classical theory wavelets are generated as translation and dilations of a mother wavelet ψ, i.e., $\psi_{jk}(x) = \psi(2^{-j}x - k)$. However, more general approaches are often used, e.g., for the construction of wavelets on the interval [20] or wavelets for finite element spaces [21]. In particular, non-uniform spline wavelets [22] will be used in our wavelet-based circuit simulation technique.

Since a wavelet basis consist of an infinite number of wavelets, in practical computations one has to consider approximations of f by partial sums of the wavelet expansion (1). A simple approach is to fix a maximal wavelet level J and approximate f by

$$f_J = \sum_{k \in \mathscr{I}} c_k \, \phi_k + \sum_{j=0}^{J} \sum_{k \in \Lambda_j} d_{jk} \, \psi_{jk}. \tag{2}$$

This approach is called linear approximation, since the approximation is determined in the linear space of wavelets with level less or equal J. For wavelets of sufficient regularity, one obtains error estimates of the form

$$\| f - f_J \|_{L_2} \leq C \, 2^{-Js} \, \| f \|_{W_2^s}, \tag{3}$$

with the Sobolev space W_2^s. However, approximation results as (3) hold also for other approximation methods, e.g., for Fourier sums (see [23]).

The real approximation power of wavelets is due to their locality, which implies that (3) holds also for small subintervals. Thus, a piecewise smooth function can be essentially approximated by some coarse scale approximation with wavelets added only at non-smooth parts to achieve a required accuracy. Doing this adaptively for any given signal leads to the notion of best n-term approximation, where the approximation is determined as linear combination of n arbitrarily chosen wavelets. This results in an essentially improved approximation for a wide class of functions, e.g., piecewise smooth function with isolated singularities. For details about this adaptive, nonlinear approximation methods we refer to [23, 24]. Usually it is not obvious which wavelets have to be chosen for optimal approximation results. In practice optimal wavelet representations can be determined by one of the two complementary strategies: coarsening or refinement.

Coarsening is used if one has already a fine, highly accurate but expensive approximation, e.g., from measurements. The goal is to throw away as much information as possible, while introducing only a small error. For a wavelet representation this can be achieved quite easily by thresholding, which means that wavelets with small expansion coefficients are removed from the representations. Inherent stability properties of wavelets ensure that this elimination of terms with small coefficients does not add up to a significant error of the wavelet expansion. For wavelets with good localization and approximation properties, one has many small wavelet coefficients for piecewise smooth signals with few local singularities (e.g., sharp transients), which will result in an essential reduction of data for the coarsened signal. A disadvantage of coarsening is that it might be too costly to acquire the fine representation. In particular, for solving operator equations, as in circuit simulation, the reason for using adaptive wavelet techniques is the reduction of computational cost, which is thwarted by computing a non-adaptive solution in advance.

In contrast, the strategy of refinement is to start with coarse approximation and introduce successively more and more degrees of freedom (e.g., wavelets) in order to improve the approximation. However, since it is not known in advance, where refinements are necessary, one has to rely on rough estimates. Therefore it is reasonable to do the refinement in several steps. This allows to check the previous steps, while acquiring more information for later steps. This approach is in particular interesting for iterative methods, where the approximation is improved in each iteration step and the number of degrees of freedom can be increased accordingly.

2 An Adaptive Wavelet Galerkin Method

We consider circuit equations in the charge/flux oriented modified nodal analysis (MNA) formulation, which yields a mathematical model in the form of an initial-value problem of differential-algebraic equations (DAEs):

$$\frac{d}{dt}\mathbf{q}\big(\mathbf{x}(t)\big) + \mathbf{f}\big(\mathbf{x}(t)\big) - \mathbf{s}(t) = 0. \tag{4}$$

Here $\mathbf{x}$ is the vector of node potentials and specific branch currents and $\mathbf{q}$ is the vector of charges and fluxes. Vector $\mathbf{f}$ comprises static contributions, while $\mathbf{s}$ contains the contributions of independent sources.

In our adaptive wavelet approach we first discretize the MNA equation (4) in terms of the wavelet basis functions, by expanding $\mathbf{x}$ as a linear combination of wavelets or related functions, i.e., $\mathbf{x} = \sum_{k=0}^{n} \mathbf{c}_k\, \varphi_k$. For such $\mathbf{x}$ we integrate the circuit equations against test functions

$$\int_0^T \left(\frac{d}{dt}\mathbf{q}(\mathbf{x}(t)) + \mathbf{f}(\mathbf{x}(t)) - \mathbf{s}(t)\right) \theta_\ell \, dt = 0, \tag{5}$$

for $\ell = 1, \ldots, n$. Together with the initial conditions $\mathbf{x}(0) = \mathbf{x}_0$, we have now $n+1$ vector valued equations, which determine the coefficients $\mathbf{c}_i$ provided that the test functions θ_ℓ are chosen suitably with respect to the basis functions φ_i.

Due to the intrinsic properties of wavelets [19] nonlinear wavelet approximation can provide an efficient representation of functions with steep transients, which often appear in a mixed analog/digital electronic circuit. However, for an efficient circuit simulation we have to take into account further properties of a wavelet system. We consider spline wavelets to be the optimal choice since spline wavelets are the only wavelets with an explicit formulation. This permits the fast computation of function values, derivatives and integrals, which is essential for the efficient numerical solution of a nonlinear problem as given in (4) (see also [25, 26]). Spline wavelets have been already used for circuit simulation [27]. However, here we use a completely new approach based on spline wavelets from [22].

With a good initial guess, Newton's method is known to converge quadratically. However, a good initial guess is usually not available. In practice we can often obtain convergence only with a slow converging initial phase of damped Newton steps, which will mainly contribute to the computational cost of the problem. On the other hand, to get a good approximation of the solution of (4), the space $X = \operatorname{span}\{\varphi_k \ : \ k = 0, \ldots, n\}$ has to be sufficiently large and the computational cost of each step depends on $n = \dim X$. Our approach is to use adaptive wavelet refinement during the Newton iteration, which leads to an efficient adaptive representation and essentially reduced computation time.

3 Interval Splitting Method

A prototype of the proposed adaptive wavelet algorithm is implemented within the framework of a productively used circuit simulator and tested on a variety of circuits. In tests on some typical RF circuits (amplifier, mixer, oscillator), we were able to reproduce the results from the transient analysis of the same circuit simulator up to high accuracy (see [28]). For all these examples, the wavelet method used a considerable smaller grid (i.e., larger stepsize) than the transient analysis, while the computation time was higher but still close to the standard method. This shows that there is a potential for wavelet methods in circuit simulation, if further optimization can be achieved.

However, in further tests with a Schmitt trigger circuit (Fig. 1, [29]) convergence could only be achieved with a highly accurate initial guess. This is of limited practical value, since we can usually not provide an initial guess of such quality. We identified inherit hysteresis of the Schmitt trigger as the main cause for this problem. In circuits exhibiting hysteresis, certain input voltages can result in different output,

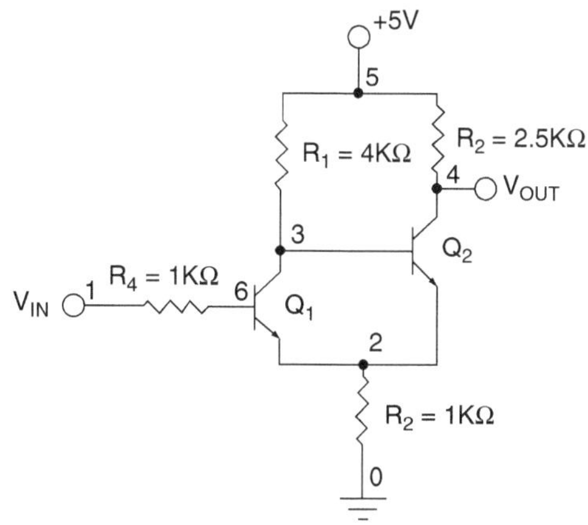

Fig. 1 Schematics of Schmitt trigger from [29]

depending on the previous behaviour of the input signal. With an insufficient initial guess Newton's method may approach locally the wrong result. This effect was observed in a Harmonic Balance simulation too, where the solution is also represented by a basis expansion over an entire period.

This convergence problem was successfully addressed by a further improvement of the basic wavelet method based on an interval splitting mechanism. Basically the wavelet method is applied to a series of smaller intervals when no convergence is detected. This is an analogous approach to the reduction of the step size in transient analysis if no convergence is encountered in the current time step. In order to preserve continuity, the initial value for each interval is obtained from the wavelet expansion of the solution on the previous interval. Furthermore, the interval size is adapted after each successful step, aiming to keep the problem size for the wavelet method in a nearly optimal range.

4 Numerical Tests

The interval splitting method was implemented as an enhancement to the basic wavelet algorithm and tested on a variety of circuits. For all examples we have compared the CPU time and the grid size (i.e., the number of spline knots or time steps) with the corresponding results from transient analysis of the underlying circuit simulator.

The error is estimated by comparison with well established high accuracy transient analysis. The estimate shown in the signal is the maximal absolute difference over all transient grid points, which gives a good approximation of the maximal error. That is, if we can obtain a small error for the wavelet analysis, this proves good agreement with the standard transient method. In particular, since we compare the solutions of two independent methods we have very good evidence that we approximate the solution of underlying DAE's with the estimated error.

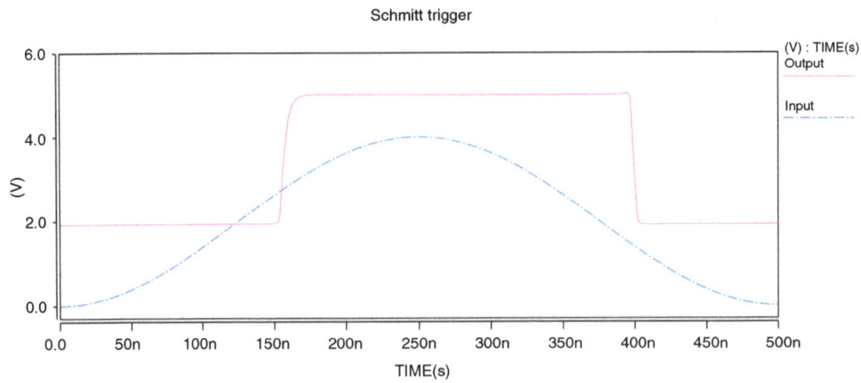

Fig. 2 Input and output signal for the Schmitt trigger from wavelet analysis

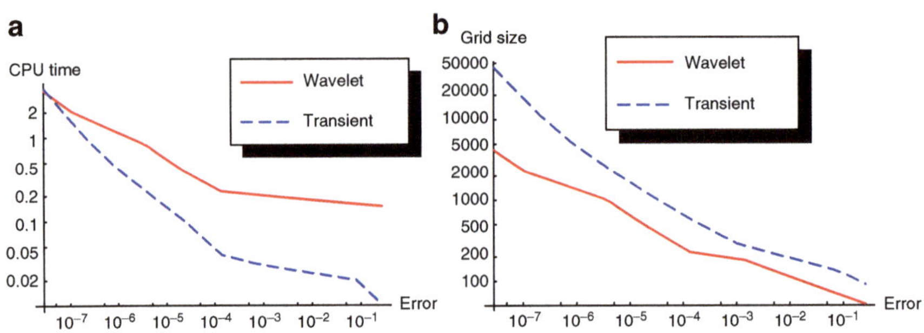

Fig. 3 Simulation results for the Schmitt trigger. Computation time versus error (*left*), and grid size versus error (*right*) for transient analysis and adaptive wavelet analysis

Schmitt trigger. The first test circuit is the Schmitt trigger [29]. As can be seen in Fig. 2 the output of the Schmitt trigger circuit signal jumps to a higher level if the input exceeds an upper threshold and jumps back to low if the input falls below a lower threshold. However, due to capacitances present in the transistor model the jumps are slightly smoothed and delayed.

Inverter chain. A further test circuit was an inverter chain consisting of 9 inverters. Therefore, the output signal represents the 9-times inverted digital input signal. However, we can observe a delay and a modification in the transition between high and low signal due to intrinsic properties of used technology. Similar to the hysteresis effect in the previous problem, the output depends strongly on the earlier behaviour of the input signal, which again requires the use of the interval splitting wavelet method to obtain the correct results (Fig. 4).

In both examples, the interval splitting wavelet method could produce the correct results and thus we have achieved robustness for the wavelet-based approach. The performance is comparable to transient analysis, although the current

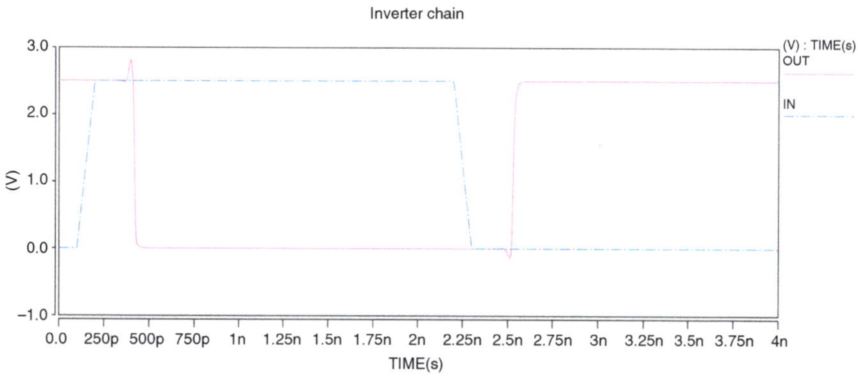

Fig. 4 Input and output signal of the inverter chain from wavelet analysis

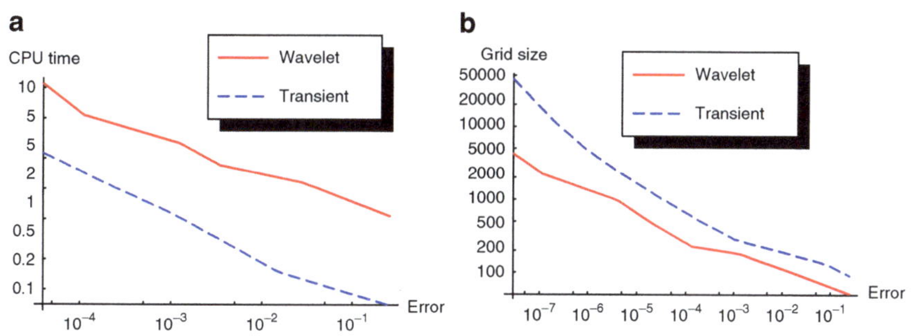

Fig. 5 Simulation results for the inverter chain. Computation time versus error (*left*), and grid size versus error (*right*) for transient analysis and adaptive wavelet analysis

implementation is not faster than the reference method (Figs. 3 and 5). However, we see a big potential for the improvement of the implemented method.

5 Conclusion

The results of the simulations indicate that the wavelet-based method is able to fulfill all accuracy requirements and may achieve the performance of the standard transient analysis. Since the relatively new wavelet approach has a large potential for optimization, we are optimistic that wavelet analysis will be a valuable tool for circuit simulation in the future. Therefore our activities on optimization and further development of the wavelet-based algorithm are continuing.

Acknowledgements This work has been supported within the EU Seventh Research Framework Project (FP7) ICESTARS with project number FP7/2008/ICT/214911.

References

1. Daubechies, I.: Orthonormal bases of compactly supported wavelets. Commun. Pure Appl. Math. **41**, 909–996 (1988)
2. Daubechies, I.: Ten Lectures on Wavelets. SIAM, Philadelphia (1992)
3. Grossmann, A., Morlet, J.: Decomposition of hardy functions into square integrable wavelets of constant shape. SIAM J. Math. Anal. **15**, 723–736 (1984)
4. Donoho, D.: Unconditional bases are optimal bases for data compression and for statistical estimation. Appl. comput. Harmonic Anal. **1**, 100–115 (1993)
5. Coifman, R., Wickerhauser, M.: Entropy based algorithms for best basis selection. IEEE Trans. Info. Theory **38**, 713–718 (1992)
6. Meyer, Y.: Wavelets: Algorithms and Applications. SIAM, Philadelphia (1993)
7. Mallat, S.: A Wavelet Tour of Signal Processing. Academic Press, Massachusetts (1998)
8. Kaiser, G.: A friendly guide to wavelets. Birkhäuser (1994)
9. Le Maitre, O., Najm, H., Ghanem, R., Knio, O.: Multi-resolution analysis of Wiener-type uncertainty propagation schemes. J. Comput. Phys. **197**, 502–531 (2004)
10. Misiti, M., Misiti, Y., Oppenheim, G., Poggi, J.M.: Wavelet Toolbox 4. Mathworks (2008)
11. Pan, G.: Wavelets in Electromagnetics and Device Modelling. Wiley-Interscience, New York (2003)
12. Vetterli, M., Kovacevic, J.: Wavelets and Subband Coding. Prentice Hall, Englewood Cliffs, NJ (1995)
13. Young, R.: Wavelet Theory and Its Applications. Kluwer Academic Publishers (1993)
14. Bartel, A., Knorr, S., Pulch, R.: Wavelet based methods for multirate partial differential-algebraic equations. Appl. Numer. Math. **59**, 495–506 (2008)
15. Christoffersen, C., Steer, M.: State-variable-based transient circuit simulation using wavelets. IEEE Microwave Wireless Components Lett. **11**, 161–163 (2001)
16. Dautbegovic, E., Condon, M., Brennan, C.: An efficient nonlinear circuit simulation technique. IEEE Trans. Microw. Theory Tech. **53**, 548–555 (2005)
17. Soveiko, N., Gad, E., Nakhla, M.: A wavelet-based approach for steady-state analysis of nonlinear circuits with widely separated time scales. IEEE Microw. Wireless Components Lett. **17**, 451–453 (2007)
18. Zhou, D., Cai, W.: A fast wavelet collocation method for high-speed circuit simulation. IEEE Trans. Circ. Syst. **46**, 920–930 (1999)
19. Dautbegovic, E.: Wavelets in circuit simulation. In: J. Roos, L. Costa (eds.) Scientific Computing in Electrical Engineering, Mathematics in Industry, vol. 14, pp. 131–142. Springer, Berlin (2010)
20. Dahmen, W., Kunoth, A., Urban, K.: Biorthogonal spline-wavelets on the interval – stability and moment conditions. Appl. Comp. Harm. Anal. **6**, 132–196 (1999)
21. Stevenson, R., Nguyen, H.: Finite element wavelets on manifolds. IMA J. Numer. Math **23**, 149–173 (2003)
22. Bittner, K.: Biorthogonal spline wavelets on the interval. In: G. Chen, M.J. Lai (eds.) Wavelets and Splines: Athens 2005, pp. 93–104. Nashboro Press, Brentwood, TN (2006)
23. DeVore, R.A.: Nonlinear approximation. Acta Numerica **7**, 51–150 (1998)
24. DeVore, R.A., Lorentz, G.G.: Constructive Approximation. Springer, New York (1993)
25. Bittner, K., Urban, K.: Adaptive wavelet methods using semiorthogonal spline wavelets: Sparse evaluation of nonlinear functions. Appl. Comput. Harmon. Anal. **24**, 94–119 (2008)
26. Dahmen, W., Schneider, R., Xu, Y.: Nonlinear functionals of wavelet expansions – adaptive reconstruction and fast evaluation. Numer. Math. **86**, 49–101 (2000)
27. Zhou, D., Cai, W.: A fast wavelet collocation method for high-speed circuit simulation. IEEE Trans. Circ. Syst. – I: Fundamental Theory Appl. **46**, 920–930 (1999)
28. Bittner, K., Dautbegovic, E.: Wavelets algorithm for circuit simulation. In: Progress in Industrial Mathematics at ECMI 2010, Mathematics in Industry. Springer, Berlin (submitted)
29. http://www.physics.ucdavis.edu/Classes/Physics116/Schmitt.html

Modeling and Simulation of Organic Solar Cells

Carlo de Falco, Antonio Iacchetti, Maddalena Binda, Dario Natali, Riccardo Sacco, and Maurizio Verri

Abstract A model for polymer Solar Cells is presented consisting of a system of nonlinear diffusion-reaction PDEs with electrostatic convection, coupled to a kinetic ODE. A proof of the existence of both stationary and transient solutions is given and an algorithm for computing them is proposed and numerically validated by comparison with experimentally measured data for a photovoltaic cell.

1 Organic Solar Cells

Third Generation solar cells [12] have recently received a lot of interest as a viable choice for a low cost renewable energy source. Roughly speaking, 3G photovoltaic devices can be divided into two main classes: electrochemical cells [11] and organic cells (OSC) [15]. While the latter are the main topic of the present contribution, some of the proposed analytical and numerical techniques can be showed to be of use when dealing with the former as well. The simplest possible structure for an organic-polymer based solar cell consists of a binary *blend* of two materials, giving rise to a bulk heterojunction (BHJ), sandwiched between

C. de Falco (✉) · R. Sacco · M. Verri
Dipartimento di Matematica, Politecnico di Milano, P.zza L. da Vinci 32, 20133 Milano, Italy
e-mail: carlo.defalco@polimi.it; riccardo.sacco@polimi.it; maurizio.verri@polimi.it

A. Iacchetti · D. Natali
Dipartimento di Elettronica e Informazione, Politecnico di Milano, P.zza L. da Vinci 32, 20133 Milano, Italy

Center for Nano Science and Technology of IIT PoliMI, Via Pascoli 70/3 20133 Milano, Italy
e-mail: iacchetti@elet.polimi.it; dario.natali@polimi.it

M. Binda
Dipartimento di Elettronica e Informazione, Politecnico di Milano, P.zza L. da Vinci 32, 20133 Milano, Italy
e-mail: binda@elet.polimi.it

B. Michielsen and J.-R. Poirier (eds.), *Scientific Computing in Electrical Engineering SCEE 2010*, Mathematics in Industry 16, DOI 10.1007/978-3-642-22453-9_35,

one transparent (e.g. indium-tin-oxide or fluorinated tin oxide) and one reflecting metal contact (usually aluminum or silver). Absorbed photons produce electron-hole pairs which, in contrast to what is usually the case in standard inorganic semiconductors, have strong binding energy (0.1–0.4 eV typically) and a distance in the sub-nanometer range. The two materials are chosen in order to display at their interface an offset in energy levels suitable to exploit photoinduced electron transfer phenomenon [21]: for pairs able to diffuse to the heterojunction before recombination, electron(hole) transfer can spontaneously occur from the excited donor(acceptor) molecule to an adjacent acceptor(donor) molecule provided that this latter has suitably high(low) electron affinity (ionization potential). Thanks to the built-in electrical field originating from the difference in metal Fermi levels, the separated charges are driven to the contacts where they are *harvested* producing a current. Model parameters of BHJs are difficult to characterize based on constituent material properties only, therefore mathematical modeling and analysis as well as numerical simulations can help fit their values by comparison to experimental measurements.

The Mathematical Model

The blend is modeled by a homogeneous material filling a bounded domain $\Omega \subset \mathbb{R}^d$, $d \geq 1$, with a Lipschitz boundary $\Gamma \equiv \partial\Omega$ divided into two disjoint subregions, Γ_D and Γ_N, representing the interface between metal and polymer blend and interior artificial boundaries, respectively. We assume that meas $(\Gamma_D) > 0$ and $\Gamma_D \cap \Gamma_N = \emptyset$, and denote by $\boldsymbol{\nu}$ the outward unit normal vector along Γ. Charge transport in the device is governed by the set of continuity equations

$$\begin{cases} \dot{n} - \mathrm{div}\,(D_n \nabla n - \mu_n n\, \nabla\varphi) = G_n - R_n\, n \\ \dot{p} - \mathrm{div}\left(D_p \nabla p + \mu_p\, p\, \nabla\varphi\right) = G_p - R_p\, p, \end{cases} \tag{1a}$$

to be solved in $\Omega_T \equiv \Omega \times (0, T)$. Using, from now on, the symbol η to indicate either of n or p, G_η are the carrier *generation rates* and $R_\eta\, \eta$ are the *recombination rates*. D_η being the charge carrier *diffusion coefficients* and μ_η the *carrier mobilities*. The *electrostatic potential* φ satisfies the Poisson equation

$$-\,\mathrm{div}(\varepsilon \nabla\varphi) = q(p - n) \qquad \text{in } \Omega_T. \tag{1b}$$

We denote by X the *volume density of geminate pairs* and we express its rate of change as

$$\dot{X} = g - r \qquad \text{in } \Omega_T. \tag{1c}$$

The geminate-pair generation and recombination rates in (1c) can be both split into two contributions as

$$g = \underbrace{G(\mathbf{x},t)}_{(a)} + \underbrace{\gamma\, p\, n}_{(b)}\,, \qquad r = \underbrace{k_{diss} X}_{(c)} + \underbrace{k_{rec} X}_{(d)}\,, \tag{2}$$

(a) accounting for the photo-generation rate, (b) accounting for the rate at which free electrons and holes are attracted to each other and recombine, (c) accounting for the rate at which free electrons and holes are produced by separation of a bound pair and (d) accounting for the rate at which geminate pairs that are not split recombine. The generation rates satisfy $G_n = G_p = k_{diss} X$ while for the recombination rates $R_n n = R_p p = \gamma p n$ holds. Boundary conditions for carriers at the contacts can be given the following Robin-type form

$$\kappa_\eta \, \mathbf{J}_\eta \cdot \boldsymbol{\nu} = \beta_\eta - \alpha_\eta \, \eta \qquad \text{on } \Gamma_D \times (0,T), \tag{3}$$

where $\mathbf{J}_\eta$ are the (particle) current densities, κ_η are non negative parameters while β_η are the rates at which charges are injected into the device and $\alpha_\eta \eta$ are the rates at which electrons and holes recombine with their image charges at the contacts, respectively. A physically sound characterization of the injection rates α_η and β_η is carried out in [7] where the classical theory of thermionic current flow at Schottky contacts is extended to the case of a metal-organic interface. Reliable models for the above parameters are, though, still subject of on-going debate and investigation (see, e.g., [3]).

2 Analysis of the Model

We introduce the following (physically plausible) simplifying assumptions:

(H1) γ, k_{diss}, k_{rec} and G are all positive constant quantities in Ω_T;
(H2) $D_\eta = V_{th}\mu_\eta$, V_{th} being the *thermal voltage* and $\mu_\eta \geq \mu_{\eta 0} > 0$ a.e. in Ω_T;
(H3) $v_n, v_p \leq v^{max} < +\infty$ where $v_\eta := \mu_\eta |\mathbf{E}|$;
(H4) $\kappa_\eta = 0$ and α_η, β_η are functions of position only in (3).

Stationary Regime

Setting $\dot{X} = 0$ in (1c), we can eliminate the dependent variable X in favor of n, p and of the input function G, so that the model unknowns are n, p and φ only. We can then prove the following (see [9].)

Theorem 1. *Let assumptions (H1)–(H4) be satisfied and let* $(\varphi_D, n_D, p_D) \in (L^\infty(\Gamma_D))^3$ *be the values of the electric potential and carrier concentrations on* Γ_D. *Then the model equations admit a weak solution* $(\varphi^*, n^*, p^*) \in (H^1(\Omega) \cap L^\infty(\Omega))^3$ *and there exist positive constants* $\underline{\mathcal{M}}, \overline{\mathcal{M}}, \underline{\mathcal{K}}, \overline{\mathcal{K}}$ *such that*

$$\underline{\mathcal{M}} \leq n^*,\ p^* \leq \overline{\mathcal{M}},\ \underline{\mathcal{K}} \leq \varphi^* \leq \overline{\mathcal{K}} \text{ a.e. in } \Omega. \tag{4}$$

Transient Regime

By approximating the integral form of (1c) as in [9], we can again eliminate the dependent variable X in favor of n, p and of the given forcing term G and initial condition X_0 so that the resulting model takes the form

$$\begin{cases} -\mathrm{div}(\varepsilon \nabla \varphi) & = q(p-n) \\ \dot{n} - \mathrm{div}\,(D_n \nabla n - \mu_n n\, \nabla \varphi) & = \widetilde{G}_n - \widetilde{R}_n n \qquad \text{in } \Omega_T \\ \dot{p} - \mathrm{div}\left(D_p \nabla p + \mu_p\, p\, \nabla \varphi\right) & = \widetilde{G}_p - \widetilde{R}_p p, \end{cases} \tag{5}$$

Theorem 2. *Let assumptions (H1)–(H4) be satisfied, and the initial data* $\mathbf{U} := (n_0, p_0)$, X_0 *and the function* Ψ *(representing a lifting of the electric potential boundary conditions) be such that* $\mathbf{U} \in (H^1(\Omega_T) \cap L^\infty(\Omega_T))^2$, *with* $\mathbf{U} > \mathbf{0}$, $X_0 \in L^\infty(\Omega)$ *with* $X_0 \geq 0$, *and* $\Psi \in H^1(\Omega_T) \cap L^\infty(\Omega_T)$. *Then, setting* $\mathbf{u} := (n, p)$, *system* (5) *admits a weak solution* $(\varphi, \mathbf{u})$ *such that:*

1. $\mathbf{u} > \mathbf{0}$ *a.e. in* Ω_T;
2. $\mathbf{u}(\mathbf{x}, 0) = \mathbf{U}(\mathbf{x}, 0)$ *and* $\mathbf{u} - \mathbf{U} \in L^2(0, T; H_0)^2$;
3. $\mathbf{u} \in \left(C(0, T; L^2(\Omega)) \cap L^\infty(\Omega_T)\right)^2$;
4. $\dfrac{\partial \mathbf{u}}{\partial t} \in L^2(0, T; H_0')^2$;
5. $\varphi - \Psi \in L^2(0, T; H_0)$ *with* $\varphi \in L^\infty(\Omega_T)$.

Numerical Discretization

To design an effective simulation algorithm, appropriate for accurately estimating the OSC photocurrent in both stationary and transient regimes, we make use of reliable numerical methods traditionally used for the spatial discretization of inorganic semiconductor devices (see, e.g., [18] Chap. 6, Sect. 4) combined with efficient time-step adaptation methods (see, e.g., [1, 13]). To this end, we first carry out a temporal semi-discretization using an implicit multistep method where the selection of the time increment is performed adaptively in such a way to minimize the time discretization error while minimizing the total number of time steps via the DAE solver software library DASPK [6, 19]. Then, the resulting sequence of differential subproblems is linearized using the Newton–Raphson method with inexact evaluation of the Jacobian. Finally, we use exponentially fitted finite elements for the spatial discretization, to ensure a stable approximation of the internal and boundary layers arising in the distribution profile of the photogenerated carriers [2, 10, 14, 20]. The adopted formulation provides a natural multidimensional extension of the classical Scharfetter-Gummel difference scheme [5, 17] and ensures that the computed carrier concentration is strictly positive under the condition that the triangulation of the domain Ω is of Delaunay type.

3 Experimental Setting and Numerical Validation

In this section, we describe the real device used for numerically validating the simulation tool implementing the method described above, and then we illustrate the obtained results by comparison with available measured data.

3.1 Description of the Device

The device chosen, schematically depicted in Fig. 1, is based on the bulk heterojunction of a squaraine molecule, acting as donor, and Phenyl-C61-Butyric-Acid-Methyl-Ester (PCBM) acting as acceptor. We chose squaraine because they are able to absorb also radiation in the 600–800 nm region: extension of the absorption spectrum towards red region for organic light harvesting devices is currently a subject of intense research. In particular, we exploited a hydrazone end-capped symmetric squaraine provided with glycolic functionalization chains, since this specific substitution pattern had been previously demonstrated to provide the appropriate phase separation between the squaraine compound and PCBM

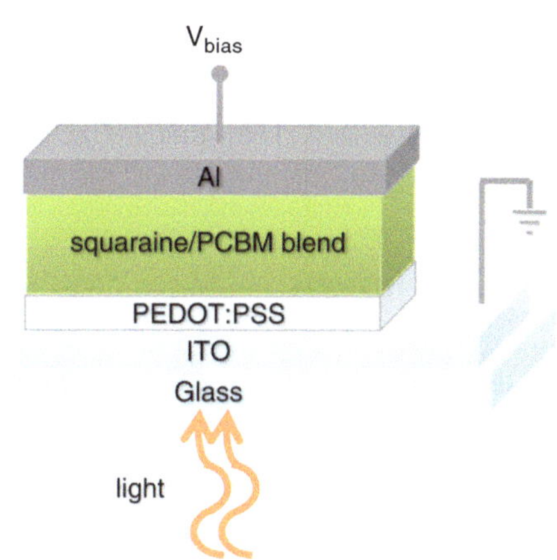

Fig. 1 Schematic representation of the vertical BHJ simulated in Sect. 3

when blended together. As far as the blend composition is concerned, we chose a squaraine:PCBM 1:3 by weight ratio in order to get a reasonable balance between hole and electron mobilities [4]. The chosen device geometry was vertical, with the active material sandwiched between a bottom transparent ITO electrode coated with PEDOT:PSS (Clevios P VP AI 4083) and a top evaporated aluminum one; device area was about 4 mm^2. PEDOT:PSS was deposited by spin-coating (from aqueous solution) at 2000 rpm onto glass-ITO substrates pre-treated with oxygen plasma and annealed at $100°C$ for 15 min under nitrogen. The blend was dissolved in chloroform (19.2 mg/ml) and deposited in a glovebox (water and oxygen content below 1 ppm) by spincoating at 100 rpm for 1 min (followed by 1 min at 1000 rpm), giving a 220 nm thick film. Electrical characterization was performed in vacuum $(P < 10^{-5} mbar)$.

3.2 Numerical Results

Figures 2a and b display the comparison between experimentally measured and numerically computed current-voltage characteristics of the device previously described, while the model parameters used for the simulation are shown in Table 1.

Figure 2a refers to the case where the device is not illuminated (dark condition) while Fig. 2b refers to the case where an incident monocromatc light source of wavelength 700 nm is applied with a power of 0.63 mW cm^{-2}.

We notice that measured and simulated data are in good agreement especially as far as the dark condition is considered. As for the results in the illuminated regime, the simulation correctly predicts the open circuit voltage $V_{oc} \simeq 0.4$ V within a tolerance of 3%, and also the short circuit current is well approximated although with slightly lesser precision. The discrepancies under illumination are likely due to the simplifications introduced in the model, in particular having disregarded the dependency of the carrier mobilities on the electric field and on the charge density, which is far higher in this regime. The former effect is already taken into account in our simulator, but was neglected due to lack of experimental data to fit the model parameters, on the other hand the latter would require extensions to the numerical algorithm (see, e.g., [16]) that are the subject of our ongoing work.

4 Conclusions

In this article, we have proposed and investigated a computational model for the study of bulk heterojunction organic polymer solar cells. The model consists of a system of drift-diffusion equations for photogenerated charge transport plus an ordinary differential equation governing the time rate of change of photoinduced excitons. Linearization of the fully coupled problem supported by suitable adaptive time stepping and stable exponentially-fitted finite elements allows to end up

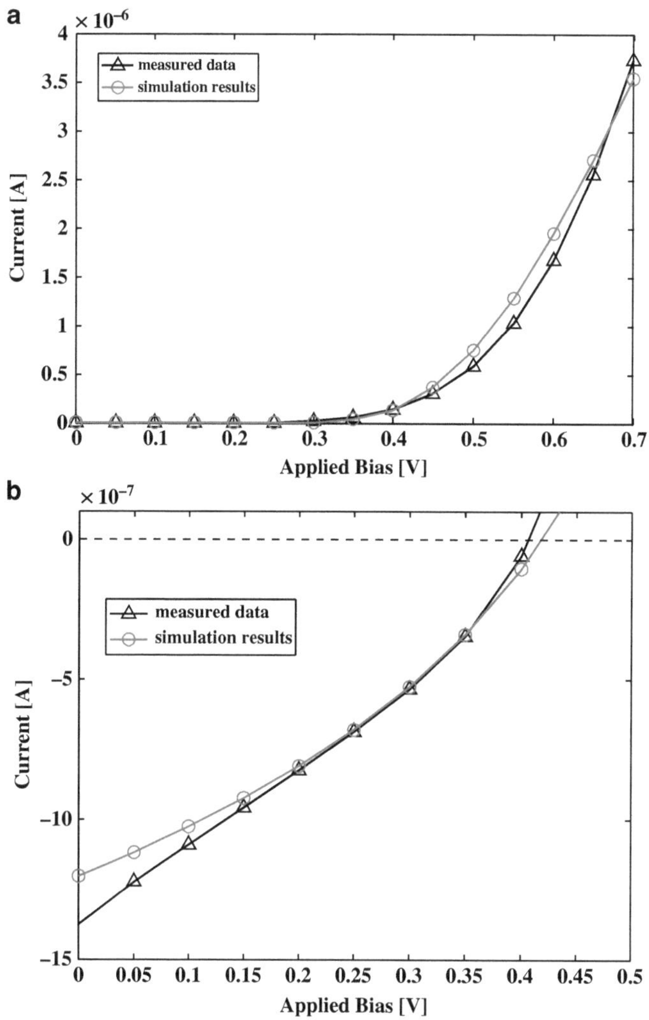

Fig. 2 (**a**) Comparison of computed and measured current-voltage characteristics, dark working conditions (**b**) Comparison of computed and measured current-voltage characteristics, illuminated working conditions

with a robust and efficient simulation tool, whose validation is carried out on the comparison with experimental data of a solar cell. The same computational model and algorithm have been applied successfully to investigate other classes of solar cells, namely, Electrochemical cells or Bilayer OSCs [8].

Acknowledgements The authors thank Professor M. Sampietro of Politecnico di Milano for stimulating and fruitful discussions.

Table 1 Parameters used for the simulations

parameter name	symbol	units	value
Electron mobility	μ_n	$cm^2\ V^{-1}\ s^{-1}$	6×10^{-6}
Hole mobility	μ_p	$cm^2\ V^{-1}\ s^{-1}$	2×10^{-6}
Electron B.C. parameters (PEDOT contact)	α_n	cm^3	2.69×10^{-13}
	β_n		1.93×10^{-22}
	κ_n	$A^{-1}\ cm^2$	0
Electron B.C. parameters (Al contact)	α_n	cm^3	2.69×10^{-13}
	β_n		1.456×10^{-6}
	κ_n	$A^{-1}\ cm^2$	0
Hole B.C. parameters (PEDOT contact)	α_p	cm^3	2.69×10^{-13}
	β_p		1.32×10^{3}
	κ_p	$A^{-1}\ cm^2$	0
Hole B.C. parameters (Al contact)	α_p	cm^3	2.69×10^{-13}
	β_p		5.47×10^{-14}
	κ_p	$A^{-1}\ cm^2$	0
Bi-molecular recombination rate	γ	$s^{-1}\ cm^3$	4.28×10^{-11}
Exciton recombination rate	k_{rec}	s^{-1}	1.0×10^{9}
Exciton dissociation rate	k_{diss}	s^{-1}	1.576×10^{8}

References

1. Ascher, U., Petzold, L.: Computer methods for ordinary differential equations and differential-algebraic equations. SIAM, Philadelphia (1998)
2. Bank, R.E., Coughran, W.M., Cowsar, L.C.: Analysis of the Finite Volume Scharfetter-Gummel Method for Steady Convection Diffusion Equations. Comput. Visualiz. Sci. **1**(3), 123–136 (1998)
3. Barker, J., Ramsdale, C., Greenham, N.: Modeling the current-voltage characteristics of bilayer polymer photovoltaic devices. Phys. Rev. B **67** (2003)
4. Binda, M., Agostinelli, T., Caironi, M., Natali, D., Sampietro, M., Beverina, L., Ruffo, R., Silvestri, F.: Fast and air stable near-infrared organic detector based on squaraine dyes. Org. Elec. **10**(7), 1314–1319 (2009)
5. Brezzi, F., Marini, L., Pietra, P.: Numerical Simulation of Semiconductor Devices. Comp. Meths. Appl. Mech. Engrg. **75**, 493–514 (1989)
6. Brown, P., Hindmarsh, A., Petzold, L.: A description of DASPK: A solver for large-scale differential-algebraic systems. Lawrence Livermore National Report UCRL (1992)
7. Campbell Scott, J., Malliaras, G.: Charge injection and recombination at the methal-organic interface. Chem. Phys. Lett. **299**, 115–119 (1999)

8. Cogliati, M., Porro, M.: Reaction-diffusion PDE/ODE nonlinear systems for the simulation of new generation solar cells (2010). Master Thesis, Politecnico di Milano
9. de Falco, C., Sacco, R., Verri, M.: Analytical and numerical study of photocurrent transients in organic polymer solar cells. Comput. Methods Appl. Mech. Engrg. (2010 (in press))
10. Gatti, E., Micheletti, S., Sacco, R.: A New Galerkin Framework for the Drift-Diffusion Equation in Semiconductors. East West J. Numer. Math. **6**(2), 101–135 (1998)
11. Grätzel, M.: Photoelectrochemical cells. Nature **414** (2001)
12. Green, M.: Third Generation Photovoltaics: Advanced Solar Electricity Generation. Springer, Berlin (2003)
13. Hairer, E., Wanner, G.: Solving Ordinary Differential Equations I. Springer Series in Computational Mathematics. Springer, Berlin (1996)
14. Lazarov, R., Zikatanov, L.: An exponential fitting scheme for general convection–diffusion equations on tetrahedral meshes. Comput. Appl. Math. (Obchysljuval'na ta prykladna matematyka, Kiev) **1(92)**, 60–69 (2005)
15. Mayer, A., Scully, S., Hardin, B., Rowell, M., McGehee, M.: Polymer-based solar cells. Mater. Today **10**(11), 28–33 (2007)
16. van Mensfoort, S.L.M., Coehoorn, R.: Effect of gaussian disorder on the voltage dependence of the current density in sandwich-type devices based on organic semiconductors. Phys. Rev. B **78** (2008)
17. Scharfetter, D., Gummel, H.: Large signal analysis of a silicon Read diode oscillator. IEEE Trans. Electron Devices **ED-16**, 64–77 (1969)
18. Selberherr, S.: Analysis and Simulation of Semiconductor Devices. Springer, Wien (1984)
19. Van Keken, P., Yuen, D., Petzold, L.: DASPK: a new high order and adaptive time-integration technique with applications to mantle convection with strongly temperature-and pressure-dependent rheology. Geophys. Astrophys. Fluid Dyn. **80**(1), 57–74 (1995)
20. Xu, J., Zikatanov, L.: A monotone finite element scheme for convection–diffusion equations. Math. Comp. **68(228)**, 1429–1446 (1999)
21. Yu, G., Gao, J., Hummelen, J., Wudl, F., Heeger, A.: Polymer photovoltaic cells: Enhanced efficiencies via a network of internal donor-acceptor heterojunctions. Science **270**, 1789 (1995)

Numerical Simulation of a Hydrodynamic Subband Model for Semiconductors Based on the Maximum Entropy Principle

G. Mascali and V. Romano

Abstract A hydrodynamic subband model for semiconductors has been formulated in (Mascali and Romano, IL NUOVO CIMENTO 33C:155163, 2010) by closing the moment system derived from the Schrödinger-Poisson-Boltzmann equations on the basis of the maximum entropy principle (MEP). Explicit closure relations for the fluxes and the production terms are obtained taking into account scattering of electrons with acoustic and non-polar optical phonons, as well as surface scattering. Here a suitable numerical scheme is presented for the above model together with simulations of a nanoscale silicon diode.

1 MEP Model for Subbands

By shrinking the dimensions of electronic devices, effects of quantum confinement are observed [2, 3], e.g. in MOSFETs at the Si-SiO_2 interface, in double gate MOSFETs, in hetero-structures like AlGa-Ga.

If electrons are quantized in the z direction and free to move in the x-y plane, one assumes the following ansatz for the electron wave function

$$\psi(\mathbf{r}) = \psi(x, y, z) = \frac{1}{\sqrt{\mathscr{A}}}\,\varphi(z)e^{i\mathbf{k}_{||}\cdot\mathbf{r}_{||}},$$

G. Mascali (✉)
Dipartimento di Matematica, Università della Calabria, Via Ponte Bucci, cubo 30 B, Arcavacata di Rende (Cs) and INFN-Gruppo c. Cosenza, Italy
e-mail: mascali@unical.it

V. Romano
Dipartimento di Matematica e Informatica, Università di Catania, Viale A.Doria 6, I-95125 Catania, Italy
e-mail: romano@dmi.unict.it

B. Michielsen and J.-R. Poirier (eds.), *Scientific Computing in Electrical Engineering SCEE 2010*, Mathematics in Industry 16, DOI 10.1007/978-3-642-22453-9_36,

with $\mathbf{k}_{||} = (k_x, k_y)$ and $\mathbf{r}_{||} = (x, y)$ denoting the longitudinal components of the wave-vector $\mathbf{k}$ and of the position vector $\mathbf{r}$, respectively, and $\mathscr{A}$ symbolizing the area of the x-y cross-section.

The Schrödinger equation in the effective mass approximation gives the following equation for φ

$$\left[-\frac{\hbar^2}{2m^*}\frac{d^2}{dz^2} + E_C\right]\varphi(z) = \left[E - \frac{\hbar^2}{2m^*}|\mathbf{k}_{||}|^2\right]\varphi(z) = \varepsilon\,\varphi(z), \quad 0 \le z \le L,$$

where $\hbar$ is the reduced Planck constant, m^* is the effective electron mass, ε is the energy associated with the confinement in the z-direction, and L is the device extension in the z-direction. One finds a countable set of eigen-pairs (subbands) $(\varphi_\nu, \varepsilon_\nu), \nu = 1, 2....$ The conduction band minimum E_C is given by $E_C = -q(V_C + V)$, where V_C is the confining potential and V is the self-consistent electrostatic potential obtained from the Poisson equation

$$\nabla \cdot (\epsilon_d \, \nabla V) = -q(C(\mathbf{r}) - n), \tag{1}$$

where q is the absolute value of the electron charge, ϵ_d is the dielectric constant, $C(\mathbf{r})$ is the doping concentration, and n is the electron density given by $n(\mathbf{r}, t) = \sum_{\nu=1}^{+\infty} \rho_\nu(x, y, t)|\varphi_\nu(z, t)|^2$, with ρ_ν the areal density of electrons of the ν-subband.

The description of the electron transport along the longitudinal direction is included by adding a system of coupled semiclassical Boltzmann equations for the distributions $f_\nu(x, y, k_x, k_y, t)$ of electrons in the subbands

$$\frac{\partial f_\nu}{\partial t} + \frac{1}{\hbar}\nabla_{\mathbf{k}_{||}} E_\nu \cdot \nabla_{\mathbf{r}_{||}} f_\nu - \frac{q}{\hbar}\mathbf{E}_\nu^{eff} \cdot \nabla_{\mathbf{k}_{||}} f_\nu = \sum_{\mu=1}^{\infty} C_{\nu,\mu}[f_\nu, f_\mu], \quad \nu = 1, 2, \ldots \tag{2}$$

where $E_\nu = \varepsilon_\nu + \frac{\hbar^2}{2m^*}|\mathbf{k}_{||}|^2$, and $\mathbf{E}_\nu^{eff} = \frac{1}{q}\nabla_{\mathbf{r}_{||}}\varepsilon_\nu(\mathbf{r}_{||})$. ρ_ν is expressed in terms of f_ν by $\rho_\nu = \int_{B_2} f_\nu(\mathbf{r}_{||}, \mathbf{k}_{||}, t) d^2\mathbf{k}_{||}$, with B_2 indicating the 2D Brillouin zone.

The relevant 2D scattering mechanisms in Si are acoustic phonon scattering, non-polar optical phonon scattering, and surface scattering, and they are taken into account by the right-hand side of (2).

The direct numerical simulation of the Schrödinger-Poisson-Boltzmann system is a daunting computational task (see for example [4–6]). Simpler macroscopic models are needed for CAD purposes. These can be obtained as moment equations of the transport Boltzmann equations under suitable closure relations. The moment of the ν-subband distribution with respect to a weight function $a(\mathbf{k}_{||})$ reads $M_a^\nu = \int_{B_2} a(\mathbf{k}_{||}) f_\nu(\mathbf{r}_{||}, \mathbf{k}_{||}, t) d^2\mathbf{k}_{||}$.

In particular we take as basic moments:

The areal density $\rho^\nu = \displaystyle\int_{B_2} f_\nu(\mathbf{r}_{||}, \mathbf{k}_{||}, t) d^2\mathbf{k}_{||}$,

The longitudinal mean velocity $\mathbf{V}^{\nu} = \frac{1}{\rho^{\nu}} \int_{B_2} \frac{\hbar \mathbf{k}_{||}}{m^*} f_{\nu}(\mathbf{r}_{||}, \mathbf{k}_{||}, t) d^2\mathbf{k}_{||},$

The longitudinal mean energy $W^{\nu} = \frac{1}{\rho^{\nu}} \int_{B_2} \frac{\hbar^2 \mathbf{k}_{||}^2}{2m^*} f_{\nu}(\mathbf{r}_{||}, \mathbf{k}_{||}, t) d^2\mathbf{k}_{||},$

The longitudinal mean energy-flux $\mathbf{S}^{\nu} = \frac{1}{\rho^{\nu}} \int_{B_2} \frac{\hbar \mathbf{k}_{||}}{m^*} \frac{\hbar^2 \mathbf{k}_{||}^2}{2m^*} f_{\nu}(\mathbf{r}_{||}, \mathbf{k}_{||}, t) d^2\mathbf{k}_{||}.$

The corresponding moment system reads

$$\frac{\partial \rho^{\nu}}{\partial t} + \nabla_{\mathbf{r}_{||}} \cdot (\rho^{\nu} \mathbf{V}^{\nu}) = \rho^{\nu} \sum_{\mu} C_{\rho}^{\nu,\mu},$$

$$\frac{\partial (\rho^{\nu} \mathbf{V}^{\nu})}{\partial t} + \nabla_{\mathbf{r}_{||}} \cdot (\rho^{\nu} \underline{\underline{\mathbf{U}}}^{\nu}) + \frac{\rho^{\nu}}{m^*} \nabla_{\mathbf{r}_{||}} \varepsilon_{\nu} = \rho^{\nu} \sum_{\mu} C_{\mathbf{V}}^{\nu,\mu},$$

$$\frac{\partial \rho^{\nu} W^{\nu}}{\partial t} + \nabla_{\mathbf{r}_{||}} \cdot (\rho^{\nu} \mathbf{S}^{\nu}) + \rho^{\nu} \nabla_{\mathbf{r}_{||}} \varepsilon_{\nu} \cdot \mathbf{V}^{\nu} = \rho^{\nu} \sum_{\mu} C_{W}^{\nu,\mu},$$

$$\frac{\partial (\rho^{\nu} \mathbf{S}^{\nu})}{\partial t} + \nabla_{\mathbf{r}_{||}} \cdot (\rho^{\nu} \underline{\underline{\mathbf{F}}}^{\nu}) + \rho^{\nu} \left[\frac{W^{\nu}}{m^*} \underline{\underline{\mathbf{I}}} + \underline{\underline{\mathbf{U}}}^{\nu} \right] \cdot \nabla_{\mathbf{r}_{||}} \varepsilon_{\nu} = \rho^{\nu} \sum_{\mu} C_{\mathbf{S}}^{\nu,\mu},$$

where

$$\begin{pmatrix} \underline{\underline{\mathbf{U}}}^{\nu} \\ \underline{\underline{\mathbf{F}}}^{\nu} \end{pmatrix} = \frac{1}{\rho^{\nu}} \int_{B_2} \begin{pmatrix} \frac{\hbar^2}{(m^*)^2} \\ \frac{\hbar^4 \mathbf{k}_{||}^2}{2(m^*)^3} \end{pmatrix} \mathbf{k}_{||} \otimes \mathbf{k}_{||} f_{\nu}(\mathbf{r}_{||}, \mathbf{k}_{||}, t) \, d^2\mathbf{k}_{||},$$

$$\begin{pmatrix} C_{\rho}^{\nu,\mu} \\ C_{W}^{\nu,\mu} \end{pmatrix} = \int_{B_2} \begin{pmatrix} 1 \\ \frac{\hbar^2 \mathbf{k}_{||}^2}{2m^*} \end{pmatrix} \left[S(\mathbf{k}_{||}^{\mu}, \mathbf{k}_{||}^{\nu}) \, f_{\mu} - S(\mathbf{k}_{||}^{\nu}, \mathbf{k}_{||}^{\mu}) \, f_{\nu} \right] d^2\mathbf{k}_{||},$$

$$\begin{pmatrix} C_{\mathbf{V}}^{\nu,\mu} \\ C_{\mathbf{S}}^{\nu,\mu} \end{pmatrix} = \int_{B_2} \begin{pmatrix} \frac{\hbar \mathbf{k}_{||}}{m^*} \\ \frac{\hbar^3 \mathbf{k}_{||}^2}{2(m^*)^2} \mathbf{k}_{||} \end{pmatrix} \left[S(\mathbf{k}_{||}^{\mu}, \mathbf{k}_{||}^{\nu}) \, f_{\mu} - S(\mathbf{k}_{||}^{\nu}, \mathbf{k}_{||}^{\mu}) \, f_{\nu} \right] d^2\mathbf{k}_{||},$$

$S(\mathbf{k}_{||}^{\mu}, \mathbf{k}_{||}^{\nu})$ being the transition rate from the longitudinal state with wave-vector $\mathbf{k}_{||}^{\mu}$ to that with wave-vector $\mathbf{k}_{||}^{\nu}$. The moment system is not closed because there are more unknowns than equations. Therefore closure relations are needed. The maximum entropy principle (MEP) leads to a systematic way for obtaining constitutive relations on the basis of information theory [7–11]. We define the entropy of the system as

$$\mathscr{S} = -k_B \sum_{\nu=1}^{+\infty} |\varphi_\nu(z,t)|^2 \int_{B_2} (f_\nu \log f_\nu - f_\nu)\, d^2\mathbf{k}_{||},$$

and, according to MEP, we estimate the f_ν's as the distributions f_ν^{MEP}'s that maximize $\mathscr{S}$ under the constraints

$$M^\nu_{a_A} = \int_{B_2} a_A(\mathbf{k}_{||})\, f_\nu^{MEP}\, d^2\,\mathbf{k}_{||},$$

where $M^\nu_{a_A}$ are the basic moments we have previously considered.

For the sake of brevity we omit the details. The interested reader is referred to [1] for the explicit expressions of the closure relations for fluxes and production terms in the case of scattering between electrons and acoustic phonons and non-polar optical phonons, along with surface scattering.

The moment system of the subbands augmented with the MEP closure relations forms a quasilinear hyperbolic system in the time direction, provided $W^\nu > 0$.

2 Numerical Simulations

The numerical method adopted for the transport part is a generalization of the Nessyahu-Tadmor (NT) scheme developed in [13] for the moment system and already adopted in [14, 15] for the semiclassical MEP hydrodynamical model. As a preliminary result, a quantum silicon diode is simulated (see Fig. 1) since in the present article the main emphasis is on the feasibility of the model rather than on the numerical issues. More complex 2d cases, like MOSFET and double gate, are under current investigation and will be presented elsewhere. We assume that the oxide gives rise to an infinitely deep potential barrier and use as bottom energies and envelope functions the analytical expressions

$$\varepsilon_\nu = \frac{\nu^2\pi^2\hbar^2}{2L^2 m^*}, \quad \varphi_\nu(z) = \sqrt{\frac{2}{L}} \sin \frac{\nu\pi}{L} z, \quad z \in [0, L], \quad \nu = 1, 2, \ldots$$

Moreover we consider as driving potential the mean electrostatic potential

$$\overline{\phi}(x) = \frac{1}{L} \int_0^L \phi(x,z)\, d\, z.$$

By taking homogeneous Neumann conditions at the Si/Si-O$_2$ interfaces, $\overline{\phi}(x)$ satisfies the 1D Poisson equation

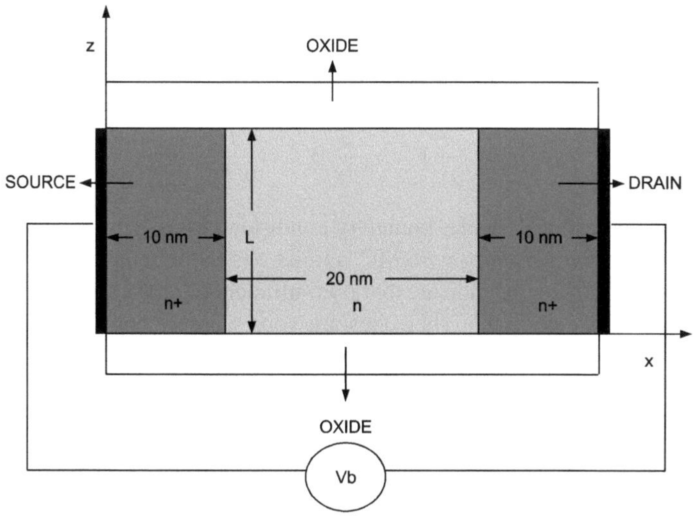

Fig. 1 Simulated diode

$$L \epsilon \overline{\phi}(x)_{xx} = -q \left(\overline{N}_D(x) - \sum_{\nu} \rho^{\nu}(x) \right)$$

where

$$\overline{N}_D(x) = \int_0^L N_D(x,z)\, d\, z$$

with N_D donor doping profile.

Six equivalent valleys are considered with a single effective mass $m^* = 0.32 m_e$, with m_e the free electron mass. A possible generalization could include both longitudinal and transverse masses.

Due to the symmetries of the problem and the boundary conditions, the transverse component of the electric field is 0. As a consequence the surface scattering vanishes and only scattering of electrons with acoustic phonons and non-polar optical phonons will be retained.

The doping in the $n+$ regions is $N_D(x) = 10^{20}$ cm^{-3} and in the n region is $N_D(x) = 10^{16}$ cm^{-3}, with a regularization at the two junctions by a hyperbolic tangent. The width of the diode is $L = 10$ nμ. A bias voltage $V_b = 0.1$ Volt between source and drain is considered.

The following initial data

$$\rho^{\nu}(x,0) = \frac{e^{-\varepsilon_{\nu}/k_B T_L}}{\sum_{\mu} e^{-\varepsilon_{\mu}/k_B T_L}} \overline{N}_D, \quad W^{\nu}(x,0) = k_B T_L, \quad V^{\nu}(x,0) = S^{\nu}(x,0) = 0$$

are taken, where T_L is the lattice temperature and V and S are the longitudinal components of $\mathbf{V}^\nu$ and $\mathbf{V}^\nu$.

Homogeneous Neumann boundary conditions at the source and drain are assumed

$$\frac{\partial}{\partial x}\rho^\nu = \frac{\partial}{\partial x}V^\nu = \frac{\partial}{\partial x}W^\nu = \frac{\partial}{\partial x}S^\nu = 0.$$

We note that imposing Dirichlet boundary conditions for the energy at source and drain leads to an inconsistency with MC simulations in the semiclassical case [16]. The numerical experiments indicate that it is sufficient to take into account only the first three subbands since the other ones are very scarcely populated. The steady state is reached after about 5 ps.

In order to establish the number of grid points to use, we compare the numerical steady state solution obtained with a mesh of 64, 128 and 256 spatial nodes. In Fig. 2 the velocity in the subband 1 is plotted. It is evident that error has a good behavior since the difference between the result with 128 and 256 grid points is about 25% of that between the results with 64 and 128 grid points. Note that the negative velocity across the first junction, present in the lower resolution case, disappears with the smaller mesh sizes. In the other figures, the results obtained with 256 nodes will be shown.

In Fig. 3 we plot the reconstruction of the electron density from the surface density and the envelope function. In Fig. 4 the densities, velocities and energies measured from the subband bottom in the first three subbands are shown. The surface density in the third subband is about 2% of the total surface density as a

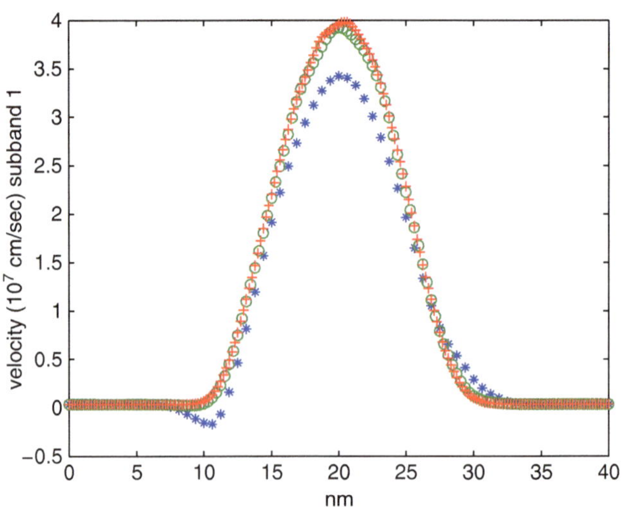

Fig. 2 Velocity in the first subband obtained with 64 (*), 128 (o) and 256 (+) grid points

Fig. 3 Steady state density reconstruction

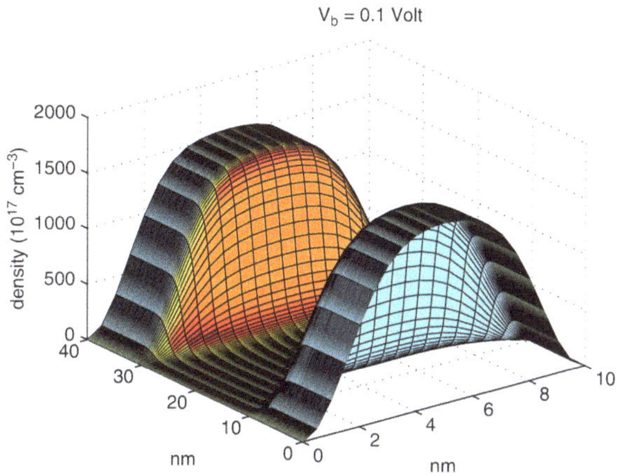

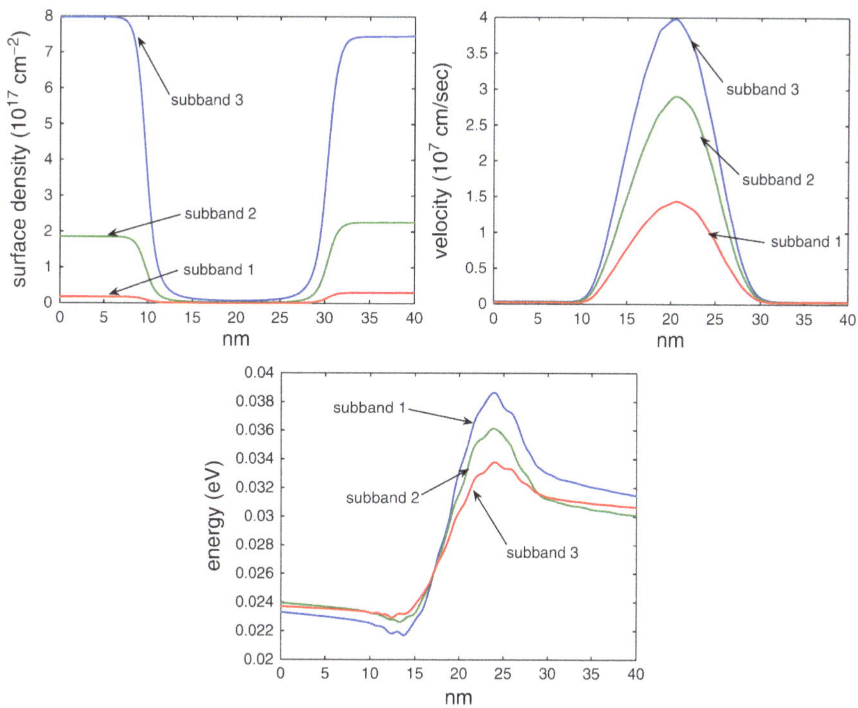

Fig. 4 Steady state densities, velocities and energies in the first three subbands

confirmation that the inclusion of further subbands has a negligible effect. It is worth to mention that the energy has an evidently different value between source and drain as happens in the semiclassical case. The use of Dirichlet conditions for the energy at the contacts misses such an effect.

3 Conclusion

A numerical integration of a hydrodynamical subband model formulated with the use of the Maximum Entropy Principle has been performed. The simulations of a quantum diode, under some simplifying assumptions, give results which capture the main confining effects. More realistic simulations require the set up of a numerical code for the solution of the moment equations coupled to the Schrödinger-Poisson block. This issue is under current investigation by the authors and will be the subject of a forthcoming article.

Acknowledgements G. M. and V. R. acknowledge the financial support by P.R.A., University of Calabria and University of Catania, respectively.

References

1. Mascali, G., Romano, V.: Hydrodynamic subband model for semiconductors based on the maximum entropy principle. IL NUOVO CIMENTO **33 C**, 155–163 (2010)
2. Datta, S.: Quantum Phenomena, vol. VIII of the Modular Series on Solid State Devices. Addison-Wesley, Reading, MA (1989)
3. Lundstrom, M.: Fundamentals of Carrier Transport. Cambridge University Press, Cambridge (2000)
4. Polizzi, E., Abdallah, N. B.: Subband decomposition approach for the simulation of quantum electron transport in nanostructures. J. Comp. Phys. **202**, 150–180 (2005)
5. Galler, M., Schuerrer, F.: A deterministic solver to the Boltzmann-Poisson system including quantization effects for silicon-MOSFETs. In: Progress in Industrial Mathematics at ECMI 2006, Series: Mathematics in Industry, pp. 531–536, Springer, Berlin (2008)
6. Abdallah, N. B., Caceres, M.J., Carrillo, J.A., Vecil, F.: A deterministic solver for a hybrid quantum-classical transport model in nanoMOSFETs. J. Comp. Phys. **228**, 6553–6571 (2009)
7. Jaynes, E.T.: Information Theory and Statistical Mechanics. Phys. Rev. **106**, 620–630 (1957)
8. Wu, N.: The Maximum Entropy Method. Springer, Berlin (1997)
9. Anile, A.M., Romano, V.: Non parabolic band transport in semiconductors: closure of the moment equations. Cont. Mech. Thermodyn. **11**, 307–325 (1999)
10. Romano, V.: Non parabolic band transport in semiconductors: closure of the production terms in the moment equations. Cont. Mech. Thermodyn. **12**, 31–51 (2000)
11. Mascali, G., Romano, V.: Hydrodynamical model of charge transport in GaAs based on the maximum entropy principle. Cont. Meh. Thermodyn. **14**, 405–423 (2002)
12. Mascali, G., Romano, V.: Si and GaAs mobility derived from a hydrodynamical model for semiconductors based on the maximum entropy principle. Physica A **352**, 459–476 (2005)
13. Liotta, S.F., Romano, V., Russo, G.: Central schemes for balance laws of relaxation type. SIAM J. Numer. Anal. **38**, 1337–1356 (2000)
14. Romano V.: 2D simulation of a silicon MESFET with a nonparabolic hydrodynamical model based on the maximum entropy principle. J. Comput. Phys. **176**, 70–92 (2002)
15. La Rosa, S., Mascali, G., Romano, V.: Exact maximum entropy closure of the hydrodynamical model for Si semiconductor: the 8-moment case. SIAM J. Appl. Math. **70**, 710–734 (2009)
16. Romano, V.: 2D numerical simulation of the MEP energy-transport model with a finite difference scheme. J. Comput. Phys. **221**, 439–468 (2007)

Inverse Doping Profile of MOSFETs via Geometric Programming

Yiming Li and Ying-Chieh Chen

Abstract In this study, we optimize one-dimensional doping profiles between the interface of semiconductor and oxide to the substrate in metal-oxide-semiconductor field-effect transistors (MOSFETs). For a set of given current-voltage curves, the problem is modelled as a geometric programming (GP) problem. The MOSFET's DC characteristics including the on- and off-state currents are simultaneously derived as functions of the doping profile in the GP problem.

1 Introduction

The channel doping profile of semiconductor devices plays an important role in determining the electrical characteristic of metal-oxide-semiconductor field-effect transistors (MOSFETs) [1]. The default engineering approach in determining a proper doping profile for specified current-voltage (I-V) curves is time-consuming and a complicated work [2]. Various computational approaches have been proposed to obtain the doping profile of MOSFETs, such as the simulation-based evolutionary technique [3], the level set method [4], and the mainfold mapping [5]. A geometric programming (GP) problem is a type of mathematical optimization problem, which is characterized by objective and constraint functions with a certain special mathematical form [6]. Recently, the optimal design of semiconductor devices and electronic circuits was found to be equivalent to a solution of a GP problem. An interior-point algorithm also was proposed to solve large-scale GP problems [7].

In this paper, we solve the inverse channel doping profile from the interface of silicon and oxide to the maximum depletion width by minimizing the subthreshold swing (SS) of MOSFETs subject to various constraints of DC characteristics including the on- and off-state currents. For extracting the corresponded doping

Y. Li (✉) · Y.-C. Chen
Department of Electrical Engineering, National Chiao-Tung University, Hsinchu 300, Taiwan
e-mail: ymli@faculty.nctu.edu.tw

B. Michielsen and J.-R. Poirier (eds.), *Scientific Computing in Electrical Engineering SCEE 2010*, Mathematics in Industry 16, DOI 10.1007/978-3-642-22453-9_37,

profile of the MOSFETs, we integrate a one-dimensional (1-D) Poisson equation [1] and derive the objective function SS in terms of the doping profile. Furthermore, the on- and off-state currents are reformulated as GP's inequalities in GP compatible constraints. The nonlinear doping profile optimization problem is transformed to a GP problem. The GP problem is transformed into a form of convex optimization problem [6, 8] and solved by using the interior-point algorithm in a global sense.

In Sect. 2, we formulate the problem. In Sect. 3, we show the numerical results. Finally, we draw the conclusions.

2 Problem Formulation and Solution Method

A GP problem can be characterized by objective and constraint functions with a certain special mathematical form [6, 8]:

$$\begin{aligned}
&\min_x f_0(x) = \sum_{t=1}^{u_0} c_{0t} \prod_{j=1}^{n} x_j^{a_{0tj}} \\
&\text{s.t } f_i(x) = \sum_{t=1}^{u_i} c_{it} \prod_{j=1}^{n} x_j^{a_{itj}} \leq 1,\ i = 1, 2, ..., m \\
&g_i(x) = \prod_{j=1}^{n} x_j^{b_{ij}} = 1,\ i = 1, 2, ..., q \\
&x_j \geq 0,\ j = 1, 2, ..., n
\end{aligned} \quad (1)$$

where the *posynomial* $f_0(x)$, containing u_0 terms, is the objective function, and the *posynomials* $f_i(x)$ for $i = 1, 2, \ldots, m$, containing u_i terms, represent m inequality constraints. By the definition of *posynomial*, all the coefficients c_{it} for $i = 0, 1, \ldots, m$ and $t = 1, 2, \ldots, u_m$ are positive, and the a_{itj} for $i = 0, 1, \ldots, m$, $t = 1, 2, \ldots, u_m$ and $j = 1, \ldots, n$ are real numbers. The $g_i(x)$ are monomial functions, where the b_{ij} for $i = 0, 1, \ldots, q$ and $j = 1, \ldots, n$ are real numbers. In the following subsections, we first write down the inverse doping profile problem and derive the object function as well as constraints in terms of the doping profile between the interface of semiconductor and oxide to the substrate. We next formulate the GP problem of 1-D doping profile for a MOSFET.

2.1 *The Inverse Doping Profile Problem*

For a set of given I-V specifications, the inverse doping profile problem can be modelled as

$$\begin{aligned}&\min SS\\&\text{s.t. } N_{min} \le N_A(x) \le N_{max},\ 0 \le x \le W_{dm}\\&I_{on} \ge I_{on-set}\\&I_{off} \le I_{off-set}\end{aligned}, \tag{2}$$

where the $N_A(x)$ is the 1-D non-uniform p-typed doping profile of n-typed MOSFETs, which is a positive function of space x ranging from the silicon/oxide interface to the maximum depletion width W_{dm}. The range of $N_A(x)$ has its lower bound of background doping level N_{min} and its upper bound of the maximum manufacturing doping concentration N_{max}. In the constraints of DC characteristics, I_{on} denotes the on-state current, I_{off} is the off-state current, and I_{on-set} and $I_{off-set}$ are the targeted specifications of I_{on} and I_{off}. For deriving the doping profile problem (2), by directly integrating the 1-D Poission equation [1], we have

$$2\psi_B = \frac{q}{\varepsilon_{si}} \int_0^{W_{dm}} x N_A(x) dx, \tag{3}$$

where $\varepsilon_{si} > 0$ is the silicon permittivity and $\psi_B > 0$ is the voltage difference between Fermi level and intrinsic level [1]. To obtain the maximum depletion width, we discretize (3) for the maximum depletion width with K uniformly spaced points, $x_j = jW_{dm} / K$, $j = 0, 1, \ldots, K$. The doping profile is then sampled at these points; we define $d_j = N_A(x_j)$, $j = 0, 1, \ldots, K$. The maximum depletion width W_{dm} can be further expressed as:

$$W_{dm} = K \sqrt{\frac{\varepsilon_{si} 2\psi_B}{q \sum_{j=0}^{K} j d_j}}, \tag{4}$$

which is a function of the discrete doping profile d_j.

2.2 The Subthreshold Swing

The subthreshold swing is calculated by the subthreshold current (I_{sub}) which is changed by the gate voltage (V_{gs}) variation of one-order magnitude [1]

$$SS \equiv \left[\frac{d(\log_{10} I_{sub})}{dV_{gs}}\right]^{-1} = 2.3\frac{mkT}{q} = 2.3\frac{kT}{q}\left(1 + \frac{3t_{ox}}{W_{dm}}\right), \tag{5}$$

where t_{ox} is the oxide thickness and m is the body-effect coefficient. Substituting the maximum depletion width W_{dm} of (4) into (5), then we can rewrite *SS* as:

$$SS = 2.3\frac{kT}{q}\left(1 + \frac{3t_{ox}}{K}\sqrt{\frac{q\sum_{j=0}^{K} jd_j}{2\varepsilon_{si}\psi_B}}\right). \tag{6}$$

The expression in (6) for *SS* satisfies the form of a *posynomial* since the coefficients of all the optimal variables d_j are positive.

2.3 The Constraint for the On-State Current

We assume the saturation current to be larger than the specification of the on-state current (I_{on-set}) when $V_{dd} = V_{ds}$ (in this work, the on-state current is defined as the targeted saturation current when the applied drain voltage V_{ds} is equal to the applied V_{dd}) and $V_{ds} = V_{gs} - V_t$, where V_{gs} is the applied gate voltage and V_t is the threshold voltage of device, we have [1]

$$\begin{aligned} I_{on} &= \frac{W}{2mL}\mu_{eff}C_{ox}\left(V_{gs} - V_t\right)^2 \\ &= \frac{W}{2mL}\mu_{eff}C_{ox}\left(V_{gs} - V_{fb} - 2\psi_B + \frac{Q_d}{C_{ox}}\right)^2 \geq I_{on-set} \end{aligned} \tag{7}$$

where L and W are device channel length and width, C_{ox} is the oxide capacitance per unit area, μ_{eff} is effective electron mobility, the V_{fb} is the flat-band voltage, and the depletion charge Q_d can be calculated by

$$Q_d = -q\int_0^{W_{dm}} N_A(x)dx. \tag{8}$$

Using a similar procedure of discretization in deriving (4), we find that

$$Q_d = -q\frac{W_{dm}}{K}\sum_{j=0}^{K} d_j. \tag{9}$$

We substitute (9) into (7) and have the following estimation

$$\frac{\sqrt{2q\psi_B\varepsilon_{si}}\sum_{j=0}^{K} d_j}{C_{ox}\left(V_{gs}-\sqrt{\frac{2mI_{on-set}}{\left(\frac{W}{L}\right)\mu_{eff}C_{ox}}}-V_{fb}-2\psi_B\right)} \leq \sqrt{\sum_{j=0}^{K} jd_j}. \tag{10}$$

Unfortunately, this inequality is not a GP compatible constraint. Because of the right-hand side of a *posynomial* inequality should be a constant or *monomial* and the body-effect coefficient is also a function of doping profile. For the body-effect coefficient m, we substitute the maximum of the manufacturing doping concentration N_{max} into the doping concentration d_j. Then m can be recalculated as m_{max} and then substituted into (10). Thus, we can ensure that for the minimum of I_{on} (according to (7), as m goes large, the I_{ds} will become small) $\geq I_{on-set}$, and the left-hand side of (10) will become a positive constant multiples a *posynomial* $\sum_{j=0}^{K} d_j$. For $\sum_{j=0}^{K} jd_j$, we use the arithmetic-geometric mean inequality to form a GP compatible approximation. Since the arithmetic mean is larger than geometric mean, we try

$$\frac{\sqrt{2q\psi_B\varepsilon_{si}}\sum_{j=0}^{K} d_j}{C_{ox}\left(V_{gs}-\sqrt{\frac{2m_{max}I_{on-set}}{\left(\frac{W}{L}\right)\mu_{eff}C_{ox}}}-V_{fb}-2\psi_B\right)} \leq \sqrt{(K+1)\left(\prod_{j=0}^{K} jd_j\right)^{\frac{1}{K+1}}}. \tag{11}$$

Therefore, if the inequality (11) holds, then the inequality (10) must also be satisfied. The inequality (11) is a GP compatible constraint, since the right-hand side is a *monomial* function with variables d_j.

2.4 The Constraint for the Off-State Current

The off-state current I_{off} is a special case of the subthreshold current I_{sub} when $V_{gs}=0$ and $V_{ds}=V_{dd}$. We assume the off-state current $I_{off} \leq I_{off-set}$, then [1]

$$I_{off}=\mu_{eff}C_{ox}\frac{W}{L}(m-1)\left(\frac{kT}{q}\right)^2\exp\left[\frac{q\left(-V_{fb}-2\psi_B+\frac{Q_d}{C_{ox}}\right)}{mkT}\right] \leq I_{off-set}. \tag{12}$$

We keep the exponential term at the left-hand side of (12), and replace m by $m_{max}>1$ to obtain the maximum I_{off} (according to (12), when m increase, the I_{off}

may consequently raise), and we can guarantee that for the maximum $I_{off} \geq I_{off-set}$. Therefore, (12) is expressed as

$$\exp\left[\frac{q\left(-V_{fb} - 2\psi_B + \frac{Q_d}{C_{ox}}\right)}{m_{max}kT}\right] \leq \frac{I_{off-set}}{\mu_{eff} C_{ox} \frac{W}{L}(m_{max} - 1)\left(\frac{kT}{q}\right)^2}. \quad (13)$$

The inequality (13) is not a GP compatible constraint because the *posynomial* function Q_d is in the exponential term. Therefore, we take the logarithm at both sides of (13) and rearrange the terms

$$Q_d \geq C_{ox}\left\langle \frac{m_{max}kT}{q} \ln\left\{\frac{\mu_{eff} C_{ox} \frac{W}{L}(m_{max} - 1)\left(\frac{kT}{q}\right)^2}{I_{off-set}}\right\} - V_f - 2\psi_B\right\rangle. \quad (14)$$

Substituting (9) into (14), we have

$$C_{ox}\sqrt{\frac{q\varepsilon_{si} 2\psi_B}{\sum_{j=0}^{K} jd_j}}\left\langle \frac{m_{max}kT}{q} \ln\left\{\frac{\mu_{eff} C_{ox} \frac{W}{L}(m_{max} - 1)\left(\frac{kT}{q}\right)^2}{I_{off-set}}\right\} - V_f - 2\psi_B\right\rangle \leq \sum_{j=0}^{K} d_j. \quad (15)$$

For obtaining the *posynomial* inequality, we again use the arithmetic-geometric mean inequality to transform the summation $\sum_{j=0}^{K} d_j$ at the right-hand side of (15). We come down to the following estimation

$$C_{ox}\sqrt{\frac{q\varepsilon_{si} 2\psi_B}{\sum_{j=0}^{K} jd_j}}\left\langle \frac{m_{max}kT}{q} \ln\left\{\frac{\mu_{eff} C_{ox} \frac{W}{L}(m_{max} - 1)\left(\frac{kT}{q}\right)^2}{I_{off-set}}\right\} - V_f - 2\psi_B\right\rangle \leq (\mathrm{K}+1)\left(\prod_{j=0}^{K} d_j\right)^{\frac{1}{K+1}}. \quad (16)$$

Thus, if the inequality (16) holds, then the inequality (15) is also achieved. The inequality (16) is a *posynomial* inequality, since the right-hand side is a *monomial* function and the left-hand side is a *posynomial* function with the variables d_j.

2.5 The Formulated GP Problem

Based on the above estimations, the 1-D optimal inverse doping profile problem in the n-type MOSFETs is expressed as

$$\min \left\{ 2.3\frac{kT}{q}\left(1+\frac{3t_{ox}}{K}\sqrt{\frac{q\sum\limits_{j=0}^{K} jd_j}{2\varepsilon_{si}\psi_B}}\right)\right\}$$

s.t $N_{min} \le d_j \le N_{max},\ j = 0,\ 1, \ldots,\ K.$

$$\frac{\sqrt{2q\psi_B\varepsilon_{si}}\sum\limits_{j=0}^{K} d_j}{C_{ox}\left(V_{gs}-\sqrt{\frac{2m_{max}I_{on-set}}{\left(\frac{W}{L}\right)\mu_{eff}C_{ox}}}-V_{fb}-2\psi_B\right)} \le \sqrt{(K+1)\left(\prod_{j=0}^{K} jd_j\right)^{\frac{1}{K+1}}},$$

$$C_{ox}\sqrt{\frac{q\varepsilon_{si}2\psi_B}{\sum\limits_{j=0}^{K} jd_j}}\left\langle \frac{m_{max}kT}{q}\ln\left\{\frac{\mu_{eff}C_{ox}\frac{W}{L}(m_{max}-1)\left(\frac{kT}{q}\right)^2}{I_{off-set}}\right\}-V_f-2\psi_B\right\rangle$$

$$\le (\mathrm{K}+1)\left(\prod_{j=0}^{K} d_j\right)^{\frac{1}{K+1}} \tag{17}$$

which is a nonlinear constrained optimization problem and is also a GP with variables $d_j, \forall j = 0,\ 1, \ldots,\ K$. Next we transfer the GP problem into a convex optimization problem. Also the corresponding dual problem can be formulated [8]. After obtaining the prime and dual problems of the GP problem in the convex form [8], we apply the logarithmic barrier function transformation to convert the constrained optimization problem into an unconstrained one [7]. Finally we employ a general search algorithm such as gradient or a Newton-method to solve this unconstrained optimization and the original solution can be inversely obtained.

3 Results and Discussion

We now test a numerical example for the optimal doping profile problem (17) for the MOSFET with a 0.35 μm device channel length and 1 μm device width.

For low-standby power (LSP) devices, the high threshold voltage to suppress the standby-power consumption due to leakage current is necessary, and therefore, we set the specification of on- and off-state currents as: $(I_{off-set}, I_{on-set}) = (5 \times 10^{-13}\ \text{A}, 1 \times 10^{-4}\text{A})$. For high performance (HP) devices, it requires low threshold voltage to increase switching speed, and the specification of HP devices is taken as: $(I_{off-set}, I_{on-set}) = (1 \times 10^{-9}\ \text{A}, 3 \times 10^{-4}\text{A})$. The optimized doping profiles for the HP (solid line) and LSP (dash line) MOSFETs are shown in Fig. 1. The corresponding I-V curves for the HP and LSP devices are shown in Fig. 2. For the LSP device, the threshold voltage is about 0.44 V, and the off-state current is about 1×10^{-13} A, which is smaller than 5×10^{-13} A; on the other hand, the HP

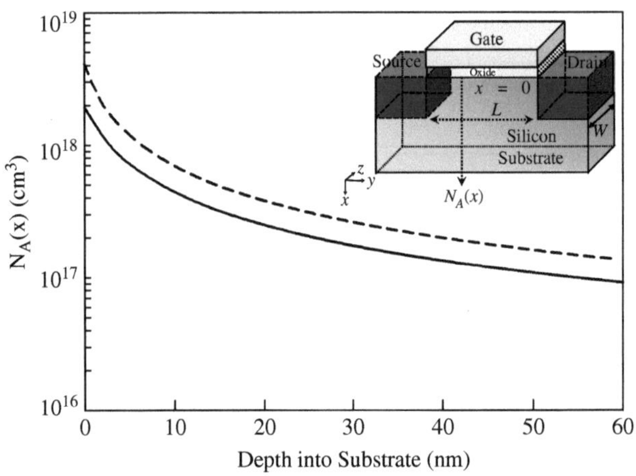

Fig. 1 The optimized doping profiles for the HP (solid line) and the LSP (dash line) MOSFETs

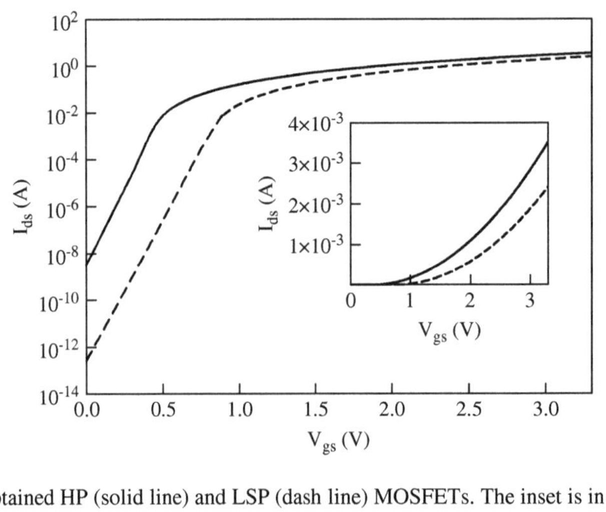

Fig. 2 The obtained HP (solid line) and LSP (dash line) MOSFETs. The inset is in linear plot

device has a low subthreshold swing, and the threshold voltage is equal to 0.13 V, which has a high switching speed as we expected. As shown in Fig. 1, the higher doping concentration near the surface contributes to a higher threshold voltage, and decrease the off-state current for the LSP device.

4 Conclusions

The 1-D inverse doping profile problem for MOSFETs was formulated and transformed into a GP problem. This study can be extended into 2-D and 3-D optimization problems. For application to deep submicron devices, channel doping profile optimization with considering the short channel and quantum mechanical effects is necessary.

Acknowledgements This work was supported in part by Taiwan National Science Council (NSC) under Contract NSC-97-2221-E-009-154-MY2 and NSC-99-2221-E-009-175.

References

1. Sze, S.M.: Physics of Semiconductor Device, 2nd edn., Wiley, New York (1981)
2. Djomehri, I.J., Antoniadis, D.A.: Inverse Modeling of Sub-100 nm MOSFETs Using I-V and C-V. IEEE Trans. Electron Dev. **49**, 568–575 (2002)
3. Li. Y., Yu, S.M.: A coupled-simulation-and-optimization approach to nanodevice fabrication with minimization of electrical characteristics fluctuation. IEEE Trans. Semiconductor Manufacturing **20**, 432–438 (2007)
4. Leitao, A., Markowich, P.A., Zubelli, J.P.: On inverse doping profile problems for the stationary voltage-current map. Inverse Problems **22**, 1071–1088 (2006)
5. Lahaye, D.J.P., Drago, C.R.: Exploiting model hierarchy in semiconductor design using manifold mapping. In: Roos, J., Costa, L. (eds.) Scientific Computing in Electrical Engineering SCEE 2008, vol. 20, pp. 445–452 (2010)
6. Boyd, S., Kim, S.J., Vandenberghe, L., Zubelli, J.P.: A tutorial on geometric programming. Opt. Eng. **8**, 67–127 (2007)
7. Kortanek, K., Xu, X., Ye, Y.: An infeasible interior-point algorithm for solving primal and dual geometric programs. Math. Prog. **76**, 155–181 (1996)
8. Boyd, S., Vandenberghe, L.: Convex Optimization, Cambridge, UK (2004)

Numerical Simulation of Semiconductor Devices by the MEP Energy-Transport Model with Crystal Heating

Vittorio Romano and Alexander Rusakov

Abstract A new numerical model of semiconductors including crystal heating effect is presented. The model equations have been obtained with the use of the maximum entropy principle. In the numerical model the iterative scheme is used for obtaining stationary solution of electro-thermal problems. Numerical simulations of a 2D nanoscale MOSFET with the self-heating effect are presented. The difference between MEP and simpler thermal expressions is analyzed.

1 Introduction

In today semiconductor technology, the miniaturization of devices is more and more progressing. As a consequence, the simulation of the modern nanoscale semiconductor devices requires advanced transport models that take into account heating effects. The analysis of device self-heating becomes more and more important for the design and validation of devices and circuits. In this paper a numerical integration of a new coupled electro-thermal model for semiconductors, developed recently in [1], is performed. At macroscopic level, several different heuristic models of lattice heating have been proposed. They are represented by the lattice energy balance equation and differ from the proposed form of thermal conductivity and energy production, e.g. [2, 3]. In [1] a macroscopic model which has been formulated starting from the semiclassical description based on the Boltzmann

V. Romano (✉)
Dipartimento di Matematica e Informatica, Università di CataniaViale A.Doria 6, I-95125 Catania, Italy
e-mail: romano@dmi.unict.it

A. Rusakov
Institute for Design Problems in Microelectronics, Sovetskaya Street 3, Moscow, Russian Federation
e-mail: rusakov@inm.ras.ru

B. Michielsen and J.-R. Poirier (eds.), *Scientific Computing in Electrical Engineering SCEE 2010*, Mathematics in Industry 16, DOI 10.1007/978-3-642-22453-9_38,

equations describing the electron-phonon system. The closure relations have been obtained with the maximum entropy principle (hereafter MEP). Explicit constitutive relations have been obtained with coefficients depending on the electron energy W and crystal temperature T_L and related to the scattering parameters (see [1, 4, 7] for more details). The case of varying lattice temperature has been numerically integrated for 1D problems in [4, 5] and for the 1D model obtained with MEP in the [6]. Here we present 2D numerical model, without radiative term which does not have clear physical meaning especially in 2D case. In comparison to [7,8] in our work we enhance our numerical model with a new iterative scheme for the efficient solution of the stationary problem. In the numerical experiment section we present and analyze solutions for a nanoscale semiconductor device with the heating source term described by the MEP and by the Joule expressions.

2 Mathematical Model

The model is represented by the system of the equations

$$\frac{\partial n}{\partial t} + \mathrm{div}\,(n\,\mathbf{V}) = -R, \tag{1}$$

$$\frac{\partial p}{\partial t} + \mathrm{div}\left(p\,\mathbf{V_p}\right) = -R, \tag{2}$$

$$\frac{\partial\,(nW)}{\partial t} + \mathrm{div}\,(n\,\mathbf{S}) + nq\mathbf{V}\cdot\nabla\phi = nC_W, \tag{3}$$

$$\rho c_V \frac{\partial T_L}{\partial t} - \mathrm{div}\,[K(T_L)\nabla T_L] = H, \tag{4}$$

$$\mathbf{E} = -\nabla\phi, \quad \varepsilon\Delta\phi = -q(N_D - N_A - n + p). \tag{5}$$

n and p are the electron and hole density respectively, W is the electron energy, T_L the lattice temperature, ϕ the electrostatic potential and $\mathbf{E} = -\nabla\phi$ the electric field. N_D and N_A are the donor and acceptor density respectively. q is the elementary charge, ρ the silicon density, c_V the specific heat, C_W the energy production term, which is in a relaxation form $C_W = -\dfrac{W - W_0}{\tau_W}$, with $W = 3/2k_BT_L$ and $\tau_W(W)$ the energy relaxation time. k_B is the Boltzmann constant and ε is the dielectric constant. R is the generation-recombination term.

The closure relations for the electron velocity $\mathbf{V}$, the energy flux $\mathbf{S}$, the thermal conductivity $K(T_L)$ and the crystal energy production term H have been obtained in [1] by employing MEP. The holes are described by a standard drift-diffusion model with constant mobility. $\mathbf{V_p}$ is the hole velocity. The expressions of the electron velocity $\mathbf{V}$ and the energy flux $\mathbf{S}$ are given in [1]

The phonon energy production is given by

$$H = -(1 + P_S)\, n\, C_W + P_S\, \mathbf{J} \cdot \mathbf{E}, \tag{6}$$

where $P_S = -c^2\, \tau_R\, c_{12}^{(p)}$ plays the role of a thermopower coefficient. On source and drain contacts the Robin boundary condition $-k_L \frac{\partial T_L}{\partial n} = R_{th}^{-1}(T_L - T_{env})$ is assumed, R_{th} being the thermal resistivity of the contact and T_{env} the environment temperature. We use no-flux condition for temperature on the lateral boundary and oxide silicon interface and Dirichlet condition on the bulk contact. The electron energy on the source, drain and bulk contact is set equal to the lattice energy. Other boundary conditions in the MOSFET model are described in detail in [9].

3 The Numerical Method

Direct integration of the numerical model for the temperature relaxation period becomes very expensive procedure as the time step of integration scheme is limited by the properties of the electrical part of the system of equations. To make our numerical simulation feasible we apply a variant of the iterative scheme. At each iteration we first integrate only the electron part of the model until it achieves stationary solution and then integrate the crystal lattice temperature diffusion equation with known source for the period needed to achieve its stationary solution. By using different integration scheme for the thermal and electron part of the model with different time steps, we achieve a significant speed-up in the computation. The iterative scheme can be written as follows:

1. **do while** $||(\mathscr{U}, T)^k - (\mathscr{U}, T)^{k-1}|| < \varepsilon$
2. Integrate the balance equations for electrons and holes, with the crystal lattice energy, frozen at the previous time step, obtaining the electron and hole density, electrical field and energy at the next time step. Schematically this step can be written as

$$\frac{\partial \mathscr{U}^k}{\partial t} + F(\mathscr{U}^k, T_L^{k-1}) = 0, \tag{7}$$

   with $\mathscr{U} = (n, p, W, \phi)$, where $k = 1, \ldots, N$ is the index of the iteration, $t \in [0, t_k]$, t_k integration period.
3. Integrate the lattice energy balance equation with n and W given by the step 1 until convergence is reached:

$$\rho c_V \frac{\partial T_L^k}{\partial t} - \operatorname{div}\left[K(T_L^k)\nabla T_L\right] = H(\mathscr{U}^k, T_L^k). \tag{8}$$

4. **set** $k = k + 1$
5. **end do**

We observed that in our numerical experiments usually only 3–5 iterations are required to obtain steady-state solution. Typically the time step for integration of (8) we can use is 100 times larger than the time step for (7). Thus the total cost of the solution of the steady state electro-thermal system is only 3-5 times larger than solution of electron part only. If we use direct integration of the (1)–(5) till the stationary solution we have to choose an integration period approximately 100–500 times larger than for electron part only. In our numerical experiment a convergence of the electron-phonon model is reached after 1500 and 5 ps for electron part only. Thus the stationary solution can be computed with the iterative scheme in 50 times faster.

For the transient analysis of the model system of equations the multi-rate integration technique has been used [8] which similar to approach presented in [11].

The numerical scheme for solution of electrical part is based on an exponential fitting like that employed in the Scharfetter-Gummel scheme for the drift-diffusion model of semiconductors. The basic idea is to split the particle and energy density currents as the difference of two terms. Each of them is written by introducing suitable mean mobilities in order to get expressions of the currents similar to those arising in other energy-transport models known in literature. A simple explicit discretization in time with constant time step proves satisfactorily efficient avoiding the problem related to the high nonlinear coupling of the discretized equations. The model equations are spatially discretized on a regular grid. The details of the numerical scheme can be found in [9].

To solve a lattice energy equation (8) a coordinate splitting technique [12] is used. For the space approximation in every direction an implicit time scheme with the three points stencil is chosen. The obtained linear system can be solved efficiently with tridiagonal matrix factorization procedure. We remark that usage of the implicit time scheme for lattice energy operator significantly improves the overall simulation time. In 1D diode model usage of implicit integration scheme gives a speed up of 100x of the total simulation time and we expect the similar effect in the MOSFET model.

4 Numerical Experiments

In the numerical experiments we consider the simulation of a 50 nm channel length MOSFET (Fig. 1). The gate length is 45 nm, source and drain are 25 nm long. The source and drain depths are 0.1 μm. The gate oxide is 5 nm thick. The substrate thickness is 0.4 μm. The environment temperature T_{env} is 300 K. In our numerical experiments we take the thermal resistivity $R_{th} = 10^{-8}$ K m^2 /W as in [5]. The doping concentration is

$$N_D(x) - N_A(x) = \begin{cases} 10^{17} cm^{-3} \text{ in the } n^+ \text{ regions} \\ -10^{14} cm^{-3} \text{in the } p \text{ region} \end{cases} \tag{9}$$

with abrupt junctions. The gate voltage is $V_{DG} = 0.8$ V.

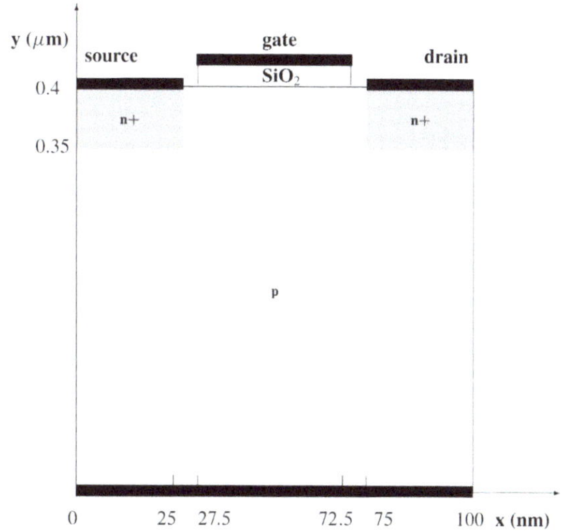

Fig. 1 Schematic representation of a bidimensional MOSFET

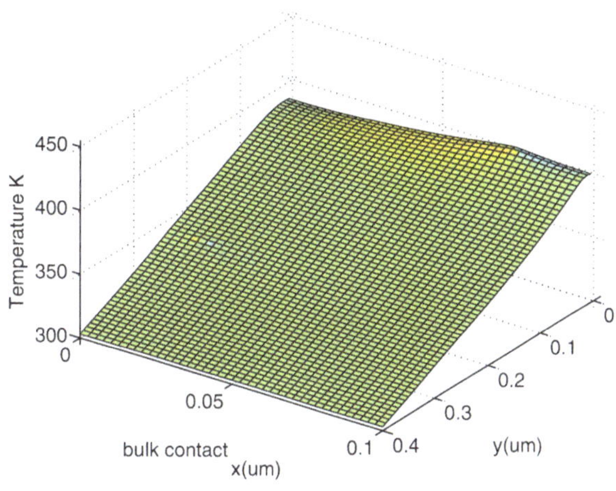

Fig. 2 Stationary solution for lattice temperature distribution in the considered MOSFET

In Fig. 2 the crystal lattice temperature obtained with MEP model for the MOSFET with is shown. In Fig. 3 electron energy is reported. The lattice temperature raises by more than 100 K in the area near the gate where there is the maximum for the electron energy. Such a strong heating effect could damage the device or change dramatically its performance.

For comparison, the simulation with only the Joule term $H = \mathbf{J} \cdot \mathbf{E}$ gives us much lower temperature raise, about 4°, as it shown in Fig. 4. The relevance of the Joule term in the MEP model is accounted for by the coefficient P_S in (6) behind the scalar product of the current and electrical field. In the pure Joule model it is about 20 times smaller than the value of P_S obtained by the MEP. Usually in the

Fig. 3 Stationary solution for electron energy in the considered MOSFET

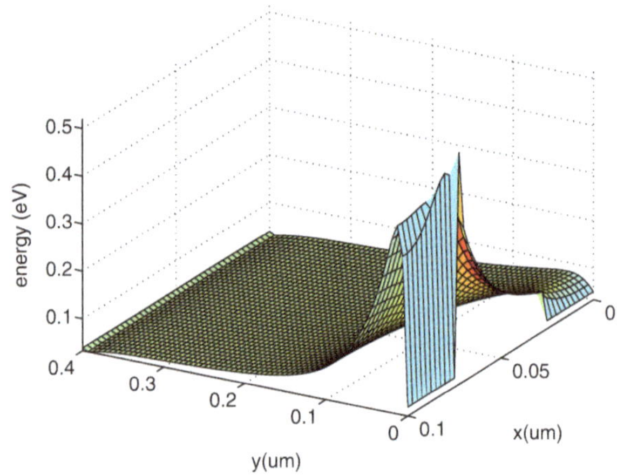

Fig. 4 Stationary solution for lattice temperature distribution computed with the pure Joule effect

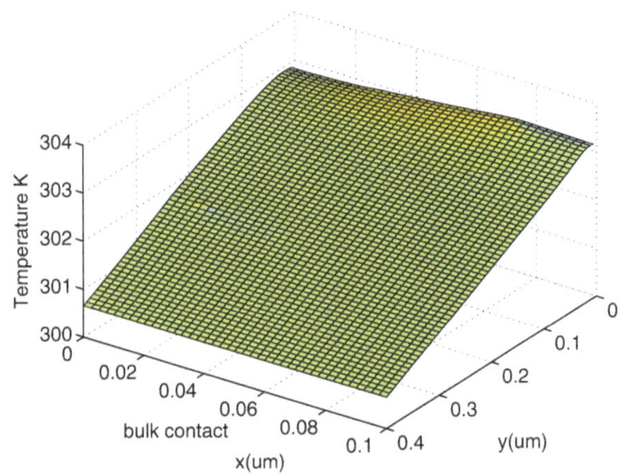

simulations this coefficient is chosen using empirical approaches but MEP gives us a way to obtain an explicit expression for this coefficient which is crucial for the electro-thermal analysis of the semiconductor devices.

References

1. Romano, V., Zwierz, M.: Electron-phonon hydrodynamical model for semiconductors. ZAMP **61**(6), 1111–1131 (2010), DOI: 10.1007/s00033-010-0089-9
2. Gaur, S.P., Navon, D.H.: Two-dimensional carrier flow in a transistor structure under non-isothermal conditions. IEEE Trans. Electron Dev. **ED-23** 50–57 (1976)
3. Wachutka, G.: Rigorous thermodynamic treatment of heat generation and conduction in semiconductor device modeling. IEEE Trans. Computer-Aided Des. **9**, 1141–1149 (1990)
4. Romano, V, Scordia, C.: Simulations of an electron-phonon hydrodynamical model based on the maximum entropy principle. Proceedings of SCEE 2008

5. Brunk, M, Jüngel, A.: Numerical coupling of electric circuit equations and energy-transport models for semiconducotrs. SIAM J. Sci. Comput. **30**, 873–894 (2008)
6. Stefano, V.: Modeling thermal effects in submicron semiconductor devices. CAIM 1(1), 110–127 (2010)
7. Romano, V, Rusakov, A.: 2d numerical simulations of an electronphonon hydrodynamical model based on the maximum entropy principle. Comput. Meth. Appl. Mech. Eng. **199**(41–44), 2741–2751 (2010), doi:10.1016/j.cma.2010.06.005
8. Romano, V., Rusakov, A.: Numerical simulation of coupled electron devices and circuits by the MEP hydrodynamical model for semiconductors with crystal heating. Nuovo Cimento C **33**(01), pp. 223–230 (2010)
9. Romano, V.: 2D numerical simulation of the MEP energy-transport model with a finite difference scheme. J. Comp. Phys. **221**, 439–468 (2007)
10. Selberherr, S.: Analysis and simulation of semiconductor devices. Springer, Wien – New York (1984)
11. Bartel, A., Guenther, M.: Multirate co-simulation of first order thermal models in electric circuit design. In: Schilders W., et al. (eds.) Scientific Computing in Electrical Engineering. Proceedings SCEE 2002, Springer, Berlin, pp. 23–28 (2002)
12. Marchuk, G.I.: Splitting and Alternating Direction Method. In: Handbook of Numerical Analysis, vol. 1, pp. 197–462 (1990)

Part V
Model Order Reduction

Introduction

In spite of the impressive data processing capacities of today's computers, it remains worthwhile to limit the number of degrees of freedom in the numerical models we use, and for a number of reasons. Of course, making models with a reduced number of unknowns allows one to handle problems of even greater complexity at the same cost. Reducing the computation time to the minimum also opens the possibility to analyse many different complex configurations in a reasonable time and thus to perform optimisations or sensitivity and statistical analysis on complex configurations.

Model order reduction (MOR) methods are designed to reduce the computational complexity of the numerical representation of a model while maintaining the accuracy needed in a given class of applications. The methods presented in this part are partly inspired by problems arising from the development of electronic circuits with a large amount of components. However, the techniques presented for coping with the huge number of degrees of freedom found in this context are also of interest in other situations. Of course, model order reduction can also be directly adapted to discretisations of a system of partial differential equations and examples of this are also presented in this part.

The sequence of the papers in this part shows a gradual evolution from circuit problems to field computation problems. Along the way we find efficient input-output modelling techniques, numerical algebra-inspired reduction techniques and network problems in macroscopic electromagnetics, where subdomain decompositions on an intermediate level (i.e., not down to the finite element level) require efficient models of extended domains considered as a "black box."

The first paper was written by J. Rommes (an invited speaker at the conference) and pesents an overview of some of the challenging problems in applied mathematics arising in the electronics industry. It lists the essential techniques of Model Order Reduction (MOR) but also shows directions of research that should be explored to solve the difficulties yet to be overcome.

The following three papers consider MOR methods for network problems. The paper by M. Ugryumova, J. Rommes and W. Schilders presents a method for dealing with large resistor networks representing interconnect systems including parasitic effects. This paper, continuing previous work by the same research group, describes a significant improvement in the model reduction while at the same time maintaining good accuracy. The paper by P. Miettinen et al. presents a method for ensuring that the model order reduction of networks with resistors, inductors and capacitors (RLC networks) does not result in a singular system matrix by excluding from the reduction the parts of the network identified beforehand as being responsible for such singularities. Their method is applicable to any netlist-in/netlist-out type of MOR method. The next paper, by the same authors, shows how their method can be extended to networks that also have mutual inductances (RLCM networks) such that the resulting reduced model is also an RLCM network.

Instead of reducing the order of a detailed model, one may also want to reduce the model to a black box model. The paper by M. Striebel and J. Rommes presents such a method for nonlinear systems. Although the starting point is again a network problem, the model order reduction now eliminates all "internal" state variables and only retains an efficient and accurate representation of the input-output relations. The paper by F. Yetkin and H. Dağ arrives at an efficient representation of a system's input-output relation by using some cleverly chosen points in the Fourier spectral representation of the system's transfer function. The equations one has to solve for finding the appropriate spectral information are provided and the method is shown to compare well with known methods both with regard to accuracy and to computational efficiency.

M. Hinze and M. Kunkel's paper examines the model order reduction of a network problem of the internals of an integrated circuit coupled to the partial differential equations for drift diffusion of charges in the device. The paper by A. Lutowska, M. Hochstenbach and W. Schilders presents a general method concerned with the modelling of a configuration of interconnected sub-systems. The sub-systems including the interconnection system all get their own reduced representation. In this way, the original block structure corresponding to the decomposition in sub-systems is preserved, including the zero blocks.

In the final paper of this part, K. Stavrakakis et al. put forward an MOR method to make efficient models for configurations in which parametric studies of three-dimensional configurations are of interest. The particular case of a filter configuration in which the electromagnetic field problem is discretised using the Finite Integration Technique is examined. The analysis of the filter impedance's sensitivity to geometrical design parameters is made practicable by using MOR in the underlying electromagnetic field problem.

Challenges in Model Order Reduction for Industrial Problems

Joost Rommes

Abstract Mathematical challenges arise in many applications in the electronics industry. Device and circuit simulation are well-known examples, and in industry these are typically crucial for circuit and layout optimization. Model order reduction is one of the available tools, and we show when and how, and when not, to use this. We will give an overview of the challenges we are facing, explain how we try to conquer these, discuss the requirements we have to deal with, and indicate where improvements are needed.

1 Introduction

The increasing demand for smaller, faster, and multi-functional electronic devices such as smart phones is one of the driving forces in semiconductor industry. Combined with requirements on power usage (low power), sustainability, and wireless functionality (RF) this is generating many challenges on several domains. There are not only many electrical design challenges; especially the increased complexity of electronic designs, under pressure by reduced time-to-market (vital to survive over your competitors), leads to mathematical challenges. These challenges arise at all levels in the design flow: at device and circuit level very large systems have to be simulated, while at system level one has to guarantee a certain overall performance, also taking into account external electro-magnetic effects [3]. Furthermore, testing final products requires state-of-the-art statistical methods [9].

In the following sections some mathematical challenges arising in the electronics industry will be described, together with the typical industrial requirements that

J. Rommes (✉)
NXP Semiconductors/Central R&D, Eindhoven, The Netherlands
e-mail: joost.rommes@nxp.com; http://www.nxp.com; http://sites.google.com/site/rommes

B. Michielsen and J.-R. Poirier (eds.), *Scientific Computing in Electrical Engineering SCEE 2010*, Mathematics in Industry 16, DOI 10.1007/978-3-642-22453-9_39,

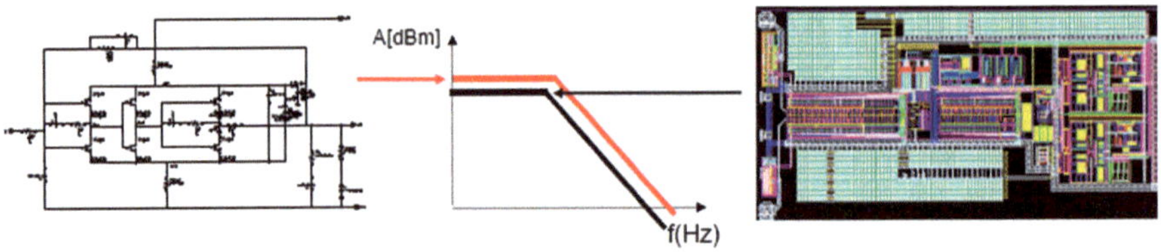

Fig. 1 The challenge in RF integrated circuit design: how to close the gap between schematic and layout simulations?

make solutions less straightforward in general. Model order reduction (MOR) [2,16] is illustrated as one of the possible solutions, but we also show that this is not always the ideal solution. We conclude with mentioning some open challenges.

2 Challenges

Many challenges in RF-analog mixed-signal design are related to the gap between simulation at schematic level and simulation at layout level, as depicted in Fig. 1. Whereas the schematic is a high level description of the design, typically assuming ideal elements and devices, the layout is a representation of the design in shapes in the physical layers (silicon, oxide, metal). Simulation of the latter is more accurate, but also more complex, due to the inclusion of several physical effects, and as a result designers will observe a gap between schematic and layout simulation. Closing this gap, by making changes to the layout, is a laborious task. As an example two related challenges are described in more detail.

2.1 Extracted Parasitics

Although big parts of modern chips are digital, still a crucial part is analog (hence *analog mixed-signal*). The analog part is for instance responsible for the conversion of an analog RF signal into a digital signal [11]. In terms of devices (e.g., transistors), the analog part may not be large, especially not when compared to the digital part (hundreds of devices vs. millions of gates). From a simulation perspective, however, the analog part is much more challenging. Simulating thousands or even millions of devices is possible, at schematic level, but eventually one has to verify the physical layout, which requires simulation including device and interconnect parasitics. Depending on the amount of interconnect and the type of extraction (translation of the layout into a form that can be simulated by a circuit simulator), thousands to millions of RLC elements and nodes can be added to the original network. Although the parasitics are linear elements in general, they do not only put a heavy demand on CPU and memory requirements, they may also influence convergence of typical RF simulations such as periodic steady-state (PSS).

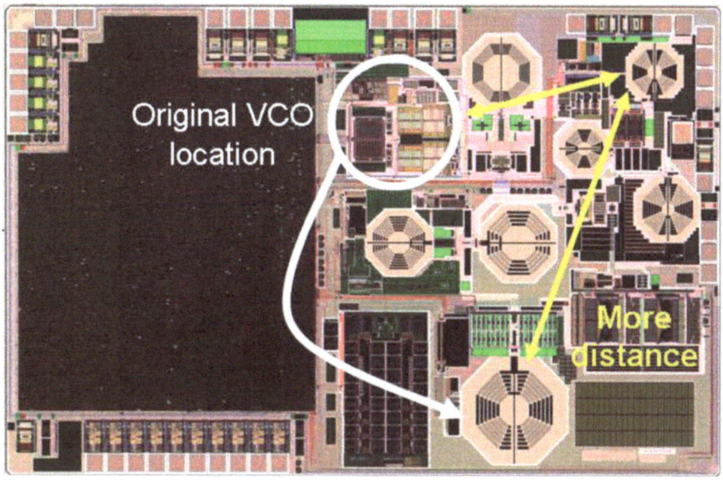

Fig. 2 Floor plan with relocation option that was considered after nonlinear phase noise analysis showed an intolerable pulling due to unintended coupling. See [6] for more details

2.2 Layout Optimization

While circuit design is usually done top-down, verification is done bottom-up. Components in the circuit, such as low-noise amplifiers and mixers are first verified individually, and at a level higher one has to verify the combined operation of these components. Again, increased complexity makes simulation a major challenge, not only due to parasitics but also due to unintended coupling of for instance oscillators [6]. What is needed in this case are models that accurately predict the phase-noise behavior of the coupled oscillators, so that, as shown in Fig. 2, an improved floor plan can be made. In practice, however, this is not always sufficient. As described above, parasitics can influence circuit performance considerably, and since parasitic effects are hard to predict, typically many layout iterations and simulations are needed. Hence, efficient ways to deal with (repeated) parasitic extraction, a mathematical challenge itself, are required.

3 Solutions

3.1 Model Order Reduction

Although electric circuits contain nonlinear devices such as transistors and hence are modeled by nonlinear dynamical systems (differential-algebraic equations) [3], at the core of many simulation types, such as transient and PSS, large-scale *linear* systems have to be solved. Furthermore, parasitics are usually modeled by linear RLC elements, resulting in linear dynamical subsystems. These linear subsystems can be separated from the nonlinear subsystems by advanced graph algorithms and reduced by MOR methods [16]. One approach (for resistor-only circuits) is to eliminate selected internal unknowns from a linear system by using the Schur complement [15]:

$$\begin{bmatrix} G_{11} & G_{12} \\ G_{12}^T & G_{22} \end{bmatrix} \begin{bmatrix} \mathbf{v} \\ \mathbf{u} \end{bmatrix} = \begin{bmatrix} B \\ 0 \end{bmatrix} \mathbf{i} \Leftrightarrow (G_{11} - G_{12} G_{22}^{-1} G_{12}^T)\mathbf{v} = B\mathbf{i},$$

where $G \in \mathbb{R}^{N \times N}$, $\mathbf{v}, \mathbf{i} \in \mathbb{R}^n$, $\mathbf{u} \in \mathbb{R}^m$, $B \in \mathbb{R}^{n \times n}$, $N = n + m$, and G_{ij}, 0 have appropriate dimensions. The effectiveness depends on which nodes are eliminated — we use graph and node reordering algorithms [1] for this selection. An example is shown in Fig. 3, where the preservation of five additional internal nodes reduces the number of resistors in the reduced network by 50%. The original network has 59 terminals, 2,300 internal nodes and 3,683 resistors. Full elimination of all internal nodes results in 1,711 resistors, keeping five internal nodes leaves just 721 resistors. For more details see [15].

An important aspect here is that if accurate reduction of parasitic extracted networks is possible, one would actually expect that the extraction itself could already have been more efficient. Furthermore, one might wonder why, with the availability of many sparse direct and iterative solvers, reduction is needed anyway. One reason is that industrial software may be affected by historical choices for datastructures and algorithms, and that limited capacity is available for software reengineering.

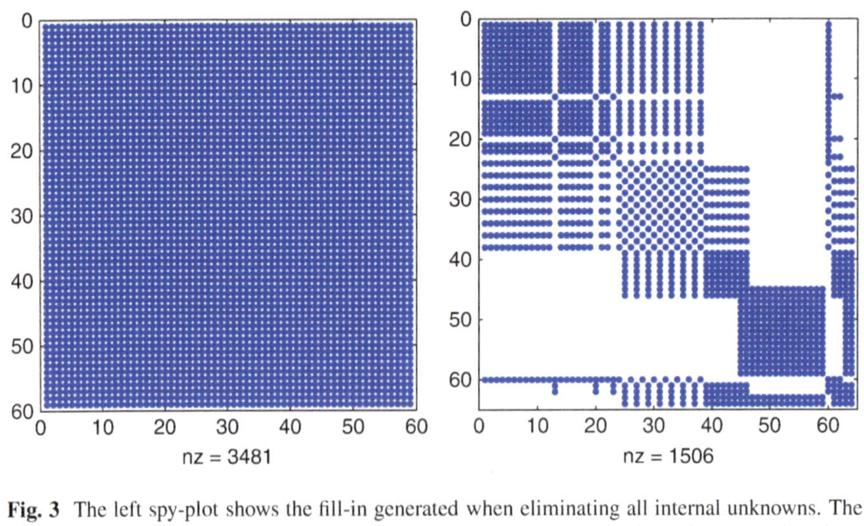

Fig. 3 The left spy-plot shows the fill-in generated when eliminating all internal unknowns. The right plot shows the result when five additional internal unknowns are kept. By preserving these unknowns the number of resistors is reduced by more than 50%

3.2 Exploit the Structure of the System

If one has knowledge of the underlying system structures, one may benefit from this. Consider the system

$$\begin{cases} E\dot{\mathbf{x}}(t) = A\mathbf{x}(t) + B\mathbf{u}(t) \\ \mathbf{y}(t) = C^T\mathbf{x}(t) + D\mathbf{u}(t) \end{cases}$$

where $\mathbf{x}(t) \in \mathbb{R}^n$ is the state vector, $\mathbf{u}(t) \in \mathbb{R}^m$ the input vector and $\mathbf{y}(t) \in \mathbb{R}^p$ the output vector; $E, A \in \mathbb{R}^{n\times n}$ are the descriptor and state matrix, and $B \in \mathbb{R}^{n\times m}$, $C \in \mathbb{R}^{n\times p}$ and $D \in \mathbb{R}^{p\times m}$ are system matrices.

The controllability gramian $P \in \mathbb{R}^{n\times n}$ [2] of state-space system ($E = I, A, B, C, D$) can be found by solving the Lyapunov equation

$$AP + PA^T = -BB^T.$$

The alternating direction implicit (ADI) method is a well known way to solve Lyapunov equations, and in the past decades efficient schemes [8, 10, 19] have been developed that exploit the fact that $P = P^T > 0$. However, applicability to large scale systems is still a challenge, because the main operation in the ADI process is the solution of linear systems

$$(A + \sigma_i I)X_i = Y_i,$$

where it must be noted that the possibly complex shifts σ_i and the right hand sides change per iteration. If A is large and sparse, one can use fill-in minimizing matrix reordering algorithms such as approximate minimum degree (AMD) [1] to limit the costs for these solves. Matters become more complicated if the original system is in descriptor form ($E \neq I, A, B, C, D$), since then one has to solve the generalized Lyapunov equation

$$APE^T + EPA^T = -P_l BB^T P_l^T,$$

where $P_l \in \mathbb{R}^{n\times n}$ are *spectral projectors* onto the deflating subspaces. There are methods that can solve this equation, see e.g., [17]. However, these require the computation of P_l and this may be unfeasible for large-scale systems. Fortunately, for systems with a certain structure, one can circumvent the computation of the spectral projectors. Suppose the structure of the descriptor system matrices is as follows

$$\begin{cases} \dot{\mathbf{x}}(t) = J_1\mathbf{x}(t) + J_2\mathbf{z}(t) + B_1\mathbf{u}(t) \\ 0 = J_3\mathbf{x}(t) + J_4\mathbf{z}(t) + B_2\mathbf{u}(t) \\ \mathbf{y}(t) = C_1^T\mathbf{x}(t) + C_2^T\mathbf{z}(t) + D_a\mathbf{u}(t) \end{cases}$$

where $\mathbf{x}(t) \in \mathbb{R}^n$, $\mathbf{z}(t) \in \mathbb{R}^{n_z}$ and $J_1, J_2, J_3, J_4, B_1, B_2, C_1, C_2, D_a$ have appropriate dimensions. Clearly, the state-space and descriptor representations are related via

$$A = J_1 - J_2 J_4^{-1} J_3, \quad B = B_1 - J_2 J_4^{-1} B_2,$$
$$C^T = C_1^T - C_2^T J_4^{-1} J_3, \quad D = D_a - C_2^T J_4^{-1} B_2.$$

So one could consider transforming the descriptor system into a state-space system and apply the standard ADI scheme. The problem here is that A will be dense in general and hence solves $(A_s + \sigma I)X = Y$ will be very expensive. Now, for implicitly solving systems of the form $(A_s + \sigma I)\mathbf{x} = \mathbf{b}$ with $A_s = J_1 - J_2 J_4^{-1} J_3$ (A_s is the Schur complement of J_4 in A) we note that [4,7]

$$\begin{bmatrix} J_1 + \sigma I & J_2 \\ J_3 & J_4 \end{bmatrix} \begin{bmatrix} X \\ * \end{bmatrix} = \begin{bmatrix} Y \\ 0 \end{bmatrix} \Longleftrightarrow (A_s + \sigma I)X = Y$$

Hence we can solve the linear equations for the state-space formulation implicitly by solving the sparse equations in terms of descriptor matrices. This is beneficial for two reasons. First, solves with the sparse descriptor matrices are in general much cheaper than (construction of and) solves with the dense state space matrices. Secondly, the ADI low-rank approximations of P, are still constructed in state space. In other words, we do not have to compute the deflating subspaces since we work in state space — implicitly when solving linear systems, and explicitly when constructing low rank approximations.

This approach has been applied successfully to MOR problems in fluid dynamics [7] and power systems [4]. Furthermore, this insight can be used in any process that needs the solution of linear systems, such as transient simulation and eigenvalue computations.

3.3 Eigenvalues and Model Order Reduction

A problem that often occurs when reducing linear systems with Krylov based projection methods such as the Arnoldi method [5], is that certain peaks in the frequency response plot of $H(s) = C^T(sE - A)^{-1}B + D$ are not matched, see Fig. 4. The problem is caused by the convergence behavior of the Arnoldi method: the eigenvalue approximations, or Ritz values, tend to approximate the eigenvalues at the outside of the spectrum [18] (and even worse, may not have converged to eigenvalues of the original system and hence cause wrong peaks). This can also be seen in Fig. 4 (right, circles), where the circles denote the poles of the moment matching model (note the inverses of the poles are shown): they match the eigenvalues at the outside. The eigenvalues that cause the peaks, also known as dominant poles [12], however, may be located anywhere in the spectrum, as can also be seen in Fig. 4 (right, squares). This explains why the Arnoldi model fails to

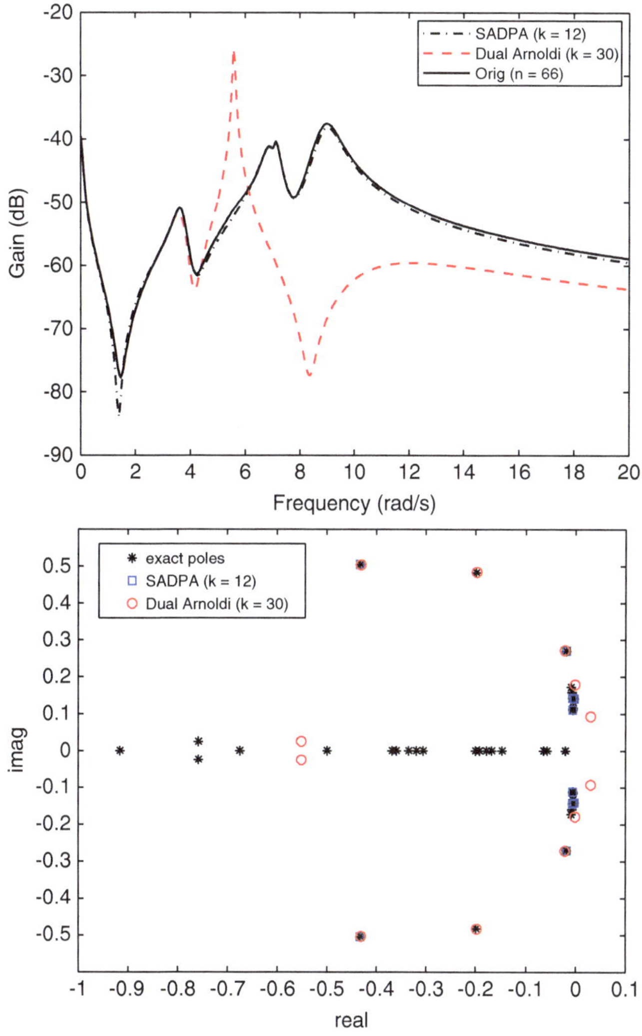

Fig. 4 Left: Original frequency response and reduced order models constructed by keeping six dominant modes (SADPA) and by preserving thirty moments (dual Arnoldi). The modal approximant matches peaks but is less accurate in between, while the Arnoldi model misses peaks. Right: corresponding eigenvalue spectra (zoom). Since Arnoldi approximates exterior eigenvalues, it may miss some of the dominant poles that cause the peaks

capture the peaks. In practice, it turns out that combining both methods may lead to better results: first one computes the dominant poles of the system, and based on these poles on chooses shifts for the (rational) Krylov subspace based reduction method. See [5, 14] for more details.

Also for parameterized MOR eigenvalues play an important role: by preserving the eigenvalues that are most sensitive to parameter changes one can make compact reduced order models that are valid for limited parameter ranges [13].

3.4 Reuse of Results

During layout optimization, many manual or automatic layout updates, parasitic extractions, and simulations have to be done. Often, per iteration the changes to a layout are minor and typically restricted to a part of the layout. An obvious idea is to reuse results from previous iterations as much as possible. For instance, by performing a partial and/or hierarchical extraction for the change part of the layout, one can gain considerably over doing a full extraction every iteration. Furthermore, one can use the generated simulation results to construct macromodels and sensitivity models, which can help to predict effective layout updates and so reduce the number of needed layout iterations. In practice this has lead to speedups of layout iterations, and increase of layout verifications, of factors 10 and more. However, since the various steps in layout simulation are typically approached individually, and solved by individual software packages, even more can be gained if steps are combined.

4 Open Challenges and Concluding Remarks

There are many open challenges. For MOR, robust methods for accurate extraction and reduction of parasitic RCLk networks with many inputs and outputs and macro modeling of large-scale nonlinear networks are only partially available. Such methods would help in circuit and layout optimization, which, however, also requires algorithms for automatically placing and routing components on a chip. Finally, there is need for methods that can deal with systems that depend on several parameters.

Even when mathematical challenges in industry are well understood, and solutions are available, practical implementation may still be a major task. Combination of several expertises — electrical engineering, mathematics, computer science — is required to successfully conquer these challenges.

Acknowledgements Part of this work was supported by EU Project O-MOORE-NICE!

References

1. Amestoy, P.R., Davis, T.A., Duff, I.S.: An approximate minimum degree ordering algorithm. SIAM J. Matrix Anal. Appl. **17**(4), 886–905 (1996)
2. Antoulas, A.C.: Approximation of large-scale dynamical systems. SIAM, Philadelphia, USA (2005)
3. Ciarlet, P.G., Schilders, W.H.A., ter Maten, E.J.W. (eds.): Numerical methods in electromagnetics, *Handbook of Numerical Analysis*, vol. 13, Elsevier (2005)
4. Freitas, F.D., Rommes, J., Martins, N.: Gramian-based reduction method applied to large sparse power system descriptor models. IEEE Trans. Power Syst. **23**(3), 1258–1270 (2008)

5. Grimme, E.J.: Krylov projection methods for model reduction. Ph.D. thesis, University of Illinois (1997)
6. Harutyunyan, D., Rommes, J., ter Maten, E.J.W., Schilders, W.H.A.: Simulation of mutually coupled oscillators using nonlinear phase macromodels. IEEE TCAD **28**(10), 1456–1466 (2009)
7. Heinkenschloss, M., Sorensen, D.C., Sun, K.: Balanced truncation model reduction for a class of descriptor systems with application to the Oseen equations. SIAM J. Sci. Comput. **30**(2), 1038–1063 (2008)
8. Li, J.R., White, J.: Low rank solution of Lyapunov equations. SIAM J. Matrix Anal. Appl. **24**(1), 260–280 (2002)
9. ter Maten, E.J.W., Doorn, T.S., Croon, J.A., Bargagli, A., Bucchianico, A.D., Wittich, O.: Importance sampling for high speed statistical Monte-Carlo simulations. CASA-report 09-37, TU Eindhoven (2009)
10. Penzl, T.: Algorithms for model reduction of large dynamical systems. Lin. Alg. Appl. **415**(2–3), 322–343 (2006)
11. Razavi, B.: Design of Analog CMOS Integrated Circuits. McGraw-Hill, New York (2001)
12. Rommes, J., Martins, N.: Efficient computation of transfer function dominant poles using subspace acceleration. IEEE Trans. Power Syst. **21**(3), 1218–1226 (2006)
13. Rommes, J., Martins, N.: Computing large-scale system eigenvalues most sensitive to parameter changes, with applications to power system small-signal stability. IEEE Trans. Power Syst. **23**(4), 434–442 (2008)
14. Rommes, J., Martins, N.: Computing transfer function dominant poles of large second-order dynamical systems. SIAM J. Sci. Comput. **30**(4), 2137–2157 (2008)
15. Rommes, J., Schilders, W.H.A.: Efficient methods for large resistor networks. IEEE TCAD **29**(1), 28–39 (2010)
16. Schilders, W.H.A., van der Vorst, H.A., Rommes, J. (eds.): Model order reduction: theory, research aspects and applications, *Mathematics in Industry*, vol. 13. Springer (2008)
17. Stykel, T.: Gramian based model reduction for descriptor systems. Math. Control Sig. Syst. **16**, 297–319 (2004)
18. van der Vorst, H.A.: Computational methods for large eigenvalue problems. In: Ciarlet P.G., Lions J.L. (eds.) Handbook of Numerical Analysis, vol. VIII, pp. 3–179. North-Holland, Elsevier, Amsterdam (2001)
19. Wachspress, E.L.: Iterative solution of the Lyapunov matrix equation. Appl. Math. Lett. **107**(1), 87–90 (1988)

On Approximate Reduction of Multi-Port Resistor Networks

M.V. Ugryumova, J. Rommes, and W.H.A. Schilders

Abstract Simulation of the influence of interconnect structures and substrates is essential for a good understanding of modern chip behavior. Sometimes such simulations are not feasible with current circuit simulators. We propose an approach to reduce the large resistor networks obtained from extraction of the parasitic effects that builds upon the work in (Rommes and Schilders, IEEE Trans. CAD Circ. Syst. 29:28–39, 2010) The novelty in our approach is that we obtain improved reductions, by developing error estimations which enable to delete superfluous resistors and to control accuracy. An industrial test case demonstrates the potential of the new method.

1 Introduction

Interconnect and substrate parasitic extraction typically lead to large resistor networks. Such networks may contain up to millions of resistors, hundreds of thousands of internal nodes and thousands of external nodes [4]. Simulations of such networks may be very time consuming or unfeasible. Model order reduction is aimed at finding smaller networks which accurately or exactly describe the port behavior of the original resistor networks.

Classical Krylov based model order reduction methods and structure preserving methods often lead to dense reduced matrices. This becomes the problem when dealing with networks with many terminals: synthesized reduced networks may

M.V. Ugryumova (✉) · W.H.A. Schilders
Eindhoven University of Technology, P.O.Box 513, 5600 MB, Eindhoven, The Netherlands
e-mail: m.v.ugryumova@tue.nl; w.h.a.schilders@tue.nl

J. Rommes
HTC 47, 5656 AE Eindhoven, The Netherlands
e-mail: joost.rommes@nxp.com

B. Michielsen and J.-R. Poirier (eds.), *Scientific Computing in Electrical Engineering SCEE 2010*, Mathematics in Industry 16, DOI 10.1007/978-3-642-22453-9_40,

have more resistors than the original networks [3]. In [4], an exact reduction technique, *ReduceR*, for resistor networks has been suggested. This approach is based on finding a special order in which internal nodes are eliminated. This allows to maximize sparsity of conductance matrix, and, therefore, the number of resistors in the reduced model. However, as it will be shown further, *ReduceR* does not always deliver a good reduction in terms of the number of resistors in the final circuit.

Therefore, in this paper, we consider an approach, called *simplification*, for reduction of resistor networks based on deleting resistors which do not affect behavior of the circuit significantly. Since such reduction is not exact, we are interested in controlling the error due to approximation. We expect that together with an existing approach for reduction, *ReduceR*, our approach will lead to improved final reduction.

This paper is organized as follows. In Sect. 2, we discuss circuit equations and goals of exact reduction for resistor networks. Also a challenging network for exact reduction is demonstrated. In Sect. 3 we suggest two independent criteria for improved reduction. For each criterion we derive error estimation which allow to control accuracy of approximation. In Sect. 4 we provide numerical examples and discuss the performance of the suggested estimations. Section 5 concludes.

2 Challenge in Exact Reduction of Resistor Networks

An n-port resistor network can be described by the Modified Nodal Analysis equation in a block form

$$\underbrace{\begin{pmatrix} G_{11} & G_{12} \\ G_{12}^T & G_{22} \end{pmatrix}}_{G} \begin{pmatrix} \mathbf{v}_e \\ \mathbf{v}_i \end{pmatrix} = \underbrace{\begin{pmatrix} B \\ 0 \end{pmatrix} \mathbf{i}_e}_{\mathbf{i}}, \tag{1}$$

where $\mathbf{v}_e \in \mathbb{R}^{n_e}$ and $\mathbf{v}_i \in \mathbb{R}^{n_i}$ are the voltages at external and internal nodes, respectively, $\mathbf{i}_e \in R_e^n$ are the currents injected in external nodes, $B \in \{-1, 0, 1\}^{n_e \times n_e}$ is the incidence matrix, and $G_{11} = G_{11}^T \in \mathbb{R}^{n_e \times n_e}$, $G_{12} \in \mathbb{R}^{n_e \times n_i}$ and $G_{22} = G_{22}^T \in \mathbb{R}^{n_i \times n_i}$. Note that $n = n_e + n_i$. System (1) must be grounded, i.e. equations corresponding to the ground nodes must be removed from the system.

In [4], the problem of reduction of large resistor networks is defined as follows: given a very large resistor network with conductance matrix G, find an equivalent network that: (a) has the same terminals; (b) has exactly the same path resistances between terminals; (c) has $\hat{n}_i \ll n_i$ internal nodes; (d) has $\hat{N} \ll N$ resistors; (e) realizable as a netlist. The existing algorithm, *ReduceR* [4], finds a subset of nodes which after being eliminated leads to a sparse conductance matrix. This is done by the use of the three strategies: graph algorithms, a reordering strategy (Approximate Minimum Degree algorithm) and a node elimination strategy. These strategies together guarantee that the reduced network is exact, i.e. no approximation

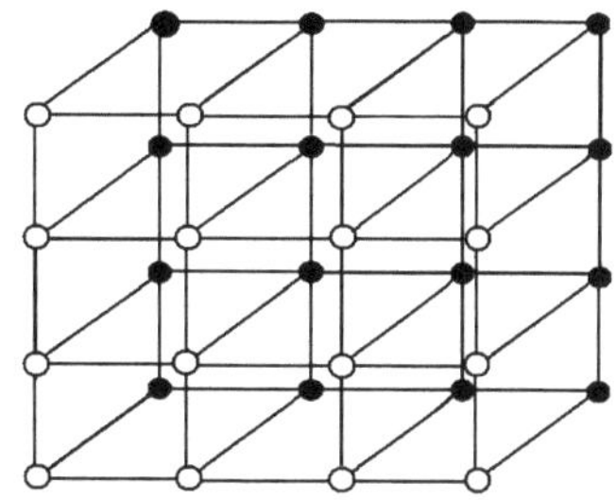

Fig. 1 Network with 16 external nodes (black dots), 16 internal nodes and 64 resistors (edges)

error is made. However the algorithm does not always deliver a good reduction. For example, networks which are obtained from substrate extraction based on Finite Element Method usually have a specific quadrilateral structure with large and sparse conductance matrices [5]. The exact reduction of such networks is challenging because elimination of internal nodes may not lead to efficient reduction. An example of such network is shown in Fig.1. Note that elimination of internal nodes in corners (they have the smallest degree) decreases the number of nodes however the number of resistors will stay unchanged.

3 Improved Approach

In order to improve the reduction of the number of resistors, we suggest a new approach that is based on finding and deleting some resistors which do not contribute to the behavior of the circuit significantly. We will call it *simplification*. Simplification does not deliver exact reduction and hence we would like to control error due to approximation. Further we suggest two types of error which can be used for controlling the quality of our approximation.

First we consider the relative error between voltages at nodes. Given a tolerance δ, the goal is to delete resistors in the network such that

$$Err_v = \frac{||\mathbf{v} - \tilde{\mathbf{v}}||}{||\mathbf{v}||} = \frac{||G^{-1}\mathbf{i} - \tilde{G}^{-1}\mathbf{i}||}{||G^{-1}\mathbf{i}||} < \delta, \tag{2}$$

where $\mathbf{v}$ is the vector of voltages at the nodes in the original network, $\tilde{\mathbf{v}}$ is the vector of voltages at the nodes after simplification, i.e. when some resistors have been deleted, G and $\tilde{G}$ are conductance matrices of the original and simplified networks. The error (2) is the most natural way of measuring the quality of approximation, because it tells us how close the output voltage of the reduced system and of the original system will be, when applying the same input currents. Since the current $\mathbf{i}_e$ is unknown in general, (2) requires knowledge of an error estimation which is independent of $\mathbf{i}_e$. In Sects. 3.1 and 3.2 we will derive estimations for Err_v and give recommendations on their use.

Secondly we consider the relative error between total path resistances. Given a tolerance δ, the goal is to delete resistors in the network such that

$$Err_{tp} = \frac{|R_{tot} - \tilde{R}_{tot}|}{|R_{tot}|} < \delta, \tag{3}$$

where R_{tot} is a total path resistance in original network (G) and $\tilde{R}_{tot}$ is a total path resistance in simplified network $(\tilde{G})$. The total path resistance is defined as [1]

$$R_{tot} = \sum_{i<j} R_{ij} = n \sum_{i=1}^{n-1} \frac{1}{\lambda_i}, \tag{4}$$

where R_{ij} is path resistance between nodes i and j, and $\lambda_1 \geq \cdots \geq \lambda_n = 0$ are eigenvalues of G. Note that close values of total path resistances do not imply that the networks have similar behavior. However, if networks have similar behavior, then corresponding total path resistances are similar. Since, in our case, the simplified network is obtained from the original one by deleting some resistors, we can expect (3) to be a measure that indicates, how well the reduced network approximates the original one. In Sect. 4 we show this fact via a numerical experiment, and indeed, the smaller δ in (3), the better the approximation to the original network. Simplification based on (3), of course, requires an efficient estimation with less computational cost than the direct computation of Err_{tp}. Derivation of such estimation will be done in Sect. 3.3.

3.1 Error Estimation for Err_v (First Version)

First approach is based on [8]. By neglecting, for instance, the smallest conductances (biggest resistors) in G, the criteria for simplification can be described as

$$||\Delta G|| \leq tol \cdot ||G||, \tag{5}$$

where $\Delta G = G - \tilde{G}$. According to [8] the error (2) can be bounded as

$$Err_v = \frac{||\mathbf{v} - \tilde{\mathbf{v}}||}{||\mathbf{v}||} \leq ||(G + \Delta G)^{-1}|| \cdot ||\Delta G|| \leq \kappa(G) \frac{||\Delta G||}{||G||} = \kappa(G) \cdot tol \equiv Err_{vc} \tag{6}$$

where $\kappa(G)$ is the condition-number of G. Thus Err_{vc} is a bound of the relative error Err_v.

Based on (5)–(6), a cheap and fast simplification procedure can be defined. For a given G and tolerance δ, one needs to compute $\kappa(G)$ and choose parameter tol such that it is less than $\delta/\kappa(G)$. After deleting a resistor (group of resistors), the condition (5) is checked. If it holds true, then the deleted resistor is confirmed and the next resistor is tried to be deleted, otherwise deleting is not confirmed and the next resistor is considered. In Sect. 3.4 we suggest an approach for deleting resistors by groups with the use of Err_{vc}.

In practice, however, the condition number, $k(G)$, is usually in the range from 10^5 till 10^8, therefore for the required accuracy, e.g. $\varepsilon = 5\%$, parameter tol must be small. As a result, condition (5) may become too strict for deleting a big amount of resistors which makes estimation not sharp enough. We remind that grounding of the network is required here in order to prevent G from being singular. If the network is not grounded one can ground an arbitrary external node and then perform simplification according to (5)–(6). After that the deleted external node is added to the network.

3.2 Error Estimation for Err_v (Second Version)

In this section we suggest an error estimation for the maximum relative error of vector of voltages which is sharper than the error estimation based on the condition number (6):

$$Err_v = \frac{||\mathbf{v} - \tilde{\mathbf{v}}||_2}{||\mathbf{v}||_2} \leq \max_{\mathbf{i} \in \mathbb{R}^n, \mathbf{i} \neq 0} \frac{||\mathbf{v} - \tilde{\mathbf{v}}||_2}{||\mathbf{v}||_2} = \max_{\mathbf{i} \in \mathbb{R}^n, \mathbf{i} \neq 0} \frac{||(I - \tilde{G}^{-1}G)\mathbf{f}||_2}{||\mathbf{f}||_2} = \sigma_1 \equiv Err_{vs}, \tag{7}$$

where σ_1 denotes the maximum singular value of $(I - \tilde{G}^{-1}G)$, $\mathbf{f} = G^{-1}\mathbf{i}$, and Err_{vs} is estimation of the error Err_v. Computation of maximum singular value can be performed, for instance, by Jacobi-Davidson type SVD method [2], or by Krylov-Schur method [7], which is used for numerical examples in Sect. 4. Note that $\tilde{G}$ has to be nonsingular, i.e. the network has to be grounded. If the network is not grounded, one can temporarily ground an arbitrary external node and after simplification unground it, i.e., to insert back the corresponding row and column in G. In Sect. 3.4 we suggest an approach for deleting resistors by groups with the use of Err_{vs}.

3.3 Error Estimation for Err_{tp}

In order to derive an estimation for (3), we consider the following theorem about perturbation of a Hermitian matrix [6]:

Theorem 1. *Let A be a Hermitian matrix with eigenvalues $\lambda_1 \geq \lambda_1 \geq \ldots \lambda_n$ and $\Delta A = A + E$ is a Hermitian perturbation of A with eigenvalues $\tilde{\lambda}_1 \geq \tilde{\lambda}_2 \ldots \tilde{\lambda}_n$. Matrix E has eigenvalues $e_1 \geq e_2 \geq \ldots e_n$. Then,* $\max\left(|\tilde{\lambda}_i - \lambda_i|\right) \leq ||E||_2$, *for $i = 1, \ldots, n$.*

We set up $\Delta A = \tilde{G}$ and $A = G$, where $\tilde{G}$ is a conductance matrix after simplification. By doing some algebra, (3), with R_{tot} and $\tilde{R}_{tot}$ defined as in (4), we obtain the estimation, Err_{tpa}, such that

$$Err_{tp} = \frac{|\tilde{R}_{tot} - R_{tot}|}{|R_{tot}|} \leq \frac{max\{|\varepsilon_1|, |\varepsilon_n|\}}{\lambda_{n-1}} \equiv Err_{tpa}, \tag{8}$$

where ε_1, ε_n are the largest and the smallest eigenvalues of $-\Delta G$, and λ_{n-1} is the second smallest eigenvalue of G. Thus deleting some resistors, one has to recompute only the largest magnitude eigenvalue of $-\Delta G$, while λ_{n-1} has to be computed once. This makes Err_{tpa} more attractive from computational point of view than a direct computing of Err_{tp}. In Sect. 3.4 we suggest an approach for deleting resistors by groups with the use of Err_{tpa}.

3.4 *Deleting Resistors by Groups*

For convenience we will use Err for denoting a generic error estimate, i.e. Err_{vc}, Err_{vs} or Err_{tpa}. Now the question is as follows. Which resistors should we delete from G and in which order should they be deleted to obtain $\tilde{G}$, such that $Err < \delta$? First we give some physical intuition. The larger resistor, the less current flows through it. Thus if a resistor is large, then almost no current goes through it and, therefore, such resistor can be neglected. This principle can be used in our case.

Further before deleting resistors we suggest to sort them in decreasing order. Since deleting resistors one by one and checking $Err < \delta$ is not efficient, we suggest to delete resistors *by groups*. To do that, choose k, which is less than the number of all resistors. (For example, k can be chosen as 10% from the whole amount of resistors.) Try to delete k resistors at once and check whether network becomes disconnected, i.e. corresponding undirected graph of the network is not connected [4]. If the network is still connected, then compute Err. If $Err < \delta$, then try to delete the next $2k$ resistors, otherwise try to delete $k/2$ resistors. If the network is disconnected, then try to delete $k/2$ resistors. If $k = 1$ and the network is disconnected, then skip the first resistor and continue the procedure from the beginning. As soon as $Err > \delta$ *and* $k = 1$, the procedure is stopped. This algorithm is just a way to select resistors that are candidates to be eliminated.

4 Numerical Results

We will show how the suggested approach for simplification with the use of error estimations and reduction by *ReduceR* work for networks from industry. The networks I,II and IV come from realistic designs of very-large-scale integration

Table 1 Results of reduction by *ReduceR* (R) and simplification (S^1 - using estimation Err_{vc}, S^2 - using estimation Err_{vs}, S^3 - using estimation Err_{tpa}). $\delta = 0.05$

Network I	Original	R	$R+S^1$	$R+S^2$	$R+S^3$
#ports	160	160	160	160	160
#resistors	23222	3315	2393	1187	1909
CPU time(s.)	–	23.8	23.9	28	24.9

Network II	Original	R	$R+S^1$	$R+S^2$	$R+S^3$
#ports	39	39	39	39	39
#resistors	2476	702	641	510	622
CPU time(s.)	–	1	1.01	1.9	1.7

Network III	Original	R	$R+S^1$	$R+S^2$	$R+S^3$
#ports	55	55	55	55	55
#resistors	70006	1485	1485	455	1480
CPU time(s.)	–	62.4	62.4	73	62.9

Network IV	Original	R	$R+S^1$	$R+S^2$	$R+S^3$
#ports	76	76	76	76	76
#resistors	1936	1397	1385	1264	1312
CPU time(s.)	–	1.1	1.2	2.6	1.8

chips [4] and the network III comes from substrate extraction which has a specific structure similar to the one in the Fig. 1. The simplification procedures have been implemented in Matlab 7.5 and have been tested on a Core 2 Duo 1.6 GHz PC.

In this paper we apply simplification *after* reduction by *ReduceR*. From the Table 1 it can be seen that simplification by Err_{vc} is faster than simplification by Err_{vs}. As was mentioned in Sect. 3.1, Err_{vc} depends on condition number of G which makes it less sharp than Err_{vs}. Noticeable reduction by Err_{vs} has been achieved for the networks I and III, where the number of resistors has been decreased by 65% and 70% respectively. Thus, estimation, Err_{vc}, can be considered for fast improvements in the amount of resistors, while Err_{vs} can be used for obtaining more advanced reduction. To compare the overall CPU simulation time, we measured time required to compute the path resistances. We first applied *ReduceR*, and then simplification by Err_{vs} to obtain the reduced network. The usage of simplification after *ReduceR* provides a further reduction in simulation time of 30% for the network *I*, and similarly, for the networks *II*, *III*, and *IV* the CPU time was reduced at 4%, 2%, and 4% respectively. Figure 2 demonstrates that the higher tolerance δ, the more resistors can be deleted. With $\delta = 20\%$, computational time is increased by 10% maximum.

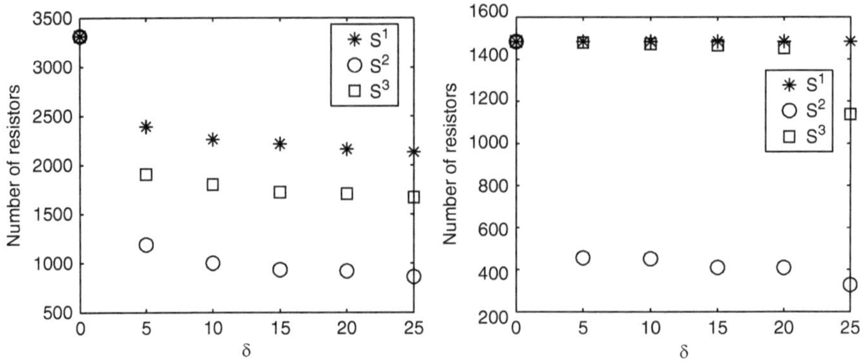

Fig. 2 Tolerance δ versus number of resistors in the networks (I-left, III-right) after simplification (S^1 – by estimation Err_{vc}, S^2 – by Err_{vs}, S^3 – by Err_{tpa}) applied after reduction by *ReduceR*

Table 2 The smaller δ, the close simplified network to the original one

δ	5%	15%	25%
Err_p	$4.38e^{-5}$	$1.64e^{-4}$	$5.79e^{-4}$

In order to show that Err_{tpa} in fact indicates how close the original and the simplified networks are, we performed the following experiment. We applied simplification by Err_{tpa} to the reduced network obtained by *ReduceR* from the network I. Then, we computed the maximum relative error, Err_p, between path resistances of the original and simplified network. From Table 2 it can be seen that smaller values of δ (with $Err_{tpa} \leq \delta$) correspond to smaller values of Err_p, which implies that the network with smaller Err_{tpa} is closer to the original network. The same tendency was observed for all other networks, and therefore we do not include these results.

5 Conclusion

By using insights from linear algebra, simplification improves the reduction of resistor networks, regarding the amount of resistors. Derived error estimations allow to keep a strict control on the accuracy of the reduced networks. Simplification, applied after reduction by *ReduceR*, improved total reduction by 70%. Since the success of simplification depends on the values of conductances, simplification can be considered as a complementary procedure to existing exact reduction techniques.

Acknowledgements The authors would like to thank M. Hochstenbach for sharing the Krylov-SVD code. The first author wants to thank P.I. Rosen Esquivel for the useful discussions.

References

1. Aldous, D.: Reversible Markov chains and random walks on graphs. In: Book in preparation, Available at www.stat.berkeley.edu/ aldous/RWG/book.html (2003)
2. Hochstenbach, M.: A Jacobi-Davidson type SVD method. SIAM J. Sci. Comput. **23**, 606–628 (2001)
3. Ionutiu, R., Rommes, J.: Circuit synthesis of reduced order models. Technical Note NXP-TN-2008/00316, NXP Semiconductors, Eindhoven, The Netherlands (2009)
4. Rommes, J., Schilders, W.H.A.: Efficient methods for large resistor networks. IEEE Trans. CAD Circ. Syst. **29**, 28–39 (2010)
5. Schrik, E., Meijs van der, N.P.: Combined BEM/FEM substrate resistance modeling. In: Proceedings of the 39th Conference on Design Automation, June 10–14. New Orleans, Louisiana, USA (2002)
6. Stewart, G.W., Sun, J.: Perturbation theory. Academic Press, INC., Boston, San Diego, New York, London, Sydney, Tokyo, Toronto (1990)
7. Stoll, M.: A Krylov-Schur approach to the truncated SVD (2010). Preprint submitted to Elsevier
8. Yang, F., Zeng, Y., Su, Y., Zhou, D.: RLC equivalent circuit synthesis method for structure-preserved reduced-order model of interconnect in VLSI. Commun. Comput. Phys. **3**, 376–396 (2008)

Improving Model-Order Reduction Methods by Singularity Exclusion

Pekka Miettinen, Mikko Honkala, Janne Roos, and Martti Valtonen

Abstract This paper presents a novel stand-alone method for overcoming a singular system matrix in Model-Order Reduction (MOR) algorithms, which would otherwise foil successful algorithm operation and thus reduction. The basic idea of the method is to locate and identify the circuit areas that generate the singularities to the system matrix prior to MOR, and exclude these from the reduction. The method is applicable to any netlist-in–netlist-out type MOR method.

1 Introduction

In order to accurately simulate transistor-level interconnect behavior, also the various non-ideal parasitic layout effects appearing at microchip and interconnect level need to be modeled. However, including these complex characteristics on top of the original circuit design often poses significant run-time and memory problems for the analysis and simulation tools. One avenue to speed up the simulations is to apply model-order reduction (MOR) algorithms (e.g., [1–6]) to the circuits, which attempt to approximate the system with a reduced-size representation.

One problem arising occasionally when using MOR methods is the singularity of the system matrices [7]. Depending on the method, the system matrices are typically derived from the y [1, 4, 5] or z-parameter [2, 3] circuit equations (see Sect. 2). The basic idea of moment-matching MOR approaches is to expand the circuit equation unknowns and the known input sources to Taylor series at some expansion point and match some of these series coefficients, i.e., moments. Since the explicit matching is numerically unstable for high number of moments, implicit moment-matching

P. Miettinen (✉) · M. Honkala · J. Roos · M. Valtonen
Department of Radio Science and Engineering, Aalto University School of Science and Technology, P.O. Box 13000, FI-00076 Aalto, Finland
e-mail: pekka.miettinen@tkk.fi; mikko.a.honkala@tkk.fi; janne.roos@tkk.fi; martti.valtonen@tkk.fi

B. Michielsen and J.-R. Poirier (eds.), *Scientific Computing in Electrical Engineering SCEE 2010*, Mathematics in Industry 16, DOI 10.1007/978-3-642-22453-9_41,

can be done via projecting the original system onto a smaller Krylov subspace. However, if the circuit equations can not be well defined at the expansion point, the MOR methods typically inherently fail. This occurs, e.g., with certain circuit structures (see Sect. 3), which cause the system matrix to become singular.

The idea of the proposed singularity exclusion method is to analyze the original netlist as a preprocessing step to the actual MOR, and exclude those parts of the circuit from the MOR that would generate the singularities to the system matrices (see Sect. 4). After reducing the remaining circuit, the excluded parts can be reconnected with the reduced circuit, to obtain the final, (partially) reduced-order model for the complete circuit. The presented method is further encouraged by the characteristic that often the problematic circuit parts that generate the singularities are located between interconnect segments, and/or consist of few elements in total. Thus, by removing the singularity-generating structures, good reduction efficiency can still be typically achieved.

It should be noted that the additional analysis required for the singularity exclusion may notably slow down a typical MOR reduction process, especially in the case of large circuits. As a trade-off, however, the method offers an automated approach of dealing with singularity-generating structures, significantly improving MOR algorithm reliability, with no loss in reduction accuracy.

2 System Matrices

Time-domain modified nodal analysis (MNA) circuit equations are commonly used to describe a circuit system. Depending on what excitation is used on the system, either y or z-parameters can then be determined: for voltage excitation at the ports, y-parameters, and for current excitation, z-parameters. In MOR methods, the y or z-parameters describing the circuit are then used to obtain a reduced model of the original circuit in terms of reduced y or z-parameters.

Using voltage sources at the ports, the MNA circuit equations for a linear N-port can be expressed as [1]

$$\begin{cases} \mathbf{C}\dfrac{\mathrm{d}\mathbf{x}_n(t)}{\mathrm{d}t} = -\mathbf{G}\mathbf{x}_n(t) + \mathbf{B}\mathbf{u}_N(t), \\ \mathbf{i}_N(t) = \mathbf{L}^{\mathrm{T}}\mathbf{x}_n(t), \end{cases} \tag{1}$$

where $\mathbf{C}$ and $\mathbf{G}$ are the susceptance and conductance matrices, respectively, and $\mathbf{x}_n$, $\mathbf{u}_N$, and $\mathbf{i}_N$ denote the MNA variables (nodal voltages, and branch currents of inductances and voltage sources), port voltages, and port currents, respectively. Here, $\mathbf{B} = \mathbf{L}$ is a selector matrix consisting of ones, minus ones, and zeros.

The y-parameters can be determined by taking Laplace transformation of (1) and solving for port currents, which results in

$$\mathbf{Y}(s) = \mathbf{L}^{\mathrm{T}}(\mathbf{G} + s\mathbf{C})^{-1}\mathbf{B}. \tag{2}$$

If current sources are used as excitation instead of voltage sources, the MNA equations for the system are given by (using the same notation for simplicity)

$$\begin{cases} \mathbf{C}\dfrac{\mathrm{d}\mathbf{x}_n(t)}{\mathrm{d}t} = -\mathbf{G}\mathbf{x}_n(t) + \mathbf{B}\mathbf{i}_N(t), \\ \mathbf{u}_N(t) = \mathbf{L}^{\mathrm{T}}\mathbf{x}_n(t). \end{cases} \tag{3}$$

where $\mathbf{C}$ and $\mathbf{G}$ are the susceptance and conductance matrices, respectively; $\mathbf{x}_n$, $\mathbf{u}_N$, and $\mathbf{i}_N$ denote the MNA variables (nodal voltages, and branch currents of inductances), port voltages, and port currents, respectively; matrix $\mathbf{B} = \mathbf{L}$ is a selector matrix consisting of ones, minus ones, and zeros.

Taking the Laplace transformation of (3) and solving for the port voltages, the z-parameter matrix is given as

$$\mathbf{Z}(s) = \mathbf{L}^{\mathrm{T}}(\mathbf{G} + s\mathbf{C})^{-1}\mathbf{B}. \tag{4}$$

3 Matrix Singularity

A square matrix that does not have a matrix inverse is called singular. For a system of linear equations, $\mathbf{A}\mathbf{x} = \mathbf{b}$, the system matrix $\mathbf{A}$ is invertible if, and only if, the number of linearly independent equations is the same as the number of unknowns in vector $\mathbf{x}$, i.e., $\mathbf{A}$ has full rank. Thus, if $\mathbf{A}$ has less linearly independent equations than unknowns, it is singular. If a system matrix is singular, the unknown vector $\mathbf{x}$ can not be solved exactly, and most numerical simulations become difficult.

Moment-matching MOR methods, such as [1–5], need to calculate $(\mathbf{G} + s\mathbf{C})^{-1}$ of (2) or (4) (where often $s = 0$) either to explicitly match the moments, or, e.g., as a part of an iterative Arnoldi process to implicitly match the moments. Here, if $\mathbf{G} + s\mathbf{C}$ is singular at the expansion point $s = s_0$, the moments can not be generated, and the MOR fails.

For circuit equations, singular matrices are often the result of violating some of the standing assumptions for allowable circuits [8]; e.g., a circuit may not contain a loop made of independent voltage sources, only, (an E loop) or a cutset made of independent current sources, only (a J cutset). Thus, the singularity-generating structures are not very common in typical RLC circuits, but do occur occasionally, especially with MOR of automatically generated circuits and/or when huge (such) circuits need to be partitioned into manageable subcircuits, and new ports are generated. Partitioning can also be used as a processing step in MOR as a powerful tool to obtain more efficient and robust reduction result [4–6], where the need to tackle the occasional singular circuit partition is emphasized. In any case, for a general RLC MOR algorithm to be reliably effective, these potentially fatal special cases need to be handled in some way.

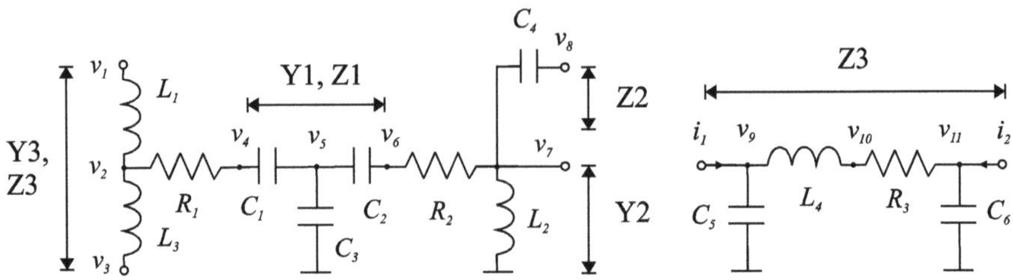

Fig. 1 Structures causing singularities for y-parameters (Y1, Y2, and Y3) and z-parameters (Z1, Z2, and Z3) at DC. The voltage and current source excitations are not included in the figure

In Fig. 1, five different circuit-topological cases are presented, which cause $\mathbf{G} + s\mathbf{C}$ to become singular at $s = s_0 = 0$ for RLC circuits. The expansion point $s_0 = 0$ (i.e., DC) is an important and typical expansion point, since low frequencies are often of major interest in a circuit application (especially so the DC). Furthermore, most interconnect circuit structures are of low-pass-filter type, and in these cases, using an expansion point near low frequencies typically gives better reduction results. However, at DC, capacitances and inductances become open and short circuits, respectively, and may more easily generate a situation where the standing assumptions for allowable circuits are violated, resulting in singular matrices.

Due to different excitations of the circuit equations (1) and (3), $\mathbf{G}+s\mathbf{C}$ calculated for the y or z-parameters become singular in slightly different cases. In Fig. 1, the circuit structures generating a singular matrix for y-parameters are noted Y1, Y2, and Y3, whereas for z-parameters, the singularities are generated by Z1, Z2, and Z3. Note that even a single occurrence of one of these structures makes the corresponding system matrix singular.

- Y1 The node v_5 is connected to capacitances, only, which become open circuits at DC. In $\mathbf{G} + s\mathbf{C}$ of (2), this translates to a row of zeros, meaning the voltage of the node, and the vector $\mathbf{x}_n$, is no longer well-defined.
- Y2 A port (voltage source) is short-circuited to the ground at DC, forming an E loop. This makes the current of inductance L_3 dependent to the current of the port at v_7, reducing the rank of $\mathbf{G} + s\mathbf{C}$ in (2) by one.
- Y3 At DC, the inductances between the two ports are reduced to a short circuit, creating an E loop. For $\mathbf{G} + s\mathbf{C}$ in (2) this translates to a linearly dependent row, and the rank of the matrix is decreased by one.
- Z1 Similarly as with y-parameters and Y1, the node v_5 is connected only to capacitances, making the system rank-deficient at DC also for z-parameters.
- Z2 The port node v_8 is connected to capacitances, only, which for z-parameters makes the system rank-deficient at DC similar to Y1 and Z1. (Note that for y-parameters, this is not a problem due to additional port current MNA stamps in the row.)
- Z3 A branch between two ports has no path to ground at DC, creating a J cutset for z-parameters. This means that the current excitations at the two ports are forced the same, (e.g., $i_1 = i_2$) and thus linearly dependent. For $\mathbf{G} + s\mathbf{C}$ in (4), this means that the matrix is again rank-deficient.

Although the above circuit cases present only the $s_0 = 0$ case, analogous situations occur at other expansion points, if the inductances and capacitances in Fig. 1 are replaced with other reactive elements: If the capacitances and inductances are replaced with inductances and capacitances, respectively, the singularities occur at $s_0 = \infty$. If the inductances and capacitances are replaced with series and parallel LC-resonators, the singularities occur when the expansion point is the same as complex resonant frequency of the resonators.

The list of singularity-generating cases presented in this section is not exhaustive, and other structures generating singularities in the system matrices exist, although they seem to be more rare for typical RLC circuits. The purpose of presenting the cases Y1, ..., Z3 is to show that the singularities are often generated by a small number of elements (except Z3) compared to the total circuit, and thus excluding them from the MOR should not generally hamper the MOR results considerably.

On the other hand, Z3 is a common structure in (partitioned) interconnects. This makes the use of z-parameters more difficult with $s_0 = 0$ in MOR approaches, especially if the many advantages of partitioning in MOR [5] are to be exploited.

4 Singularity Exclusion

A possible method of overcoming the singular system matrices is to introduce new small parasitic elements to the circuit [8]. This means adding a small conductive element in parallel and a small resistive element in series to each original reactive element in the circuit. This prevents the formation of any possible E loops, J cutsets, or nodes with no connection to other parts of the circuit. However, it is obvious that if thus processed, the original system may increase considerably in size. Furthermore, in case of large systems, even if the newly introduced parasitics are small in value, the cumulative impact of the generated error may become substantial, and worsen the reduction accuracy. In the following, a more sophisticated method for dealing with singular system matrices in MOR is presented.

The basic idea of the singularity exclusion method is to locate and isolate the areas in the original circuit that render the system matrices singular. Once the singularity-generating regions are found, the algorithm removes these parts of the circuit from the MOR process, and after all such regions are removed, MOR is performed on the non-singular part of the circuit. As a final step, the singularity-generating portion is combined to the reduced circuit.

The singularity of a matrix can be monitored, e.g., by calculating the condition number of the matrix. This measures the sensitivity of the solution of a linear equations to errors in the data, and gives an indication of the accuracy of the results from matrix inversion. In this paper, a MATLAB reciprocal condition number estimate `rcond` was used, which gives the reciprocal of the condition of the matrix in 1-norm [9]. If the matrix is near singular, `rcond` returns a value close to 0, and a value near 1 for a well-conditioned matrix. Even if the system is not strictly singular, it may be ill-conditioned. For example, a nearby singularity-generating region in the complex plane may still result in numerical unstability and poor accuracy for MOR, making it a valid target for further processing to avoid the numerical problems.

The proposed method can be divided into the following steps:

1. The condition number of the current system matrix is estimated using `rcond`.
2. If the `rcond` value is lower than a preset threshold, the matrix is partitioned into two submatrices .
3. The two new submatrices are analyzed again as in step 1. If either of the submatrices has a low `rcond` value that matrix is partitioned again into two new submatrices, and the steps 1–2 are repeated in a recursive manner.

   This partitioning process is continued until the size of a new submatrix decreases below the threshold for minimum submatrix (or corresponding subcircuit) size. At this point, the singularity-generating subcircuit is removed from the original circuit netlist to be reduced.
4. After all singularity-generating subcircuits have been localized and removed, the remaining subcircuits (i.e., partitions with acceptable `rcond`) are recombined.
5. The recombined circuit (without the singularity generating subcircuits) is reduced with the MOR method of choice.
6. Lastly, the previously excluded circuit parts are added to the reduced circuit.

If the MOR method uses partitioning as a natural part of the MOR algorithm (e.g., partitioning-based MOR methods, [4–6]), the singularity exclusion may be included in the MOR process. Here, the analysis and possible recursive partitioning described above may be done for each (MOR) partition at a time, in tandem with the MOR method. Furthermore, if the initial partition size is small (equal to the minimum subcircuit size in step 3), no recursive partitioning needs to be done, and the condition number calculations for the whole circuit become notably faster. For faster condition number calculations, the fast reciprocal condition number estimation `klu_rcond` provided by the KLU algorithm package [10] can also be used.

It should be noted that the singularity-generating structures described in Sect. 3 are mainly a problem for the MOR process; e.g., in transient analysis, z or y-parameters are not typically needed.

5 Simulations

The singularity exclusion method presented in this paper was implemented in C and MATLAB, using the hMETIS [11] algorithm-package for partitioning. Table 1 shows the transient analysis results of circuit `tree3`, first as original circuit, then the reduced circuit using PRIMA ($s_0 = 0$ and order of the reduced model, $q = 20$) and Matsumoto realization [12], with and without the singularity exclusion.

In the partitioning and `rcond`-based analysis, two singularity-generating areas (Y1) were located and removed. The threshold for `rcond` was 1×10^{-12} and the threshold for the minimum partition size was 10 circuit elements (i.e., 20 elements

Table 1 Comparison of MOR results for circuit `tree3` with and without singularity exclusion, showing the speedup in transient simulation results ($s_0 = 0$)

Method	*R*	*L*	*C*	*VCCS*	Error/%	Speedup
Original circuit (no reduction)	363	360	375	0	0	1.00
Reduced — no singularity exclusion	Singular matrix; MOR algorithm stopped, no results					
Reduced — singularity exclusion	27	7	43	160	0.07	4.72

Table 2 Transient analysis for reduced `tree3` with different expansion points s_0

s_0 / rad/s	*R*	*L*	*C*	*VCCS*	Error/%	Speedup
Original circuit	363	360	375	0	0	1.00
$\{100, 10^4, 10^6\}$	20	0	25	120	$\geq 10^3$	NaN
10^7	20	0	27	120	0.80	8.05
10^9	20	0	27	120	0.06	8.05

in total were excluded from the MOR). If the singularities were not removed, the MOR process could not continue, and no results could be obtained. As can be seen from the results, by using the singularity exclusion method, accurate and efficient reduction was still possible even with the singularity-generating structures in the original circuit.

One alternative approach to the singularity exclusion method is to change the expansion point of the MOR. However, $(\mathbf{G} + s\mathbf{C})^{-1}$ may be inaccurate for nearby expansion points that are close to a singularity-generating point in the complex plane, depending on the sensitivity of the system. Table 2 shows the transient analysis results for the circuit `tree3`, before and after MOR, with different expansion points. Here, the singularity at $s = 0$ caused numerical inaccuracies for MOR attempts in a wide frequency range, and shows that the choosing of a new s_0 may be non-trivial without further analysis.

It should be noted that the complete singularity exclusion algorithm with the recursive partitioning is relatively time-consuming, especially in the case of large circuits. Thus, although usable by any netlist-in–netlist-out MOR algorithm, the presented method is best suited to be used with partitioning-based MOR, where the partitioning step of the singularity exclusion method may be obtained at low computational cost alongside the MOR partitioning process.

6 Conclusions

In this paper, a method for overcoming singularity-generating circuit structures in MOR was presented. If a circuit contains areas that produce singularities to the system matrices, the MOR may inherently fail if additional precautions are not taken. By locating and excluding these areas from the MOR with automated processing, high reduction ratio can still be ensured, with no loss in accuracy. Alternative approaches to dealing with the singularities, such as introducing new small parasitics or switching the expansion point, may easily generate additional error to the reduction and/or require non-trivial heuristics of their own to be of similar use.

On the downside, the additional analysis needed may slow down the MOR considerably. Thus, the method is best suited to be used with partitioning-based MOR algorithms, where the computational cost of the partitioning step becomes small. If the reduction speed is not of vital interest, the presented method offers any netlist-in–netlist-out MOR algorithm notably improved method reliability.

Acknowledgements This work was partially funded by the Graduate School in Electronics, Telecommunications and Automation (GETA). Financial support from the Nokia Foundation and the Foundation of Walter Ahlström is acknowledged.

References

1. Odabasioglu, A., Celik, M., Pileggi, L.T.: PRIMA: passive reduced-order interconnect macromodeling algorithm. IEEE Trans. CAD, **17**, 645–654 (1998)
2. Su, Y.F., Wang, J., Zeng, X., Bai, Z.: SAPOR: second-order arnoldi method for passive order reduction of RCS circuits. In: Proceedings of ICCAD'04, pp. 74–79. San Jose, California, Nov (2004)
3. Freund, R.W.: SPRIM: structure-preserving reduced-order interconnect macromodeling. In: Proceedings of ICCAD'04, pp. 80–87. Nov (2004)
4. Liao, H., Dai,W.W.-M.: Partitioning and reduction of RC interconnect networks based on scattering parameter macromodels. In: Proceedings of ICCAD 1995, pp. 704–709. (1995)
5. Miettinen, P., Honkala, M., Roos, J., Valtonen, M.: PartMOR: Partitioning-based realizable model-order reduction method for RLC circuits. IEEE Trans. CAD, **30**(3), 374–387, (2011)
6. Yu, H., Shi, Y., He, L., Smart, D.: A fast block structure preserving model order reduction for inverse inductance circuits. In: Proceedings of ICCAD'06, pp. 7–12. San Jose, California, Nov (2006)
7. Aaltonen, S., Order reduction of interconnect circuits. Licentiate Thesis, Helsinki University of Technology (2003)
8. Chua, L.O., Lin, P.-M.: Algorithms and computational techniques: computer-aided analysis of electronic circuits. Prentice-Hall, Englewood Cliffs (1975)
9. Anderson, E., Bai, Z., Bischof, C., Blackford, S., Demmel, J., Dongarra, J., Du Croz, J., Greenbaum, A., Hammerling S., McKenney A., Sorensen, D.: LAPACK user's guide, 3rd edn. SIAM, Philadelphia (1999)
10. Davis, T. A., Palamadai Natarajan, E.: Algorithm 907: KLU, a direct sparse solver for circuit simulation problems. ACM Trans. Math. Softw., **37**, 36:1–36:17 (2010)
11. Karypis, G., Kumar, V.: hMETIS, A Hypergraph Partitioning Package (Version 1.5.3). (2007) http://glaros.dtc.umn.edu/gkhome/metis/hmetis/
12. Matsumoto, Y., Tanji, Y., Tanaka, M.: Efficient SPICE-Netlist representation of reduced-order interconnect model. In: Proceedings of ECCTD 2001, vol. 2, pp. 145–148. (2001)

Partitioning-Based Reduction of Circuits with Mutual Inductances

Pekka Miettinen, Mikko Honkala, Janne Roos, and Martti Valtonen

Abstract This paper describes a novel model-order reduction (MOR) method to reduce the number of mutual inductances in conjunction with a recently proposed MOR algorithm, PartMOR. As the method produces passive mutual inductances as a reduction realization, it extends the existing RLC-in–RLC-out PartMOR to a RLCM-in–RLCM-out MOR method. The method is verified and compared to a well-known MOR method with test simulations and is shown to produce good reduction results in terms of CPU speed-up and generated error.

1 Introduction

In order to accurately simulate transistor-level interconnect behavior, also the various non-ideal parasitic layout effects appearing at microchip and interconnect level need to be modeled. However, including these complex characteristics on top of the original circuit design often poses significant run-time and memory problems for the analysis and simulation tools. One avenue to speed up the simulations is to apply model-order reduction (MOR) algorithms (e.g., [1–6]) to the circuits, which attempt to approximate the system with a reduced-size representation.

Crosstalk and other coupling phenomena between neighboring interconnect lines can be divided into three types of coupling – parallel coupling, forward mutual coupling, and forward self coupling – which are typically modeled with inductances and/or capacitances [7]. Parallel coupling describes capacitive or inductive coupling oriented mostly orthogonal to two interconnect segments. Forward mutual

P. Miettinen (✉) · M. Honkala · J. Roos · Martti Valtonen
Department of Radio Science and Engineering, Aalto University School of Science and Technology, P.O. Box 13000, FI-00076 Aalto, Finland
e-mail: pekka.miettinen@tkk.fi; mikko.a.honkala@tkk.fi; janne.roos@tkk.fi; martti.valtonen@tkk.fi

B. Michielsen and J.-R. Poirier (eds.), *Scientific Computing in Electrical Engineering SCEE 2010*, Mathematics in Industry 16, DOI 10.1007/978-3-642-22453-9_42,

coupling describes a coupling effect that is not orthogonal to the interconnects, e.g., a coupling between interconnect segments that are further apart. Finally, forward self coupling is between two segments of the same interconnect line.

Capacitive coupling is a short-range effect, and the coupling terms are typically small in relative magnitude between long interconnect lines. Thus, in typical applications, capacitive coupling beyond the immediate neighboring capacitors can be discarded as negligible [7]. However, unlike capacitive coupling, inductive coupling has a wide-ranging area of effect. As a result, the mutual inductances can easily generate a dense mesh of elements between the interconnect lines.

This paper presents a novel method to reduce the number of mutual inductances in an RLCM interconnect circuit as a part of a recently proposed partitioning-based RLC-in–RLC-out MOR method, PartMOR [1] or the RL-in–RL-out MOR proposed in [5], in case of RLM circuits. The basic idea is that the RLCM (RLM) circuit is first treated for the RLC (RL) interconnects separately, and after the interconnects are reduced, the mutual inductive coupling between the reduced interconnect lines can be added in the same proportions as in the original interconnects.

2 PartMOR

PartMOR is a partitioning-based RLC-in–RLC-out MOR method that generates passive reduced-order circuits with positive-valued RLC elements [1]. The method first divides the original circuit into small partitions, which can then be approximated with low-order RLC macromodels. The approximation is performed by generating the y-parameter moment series at DC and infinity,

$$\mathbf{Y}(s) = \mathbf{M}_0 + \mathbf{M}_1 s + \mathbf{M}_2 s^2 + \cdots , \tag{1}$$

$$\mathbf{Y}(s) = \mathbf{N}_0 + \mathbf{N}_1 \frac{1}{s} + \mathbf{N}_2 \frac{1}{s^2} + \cdots , \tag{2}$$

and matching the first few moments from both of the series with one of the presented RLC macromodels. Since the method matches moments at both DC and infinity, good approximation can be achieved over a wide frequency band.

Using partitioning in MOR provides the method numerous beneficial assets, such as (block-level) sparsity, economical memory use, natural parallel processing, and facilitated port reduction. In PartMOR, specifically, the main advantage of partitioning is that small enough RLC interconnect circuit partitions can be typically approximated using few moments with still sufficient accuracy. The use of few moments, only, enables using numerically stable explicit matching with low-order macromodels. As the macromodels in turn can be relatively simple, it is possible to generate them using positive-valued RLC elements, only.

Regardless whether positive-valued RLC elements are specifically required by the design flow, PartMOR achieves excellent reduction results for various types of

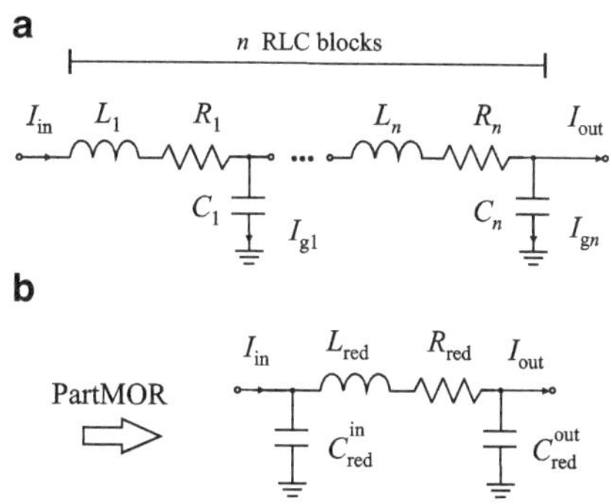

Fig. 1 (**a**) A typical interconnect section, (**b**) reduced interconnect section after PartMOR. Note that I_{in} and I_{out} should remain approximately the same before and after reduction

RLC, RC and RL circuits, outperforming an alternative RLC-in–RLC-out MOR method, SPRIM [3] using RLCSYN [8], for the cases shown in [1].

A typical PartMOR reduction of an interconnect partition is shown in Fig. 1.

3 Reducing Mutual Inductances

Reducing circuits containing mutual inductances presents partitioning-based MOR methods (such as PartMOR) several problems. Since a mutual inductance between two self inductances essentially generates a 4-port out of two 2-ports, in case of a dense mesh of inductances between interconnects, a partitioning may generate a huge number of port nodes, and lead to poor reduction efficiency. Dealing with circuits with dense meshes of elements is, thus, often difficult for partitioning-based approaches in general [1].

For low-order approximations, mutual inductances are also tricky, since it appears (as shown by preliminary simulations) that the dynamic behavior of even a small mesh of mutual inductances may be relatively complex, and thus it can not be always reliably approximated using low-order approximations.

In, e.g., [6] the reduction of mutual inductances is achieved by converting each pair of mutual inductances into a mesh of self inductances, some of which are of negative value. However, this approach has certain shortcomings, such as that the conversion model is highly inaccurate at DC, and in practice generates new 4-ports, which are difficult to partition efficiently. The method presented here relies on the physical characteristics of a transmission line, is accurate at DC, and is efficient for reducing even a dense mesh of mutual inductances using partitioning-based MOR.

Consider a typical transmission line discretization presented in Fig. 1a. If the currents $I_{g1}, \ldots, I_{gn}$ are small compared to I_{out}, it can be approximated that $I_{\text{in}} \approx I_{\text{out}}$. This holds for most interconnect models, especially when n is small.

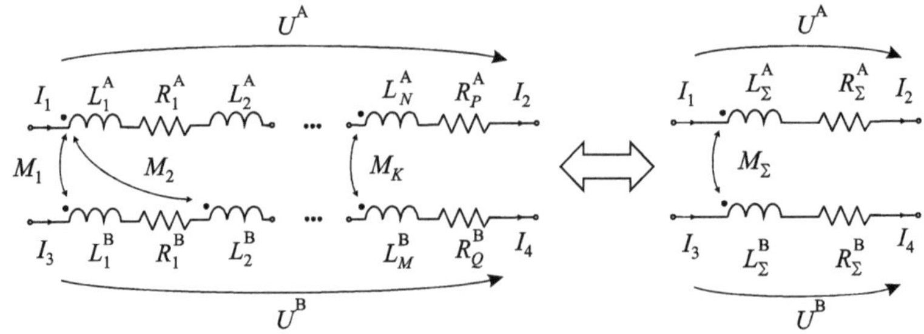

Fig. 2 Two interconnect lines, with negligible currents to ground: $I_1 \approx I_2$, $I_3 \approx I_4$

If the currents to the ground are negligible, the voltages over the two branches, U^A and U^B (see Fig. 2), where the branches have K mutual inductances, N and M self inductances, and P and Q resistances, respectively, are given by

$$\begin{aligned} U^A &= L_1^A I_1 s + R_1^A I_1 + M_1 I_3 s + \cdots + L_N^A I_1 s + R_P^A I_1 + M_K I_3 s \\ &= \sum\nolimits_{i=1}^{N} L_i^A I_1 s + \sum\nolimits_{i=1}^{P} R_i^A I_1 + \sum\nolimits_{i=1}^{K} M_i I_3 s \\ &\equiv L_\Sigma^A I_1 s + R_\Sigma^A I_1 + M_\Sigma I_3 s, \end{aligned} \tag{3}$$

$$\begin{aligned} U^B &= L_1^B I_3 s + R_1^B I_3 + M_1 I_1 s + \cdots + L_M^B I_3 s + R_Q^B I_3 + M_K I_1 s \\ &= \sum\nolimits_{i=1}^{M} L_i^B I_3 s + \sum\nolimits_{i=1}^{Q} R_i^B I_3 + \sum\nolimits_{i=1}^{K} M_i I_1 s \\ &\equiv L_\Sigma^B I_3 s + R_\Sigma^B I_3 + M_\Sigma I_1 s. \end{aligned} \tag{4}$$

Here, the mutual inductance between each pair of inductances, L^A and L^B, is

$$M = k\sqrt{L^A L^B}, \tag{5}$$

where k is the coupling coefficient. Note that the mutual inductive coupling may as well be of parallel (M_1 in Fig. 2) or forward mutual coupling (M_2 in Fig. 2). In case of forward self coupling, the mutual inductance can be treated as a self inductance, since $I_{in} \approx I_{out}$.

Now, consider that two transmission line partitions, named A and B (as shown for one line in Fig. 1a), have mutual inductances. The RLCM partitions are first reduced similarly as RLC partitions (into Fig. 1b). Since the current $I_{in} \approx I_{out}$ should be approximately the same before and after reduction (depending on the accuracy of the MOR), the mutual coupling of the currents between the two lines, described with M_Σ, is also the same after MOR. In case of, e.g., SPICE K elements are used to realize the mutual inductive coupling, the new coupling coefficient between the inductances of the reduced partitions can be calculated with

$$k_{\text{red}} = \frac{M_\Sigma}{\sqrt{L^{\text{A}}_{\text{red}} L^{\text{B}}_{\text{red}}}}, \tag{6}$$

where $L^{\text{A}}_{\text{red}}$ and $L^{\text{B}}_{\text{red}}$ are the reduced inductances of the two partitions, respectively, after MOR.

The general MOR algorithm flow can be summarized as follows:

1. By first ignoring the mutual inductance elements M, a partitioning is generated for the remaining RLC interconnect circuit.
2. For each M in the original circuit, the inductances connected by M (L^{A} and L^{B}) and the partition they belong to, are noted.
3. Using the information from step 2, $M_\Sigma^{(i,j)}$ is calculated using (3) and (5) between each pair of partitions, (i, j), that have connecting mutual coupling.
4. The RLC circuit is reduced using PartMOR (or [5], in case of RLM circuits).
5. The original mutual couplings $M_\Sigma^{(i,j)}$ are re-introduced to the circuit and new elements are generated between the reduced partitions (i, j) (using, e.g., (6)).

In step 1, the partitioning should be done such that the partitions generated are 2-ports. If a partition ends up with more than two ports, the approximation $I_{\text{in}} \approx I_{\text{out}}$ may become highly erroneous. Here, recursive partitioning similar to, e.g., [1], can be applied to reduce the size of the partition, and hopefully the number of ports. If obtaining a 2-port partitioning is not feasible, a partitioning with more than two ports may be left out from the MOR after a certain treshold for minimum partition size in order to avoid error. In practise, for typical interconnect discretizations, this should not become a problem.

Since the reduction relies on the approximation $I_{\text{in}} \approx I_{\text{out}}$, it is important that the currents to the ground remain small compared to the current in the main branch. In general, this applies to interconnect applications with large risetimes and small propagation delays [7]. For typical situations, it is important that the partitions are small enough to ensure that the difference between I_{in} and I_{out} – and the error generated by the approximation – remains small.

4 Passivity and Stability

A linear RLC circuit is passive and thus stable, if all element values in the circuit are non-negative [4] – this is a sufficient (but not necessary) condition for passivity. Reduced RLC circuits generated by PartMOR contain only positive valued RLC elements and are thus passive and stable.

A mutual inductance M between two self inductances L_1 and L_2 is passive, if [9]

$$M^2 \leq L_1 L_2. \tag{7}$$

However, in a general case, a self inductance may have multiple mutual inductances connected to several other self inductances (see, e.g., L_1 in Fig. 2). If the single self inductance L_2 in (7) is replaced with n self inductances, $L_2 \equiv L_{21} + L_{22} + \ldots + L_{2n}$, and similarly, the mutual inductance M is divided between L_1 and $L_{21}, L_{22}, \ldots, L_{2n}$, respectively, such that $M \equiv M_{21} + \ldots + M_{2n}$, from (7) it follows

$$(M_{21} + M_{22} + \ldots + M_{2n})^2 \leq L_1(L_{21} + L_{22} + \ldots + L_{2n}), \tag{8}$$

$$1 \leq \frac{L_1 L_{21} + L_1 L_{22} + \ldots + L_1 L_{2n}}{(M_{21} + M_{22} + \ldots + M_{2n})^2}. \tag{9}$$

Thus, if (9) applies to the mutual inductances of the reduced circuit, and all other elements are passive, the whole circuit is passive and stable.

The passivity criterion (9) can be easily checked in the presented MOR algorithm flow step 5 (see Sect. 3). Typically, if the original circuit is passive, the reduction produces a passive circuit without problems. In a rare case, e.g., if $M^2 = L_1 L_2$, the numerical noise generated by the MOR might produce a reduced circuit where $M^2 > L_1 L_2$ by a slight margin, and the reduced circuit would lose passivity (and stability). To prevent this, if a violation of (9) is observed in step 5, the total mutual inductive coupling from one partition, i, can be forced to $M = \sqrt{L_1 L_2}$, i.e.,

$$\sum_{j=1}^{n} M_{\Sigma}^{(i,j)} = \sqrt{\sum_{j=1}^{n} L_{\text{red}}^{i} L_{\text{red}}^{j}}, \tag{10}$$

where n is the number of M_{Σ} connected to partition i. Using (10), the generated new mutual inductance elements (using, e.g., (6)) can be scaled down to ensure passivity.

5 Simulations

The reduction method presented in this paper was verified and simulated with several RLCM circuits, of which `RLCMbuses3` and `RLCMbuses5` are shown as representative samples. The circuit `RLCMbuses3` is shown in Fig. 3 and consist of 11 varying interconnect segments, of which five have parallel mutual coupling. The circuit `RLCMbuses5` consist of two parallel interconnects, with heavy parallel and forward coupling, and is shown as a case with a relatively dense mesh of mutual inductances. For both circuits, the output voltage is read from a victim line.

The circuits were first reduced with PartMOR as RLC circuits and then further processed for mutual inductances with the method described in this paper.

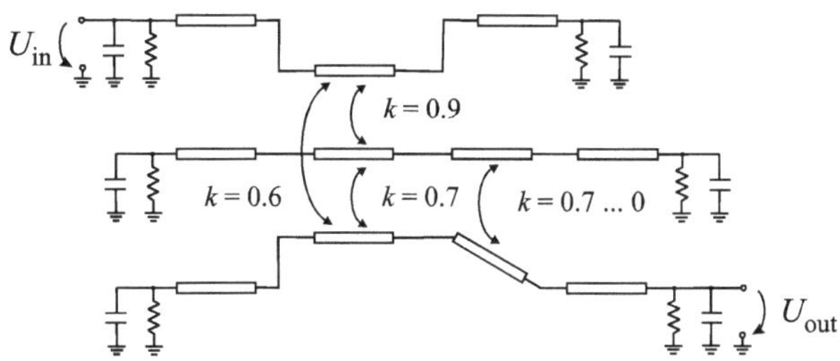

Fig. 3 Circuit `RLCMbuses3`. The inductive coupling between the interconnects is parallel coupling (with a coupling coefficient k) between neighboring interconnects. Each interconnect consist of 200–500 RLC elements

Table 1 Transient simulation results for original and reduced `RLCMbuses3`. For PartMOR+M, the partition size was 200 elements, and for PRIMA+Matsumoto $q = 40$ and $s_0 = 0$

MOR method	Nodes	R	L	C	K	$VCCS$	Error/%	CPU/s	Speed-up
Original circuit	4,510	1,507	1,502	1,507	400	0	–	30.65	1.00
PartMOR+M	74	37	24	25	18	0	0.90	0.24	127.71
PRIMA+Matsumoto	43	40	0	57	0	160	0.84	0.28	109.46

Table 2 Transient simulation results for original and reduced `RLCMbuses5`. For PartMOR+M, the partition size was 75 elements, and for PRIMA+Matsumoto $q = 20$ and $s_0 = 0$

MOR method	Nodes	R	L	C	K	$VCCS$	Error/%	CPU/s	Speed-up
Original circuit	606	205	201	203	1,910	0	–	14.21	1.00
PartMOR+M	27	14	8	10	10	0	0.48	0.13	109.31
PRIMA+Matsumoto	23	20	0	29	0	80	0.51	0.18	78.94

Transient-analysis simulations were then performed on the original and reduced circuits, and are shown in Tables 1 and 2 (where the reduction results are noted as PartMOR+M) and in Fig. 4. Here, the mutual inductances were realized using SPICE K elements. In the simulations, a short iteration (2–5 steps) was done to obtain a suitable partition size for the partitioning. The error is calculated as in [1]. As can be seen from the tables and the figure, the simulations showed good reduction results.

For comparison purposes, the two circuits were also reduced with PRIMA [2] using an efficient Matsumoto's realization method [10]. The results of the transient simulations for the PRIMA+Matsumoto reduced circuits are also shown in Tables 1 and 2. Similarly as for PartMOR+M, a short experimentation with parameter values was done to obtain a suitable order of the reduced model, q, and expansion point, s_0. By comparing the two MOR approaches, it seems that the presented method is equal to or even slightly better than PRIMA+Matsumoto in terms of reduction results.

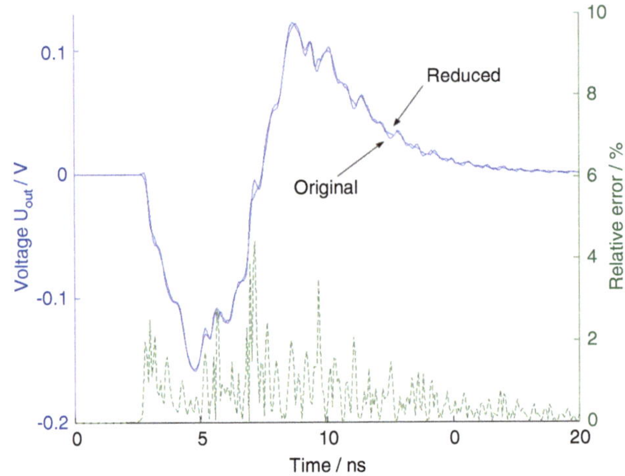

Fig. 4 Transient simulations of original and reduced `RLCMbuses3` using PartMOR and the method presented in this paper. The relative error is shown with a dashed line

6 Conclusions

In this paper, a method to reduce the number of mutual inductances in conjunction with a recently proposed MOR method, PartMOR, was presented. The method produces new passive mutual inductances as a reduction realization, thus extending the existing RLC-in–RLC-out PartMOR to a RLCM-in–RLCM-out MOR method. Additionally, since the mutual inductances are not mapped to, e.g., MNA matrices prior to MOR, even dense meshes of mutual inductances can be processed with good results. The method was verified and compared to PRIMA+Matsumoto with test simulations, and was shown to produce good reduction results. However, PRIMA+Matsumoto needs controlled sources ($VCCS$'s) in reduction realization, which may cause problems for some analysis tools. As the presented method uses only standard RLCM elements, it is readily usable in any typical design flow using mutual inductances.

Acknowledgements This work was partially funded by the Graduate School in Electronics, Telecommunications and Automation (GETA). Financial support from the Nokia Foundation and the Foundation of Walter Ahlström is acknowledged.

References

1. Miettinen, P., Honkala, M., Roos, J., Valtonen, M.: PartMOR: Partitioning-based realizable model-order reduction method for RLC circuits. IEEE Trans. CAD, **30**(3), 374–387 (2011)
2. Odabasioglu, A., Celik, M., Pileggi, L.T.: PRIMA: passive reduced-order interconnect macro-modeling algorithm. IEEE Trans. CAD, **17**, 645–654 (1998)

3. Freund, R.W.: SPRIM: structure-preserving reduced-order interconnect macromodeling, In: Proceedings of ICCAD'04, pp. 80–87. San Jose, California, Nov (2004)
4. Liao, H., Dai,W.W.-M.: Partitioning and reduction of RC interconnect networks based on scattering parameter macromodels. In: Proceedings of ICCAD'95, San Jose, California, pp. 704–709 (1995)
5. P. Miettinen, Honkala, M., Roos, J.: Partitioning-based RL-in–RL-out MOR method. In: Roos, J., Costa L.R.J. (eds.) Scientific Computing in Electrical Engineering SCEE 2008 *Mathematics in Industry* vol. 14, pp. 547–554. Springer, Berlin Heidelberg (2010)
6. Qin, Z., Cheng, C. K.: Realizable parasitic reduction using generalized Y-Δ transformation. In: Proceedings of DAC'03, pp. 220–225. Anaheim, Nov (2003)
7. Lin, T., Beattie, M.W., Pileggi, L.T: On the efficacy of simplified 2D on-chip inductance models. In: Proceedings of DAC'02, pp. 757–762. New Orleans, June (2002)
8. Yang, F., Zeng, X., Su, Y., Zhou, D.: RLCSYN: RLC equivalent circuit synthesis for structure-preserved reduced-order model of interconnect. In: Proceedings of ISCAS'07, New Orleans, pp. 2710–2713 (2007)
9. DeCarlo, R. A., Lin, P.-M.: Linear Circuit Analysis: Time Domain, Phasor, and Laplace Transformation Approaches. Oxford University Press, New York (2001)
10. Matsumoto, Y., Tanji, Y., Tanaka, M.: Efficient SPICE-netlist representation of reduced-order interconnect Model. In: Proceedings of ECCTD 2001, vol. 2, pp. 145–148 (2001)

Model Order Reduction of Parameterized Nonlinear Systems by Interpolating Input-Output Behavior

Michael Striebel and Joost Rommes

Abstract In this paper we propose a new approach for model order reduction of parameterized nonlinear systems. Instead of projecting onto the dominant state space, an analog macromodel is constructed for the dominant input-output behavior. This macromodel is suitable for (re)use in analog circuit simulators. The performance of the approach is illustrated for a benchmark nonlinear system.

1 Introduction

Simulation of VLSI chips is becoming CPU and memory intensive, or even infeasible, due to the increasing amount of layout parasitics and devices in analog designs. A popular method for speeding up and/or enabling simulation of large-scale dynamical systems is model order reduction [1]. For linear systems, several methods [2, 4, 5] have been developed that are now used in industrial circuit simulators.

Well-known methods for nonlinear systems in circuit simulation are Proper Orthogonal Decomposition (POD) based methods [6] and piecewise-linearization (PWL) methods [7]. Both approaches try to obtain reduction by projection on the dominant dynamics. However, both approaches may suffer from difficulties that may limit their practical use [8]. Robust and efficient resimulation of POD models is still a challenge, while PWL based approaches require application-dependent selections strategies for linearization points and weights.

M. Striebel (✉)
Bergische Universität Wuppertal, Wuppertal, Germany
e-mail: striebel@math.uni-wuppertal.de

J. Rommes
NXP Semiconductors, Eindhoven, The Netherlands
e-mail: joost.rommes@nxp.com

B. Michielsen and J.-R. Poirier (eds.), *Scientific Computing in Electrical Engineering SCEE 2010*, Mathematics in Industry 16, DOI 10.1007/978-3-642-22453-9_43,

We present a new method for the reduction of large nonlinear systems. The most significant difference with respect to existing methods is that instead of focusing on the dominant state dynamics, the proposed method tries to capture the dominant input-output behavior. Another novelty is that the resulting analog model behaves like a circuit element and can easily be used by circuit simulators. Table models have been used before, for instance for device modelling [3]; in this paper, however, we use table models for complete circuit blocks.

2 Circuit Modeling

Complex electrical systems are designed in a modular way.To enable communication with other circuit blocks, some nodes of each unit act as terminals, or pins. At these, say n_P pins, information in terms of pin voltages and pin currents, $\mathbf{v}_{\text{pin}}, \mathbf{i}_{\text{pin}} \in \mathbb{R}^{n_P}$, respectively, is exchanged. Applying Modified Nodal Analysis (MNA), a block is described by

$$0 = \mathbf{A}_C \frac{d}{dt}\mathbf{q}_C(\mathbf{A}_C^T\mathbf{e}) + \mathbf{A}_R\mathbf{r}(\mathbf{A}_R^T\mathbf{e}) + \mathbf{A}_L\mathbf{j}_L + \mathbf{A}_V\mathbf{j}_V + \mathbf{A}_I\mathbf{i}(t) - \mathbf{A}_{\text{pin}}\mathbf{i}_{\text{pin}}, \qquad (1a)$$

$$0 = \frac{d}{dt}\boldsymbol{\Phi}_L(\mathbf{j}_V) - \mathbf{A}_L^T\mathbf{e}, \qquad (1b)$$

$$0 = \mathbf{v}(t) - \mathbf{A}_V^T\mathbf{e}, \qquad (1c)$$

$$0 = \mathbf{v}_{\text{pin}} - \mathbf{A}_{\text{pin}}^T\mathbf{e}, \qquad (1d)$$

where $\mathbf{e}(t) \in \mathbb{R}^{n_e}$, $\mathbf{j}_L(t) \in \mathbb{R}^{n_L}$, $\mathbf{j}_V(t) \in \mathbb{R}^{n_V}$ denote the unknown node voltages and currents through inductors and voltage sources, respectively. The incidence matrices $\mathbf{A}_\Omega \in \{0, \pm 1\}^{n_e \times n_\Omega}$, describe the placement of the basic network elements resistor (Ω = R), capacitor (C), inductor (L), voltage (V) and current (I) source, respectively. The, in general nonlinear, characteristics of the network elements are represented by $\mathbf{q}_C(\cdot)$, $\boldsymbol{\Phi}(\cdot)$, $\mathbf{r}(\cdot)$, $\mathbf{i}(\cdot)$, $\mathbf{v}(t)$. The incidence matrix $\mathbf{A}_{\text{pin}} \in \{0, \pm 1\}^{n_e \times n_P}$ addresses the circuit nodes acting as pins. Injecting, i.e., prescribing the pin voltages $\mathbf{v}_{\text{pin}}$, the pin currents $\mathbf{j}_{\text{pin}}$ become additional unknowns, meant to be passed back to the system the block is embedded in, or vice versa. By this, a circuit unit turns into an input-output system, represented in the compact form

$$\mathbf{0} = \frac{d}{dt}\mathbf{q}(\mathbf{x}) + \mathbf{j}(\mathbf{x}) + \mathbf{s}(t) + \mathbf{B}\mathbf{u}; \quad \mathbf{y} = \mathbf{B}^T\mathbf{x}, \qquad (2)$$

where $\mathbf{u}(t), \mathbf{y}(t) \in \mathbb{R}^{n_P}$ represent the input and output of the system and $\mathbf{x}(t) \in \mathbb{R}^n$ ($n = n_e + n_V + n_L + n_P$) denotes the internal states. Note, that in the following we will omit the excitation $\mathbf{s}(t)$.

Frequently, design parameters, e.g., width and length of transistor channels, are kept variable, in order to optimize them in the design process. We take this into account by including a parameter vector $\boldsymbol{\rho} \in \mathbb{R}^{n_{\text{par}}}$ in the element functions, i.e., by extending $\mathbf{q}(\mathbf{x};\boldsymbol{\rho})$ and $\mathbf{j}(\mathbf{x};\boldsymbol{\rho}) \in \mathbb{R}^n$ in (2).

2.1 Model Order Reduction

A compound of subsystems, described by (2), arises e.g., in full system verification and post-layout simulation. The arising overall system usually is very large.

Focusing on the interaction in a compound of systems one is often not interested in the individual internal states $\mathbf{x}(t)$ but merely in the way a subsystem translates $\mathbf{u}(t)$ to $\mathbf{y}(t)$. Classically, Model Order Reduction (MOR) aims at replacing (2) by a dynamical system of reduced dimension $r \ll n$. The idea is, that, given the same input $\mathbf{u}(t)$, the substitute dynamical system with internal states $\mathbf{z}(t) \in \mathbb{R}^r$ produces (almost) the same output $\mathbf{y}(t)$ as the full system (2). Hence, replacing individual blocks by models of reduced order, the dimension of the compound system is kept small, enabling the overall system to be simulated at reasonable computational costs.

MOR for linear systems, arising from parasitic extraction, used in post-layout simulation, reached a high level of maturity. Several methods are now used in industrial circuit simulators. For an overview we refer to [1, 9]. MOR for linear problems bases upon the transfer function, i.e., the representation of the dynamical system under consideration in the frequency domain and is usually combined with projecting (2) onto a lower dimensional subspace.

For nonlinear problems the situation is somewhat different. Here, in general no transfer function can be specified and also projection to a lower dimensional subspace may reduce the dimension of the system but not the computational costs evaluating the system since still (dense) right-hand side and jacobian evaluations are needed. We propose an approach to reproduce the input-output mapping, starting from time-domain considerations.

2.2 Numerical Time Integration

Systems of type (2) usually can not be solved analytically for $\mathbf{x}(t), \mathbf{y}(t)$. Numerical integration is carried out instead. Both onestep and multistep methods discretize the system. For the backward Euler as a showcase this amounts to

$$\mathbf{0} = \frac{1}{h}\left[\mathbf{q}(x_n) - \mathbf{q}(x_{n-1})\right] + \mathbf{j}(\mathbf{x}_n) + \mathbf{B}\mathbf{u}_n; \quad \mathbf{y}_n = \mathbf{B}^T\mathbf{x}_n. \tag{3}$$

Given $\mathbf{u}_n$ and $\mathbf{x}_{n-1}$, (3) defines $\mathbf{x}_n$ and $\mathbf{y}_n$, i.e., approximations to $\mathbf{x}(t_n)$ and $\mathbf{y}(t_n)$ at $t_n = t_{n-1} + h$. Applying a Newton–Raphson technique to solve this problem, a series of linear equations have to be solved. The main ingredients for setting up the corresponding linear system are

$$\alpha\mathbf{C}(\bar{\mathbf{x}}) + \mathbf{G}(\bar{\mathbf{x}}); \quad \alpha\mathbf{q}(\bar{\mathbf{x}}) + \mathbf{j}(\bar{\mathbf{x}}); \quad \mathbf{q}(\mathbf{x}_{n-1}), \tag{4}$$

Table 1 Macromodel using tabulated data: mapping $\boldsymbol{\tau}_\Omega$ and derivative evaluated at inputs $\mathbf{u}^{(i)}$.

$\mathbf{u}$	$\mathbf{u}^{(1)}$	$\cdots$	$\mathbf{u}^{(k)}$
$\boldsymbol{\tau}_\Omega$	$\boldsymbol{\tau}_\Omega^{(1)}$	$\cdots$	$\boldsymbol{\tau}_\Omega^{(k)}$
$\mathbf{T}_\Omega$	$\mathbf{T}_\Omega^{(1)}$	$\cdots$	$\mathbf{T}_\Omega^{(k)}$

with $\mathbf{C}(\cdot) = \frac{d}{dx}\mathbf{q}(\cdot)$, $\mathbf{G}(\cdot) = \frac{d}{dx}\mathbf{j}(\cdot)$, evaluated at some intermediate points $\bar{\mathbf{x}}$. For the backward Euler we have $\alpha = h^{-1}$. The term $\mathbf{q}(\mathbf{x}_{n-1})$ reflects the history of the dynamic elements.

Note, that for didactic reasons only we stick to the Euler discretisation during the rest of this paper. For an overview of schemes applicable to DAEs we refer to [10].

3 Input-Output Behavior Macromodeling

Being interested in the translation of the input to the output reads, in terms of the discretised problem (3): we are interested in $\mathbf{y}_n$ and $\mathbf{x}_n$ as an auxiliary quantity only. Hence, ideally we are able to replace the system (3) by an input-output mapping

$$\boldsymbol{\tau} : \mathbb{R}^{n_P} \to \mathbb{R}^{n_P}, \quad \mathbf{u}_n \mapsto \mathbf{y}_n = \boldsymbol{\tau}(\mathbf{u}_n). \tag{5a}$$

At first glance it seems that this is not realizable. From (4), not only a combined evaluation of $\{\mathbf{q}, \mathbf{j}\}$ and $\{\mathbf{C}, \mathbf{G}\}$ is needed but also the dynamics' history $\mathbf{q}(\mathbf{x}_{n-1})$.

However, for homogeneous structures, i.e., blocks comprising only resistive (R), capacitive (C) or inductive (L) elements, the mapping $\boldsymbol{\tau}_\Omega$ ($\Omega = R, C, L$), can be derived. Still, in general, no analytic expression can be specified. The idea is to replace function evaluation with interpolation from tabulated data (see Table 1).

This table includes also the derivative of $\boldsymbol{\tau}$ w.r.t. the input, i.e.,

$$\mathbf{T} : \mathbb{R}^{n_P} \to \mathbb{R}^{n_P \times n_P} : \mathbf{u}_n \mapsto \mathbf{T}(\mathbf{u}_n) = \left.\frac{\partial \boldsymbol{\tau}(\mathbf{u})}{\partial \mathbf{u}}\right|_{\mathbf{u}=\mathbf{u}_n}. \tag{5b}$$

The basic concept is to replace homogeneous structures by a macromodel or macroelement with the same characteristics. Resistors turn voltages to currents, capacitors answer with charges when a voltage is applied and inductors show a current-flux relation. These facts are to be preserved by the macromodel.

In the following we give some details for purely resistive and purely capacitive structures, i.e., for static and dynamic blocks.

3.1 Models for Resistive Structures

A circuit block consisting of resistors only is described by

$$\begin{aligned} \mathbf{0} &= \mathbf{A}_R \mathbf{r}(\mathbf{A}_R^T \mathbf{e}) - \mathbf{A}_{\text{pin}} \mathbf{j}_{\text{pin}}, \\ \mathbf{0} &= \mathbf{v}_{\text{pin}} - \mathbf{A}_{\text{pin}}^T \mathbf{e}. \end{aligned} \tag{6a}$$

We choose the pin voltages $\mathbf{v}_{\text{pin}}$ as input parameters. Assuming sufficient regularity of the conductance matrix $\mathbf{G}_r(\mathbf{w}) := \frac{\partial}{\partial \mathbf{w}}\mathbf{r}(\mathbf{w})$, (6a) implicitly defines the node voltages and pin currents as functions of the pin voltages, i.e., $\mathbf{e} = \mathbf{e}(\mathbf{v}_{\text{pin}})$ and $\mathbf{j} = \mathbf{j}(\mathbf{v}_{\text{pin}})$, respectively. We differentiate (6a) with respect to $\mathbf{v}_{\text{pin}}$ to get:

$$\begin{aligned} \mathbf{0} &= \mathbf{A}_R \mathbf{G}_r(\mathbf{A}_R^T \mathbf{e}) \mathbf{A}_R^T \frac{\partial \mathbf{e}}{\partial \mathbf{v}_{\text{pin}}} - \mathbf{A}_{\text{pin}} \frac{\partial \mathbf{j}_{\text{pin}}}{\partial \mathbf{v}_{\text{pin}}}, \\ \mathbf{0} &= \mathbf{I}_{n_P} - \mathbf{A}_{\text{pin}}^T \frac{\partial \mathbf{e}}{\partial \mathbf{v}_{\text{pin}}} \end{aligned} \tag{6b}$$

where $\mathbf{I}_{n_P}$ is the $n_P \times n_P$ identity matrix.

For purely resistive structures we construct Table 1, describing the mapping "pin voltages" to "pin currents" in the following way:

1. Choose a discrete set of $k \in \mathbb{N}$ terminal voltages $\mathbf{v}_{p,1}, \ldots, \mathbf{v}_{p,k}$ with $\mathbf{v}_{p,i} \in \mathbb{R}^{n_P}$
2. For each $i \in \{1, \ldots, k\}$
   a. compute $\mathbf{e}_i = \mathbf{e}(\mathbf{v}_{p,i})$ and $\mathbf{j}_{p,i} = \mathbf{j}_{\text{pin}}(\mathbf{v}_{p,i})$ by solving (6a) for $\mathbf{v}_{\text{pin}} = \mathbf{v}_{p,i}$
   b. Solve the linear system (6b) for $\frac{\partial \mathbf{e}}{\partial \mathbf{v}_{\text{pin}}}|_{\mathbf{v}_{p,i}}$ and $\frac{\partial \mathbf{j}_{\text{pin}}}{\partial \mathbf{v}_{\text{pin}}}|_{\mathbf{v}_{p,i}} =: \mathbf{J}_{p,i}$. Here, $\mathbf{G}_r(\cdot)$ is evaluated at $\mathbf{A}_R^T \mathbf{e}_i$. This amounts to computing the Schur complement

$$\mathbf{J}_{p,i} = \frac{\partial \mathbf{j}_{\text{pin}}}{\partial \mathbf{v}_{\text{pin}}} = \left(\mathbf{A}_{\text{pin}}^T \left(\mathbf{A}_R \mathbf{G}_r(\mathbf{A}_R^T \mathbf{e}_i) \mathbf{A}_R^T\right)^{-1} \mathbf{A}_{\text{pin}}\right)^{-1}. \tag{6c}$$

3. The parameters for the resistive macromodel from Table 1 are

$$\mathbf{u}^{(i)} = \mathbf{v}_{p,i}\,, \quad \boldsymbol{\tau}_R^{(i)} = \mathbf{j}_{p,i}\,, \quad \mathbf{T}_R^{(i)} = \mathbf{J}_{p,i}$$

for $i = 1, \ldots, k$ where $\mathbf{v}_{p,i} \in \mathbb{R}^{n_P}$, $\mathbf{j}_{p,i} \in \mathbb{R}^{n_P}$, $\mathbf{J}_{p,i} \in \mathbb{R}^{n_P \times n_P}$

3.2 Models for Capacitive Structures

The distribution of charges and voltages in a network of capacitors is described by

$$\begin{aligned} \mathbf{0} &= \mathbf{A}_C \mathbf{q}(\mathbf{A}_C^T \mathbf{e}) - \mathbf{A}_{\text{pin}} \mathbf{q}_{\text{pin}}, \\ \mathbf{0} &= \mathbf{v}_{\text{pin}} - \mathbf{A}_{\text{pin}}^T \mathbf{e}, \end{aligned} \tag{7a}$$

where $\mathbf{q}_{\text{pin}}$ are point charges at the structure's pins. In other words: we map a large number of charges $\mathbf{q}(\cdot)$ to n_P point charges $\mathbf{q}_{\text{pin}}$. Here the voltage $\mathbf{v}_{\text{pin}} \in \mathbb{R}^{n_P}$ is prescribed at the pins.

Analog to the procedure for resistive structures (Sect. 3.1) we construct Table 1 for purely capacitive structures. For different voltages $\mathbf{v}_{\text{pin}} \in \{\mathbf{v}_{p,1}, \ldots, \mathbf{v}_{p,k}\}$ we solve the nonlinear system (7a) for $\mathbf{q}_{\text{pin}}(\mathbf{v}_{p,i}) =: \mathbf{q}_{p,i}$ and $\mathbf{e}(\mathbf{v}_{p,i})$.

The column in Table 1 reflecting the charge replies is made up of $\{\mathbf{q}_{p,1},\ldots,\mathbf{q}_{p,k}\}$. Items for the column in Table 1 describing the Jacobians $\mathbf{T}_C$ are found by solving

$$\begin{aligned}\mathbf{0} &= \mathbf{A}_C\mathbf{C}_q(\mathbf{A}_C^T\mathbf{e})\mathbf{A}_C^T\frac{\partial\mathbf{e}}{\partial\mathbf{v}_{\text{pin}}} - \mathbf{A}_{\text{pin}}\frac{\partial\mathbf{q}_{\text{pin}}}{\partial\mathbf{v}_{\text{pin}}},\\ \mathbf{0} &= \mathbf{I}_{np} - \mathbf{A}_{\text{pin}}^T\frac{\partial\mathbf{e}}{\partial\mathbf{v}_{\text{pin}}}\end{aligned} \tag{7b}$$

for $\frac{\partial\mathbf{e}}{\partial\mathbf{v}_{\text{pin}}}|_{v_{p,i}}$ and $\frac{\partial\mathbf{q}_{\text{pin}}}{\partial\mathbf{v}_{\text{pin}}}|_{v_{p,i}} =: \mathbf{Q}_{p,i} =: \mathbf{T}_C^{(i)}$, i.e., from the Schur complement

$$\mathbf{Q}_{p,i} = \mathbf{T}_C^{(i)} = \left(\mathbf{A}_{\text{pin}}^T\left(\mathbf{A}_C\mathbf{C}_q(\mathbf{A}_C^T\mathbf{e}_i)\mathbf{A}_C^T\right)^{-1}\mathbf{A}_{\text{pin}}\right)^{-1}. \tag{7c}$$

3.3 *Parameterized Structures*

A purely resistive structure that contains parameterized elements is modeled by

$$\begin{aligned}\mathbf{0} &= \mathbf{A}_R\mathbf{r}(\mathbf{A}_R^T\mathbf{e};\boldsymbol{\rho}) - \mathbf{A}_{\text{pin}}\mathbf{j}_{\text{pin}},\\ \mathbf{0} &= \mathbf{v}_{\text{pin}} - \mathbf{A}_{\text{pin}}^T\mathbf{e},\end{aligned} \tag{8}$$

with the vector $\boldsymbol{\rho} \in \mathbb{R}^{n_{\text{par}}}$ of parameters. The task is now to not only cover a range of terminal voltages $\mathbf{v}_{\text{pin}}$ but also a parameters $\boldsymbol{\rho}$ in a reasonable range.

Therefore, the procedure from Sect. 3.1 has to be adapted: besides sweeping over a range of pin voltages $\mathbf{v}_{\text{pin}} \in \{\mathbf{v}_{p,1},\ldots,\mathbf{v}_{p,k}\}$ we also scan the input-output behavior for different parameters $\boldsymbol{\rho} \in \{\boldsymbol{\rho}_1,\ldots,\boldsymbol{\rho}_l\}$. This leads to an extended macromodel

$$\boldsymbol{\tau}: \mathbb{R}^{nP}\times\mathbb{R}^{n_{\text{par}}} \to \mathbb{R}^{nP}, \quad (\mathbf{u}_n,\boldsymbol{\rho}) \mapsto \mathbf{y}_n = \boldsymbol{\tau}(\mathbf{u}_n,\boldsymbol{\rho}), \tag{9}$$

realized by a table with datapoints $\left[(\boldsymbol{\rho}^{(\mu)},\mathbf{u}^{(\nu)}),(\boldsymbol{\tau}^{(\mu,\nu)},\mathbf{T}^{(\mu,\nu)})\right]$ for $\mu = 1,\ldots,l$ and $\nu = 1,\ldots,k$.

3.4 *Using the Macromodels*

A system containing purely resistive and capacitive subblocks can be modeled by

$$\mathbf{0} = \frac{d}{dt}\mathbf{q}(\mathbf{x}) + \mathbf{j}(\mathbf{x}) + \mathbf{s}(t) + \mathbf{B}_R\boldsymbol{\tau}_R(\mathbf{B}_R^T\mathbf{x}) + \mathbf{B}_C\frac{d}{dt}\boldsymbol{\tau}_C(\mathbf{B}_C^T\mathbf{x}), \tag{10}$$

with incidence matrices $\mathbf{B}_R, \mathbf{B}_C$ describing the interfaces. In this way, we accommodate the characteristics of a subblock being reactive or nonreactive. Macromodels of inductive nature can be added to (10) in a similar way.

Applying any numerical time integration technique to (10), we see, that the basic ingredients for the systems to be solved in this process are (cf. (4))

$$\begin{aligned}&\alpha\left[\mathbf{C}+\widetilde{\mathbf{T}}_C\right](\bar{\mathbf{x}}_n)+\left[\mathbf{G}+\widetilde{\mathbf{T}}_R\right](\bar{\mathbf{x}}_n) \text{ and } \alpha\left[\mathbf{q}+\widetilde{\boldsymbol{\tau}}_C\right](\bar{\mathbf{x}}_n)+\left[\mathbf{j}+\widetilde{\boldsymbol{\tau}}_R\right](\bar{\mathbf{x}}_n);\\&\mathbf{q}(\mathbf{x}_{n-1}) \text{ and } \widetilde{\boldsymbol{\tau}}_C(\mathbf{x}_{n-1}),\end{aligned}\tag{11}$$

where $\widetilde{\boldsymbol{\tau}}_\Omega(\cdot) = \mathbf{B}_\Omega\boldsymbol{\tau}_\Omega(\mathbf{B}_\Omega^T\cdot)$ and $\widetilde{\boldsymbol{T}}_\Omega(\cdot) = \mathbf{B}_\Omega\boldsymbol{T}_\Omega(\mathbf{B}_\Omega^T\cdot)$ for $\Omega \in \{R, C\}$. We clearly see that the Jacobians (5b) are necessary as well.

Recall, that evaluation of the macromodel-functions and the corresponding Jacobians are realized by interpolation from the corresponding Table 1.

4 Numerical Experiments

The proposed approach has been implemented in `matlab` with interpolation done using the `interpn`-functions. The integrator used is an ROW-scheme. Although one may expect a large dependence on the quality of interpolation of the derivatives, in several testcases we neither recognized an increase in iterations done for calculating the DC-solution nor in timesteps rejected during simulation.

An extended, parameterized version of the transmission line [7] serves as a test example. This circuit, displayed in Fig. 1, consists of a series of N blocks, each containing M pairs of resistor and diode (modeled by $i_d(u) = \exp(\rho\cdot u) - 1$) in parallel. This leads to a system of dimension $N\cdot M + 1$. For the nonlinear resistor-diode block (with $M = 100$) a compact model is derived by sweeping $v_{\text{pin}} = \{0.0, \pm 0.01, \pm 0.02\}$ and $\rho = \{35, 55\}$, i.e., by solving (6a), (6b) $5\cdot 2 = 10$ times for each combination of (v_{pin}, ρ). For testing, the block was instantiated $N = 10$ times and a current source $i(t) = 0.5(\cos(2\pi\cdot 0.1\cdot t) + 1)$ was chosen. To test the accuracy of the reduced model, each of the $N = 10$ blocks was replaced by a tablemodel. Hence, the full system of dimension $N\cdot M + 1 = 1001$ is replaced by a model of dimension $N + 1 = 11$. From Fig. 2 a speedup of about 6 for each choice of the parameter ρ and an almost perfect match with full system simulation can be observed.

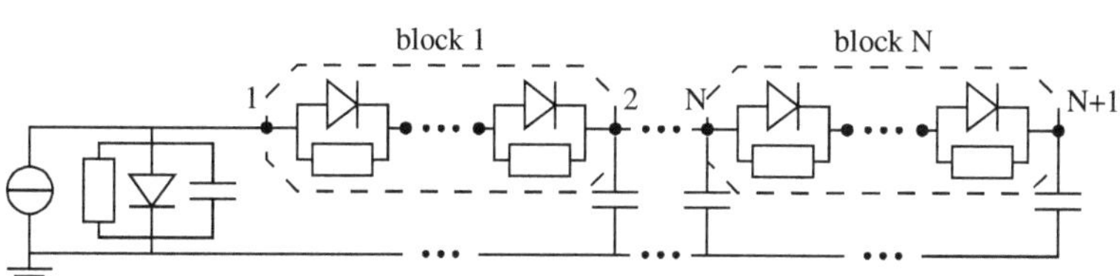

Fig. 1 Transmission line

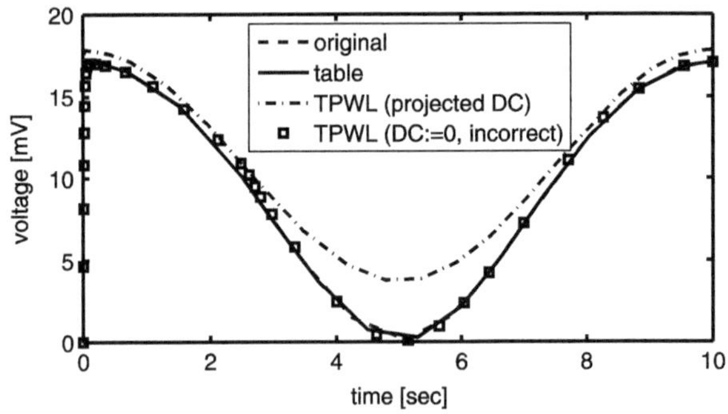

Fig. 2 Parameterized transmission line: Results for $\rho = 40$ in comparison with TPWL, node 1

	full system	reduced system extraction time: 1 .78 sec
$\rho = 40$	28 .45 sec	4 .18 sec
$\rho = 45$	26 .65 sec	4 .18 sec
$\rho = 50$	27 .86 sec	4 .18 sec

Resimulation with a TPWL-reduced model of similar size, replacing all the elements apart from the current source, was faster (about 1.5 s with 80 s for training) but less accurate. Furthermore TPWL depends on several heuristics and may be sensitive to the choice of the initial value as we see in Fig. 2. For details see [8].

5 Conclusion

We have presented a method that directly approximates the input-output behavior of large parameterized nonlinear circuits by interpolating precomputed contributions to the network equations. Numerical results confirm that significant speedups can be obtained while maintaining accuracy and not requiring heursitic model tuning. Extensions to mixed static/dynamic circuits and advanced interpolation methods that can deal with large numbers of inputs and outputs are subject to future research.

References

1. Antoulas, A.C.: Approximation of Large-Scale Dynamical Systems. SIAM, Philadelphia, (2005)
2. Odabasioglu, A., Celik, M., Pileggi, T.: PRIMA: Passive Reduced-order Interconnect Macromodeling Algorithm. IEEE Trans. Computer-Aided Des. Int. Circ. Systems **17**(8), 645–654 (1998)
3. Meijer, P.B.L.: Table models for device modelling. Proc. ISCAS **3**, 2593–2596 (1988)
4. Kerns, K.J., Yang, A.T.: Stable and efficient reduction of large, multiport networks by pole analysis via congruence transformations. IEEE Trans. Computer-Aided Des. Int. Circ. Syst. **16**(7), 734–744 (1997)

5. Rommes, J., Schilders, W.H.A.: Efficient methods for large resistor networks. IEEE Trans. CAD Int. Circ. Syst. **29**(1), 28–39 (2010)
6. Verhoeven, A., ter Maten, J., Striebel, M., Mattheij, R.: Model order reduction for nonlinear IC models. CASA-Report 07-41, TU Eindhoven (2007)
7. Rewieński, M.J., White, J.: A trajectory piecewise-linear approach to model order reduction and fast simulation of nonlinear circuits and micromachined devices. IEEE Trans. CAD Int. Circ. Syst. **22**(2), 155–170 (2003)
8. Striebel, M., Rommes, J.: Model order reduction of nonlinear systems: status, open issues, and applications. Tech. Rep. CSC/08-07, Technische Universität Chemnitz (2008)
9. Schilders, W.H.A., van der Vorst, H.A., Rommes, J. (eds.): Model Order Reduction: Theory, Research Aspects and Applications, *Mathematics in Industry*, vol. 13. Springer, Berlin (2008)
10. Hairer, E., Nørsett, S.P., Wanner, G.: Solving Ordinary Differential Equations I – Nonstiff Problems, 2nd revised edn., Springer, Berlin (2000)

On the Selection of Interpolation Points for Rational Krylov Methods

E. Fatih Yetkin and Hasan Dağ

Abstract We suggest a simple and an efficient way of selecting a suitable set of interpolation points for the well-known rational Krylov based model order reduction techniques. To do this, some sampling points from the frequency response of the transfer function are taken. These points correspond to the places where the sign of the numerical derivation of transfer function changes. The suggested method requires a set of linear system's solutions several times. But, they can be computed concurrently by different processors in a parallel computing environment. Serial performance of the method is compared to the well-known H_2 optimal method for several benchmark examples. The method achieves acceptable accuracies (the same order of magnitude) compared to that of H_2 optimal methods and has a better performance than the common selection procedures such as linearly distributed points.

1 Introduction

Model order reduction (MOR) techniques are getting more important in large scale computational tasks, such as large scale electronic circuit simulations. Models of the interconnect structure of the very large scale integrated (VLSI) circuits can be given in general as a linear state space system:

$$E\dot{x}(t) = Ax(t) + Bu(t), \quad y(t) = C^T x(t) + Du(t) \tag{1}$$

E.F. Yetkin (✉)
Istanbul Technical University, Informatics Institute, Istanbul, Turkey
e-mail: e.fatih.yetkin@be.itu.edu.tr

H. Dağ
Kadir Has University, Information Technologies Department, Istanbul, Turkey
e-mail: hasan.dag@khas.edu.tr

B. Michielsen and J.-R. Poirier (eds.), *Scientific Computing in Electrical Engineering SCEE 2010*, Mathematics in Industry 16, DOI 10.1007/978-3-642-22453-9_44,

where $A,\ E \in \Re^{nxn}$, $B \in \Re^{nxm}$, $C \in \Re^{nxm}$, $D \in \Re^{mxm}$, and n is the order of the system at hand.

If a system quadraple is given as $\Sigma = (E, A, B, C)$, one can produce the projection matrices $V \in \mathbb{C}^{nxk}$ and $W \in \mathbb{C}^{kxn}$ to obtain a kth order reduced system $\hat{\Sigma} = (\hat{E}, \hat{A}, \hat{B}, \hat{C})$. These projection matrices have to satisfy the condition $W^*V = I_k$ where I_k is the kth order identity matrix. In this work we assume that we work on a standard single input single output (SISO) system, and thus $E = I$. In SISO systems, B and C become vectors, b and c respectively. Thus, the reduced order system matrix and vectors are given below [1].

$$\hat{A} = W^*AV, \quad \hat{b} = W^*b, \quad \hat{c} = cV \tag{2}$$

Basically rational Krylov based methods match the transfer function at the selected different interpolation points [2].

Assume that k distinct points in complex plane are selected for interpolation. Then the interpolation matrices, $\hat{V}$ and $\hat{W}$, can be built as below.

$$\begin{aligned} \hat{V} &= [(s_1 I - A)^{-1}b \quad (s_2 I - A)^{-1}b \ldots (s_k I - A)^{-1}b] \\ \hat{W} &= [(s_1 I - A)^{-T}c^T \quad (s_2 I - A)^{-T}c^T \ldots (s_k I - A)^{-T}c^T] \end{aligned} \tag{3}$$

Assuming that $det(\hat{W}^*V) \neq 0$, then the projected reduced system can be built as, $\hat{A} = W^TAV, \hat{b} = W^Tb, \hat{c} = cV, \hat{D} = D$ where $V = \hat{V}$ and $W = \hat{W}(\hat{V}^*W)^{-1}$ to ensure $W^*V = I_k$. The basic problem is then to find a strategy to select the interpolation points. There are several studies for the selection of interpolation points in the literature [3].

In this work, a new approach is suggested for the selection of interpolation points. In the suggested method, the frequency response of the transfer function is sampled at some selected points. Then, numerical derivative of the sample array of these points, is computed. Obviously, the peaks of the transfer function can be determined by the sign changes of the derivative. These peaks correspond to the dominant poles of the system at hand. There are quite a few ways to find the dominant poles of a system [4, 5]. In literature there are also some ways of using these dominant pole approximation methods with spectral zeros of a system [6]. If these peaks are determined then the corresponding frequency values can be used as the interpolation points for producing the rational Krylov projectors. Finally, a reduced order model can be formed by using the projectors given in (3). Unfortunately, the reduced model may loose its stability in most cases. Therefore, one can employ the sign function based spectral projectors to neglect the unstable poles of the reduced system. Although the suggested method requires several factorizations to compute $(s_i I - A)^{-1}b$, these factorizations can be computed on different processors concurrently. Also the matrix-matrix and matrix-vector multiplications in the algorithm are amenable to parallel processing [7]. The selection procedure of the method is completely based on the information from the transfer function itself and the reduction order is automatically determined. This is the main difference from the methods given in [3].

Remainder of the paper is organized as follows. In the second part, the method is introduced. Some numerical results are given in the third section. In the last part, the conclusions and the future work are given.

2 Method and Algorithm

To produce the projectors given in (3) one has to select a suitable set of interpolation points. The common selection procedure for the interpolation points is to select either logarithmically or linearly distributed points from the working frequency. Then these initial selections can be used either in an iterative scheme or by a direct method to produce a reduced model. Our suggestion for the selection is based on the knowledge of the frequency behaviour of the system at hand. To obtain this knowledge one can select sufficient number of frequency points and compute the frequency response of the system in distinct points. Let a single input single output transfer function of the system in (1) with $E = I$ and $D = 0$ be given as, $H(s) = c^T(sI - A)^{-1}b$.

If N points are selected from the working frequency $[w_{min}, w_{max}]$, the sampled frequency response of the transfer function can be obtained as

$$H_i(s_i) = c^T(s_i I - A)^{-1}b \qquad i = 1, 2, \ldots, N. \tag{4}$$

Selection of the number of the points depends on a-priori knowledge about the frequency behaviour of the system if available. After obtaining the array, one can compute the numerical derivation of these data. The sign changes of the derivative means that there is a negative or positive peak on the Bode diagram. Then one can use these frequency points as the interpolation points.

The idea is illustrated in Fig.1. Frequency response of the building example given in [9] is computed for 20 different sample frequency points and sign changes of the numerical derivation of the frequency response are computed. It can be easily seen from the Fig. 1 that there are only three peak points for this example. If the number of frequency points is increased the frequency response curve will get more accurate and the number of peaks will also be increased until it achieves its exact value. But, we observed that it is not necessary to take all peak points of the frequency response to obtain accurate reduced models.

After the number of the peaks is determined, it can be used as the order of the reduced model and the frequencies corresponding to these peaks can be used as the interpolation points. Unfortunately, there is no guarantee for the stability of the reduced system obtained. Moreover, most of the time, the reduced model is unstable. Therefore, the unstable poles of the reduced system have to be eliminated. For this reason, spectral projectors can be employed [8]. On the other hand, accuracy of the reduced model decreases when the unstable poles eliminated.

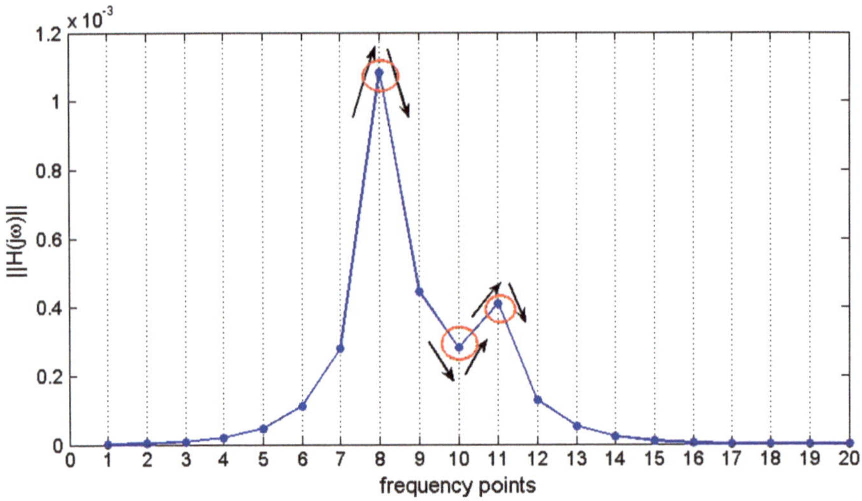

Fig. 1 Frequency response of the building example given in [9] for 20 different sample frequency points. There are only three peak points for this example

Computational cost of the rational Krylov methods is given as $\mathcal{O}(kn^3)$ for dense problems where k is the number of interpolation points. Iterative rational Krylov methods are used iteratively and the computational complexity has to be multiplied by the iteration number r [1]. In the suggested method, the main computational cost is the LU decomposition for the computation of frequency response of the transfer function in each sample point. This decomposition can also be used to form the projectors given in (3). On the other hand, to eliminate the unstable poles spectral projectors are used and a Sylvester equation is built and solved. Although these tools are computationally expensive, they are implemented on the reduced system. Therefore, the computational cost of them is negligible. As a result, the computational cost of the suggested method can be said to be comparable to that of the rational Krylov methods. The algorithm of the suggested method is given in Alg. 5.

3 Numerical Results

To test our algorithm, we used some well-known benchmark examples from the SLICOT suite [9]. We compare our algorithm with the H_2 optimal method given in [10]. The error is defined as $err = ||H(j\omega) - H_k(j\omega)||/||H(j\omega)||$ where $||H(j\omega)||$, $||H_k(j\omega)||$ are 2-norm of the frequency response of the original and the reduced system respectively. In the first experiment, relationship between the number of frequency points and the number of the peaks captured is investigated.

Algorithm 5 SUGGESTED METHOD

Require: System matrix A and the vectors b, c.
Ensure: Reduced system matrix $\hat{A}$, and the vectors $\hat{b}, \hat{c}$.

1: Select N points from $[w_{min}, w_{max}]$.
2: Compute $H_i(s_i) = c^T(s_i I - A)^{-1}b$ with appropriate linear solver for $i = 1 \ldots N$.
3: Compute the numerical derivation of the $H_i(s_i)$ data to obtain the k peak frequencies.
4: Add the w_{min} and w_{max} to the peak frequencies and obtain the $k + 2$ interpolation points.
5: Compute $\hat{W}$ and $\hat{V}$ using (3).
6: Compute $W = \hat{W}(\hat{V}^* W)^{-1}$ and assign $V = \hat{V}$ in order to ensure $W^* V = I_k$.
7: Compute $\hat{A}_1 = W^* AV, \hat{b}_1 = W^* b, \hat{c}_1 = cV$.
8: Compute the matrix sign function $S = sign(\hat{A}_1)$ and the rank revealing QR decomposition of

$$T = \frac{1}{2}(I_n - S)$$

matrix which is the spectral projector onto the stable part of the $\hat{A}_1$ as $T = QR\Pi$.
9: Compute

$$Q^T \hat{A}_1 Q = \begin{bmatrix} \hat{A}_{11} & \hat{A}_{12} \\ 0 & \hat{A}_{22} \end{bmatrix}, \quad Q^T \hat{b}_1 = \begin{bmatrix} \hat{b}_{11} \\ \hat{b}_{12} \end{bmatrix}, \quad \hat{c}_1 Q = \begin{bmatrix} \hat{c}_{11} \\ \hat{c}_{12} \end{bmatrix}$$

10: Solve the Sylvester equation

$$(\hat{A}_{11} - \beta I_k)Y - Y(\hat{A}_{22} - \beta I_{n-k}) + \hat{A}_{12} = 0$$

where $\beta \geq \max_{\lambda \in \sigma(\hat{A}_{22})}(Re(\lambda))$
11: The reduced order model is, $\hat{A} = \hat{A}_{11}, \hat{b} = \hat{b}_{11} - Y\hat{b}_{12}$ and $\hat{c} = \hat{c}_{11}$.

The change of the number of peaks and relative error according to the number of sample frequency point are given in Table 1.

In all the test cases, there is a limit for sample frequency points. After that value, there is no change in either the number of peak points or in the relative error for the reduced model. On the other hand, for small number of sample frequency points it is also possible to obtain satisfactory results with the method.

The suggested method has a comparable accuracy with iterative rational Krylov method with the same reduction order. Main problem of the method is, there is no guarantee for producing a stable and a passive reduced system with it. On the other hand, spectral projectors can be used as a post-processor to eliminate the unstable poles from the reduced system. Some comparisons of the error and computational cost for various methods can be found in Table 2.

Here, we only consider the full size linear equation solutions for computational comparison. In Fig. 2, Bode amplitude plots of the original and the reduced systems of transmission line benchmark example are given.

As the last example, we use a ladder RLC network given in [11]. Minimal realization of the circuit is given in Fig. 3. For this circuit order of the system $n = 5$. On the other hand, system matrices of this circuit can easily be extended. The order

Table 1 The change of the number of peaks and relative error according to the number of sample frequency points

# of sample points	Case	# of peaks captured	relative error
50	Beam	9	3.2×10^{-3}
	build	5	0.234
	CDplayer	11	0.173
100	Beam	15	6.4×10^{-4}
	build	7	0.196
	CDplayer	19	7×10^{-3}
200	Beam	19	2.1×10^{-4}
	build	17	9×10^{-3}
	CDplayer	25	3×10^{-3}
500	Beam	25	1.9×10^{-4}
	build	23	4.2×10^{-4}
	CDplayer	37	3×10^{-3}
1000	Beam	29	4×10^{-4}
	build	25	5.9×10^{-4}
	CDplayer	37	3×10^{-3}
5000	Beam	29	4×10^{-4}
	build	25	5.9×10^{-4}
	CDplayer	37	3×10^{-3}

Table 2 Comparison of the test results of the suggested and those of H_2 method

Case	n	k	Method	Error	# of LU
CDplayer	120	21	sugg.	7.03×10^{-3}	100
			H_2 opt.	3.9×10^{-3}	9×42
			linear	0.11	42
Build	48	6	sugg.	9.0×10^{-3}	200
			H_2 opt.	8.2×10^{-2}	10×12
			linear	6.6×10^{-3}	18

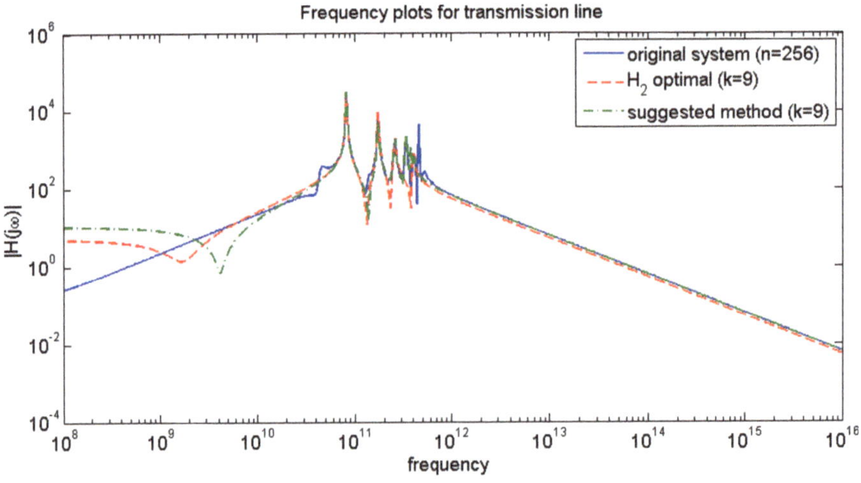

Fig. 2 Bode amplitude plots for transmission line system from [9]

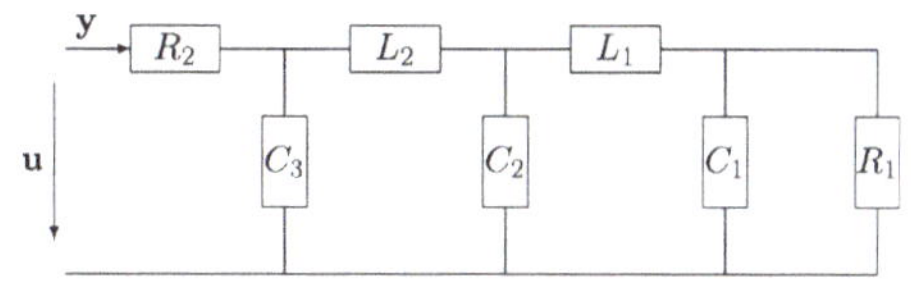

Fig. 3 Fifth order minimal realization of the RLC circuit used in experiments

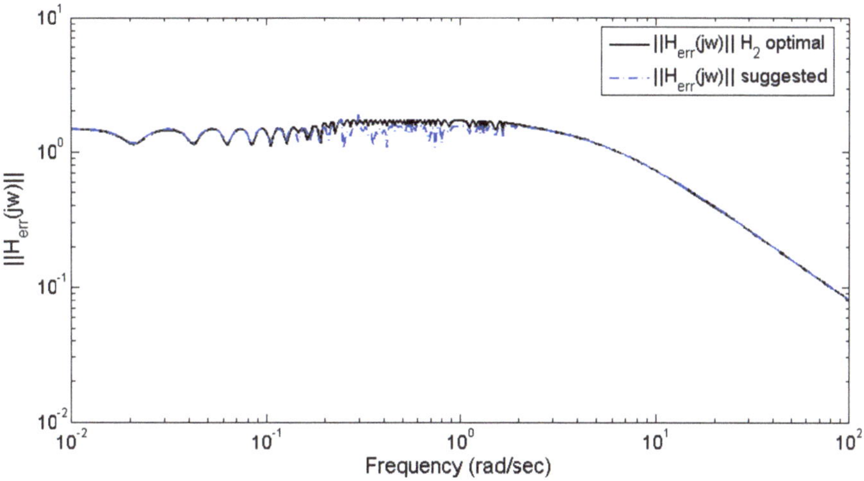

Fig. 4 Bode amplitude plots for error systems related to H_2 optimal and the suggested method

of the system n is taken as 301 in experiments. The number of frequency points is selected as 500 and the suggested method produces a 27th order reduced system. Hence, we select the number of interpolation points in H_2 optimal reduction as 28. The sigma plots of the error system for both reduction process are given in Fig. 4.

3.1 Discussions

In the suggested method, while projector matrices (V and W) are computed in step 5 of algorithm, one can use the results already computed in step 2 of the same algorithm. So there is no need for any extra computational effort to build the projection matrices. Moreover, from Table 1, we can say that satisfactory results can be achieved by relatively small number of sampling points. On the other hand, building blocks of the algorithm are based on well known procedures like matrix multiplications, linear system solutions and QR decomposition. So all these computations can be done efficiently on several type of architectures (multi-processor, multi-core or GPU systems).

4 Conclusion and Future Work

In this work, we suggest a new approach for the selection of the interpolation points for rational Krylov method. The method is based on the idea of determining the frequency response peaks of the transfer function. The frequencies of the peak points are selected as the interpolation points. The reduced order model size is determined automatically via the suggested algorithm. In the cases tested, it is observed that the method finds suitable set for using in rational Krylov method without iteration. The method needs linear equation solutions several times and this is the main computational cost of the suggested method. But these computations are completely independent from each other and the method can be easily run on multi processor systems. In MIMO cases, the method can be implemented for every input/output pairs independently and the reduced system can be determined with same approach for each input/output pair. In our future work, we plan to implement and test the suggested algorithm on parallel environments with more realistic multiple input multiple output examples.

References

1. Antoluas, A. C.: Approximation of Large-Scale Dynamical Systems. SIAM, Philadelphia (2006)
2. Gugercin, S., Antoulas, A. C.: A comparative study of 7 model reduction algorithms. In: Proceedings of the 39th IEEE Conference on Dec. and Control, Sydney, Australia, Dec. (2000)
3. Grimme, E.J.: Krylov Projection Methods for Model Reduction. Ph.D. Thesis, U. of Illinois, Urbana-Champaign, (1997)
4. Rommes, J.: Methods for Eigenvalue Problems with Applications in Model Reduction, Ph.D. Thesis, Utrecht University, (2007)
5. Rommes, J., Martins N.: Efficient computation of transfer function dominant poles using subspace acceleration. IEEE Trans. Power Syst. **21**(3), 1218–1226 (2006)
6. Ionutiu, R., Rommes J., Antoulas, A.C.: Passivity-preserving model reduction using dominant spectral-zero interpolation. IEEE Trans. Computer-Aided Des. Int. Circ. Syst. **28**(10) 1456–1466 (2009)
7. Yetkin, E. F., Dag H.: Parallel implementation of iterative rational Krylov methods for model order reduction. In: Proceedings of the ICSCCW 2009, Famagusa, Cyprus, Sept. 2–4 (2009)
8. Benner, P., Quintana-Orti E.S.: Model reduction based on spectral projection methods. In: Benner, P., Mehrmann, V., Sorensen, D.C. (eds.) Dimension Reduction of Large-Scale Systems, pp. 5–48, Springer, Berlin (2005)
9. Chahlaoui Y., Van Dooren P.: A Collection of Benchmark Examples for Model Order Reduction of Linear Time Invariant Dynamical Systems, SLICOT working note, 2002-2, Feb. (2002).
10. Gugercin, S.: An iterative SVD-Krylov based method for model reduction of large-scale dynamical systems. In: Proceedings of the 44th IEEE Conference on Dec. and Control, Seville, Spain, Dec. (2005)
11. Sorensen, D.C.: Passivity preserving model reduction via interpolation of spectral zeros. Syst. Control Lett. **54**, 347–360 (2005)

Discrete Empirical Interpolation in POD Model Order Reduction of Drift-Diffusion Equations in Electrical Networks

Michael Hinze and Martin Kunkel

Abstract We consider model order reduction of integrated circuits with semiconductors modeled by modified nodal analysis and drift-diffusion (DD) equations. The DD-equations are discretized in space using a mixed finite element method. This discretization yields a high dimensional, nonlinear system of differential-algebraic equations. Proper orthogonal decomposition is used to reduce the dimension of this model. Since the computational complexity of the reduced order model through the nonlinearity of the DD equations still depends on the number of variables of the full model we apply the discrete empirical interpolation method to further reduce the computational complexity. We provide numerical comparisons which demonstrate the performance of this approach.

1 Introduction

In this article we investigate model order reduction (MOR) based on proper orthogonal decomposition (POD) for semiconductors in electrical networks using discrete empirical interpolation method (DEIM) to treat the reduction of nonlinear components. Electrical networks can be modeled efficiently by a differential-algebraic equation (DAE) which is obtained from modified nodal analysis. Often semiconductors themselves are modeled by electrical networks. These models are stored in a library and are stamped into the surrounding network in order to create a complete model of the integrated circuit. In [7] POD-based MOR (POD-MOR)

M. Hinze (✉)
Fachbereich Mathematik, Universität Hamburg, Bundesstr. 55, 20146 Hamburg, Germany
e-mail: michael.hinze@uni-hamburg.de

M. Kunkel
Institut für Mathematik, Universität Würzburg, Am Hubland, 97074 Würzburg, Germany
e-mail: martin.kunkel@mathematik.uni-wuerzburg.de

B. Michielsen and J.-R. Poirier (eds.), *Scientific Computing in Electrical Engineering SCEE 2010*, Mathematics in Industry 16, DOI 10.1007/978-3-642-22453-9_45,

is proposed to obtain a reduced surrogate model conserving as much of the drift-diffusion (DD) structure as possible in the reduced order model (ROM). This approach in [6] is extended to parametrized electrical networks using the greedy sampling proposed in [9]. Advantages of the POD approach are the higher accuracy of the model and fewer model parameters. On the other hand, numerical simulations are more expensive. For a comprehensive overview of the DD equations we refer to [2, 8, 11].

This paper is organized as follows. We describe the unreduced model in Sect. 2. In Sect. 3, we present the MOR method based on snapshot POD combined with DEIM. In Sect. 4 we present numerical experiments, and also discuss advantages and shortcomings of our approach.

2 Discretized Coupled Model

Using the notation introduced in [5, 13] the finite element method (FEM) discretization of one semiconductor with domain $\Omega \subset \mathbb{R}^d$ $(d = 1, 2, 3)$ in an electrical network leads to a nonlinear, fully coupled DAE system of the form

$$A_C \frac{d}{dt} q_C(A_C^\top e(t), t) + A_R g(A_R^\top e(t), t) + A_L j_L(t) + A_V j_V(t) + A_S j_S(t) + A_I i_s(t) = 0, \quad (1)$$

$$\frac{d}{dt}\phi_L(j_L(t), t) - A_L^\top e(t) = 0, \quad (2)$$

$$A_V^\top e(t) - v_s(t) = 0, \quad (3)$$

$$q_S(t) - \frac{dg_\psi}{dt}(t) = 0, \quad (4)$$

$$j_S(t) - C_1 J_n(t) - C_2 J_p(t) - C_3 q_S(t) = 0, \quad (5)$$

$$\begin{pmatrix} 0 \\ -M_L \frac{dn}{dt}(t) \\ M_L \frac{dp}{dt}(t) \\ 0 \\ 0 \\ 0 \end{pmatrix} + A_{FEM} \begin{pmatrix} \psi(t) \\ n(t) \\ p(t) \\ g_\psi(t) \\ J_n(t) \\ J_p(t) \end{pmatrix} + F(n(t), p(t), g_\psi(t)) - b(e(t)) = 0, \quad (6)$$

compare Fig. 1, and see [6, 7]. Here, (1)–(3) describe the electrical network with unknown node potentials e, and branch currents j_L of inductive, and j_V of voltage source branches, respectively. Equations (4)–(5) are discretized coupling conditions.

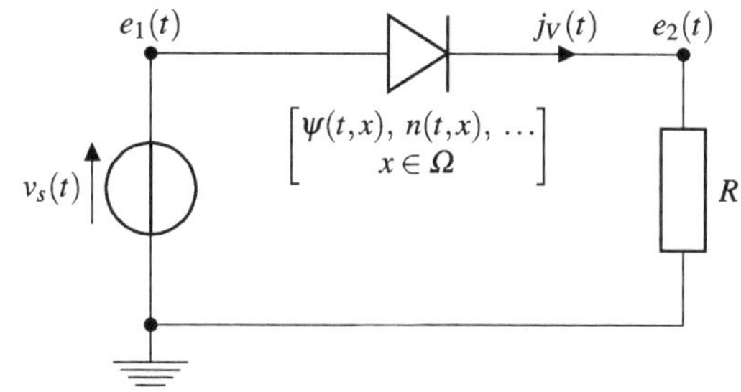

Fig. 1 Basic test circuit with one diode. The network is described by

$$\begin{aligned} A_V &= (\ \ 1,\ \ 0)^\top, \\ A_S &= (-1,\ \ 1)^\top, \\ A_R &= (\ \ 0,\ \ 1)^\top, \\ g(A_R^\top e, t) &= \frac{1}{R} e_2(t). \end{aligned}$$

The vector-valued function ψ contains the weights for the ansatz functions φ_i in the Galerkin ansatz

$$\psi^h(t,x) = \sum_{i=1}^{N} \psi_i(t)\varphi_i(x), \qquad x \in \Omega, \tag{7}$$

for the discretized potential of the semiconductor. Here, h denotes the discretization parameter and N denotes the number of finite elements. The discretized electron and hole concentrations $n^h(t,x)$ and $p^h(t,x)$, the electric field $-g_\psi^h(t,x)$ and the current densities $J_n^h(t,x)$ and $J_p^h(t,x)$ are defined likewise. The incidence matrix $A = [A_R, A_C, A_L, A_V, A_I, A_S]$ represents the network topology and is defined as usual. The matrices A_{FEM} and M_L are large and sparse. The voltage sources v_s and current sources i_s are considered as inputs of the network.

3 Model Order Reduction

We use POD-MOR applied to the DD part (6) to construct a dimension-reduced surrogate model for (1)–(6). For this purpose we run a simulation of the unreduced system and collect l snapshots $\psi^h(t_k,\cdot)$, $n^h(t_k,\cdot)$, $p^h(t_k,\cdot)$, $g_\psi^h(t_k,\cdot)$, $J_n^h(t_k,\cdot)$, $J_p^h(t_k,\cdot)$ at time instances $t_k \in \{t_1,\dots,t_l\} \subset [0,T]$. The optimal selection of the time instances is not considered here. We use the time instances delivered by the DAE integrator. The snapshot variant of POD introduced in [12] finds a best approximation of the space spanned by the snapshots w.r.t. to the considered scalar product.

Since every component of the state vector $z := (\psi, n, p, g_\psi, J_n, J_p)$ has its own physical meaning we apply POD-MOR to each component separately. Among other things this approach has the advantage of yielding a block-dense model and the approximation quality of each component is adapted individually.

The time-snapshot POD procedure delivers Galerkin ansatz spaces for ψ, n, p, g_ψ, J_n and J_p and we set $\psi^{POD}(t) := U_\psi \gamma_\psi(t)$, $n^{POD}(t) := U_n \gamma_n(t), \ldots$. The injection matrices $U_\psi \in \mathbb{R}^{N \times s_\psi}$, $U_n \in \mathbb{R}^{N \times s_n}, \ldots$, contain the (time independent) POD basis functions, and the vectors $\gamma_{(\cdot)}$ the corresponding time-variant coefficients. The numbers $s_{(\cdot)}$ denote the respective number of POD basis functions included. Assembling the POD system yields the ROM

$$A_C \frac{d}{dt} q_C(A_C^\top e(t), t) + A_R g(A_R^\top e(t), t) + A_L j_L(t) + A_V j_V(t) + A_S j_S(t) + A_I i_s(t) = 0, \quad (8)$$

$$\frac{d}{dt} \phi_L(j_L(t), t) - A_L^\top e(t) = 0, \quad (9)$$

$$A_V^\top e(t) - v_s(t) = 0, \quad (10)$$

$$q_S(t) - U_{g_\psi} \frac{d g_\psi}{dt}(t) = 0, \quad (11)$$

$$j_S(t) - C_1 U_{J_n} \gamma_{J_n}(t) - C_2 U_{J_p} \gamma_{J_p}(t) - C_3 q_S(t) = 0, \quad (12)$$

$$\begin{pmatrix} 0 \\ -\frac{d\gamma_n}{dt}(t) \\ \frac{d\gamma_p}{dt}(t) \\ 0 \\ 0 \\ 0 \end{pmatrix} + A_{POD} \begin{pmatrix} \gamma_\psi(t) \\ \gamma_n(t) \\ \gamma_p(t) \\ \gamma_{g_\psi}(t) \\ \gamma_{J_n}(t) \\ \gamma_{J_p}(t) \end{pmatrix} + U^\top F(U_n \gamma_n(t), U_p \gamma_p(t), U_{g_\psi} \gamma_{g_\psi}(t)) - U^\top b(e(t)) = 0, \quad (13)$$

with $A_{POD} = U^\top A_{FEM} U$ and $U = \operatorname{diag}(U_\psi, U_n, U_p, U_{g_\psi}, U_{J_n}, U_{J_p})$. All matrix-matrix multiplications are calculated in an offline-phase. The nonlinear function F has to be evaluated online which means that the computational complexity of the ROM still depends on the number of unknowns of the unreduced model. The nonlinearity in (13) is given by

$$U^\top F(U\gamma(t)) = \begin{pmatrix} 0 \\ U_n^\top F_n(U_n \gamma_n(t), U_p \gamma_p(t)) \\ U_p^\top F_p(U_n \gamma_n(t), U_p \gamma_p(t)) \\ 0 \\ U_{J_n}^\top F_{J_n}(U_n \gamma_n(t), U_{g_\psi} \gamma_{g_\psi}(t)) \\ U_{J_p}^\top F_{J_p}(U_n \gamma_p(t), U_{g_\psi} \gamma_{g_\psi}(t)) \end{pmatrix},$$

see e.g. [6]. The subsequent considerations apply for each block component of F. For the sake of presentation we only consider the second block

$$\underbrace{U_n^\top}_{\text{size } s_n \times N} \underbrace{F_n}_{N \text{ evaluations}} (\underbrace{U_n}_{\text{size } N \times s_n} \gamma_n(t), \underbrace{U_p}_{\text{size } N \times s_p} \gamma_p(t)), \tag{14}$$

and its derivative with respect to γ_p,

$$\underbrace{U_n^\top}_{\text{size } s_n \times N} \underbrace{\frac{\partial F_n}{\partial p}(U_n \gamma_n(t), U_p \gamma_p(t))}_{\text{size } N \times N, \text{ sparse}} \underbrace{U_p}_{\text{size } N \times s_p}.$$

Here, the matrices $U_{(\cdot)}$ are dense and the Jacobian of F_n is sparse. The evaluation of (14) is of computational complexity $O(N)$. Furthermore, we need to multiply large dense matrices in the evaluation of the Jacobian. Thus, the POD-MOR may become inefficient.

To overcome this problem, we apply DEIM, proposed in [3], which we now describe briefly. The snapshots $\psi^h(t_k, \cdot)$, $n^h(t_k, \cdot)$, $p^h(t_k, \cdot)$, $g_\psi^h(t_k, \cdot)$, $J_n^h(t_k, \cdot)$, $J_p^h(t_k, \cdot)$ are collected at time instances $t_k \in \{t_1, \ldots, t_l\} \subset [0, T]$ as before. Additionally, we collect snapshots $\{F_n(n(t_k), p(t_k))\}$ of the nonlinearity. DEIM approximates the projected function (14) such that

$$U_n^\top F_n(U_n \gamma_n(t), U_p \gamma_p(t)) \approx U_n^\top V_n (P_n^\top V_n)^{-1} P_n^\top F_n(U_n \gamma_n(t), U_p \gamma_p(t)),$$

where $V_n \in \mathbb{R}^{N \times \tau_n}$ contains the first τ_n POD basis functions of the space spanned by the snapshots $\{F_n(n(t_k), p(t_k))\}$ associated with the largest singular values. The selection matrix $P_n = \left(e_{\rho_1}, \ldots, e_{\rho_{\tau_n}}\right) \in \mathbb{R}^{N \times \tau_n}$ selects the rows of F_n corresponding to the so-called DEIM indices $\rho_1, \ldots, \rho_{\tau_n}$ which are chosen such that the growth of a global error bound is limited and $P_n^\top V_n$ is regular, see [3] for details.

The matrix $W_n := U_n^\top V_n (P_n^\top V_n)^{-1} \in \mathbb{R}^{s_n \times \tau_n}$ as well as the whole interpolation method is calculated in an offline phase. In the simulation of the ROM we instead of (14) evaluate:

$$\underbrace{W_n}_{\text{size } s_n \times \tau_n} \underbrace{P_n^\top F_n}_{\tau_n \text{ evaluations}} (\underbrace{U_n}_{\text{size } N \times s_n} \gamma_n(t), \underbrace{U_p}_{\text{size } N \times s_p} \gamma_p(t)), \tag{15}$$

with derivative

$$\underbrace{W_n^\top}_{\text{size } s_n \times \tau_n} \underbrace{\frac{\partial P_n^\top F_n}{\partial p}(U_n \gamma_n(t), U_p \gamma_p(t))}_{\text{size } \tau_n \times N, \text{ sparse}} \underbrace{U_p}_{\text{size } N \times s_p}.$$

In the applied FEM a single functional component of F_n only depends on a small constant number $c \in \mathbb{N}$ components of $U_n\gamma_n(t)$. Thus, the matrix-matrix multiplication in the derivative does not really depend on N since the number of entries per row in the Jacobian is at most c.

But there is still a dependence on N, namely the calculation of $U_n\gamma_n(t)$. To overcome this dependency we identify the required components of the vector $U_n\gamma_n(t)$ for the evaluation of $P_n^\top F_n$. This is done by defining selection matrices $Q_{n,n} \in \mathbb{R}^{c\tau_n \times s_n}$, $Q_{n,p} \in \mathbb{R}^{c\tau_p \times s_p}$ such that

$$P_n^\top F_n(U_n\gamma_n(t), U_p\gamma_p(t)) = \hat{F}_n(Q_{n,n}U_n\gamma_n(t), Q_{n,p}U_p\gamma_p(t)),$$

where $\hat{F}_n$ denotes the functional components of F_n selected by P_n restricted to the arguments selected by $Q_{n,n}$ and $Q_{n,p}$.

Supposed that $\tau_n \approx s_n \ll N$ we obtain a ROM which does not depend on N any more.

4 Numerical Investigation

The discussed method is implemented in C++ based on the FEM library deal.II [1]. The high dimensional DAE is integrated using the DASPK software package [10]. The derivative of the nonlinear functional is hard to compute and thus we calculate the Jacobians by automatic differentiation with the package ADOL-C [14]. The Newton systems which arise from the BDF method are solved with the direct sparse solver SuperLU [4].

A basic test circuit with a single 1-dimensional diode is depicted in Fig. 1. The parameters of the diode are summarized in [6]. The input $v_s(t)$ is chosen to be sinusoidal with amplitude 5 V. In the sequel the frequency of the voltage source will be considered as a model parameter.

We first validate the ROM at a fixed reference frequency of $5 \cdot 10^9$ Hz. Figure 2 shows the development of the relative error between the POD reduced, the POD-

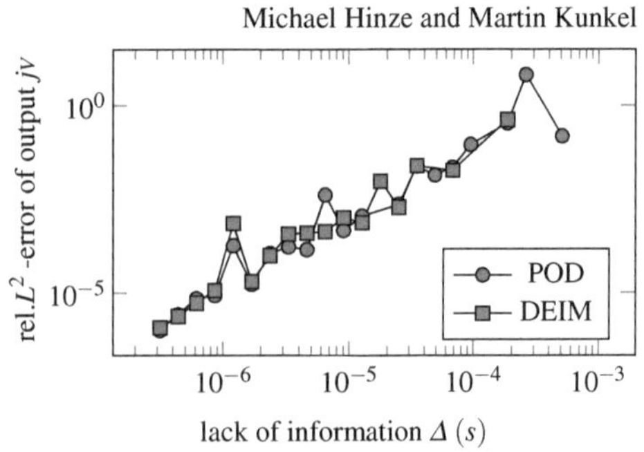

Fig. 2 Relative error between reduced and unreduced problem at the fixed frequency $5 \cdot 10^9$ Hz

DEIM reduced and the unreduced numerical solutions, plotted over the lack of information Δ of the POD basis functions with respect to the space spanned by the snapshots. The figure shows that the approximation quality of the POD-DEIM reduced model is comparable with the more expensive POD reduced model. The number of POD basis functions $s_{(\cdot)}$ for each variable is chosen such that the indicated approximation quality is reached, i.e. $\Delta := \Delta_\psi \simeq \Delta_n \simeq \Delta_p \simeq \Delta_{g_\psi} \simeq \Delta_{J_n} \simeq \Delta_{J_p}$. The numbers $\tau_{(\cdot)}$ of POD-DEIM basis functions are chosen likewise.

In Fig. 3 the simulation times are plotted versus the lack of information Δ. The POD ROM does not reduce the simulation times significantly for the chosen parameters. The reason for this is the dependency on the number of variables of the unreduced system. Here, the unreduced system contains 1000 finite elements which yields 12012 unknowns. The POD-DEIM ROM behaves very well and leads to a reduction in simulation time of about 90% without reducing the accuracy of the ROM. However, we have to report a minor drawback; not all tested ROMs converge for large $\Delta(s) \geq 3 \cdot 10^{-5}$. This is indicated in the figures by missing squares.

In Fig. 4 we plot the corresponding total number of required POD basis functions. It can be seen that with the number of POD basis functions increasing linearly, the lack of information tends to zero exponentially. Furthermore, the number of DEIM interpolation indices behaves in the same way.

In Fig. 5 we investigate the dependence of the ROMs on the number of finite elements N. One sees that the simulation times of the unreduced model depends linearly on N. The POD ROM still depends on N linearly with a smaller constant. The dependence on N of our POD-DEIM implementation is negligible.

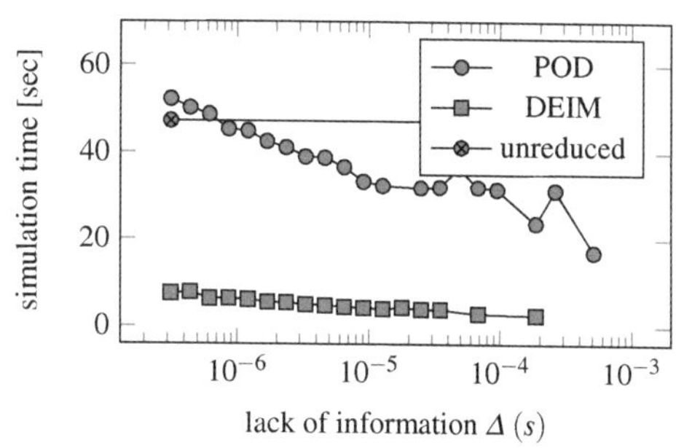

Fig. 3 Time consumption for simulation runs of Fig. 2. The horizontal line indicates the time consumption for the simulation of the original full system

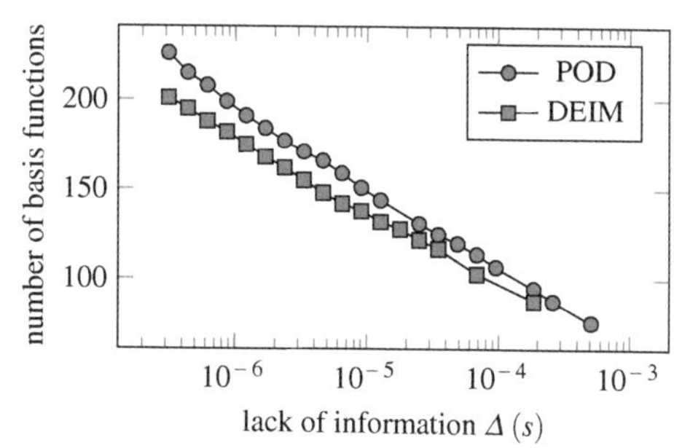

Fig. 4 The number of required POD basis function and DEIM interpolation indices grows only logarithmically with the requested information content

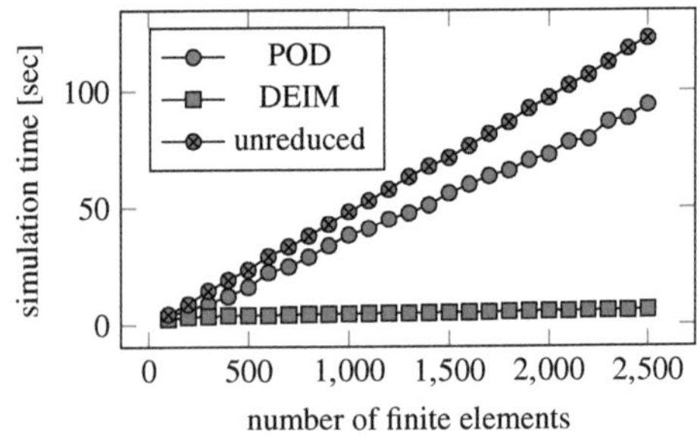

Fig. 5 Computation times of the unreduced and the reduced order models plotted versus the number of finite elements

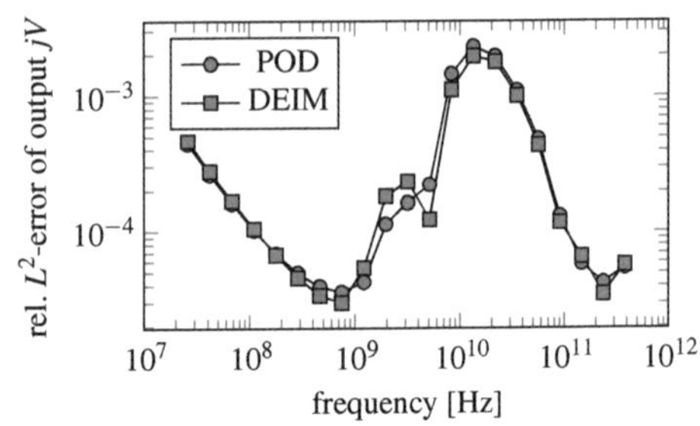

Fig. 6 The reduced models are compared with the unreduced model at various input frequencies

Finally, we in Fig. 6 analyze the behaviour of the models with respect to parameter changes. We consider the frequency of the sinusoidal input voltage as model parameter. The ROMs are created based on snapshots gathered in a full simulation at a frequency of $5 \cdot 10^9$ Hz. We see that the POD model and the POD-DEIM model behave very similarly. The adaptive refinement of the ROM is discussed in [6].

Summarizing all numerical results we conclude that the significantly faster POD-DEIM reduction method yields a ROM with the same qualitative behaviour as the ROM obtained by classical POD-MOR.

Acknowledgements The work reported in this paper was supported by the German Federal Ministry of Education and Research (BMBF), grant no. 03HIPAE5. Responsibility for the contents of this publication rests with the authors.

References

1. Bangerth, W., Hartmann, R., Kanschat, G.: deal.II – a General Purpose Object Oriented Finite Element Library. ACM Trans. Math. Softw. **33**(4) (2007)
2. Brezzi, F., Marini, L., Micheletti, S., Pietra, P., Sacco, R., Wang, S.: Discretization of Semiconductor Device Problems. I. Schilders, W. H. A. et al. (ed.) Handbook of Numerical Analysis, vol. XIII, pp. 317–441, Elsevier/North Holland, Amsterdam (2005)
3. Chaturantabut, S., Sorensen, D.C.: Nonlinear Model Reduction via Discrete Empirical Interpolation. SIAM J. Scientific Comput. **32**(5), 2737–2764 (2010)

4. Demmel, J.W., Eisenstat, S.C., Gilbert, J.R., Li, X.S., Liu, J.W.H.: A Supernodal Approach to Sparse Partial Pivoting. SIAM J. Matrix Anal. Appl. **20**(3), 720–755 (1999)
5. Günther, M., Feldmann, U., ter Maten, J.: Modelling and Discretization of Circuit Problems. Schilders, W.H.A. et al. (ed.) Handbook of Numerical Analysis, vol. 13, pp. 523–629, Elsevier/North Holland, Amsterdam (2005)
6. Hinze, M., Kunkel, M.: Residual Based Sampling in POD Model Order Reduction of Drift-Diffusion Equations in Parametrized Electrical Networks. J. Appl. Math. Mech. **91**(9), 1–14 (2011) URL http://arxiv.org/abs/1003.0551
7. Hinze, M., Kunkel, M., Vierling, M.: POD model order reduction of drift-diffusion equations in electrical networks. In: ter Maten, E. Jan W. (eds.) Lecture Notes in Electrical Engineering, vol. 74, Benner, Peter; Hinze, Michael (2011)
8. Markowich, P.: The Stationary Semiconductor Device Equations. Computational Microelectronics. Springer, New York (1986)
9. Patera, A., Rozza, G.: Reduced Basis Approximation and A Posteriori Error Estimation for Parametrized Partial Differential Equations. Version 1.0. Copyright MIT 2006–2007, to appear in (tentative rubric) MIT Pappalardo Graduate Monographs in Mechanical Engineering (2007)
10. Petzold, L.R.: A Description of DASSL: A Differential/Algebraic System Solver. IMACS Trans. Scientific Comput. **1**, 65–68 (1993)
11. Selberherr, S.: Analysis and Simulation of Semiconductor Devices. Springer, New York (1984)
12. Sirovich, L.: Turbulence and the Dynamics of Coherent Structures I: Coherent Structures. II: Symmetries and Transformations. III: Dynamics and Scaling. Q. Appl. Math. **45**, 561–590 (1987)
13. Tischendorf, C.: Coupled Systems of Differential Algebraic and Partial Differential Equations in Circuit and Device Simulation. Habilitation thesis, Humboldt-University of Berlin (2003)
14. Walther, A., Griewank, A.: A Package for Automatic Differentiation of Algorithms written in C/C++. URL https://projects.coin-or.org/ADOL-C

Model Order Reduction for Complex High-Tech Systems

Agnieszka Lutowska, Michiel E. Hochstenbach, and Wil H.A. Schilders

Abstract This paper presents a computationally efficient model order reduction (MOR) technique for interconnected systems. This MOR technique preserves block structures and zero blocks and exploits separate MOR approximations for the individual sub-systems in combination with low rank approximations for the interconnection blocks. The reduction is demonstrated to be accurate and efficient for a beam-controller system.

1 Introduction

Modeling and simulation of the behavior of complex multi-physical high-tech systems is an important and widely used part of production processes. Accurate simulations with many degrees of freedom are possible but can be inadmissibly time- and memory-consuming: A simulation of all of the electromagnetic, mechanical and acoustic effects of a magnetic resonance imaging (MRI) scanner (see Fig. 1) may take up to a few days. A manner to reduce the required time and computer resources is model order reduction which considerably reduces the size of the system – the amount of degrees of freedom – but preserves the model's characteristics and required accuracy.

MOR techniques for generic systems are well developed and widely used, see for instance [1] for an overview. Specialized MOR techniques for specifically structured systems – such as coupled or interconnected systems – recently receive more and more attention, for instance [2–4] present MOR techniques for block-structured systems which reduce the systems but keep the block structure and the location of the zero blocks.

A. Lutowska (✉) · M.E. Hochstenbach · W.H.A. Schilders
Dept. of Mathematics and Computer Science, TU Eindhoven, 5600 MB Eindhoven, The Netherlands
e-mail: a.lutowska@tue.nl; m.e.hochstenbach@tue.nl; w.h.a.schilders@tue.nl

B. Michielsen and J.-R. Poirier (eds.), *Scientific Computing in Electrical Engineering SCEE 2010*, Mathematics in Industry 16, DOI 10.1007/978-3-642-22453-9_46,

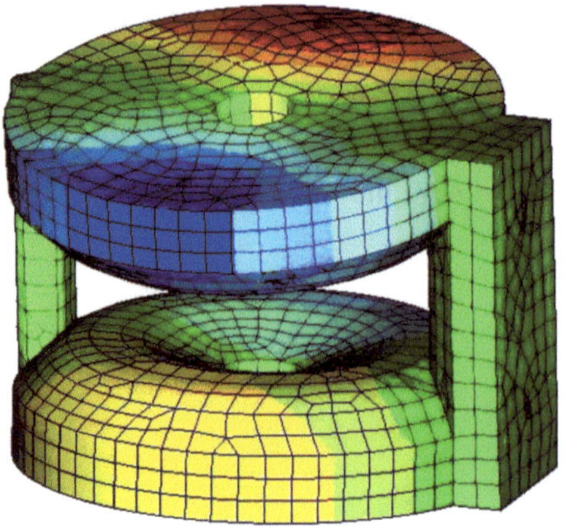

Fig. 1 A vibration analysis MRI scanner model (courtesy of Bert Roozen)

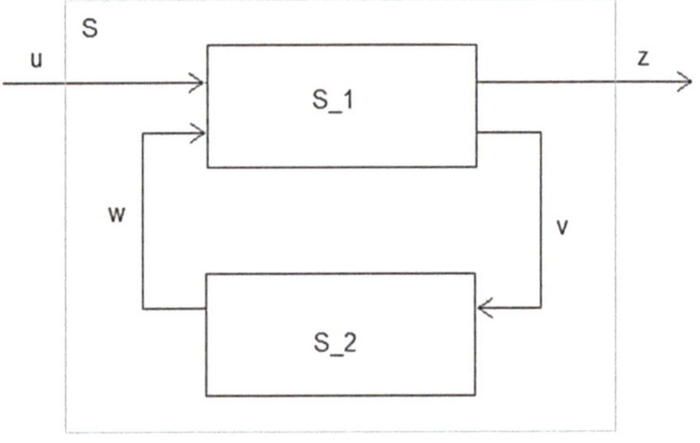

Fig. 2 Schematic representation of the considered systems

In this paper, the authors present a new technique which preserves the block structure of the system matrices and which in addition is computationally more efficient than the previously mentioned methods. The technique is based on the use of MOR approximations for the coupled systems' different sub-systems in combination with singular value decomposition based lower rank approximations for the coupling blocks.

2 Coupled Systems

For the sake of simplicity we focus on a system of two sub-systems where one sub-system's output is used as part of another sub-system's input and vice versa (see for instance Fig. 2). The time domain behavior of the each of the sub-systems S_1 and S_2 is modeled by a system of first order differential-algebraic equations after which the frequency domain behavior is obtained via a Laplace transformation. For the two sub-system example in Fig. 2, along with $N \in \mathbb{N}$ being the number of degrees of freedom of each sub-system, $m \in \mathbb{N}$ – the amount of inputs/outputs of each subsystem, A_{11}, $A_{22} \in \mathbb{R}^{N\times N}$, B_u, B_v, B_w, C_v, C_w, $C_z \in \mathbb{R}^{N\times m}$, $\mathbf{u}$, $\mathbf{v}$, $\mathbf{w}$, $\mathbf{z} \in \mathbb{R}^m$, and $\mathbf{x}_1$, $\mathbf{x}_2 \in \mathbb{R}^N$, this procedure leads to the frequency domain systems:

$$S_1 : \begin{cases} sI\mathbf{x}_1 = A_{11}\mathbf{x}_1 + B_u u + B_w w, \\ z = C_z^T \mathbf{x}_1, \\ v = C_v^T \mathbf{x}_1, \end{cases} \tag{1}$$

$$S_2 : \begin{cases} sI\mathbf{x}_2 = A_{22}\mathbf{x}_2 + B_v v, \\ w = C_w^T \mathbf{x}_2. \end{cases} \tag{2}$$

When the output w of S_2 is used as a part of the input of S_1 and the output v of S_1 is used as a part of the input of S_2, (1) and (2) reduce to an interconnected frequency domain system:

$$S_c\!: \begin{cases} sI\begin{bmatrix}\mathbf{x}_1\\ \mathbf{x}_2\end{bmatrix} = \begin{bmatrix} A_{11} & B_w C_w^T \\ B_v C_v^T & A_{22}\end{bmatrix}\begin{bmatrix}\mathbf{x}_1\\ \mathbf{x}_2\end{bmatrix} + \begin{bmatrix} B_u \\ 0\end{bmatrix} u, \\ z = [C_z^T \quad 0]\begin{bmatrix}\mathbf{x}_1\\ \mathbf{x}_2\end{bmatrix}, \end{cases} \tag{3}$$

where the sub-systems' matrices A_{11} and A_{22} form the block-diagonal of the system matrix of S_c. We assume that the off-diagonal blocks $A_{12} = B_w C_w^T$ and $A_{21} = B_v C_v^T$ are of low rank and show how this property can be exploited in a MOR context. The off-diagonal blocks need not to be sparse. In this paper, the $\mathbf{v}$ and $\mathbf{w}$ are of the same dimension, but this assumption is not necessary. Different dimensions of these vectors will only influence the rank of the off-diagonal blocks $B_w C_w^T$ and $B_v C_v^T$ and not their size, and they would not limit the methods proposed in this paper.

We assume

$$A_c = \begin{bmatrix} A_{11} & B_w C_w^T \\ B_v C_v^T & A_{22}\end{bmatrix}, \quad B_c = \begin{bmatrix} B_u \\ 0\end{bmatrix}, \quad C_c = [C_z^T \quad 0]. \tag{4}$$

Let $\mathbf{x} = [\mathbf{x}_1 \quad \mathbf{x}_2]^T$ and $V, W \in \mathbf{R}^{N\times n}$ and recall that

$$\hat{A}_c\mathbf{x} = \hat{B}_c u, \quad z = \hat{C}_c\mathbf{x}, \tag{5}$$

where

$$\hat{A}_c = W^T(sI - A_c)V, \quad \hat{B}_c = W^T B_c, \quad \hat{C}_c = C_c^T V, \tag{6}$$

is the reduced system related to (3).

3 The Separate Bases Reduction Approach

For the reduction of the linear system (3), i.e., the determination of V and W in (5), one can use any of the generic MOR methods referred to in Sect. 1. Our method reduces each of the sub-systems bases separately and therefore is called separate bases reduction (SBR). We start to show how it differs from the SPRIM approach.

Define the Krylov subspace $K_n(E, v) = [\![v, Ev, E^2 v, \ldots, E^{n-1}v]\!]$. The SPRIM block-structure Krylov subspace technique [4] constructs for $K_n((sI - A_c)^{-1}, (sI - A_c)^{-1}B_c)$ an orthonormal basis of vectors which constitute the columns of

$$\begin{bmatrix} V_1 \\ V_2 \end{bmatrix} \in \mathbb{R}^{2N \times n} \tag{7}$$

and thereafter uses the related block-diagonal matrices

$$V, W = \begin{bmatrix} V_1 & 0 \\ 0 & V_2 \end{bmatrix} \in \mathbb{R}^{2N \times 2n} \tag{8}$$

in order to obtain the reduced system in (5). We propose for the individual systems S_1 in (1) and S_2 in (2) to independently construct orthonormal bases of vectors which constitute the columns of $V_1, V_2 \in \mathbb{R}^{N \times n}$ related to respectively

$$K_n((sI - A_{11})^{-1}(sI - A_{11})^{-1}[B_u B_w]) \text{ and } K_n((sI - A_{22})^{-1}(sI - A_{22})^{-1}B_v) \tag{9}$$

and thereafter to use the SBR matrices

$$V, W = \begin{bmatrix} V_1 & 0 \\ 0 & V_2 \end{bmatrix} \in \mathbb{R}^{2N \times 2n} \tag{10}$$

in order to obtain the reduced system in (5). This method preserves the block structure of the coupled system as well as the zero blocks and can be applied also to the case of more than two coupled sub-systems.

The application of the components of the block-diagonal matrix V, namely V_1 and V_2, to each of the corresponding sub-systems separately, would result in the MOR procedure that exhibits the moment matching property. However, in case of the separate bases reduction, the moment matching property of the reduced coupled system has not been proved. Further studies on the relation between moment expansion coefficients of the sub-systems and those of the coupled system are in the scope of the future work.

4 SBR in Combination with Low-Rank Approximation

In this section, we show how to construct low-rank approximations for the A_{12} and A_{21} blocks which can used for an efficient calculation of the inverse in (5) – based on the Sherman–Morrison formula (see [5, (2.1.4)]).

For the sake of simplicity assume that A_{11} and A_{22} are scaled, i.e., that for instance $||A_{11}||_2 = 1$ and $||A_{22}||_2 = 1$ (a pre- and post-multiplication of A_c with a suitable diagonal matrix leads to the desired result). If the blocks are not of low rank by nature (i.e., if it does not hold that $m << N$) one can construct a low rank approximation with the use of generalized singular value decomposition (GSVD) – see [5, Theorem 8.7.4] for details – as follows.

We use GSVD to factor the two block-column matrices

$$\begin{bmatrix} A_{11} \\ A_{21} \end{bmatrix} = \begin{bmatrix} U_1 S_1 X_1' \\ V_1 C_1 X_1' \end{bmatrix}, \quad \begin{bmatrix} A_{22} \\ A_{12} \end{bmatrix} = \begin{bmatrix} U_2 S_2 X_2' \\ V_2 C_2 X_2' \end{bmatrix} \tag{11}$$

where S_i and C_i are diagonal matrices. We approximate $V_i C_i X_i'$ by a low(er)-rank blocks $\hat{V}_i \hat{C}_i \hat{X}_i'$ where $\hat{V}_i$ and $\hat{X}_i$ consist of the first $k_i < m$ columns of V_i and X_i, respectively, and $\hat{C}_i$ is the top-left $k_i \times k_i$ sub-matrix of the matrix C_i. This provides us with a low-rank approximation

$$\begin{bmatrix} \hat{A}_{21} \\ \hat{A}_{12} \end{bmatrix} := \underbrace{\begin{bmatrix} \hat{V}_1 \hat{C}_1 \hat{X}_1' \\ \hat{V}_2 \hat{C}_2 \hat{X}_2' \end{bmatrix}}_{\hat{B}_i \hat{C}_i^T} \approx \underbrace{\begin{bmatrix} V_1 C_1 X_1' \\ V_2 C_2 X_2' \end{bmatrix}}_{B_i C_i^T} = \begin{bmatrix} A_{21} \\ A_{12} \end{bmatrix}.$$

The low-rank property of the coupling blocks of the matrix A_c may be advantageous, e.g., while calculating the transfer function of the system (3), which is given by

$$H(s) = \begin{bmatrix} C_z^T & 0 \end{bmatrix} \begin{bmatrix} sI - A_{11} & -B_w C_w^T \\ -B_v C_v^T & sI - A_{22} \end{bmatrix}^{-1} \begin{bmatrix} B_u \\ 0 \end{bmatrix}. \tag{12}$$

If the products $B_v C_v^T$ and $B_w C_w^T$ are approximated by lower-rank ones, $\hat{B}_v \hat{C}_v^T$ and $\hat{B}_w \hat{C}_w^T$, respectively, one can evaluate the inverse in a computationally cheaper way. This can be done by applying a generalized Sherman-Morrison formula (see [5, (2.1.4)]), which for an arbitrary nonsingular matrix M expressed as a sum of a nonsingular matrix Z and its low-rank update UV^T reads

$$M^{-1} = (Z + UV^T)^{-1} = Z^{-1} - Z^{-1}U(I + V^T Z^{-1} U)^{-1} V^T Z^{-1}. \tag{13}$$

In our case, the matrix to be inverted in the transfer function (12) can be decomposed into

$$\begin{bmatrix} sI - A_{11} & -B_w C_w^T \\ -B_v C_v^T & sI - A_{22} \end{bmatrix} = \begin{bmatrix} sI - A_{11} & 0 \\ 0 & sI - A_{22} \end{bmatrix} - \begin{bmatrix} \hat{B}_w \\ 0 \end{bmatrix} [0 \quad \hat{C}_w^T] - \begin{bmatrix} 0 \\ \hat{B}_v \end{bmatrix} [\hat{C}_v^T \quad 0] \tag{14}$$

and the Sherman-Morrison formula will have to be applied twice. With this procedure, the calculation of the inverse of the full size matrix is replaced by calculation of the inverses of the smaller sub-blocks, making it computationally more efficient.

5 Numerical Results

As benchmark we use two models of a coupled beam-controller system, one with 120 and the other one with 80 degrees of freedom. Each model consists of two sub-systems, the first sub-system S_1 is a linear beam subjected to mechanical vibrations. As outputs, its displacement is measured at certain points and is used as input for the second system S_2, the controller. Part of the output of the controller is the input of the beam.

The first system consists of 120 degrees of freedom, 60 of them correspond to the beam and 60 to the controller. The reduced systems, after applying SPRIM and, on the other hand, SBR with reduced-rank approximation, have 30 and 31 degrees of freedom, respectively. In both cases, the multi point based projection space was built, with the same sample frequency values used to create the bases for the beam, the controller, and the coupled system. Figure 3 shows three plots of the magnitudes of the transfer functions of the coupled systems as the function of the frequency. The three transfer functions are calculated for the original system, the system reduced after applying SPRIM method, and the one reduced using the uncoupled bases approach. The vertical dashed lines mark the frequency values that were used as sampling frequency points. We can conclude that the reduced and the unreduced transfer functions match well in the region of interest. We independently reduce the sub-systems which is more efficient than in case of SPRIM algorithm.

Table 1 presents the accumulated computational times in seconds needed to construct the reduction bases a 1000 subsequent times. Column 1 shows the time

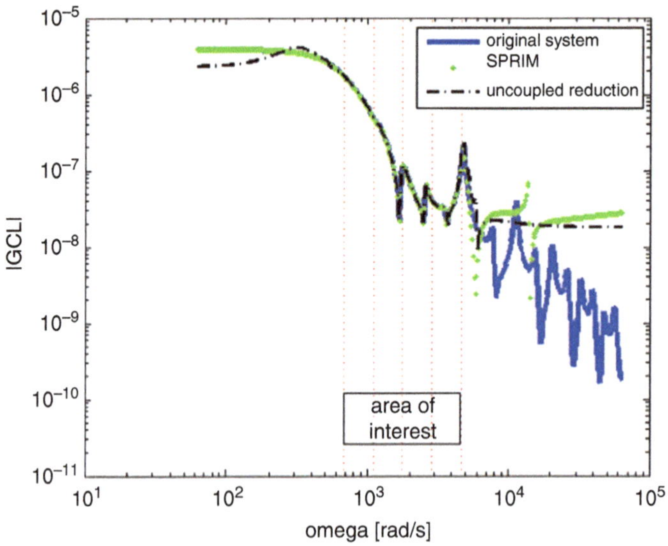

Fig. 3 Comparison of the transfer functions reduced by SPRIM and SBR

Table 1 The reduction basis construction time (in [s])

SPRIM	V_1 and V_2	V_1	V_2
0.034	0.183	0.103	0.095
64.003	62.135	37.224	30.965

required to construct the reduction bases (8). Next, column 2 shows the time required to, serially, calculate the reduction bases (9) and (10) (first V_1, next V_2). Finally, columns 3 and 4 show the time expected for the calculation of same bases, (9) and (10), parallel. In column 3 only V_1 is calculated (and the computation of V_2 is assumed to be equally fast) and in column 4 vice versa. From the results it can be observed, that building the reduction bases V_1 or V_2 is much faster that building the basis in the way suggested by SPRIM algorithm. It should be noted, that the considered system is of a relatively small size. For higher- dimensional systems, even larger advantage (with respect to the computational cost) of using the SBR algorithm is expected. It is caused by the fact, that the computational cost of the SPRIM and SBR methods is mainly influenced by the cost of computing the matrix inverse.

Figure 4 compares the magnitude of the unreduced transfer function of the second system with 80 degrees of freedom (40 for beam and 40 for the controller), as a function of the frequency, to the magnitude of the transfer function obtained after calculating a low-rank approximation of the coupling blocks. Originally, the coupling blocks are of rank 10 and, after application of our algorithm, they can be well approximated by a rank 6 matrix for the sub-block A_{12} and rank 4 matrix for the sub-block A_{21}. In this example, no reduction in size of the system matrices of S_c in (3) is done.

Table 2 shows the accumulated computational times of calculating the inverse of the matrix $\tilde{A}_c$ being the low-rank approximation of matrix A_c (negative of the left hand side of (14) for $s = 0$) 10000 subsequent times. Four cases are considered, each in therein related column

1. $\tilde{A}_c^{-1}$: The matrix inverse is based on the MATLAB "\" operator.
2. S-M (1) : The Sherman -Morrison formula is applied to the matrix $\tilde{A}_c$ and the MATLAB "\" operator is used to calculate the inverse of the corresponding block-diagonal matrix on the right hand side of (14) .
3. S-M (2) : same as S-M (1), except that the inverse of the block-diagonal matrix on the right hand side of (14) is calculated by applying the MATLAB "\" operator to each diagonal block individually.
4. S-M (3) same as S-M (2), except that to simulate parallelism, only the inverse of the diagonal sub-block A_{11} is calculated by the MATLAB "\" operator.

The first and the last approach give similar results. This can be again explained by the small size of the studied system. For larger systems, the last approach is expected to have much smaller computational cost than the first one.

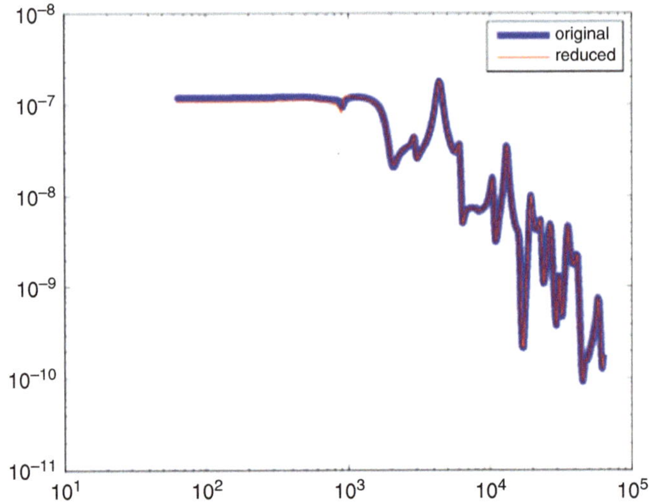

Fig. 4 Transfer function after applying a low-rank approximation

Table 2 The time to compute the inverse of the matrix $\tilde{A}_c$ in [s]

$\tilde{A}_c^{-1}$	S-M (1)	S-M (2)	S-M (3)
0.0005	0.0008	0.0010	0.0012
3.878	5.319	4.581	3.583

6 Conclusion

Tables 1 and 2 show that the presented SBR algorithm in combination with low rank approximation of the off-diagonal blocks is promising and can serve as a fast alternative to the existing MOR methods.

References

1. Antoulas, A.: Approximation of large-scale dynamical systems. Advances in Design and Control, SIAM, Philadelphia (2005)
2. Bai, Z., Li, R., Su Y.: A unified Krylov projection framework for structure-preserving model reduction. In: Model Order Reduction: Theory, Research Aspects and Applications, pp. 77–95 (2008)
3. Fernández Villena, J., Schilders, W.H.A., Silveira, L.M.: Block oriented model order reduction of interconnected systems, Technical Report (2009)
4. Freund, R.W.: SPRIM: Structure-preserving reduced-order interconnect macromodelling. In: Proceedings of the 2004 IEEE/ACM International conference on Computer-aided design, pp. 80–87, IEEE Computer Society, Washington, DC (2004)

5. Golub, G.H., Van Loan, C.F.: Matrix Computations. 3rd edn., Johns Hopkins University Press, Baltimore, MD (1996)
6. Salimbahrami, B., Lohmann, B.: Krylov Subspace Methods in Linear Model Order Reduction: Introduction to Invariance Properties. Technical report (2002)

Parametric Model Order Reduction by Neighbouring Subspaces

Kynthia Stavrakakis, Tilmann Wittig, Wolfgang Ackermann, and Thomas Weiland

Abstract Electrodynamic field simulations in the frequency domain typically require the solution of large linear systems. Model Order Reduction (MOR) techniques offer a fast approach to approximate the system impedance with respect to the frequency parameter. Most commonly, MOR via projection is applied associated with certain Krylov projection matrices. During the design process it is desirable to vary specified parameters like the frequency, geometry details as well as material parameters, giving rise to multivariate dynamical systems. In this work, a multivariate MOR method is presented for parameterized systems based on the Finite Integration Technique (FIT). It utilizes the observation, that for small parameter variations the matrices associated with the univariate MOR differ only slightly. Thus, the multivariate MOR method is deduced from the usage of specified univariate subspaces.

1 Introduction to Model Order Reduction for Large Dynamical Systems

In electrodynamic field computations the continuous Maxwell equations are typically discretized in the space variables, i.e the continuous space is mapped to a finite set of discrete elements leading to a system of differential equations constituting the

K. Stavrakakis (✉) · W. Ackermann · T. Weiland
Institut für Theorie Elektromagnetischer Felder, Technische Universität Darmstadt, Schlossgartenstr. 8, 64289 Darmstadt, Germany
e-mail: stavrakakis@temf.tu-darmstadt.de; ackermann@temf.tu-darmstadt.de; weiland@temf.tu-darmstadt.de

T. Wittig
Computer Simulation Technology AG, Bad Nauheimer Str. 19, 64289 Darmstadt, Germany
e-mail: tilmann.wittig@cst.com

B. Michielsen and J.-R. Poirier (eds.), *Scientific Computing in Electrical Engineering SCEE 2010*, Mathematics in Industry 16, DOI 10.1007/978-3-642-22453-9_47,

Maxwell Grid Equations. On the basis of these equations, in this work we consider dynamical systems, denoted by $\mathbf{\Sigma}$, where in- and output vectors **i** and **u** respectively and an auxiliary vector **x** are defined. For a thorough introduction to mathematical systems see [1]. As the size of this system can be very large, due to limited computational, accuracy and storage capabilities, simplified models which capture the main features of the original model are needed. The simplified models are then used instead of the original models. In the past years, methods of Model Order Reduction (MOR) have been developed to determine an approximate dynamical system $\hat{\mathbf{\Sigma}}$ by appropriately reducing the number of equations describing the initial system. The approximation error should be small, important system properties as stability and passivity should be preserved and the procedure should be stable and efficient.

In order to show the basic idea of MOR, let $\mathbf{\Sigma}$ be a linear, time-invariant (LTI) system consisting, for simplicity, of first-order ODEs, i.e let $\mathbf{\Sigma}$ be in the classical state-space form. Then, in the frequency domain, with the complex frequency parameter s, it is:

$$\mathbf{\Sigma} : \begin{cases} s\mathbf{x}(s) = \mathbf{A}\mathbf{x}(s) + \mathbf{B}\mathbf{i}(s), \\ \mathbf{u}(s) = \mathbf{C}\mathbf{x}(s) + \mathbf{D}\mathbf{u}(s), \end{cases} \tag{1}$$

with **i** and **u** the in- and output vectors and **x** the state vector of the system. **A** is the system matrix and the **B** and **C** are matrices related to the in- and the output vector, respectively. The MOR-step consists in approximating $\mathbf{\Sigma}$ by a system $\hat{\mathbf{\Sigma}}$

$$\hat{\mathbf{\Sigma}} : \begin{cases} s\hat{\mathbf{x}}(s) = \hat{\mathbf{A}}\hat{\mathbf{x}}(s) + \hat{\mathbf{B}}\mathbf{i}(s), \\ \mathbf{u}(s) = \hat{\mathbf{C}}\hat{\mathbf{x}}(s) + \hat{\mathbf{D}}\mathbf{i}(s), \end{cases} \tag{2}$$

by an appropriate reduction of the number of equations of $\mathbf{\Sigma}$. Details about the sense of this reduction will be given in the next sections. As s is the only variable in this case, we refer to the univariate case. The most common MOR methods are based on projection in an appropriate subspace, as explained in the following.

2 Univariate Moment-Matching MOR via Projection

MOR methods by projection correspond to truncation in an appropriate subspace. Let **x** live in $\mathbb{R}^{n\times 1}$ and consider the change of basis $\mathbf{T} \in \mathbb{R}^{n\times n}$ in the state space $\bar{\mathbf{x}} = \mathbf{T}\mathbf{x}$. The quantities $\mathbf{x}, \mathbf{T}$ and $\mathbf{T}^{-1}$ are partitioned as follows

$$\bar{\mathbf{x}} = \begin{pmatrix} \hat{\mathbf{x}} \\ \tilde{\mathbf{x}} \end{pmatrix}, \quad \mathbf{T}^{-1} = \begin{bmatrix} \mathbf{V}\ \mathbf{T}_1 \end{bmatrix}, \quad \mathbf{T} = \begin{bmatrix} \mathbf{W}^* \\ \mathbf{T}_2^* \end{bmatrix}, \tag{3}$$

where $\hat{\mathbf{x}} \in \mathbb{R}^m, \tilde{\mathbf{x}} \in \mathbb{R}^{n-m}, \mathbf{V}, \mathbf{W} \in \mathbb{R}^{n\times m}$. Substituting for **x** in (1) and retention of only the first m differential equations leads to

$$\begin{aligned} s\hat{\mathbf{x}} &= \mathbf{W}^*\mathbf{A}(\mathbf{V}\hat{\mathbf{x}} + \mathbf{T}_1\tilde{\mathbf{x}}) + \mathbf{W}^*\mathbf{B}\mathbf{i}, \\ \mathbf{u} &= \mathbf{C}(\mathbf{V}\hat{\mathbf{x}} + \mathbf{T}_1\tilde{\mathbf{x}}) + \mathbf{D}\mathbf{i}, \end{aligned} \tag{4}$$

The approximation occurs by neglecting the term $\mathbf{T}_1\tilde{\mathbf{x}}$. The resulting approximant is defined by the following matrices:

$$\hat{\mathbf{A}} = \mathbf{W}^*\mathbf{A}\mathbf{V}, \quad \hat{\mathbf{B}} = \mathbf{W}^*\mathbf{B}, \quad \hat{\mathbf{C}} = \mathbf{C}\mathbf{V}. \tag{5}$$

The determination of $\mathbf{V}$ and $\mathbf{W}$ depends on the system requirements. In [2] a detailed description of univariate projection methods is given. As the systems treated in this work are uniquely determined by their transfer function $\mathbf{Z}$, which is defined by $\mathbf{u} = \mathbf{Z}\mathbf{i}$, and are rational functions, [2], one way to approximate the original system is to approximate its transfer function by a rational function of lower degree. Consider the Laurent series expansion of $\mathbf{Z}$ around s_0:

$$\mathbf{Z}(s) = \sum_{k=0}^{\infty} \underbrace{\mathbf{C}(-(\mathbf{A} + s_0\mathbf{I}))^{-k}\mathbf{B}}_{\mathbf{M}_k} (s - s_0)^k, \tag{6}$$

where $\mathbf{M}_k$ is the kth moment of the system. The approximation of $\mathbf{Z}$ can be achieved by matching some of the moments of the series expansion with the transfer function of the reduced system $\hat{\mathbf{Z}}$. As the direct calculation of the moments is an ill-conditioned problem, implicit methods are recommended which iteratively build up the matrices $\mathbf{V}$ and $\mathbf{W}$. In general, $\mathbf{V}$ and $\mathbf{W}$ can be different, but, by choosing $\mathbf{V} = \mathbf{W}$, stability and passivity are preserved in the reduced model, [3]. The iteration is accomplished by means of the Arnoldi algorithm, which produces $\mathbf{V}$, such that $\text{colsp}\{\mathbf{V}\} \supseteqq \mathscr{K}_q(\mathbf{A}, \mathbf{x})$, where $\text{colsp}\{\mathbf{V}\}$ denotes the column space of $\mathbf{V}$ and $\mathscr{K}_q(\mathbf{A}, \mathbf{x})$ is the Krylov subspace related to $\mathbf{A}$ with dimension q. In this way, the original system $\mathbf{\Sigma}$ is reduced to the system $\hat{\mathbf{\Sigma}}$.

3 Multivariate Parameterized Systems Resulting from the Maxwell Grid Equations in the Finite Integration Theory

Often, it is desirable to vary specified model parameters, for example the frequency, geometry details or material parameters. Let $\boldsymbol{\xi} = (\xi_1, \xi_2, \ldots, \xi_r)$ be the vector containing the variable parameters, excluding the frequency parameter s. The multivariate dynamical system is described by:

$$\mathbf{\Sigma}_{\text{param}} : \begin{cases} s\mathbf{x}(s) = \mathbf{A}(\boldsymbol{\xi})\mathbf{x}(s) + \mathbf{B}(\boldsymbol{\xi})\mathbf{i}(s), \\ \mathbf{u}(s) = \mathbf{C}(\boldsymbol{\xi})\mathbf{x}(s) + \mathbf{D}(\boldsymbol{\xi})\mathbf{i}(s), \end{cases} \tag{7}$$

Fig. 1 Model Order Reduction of a multivariate parameterized system

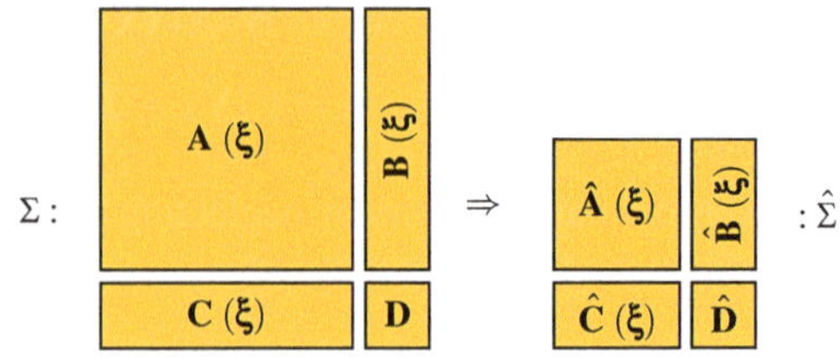

Analogously to the one-parameter case, a multivariate MOR method consists in developing a system $\hat{\boldsymbol{\Sigma}}$ by appropriately reducing the number of initial differential equations. We require that the dependence on the parameter-vector $\boldsymbol{\xi}$ remains also in the reduced system, as visualized in Fig. 1. Notice, that to each parameter vector $\boldsymbol{\xi}$ corresponds one system, which will denoted by $\boldsymbol{\Sigma}_{\boldsymbol{\xi}}$ in the following.

In this work, the Maxwell Grid Equations are obtained from the continuous Maxwell equations with the help of the Finite Integration Technique (FIT) [4,5]:

$$\begin{aligned} \mathbf{C}_{\mathrm{FIT}}\widehat{\mathbf{e}} &= -\tfrac{\mathrm{d}}{\mathrm{dt}}\mathbf{M}_{\mu}\widehat{\mathbf{h}}, & \mathbf{S}\mathbf{M}_{\mu}\widehat{\mathbf{h}} &= 0, \\ \widetilde{\mathbf{C}}_{\mathrm{FIT}}\widehat{\mathbf{h}} &= (\tfrac{\mathrm{d}}{\mathrm{dt}}\mathbf{M}_{\varepsilon} + \mathbf{M}_{\Sigma})\widehat{\mathbf{e}} + \widehat{\widehat{\mathbf{j}}}_{s}, & \widetilde{\mathbf{S}}\mathbf{M}_{\varepsilon}\widehat{\mathbf{e}} &= 0. \end{aligned} \tag{8}$$

Here, $\mathbf{M}_{\varepsilon}, \mathbf{M}_{\Sigma}$ and $\mathbf{M}_{\mu}$ are diagonal matrices, which express the mesh geometry and the material property of each meshcell. The matrices $\mathbf{S}$ and $\mathbf{C}_{\mathrm{FIT}}$ (as well as $\widetilde{\mathbf{S}}$ and $\widetilde{\mathbf{C}}_{\mathrm{FIT}}$) are topology matrices representing the divergence and the curl operator, respectively. In the following, losses are discarded ($\mathbf{M}_{\Sigma} = \mathbf{0}$).

We apply a Laplace transformation and consider the system in the Laplace-domain with the complex frequency parameter s. One way to obtain a dynamical system that relates input to output is to eliminate one of the vectors $\widehat{\mathbf{e}}$ or $\widehat{\mathbf{h}}$. For example, elimination of $\widehat{\mathbf{h}}$ leads to the discrete Helmholtz Equation

$$\begin{aligned} \mathbf{M}_{\varepsilon}s^{2}\widehat{\mathbf{e}} + \mathbf{C}_{\mathrm{FIT}}{}^{\mathrm{T}}\mathbf{M}_{\mu}^{-1}\mathbf{C}_{\mathrm{FIT}}\widehat{\mathbf{e}} &= s\mathbf{B}\mathbf{i}, \\ \mathbf{u} &= \mathbf{C}\widehat{\mathbf{e}}, \end{aligned} \tag{9}$$

where the matrices $\mathbf{B}$ and $\mathbf{C}$ have been introduced. If the considered structure is excited at m ports, we can define the input at the ports in terms of the matrix $\mathbf{B}$ and the generalized current $\mathbf{i}$, i.e. $-\widehat{\widehat{\mathbf{j}}}_{s} = \mathbf{B}\mathbf{i}$. Analogously, we can define the output voltages in terms of the vector $\widehat{\mathbf{e}}$ and the matrix $\mathbf{C}$. Notice, that (9) does not correspond to a first-order differential equation, nevertheless the same MOR methods as for first-order differential equations can be applied, as every higher-order differential equation can be transformed in a first-order one, [3]. Also in [3], a detailed description about univariate MOR methods using the FIT can be found.

When parameter variations, such as material or geometry parameters, come into play, the matrices $\mathbf{M}_{\varepsilon}, \mathbf{M}_{\mu}$ are affected and (9) becomes:

$$\begin{aligned} \mathbf{M}_{\varepsilon}(\boldsymbol{\xi})s^{2}\widehat{\mathbf{e}} + \mathbf{C}_{\mathrm{FIT}}{}^{\mathrm{T}}\mathbf{M}_{\mu}^{-1}(\boldsymbol{\xi})\mathbf{C}_{\mathrm{FIT}}\widehat{\mathbf{e}} &= s\mathbf{B}(\boldsymbol{\xi})\mathbf{i}, \\ \mathbf{u} &= \mathbf{C}(\boldsymbol{\xi})\widehat{\mathbf{e}}. \end{aligned} \tag{10}$$

4 MOR-Techniques for Multivariate Dynamical Systems

In this section, a multivariate MOR method is presented that assigns a reduced order system $\hat{\mathbf{\Sigma}}_{\boldsymbol{\xi}}$ to the system $\mathbf{\Sigma}_{\boldsymbol{\xi}}$ described by (10), that keeps the parameter dependence on the parameter-vector $\boldsymbol{\xi}$. According to the projection framework, for the multivariate case, this is achieved by assigning a common projection matrix $\mathbf{V}$ to all multivariate systems $\mathbf{\Sigma}_{\boldsymbol{\xi}}$ such that $\hat{\mathbf{Z}}_{\xi}$ approximates $\mathbf{Z}_{\xi}$.

Among the several approaches existing to calculate $\mathbf{V}$, we point out [6] and [7], in which multivariate moment matching techniques have been developed that not only match some of the first moments with respect to s, but also with respect to the parameter-vector $\boldsymbol{\xi}$. Their application requires an explicit dependence on the parameters, which is not given in (10). There, only the frequency parameter s appears explicitly. Other parameters, e.g. the material parameters ε and μ, or geometry parameters are implicitly included in the matrices $\mathbf{M}_{\varepsilon}$ and $\mathbf{M}_{\mu}$ in a nonlinear dependence. In [9], a linearization method is presented for FIT-systems depending on frequency and rectilinear length variation, which results in an explicit specification of the parameters. A three-dimensional version of this method has been also implemented. This intermediate step, besides being calculational demanding and adding further error to the subsequent order reduction, can only be accomplished when the topology of the mesh remains the same for all parameter changes. Obviously, this is a strong limiting factor for geometrical variations, as the systems often result from FIT-models with automatically created meshes, which are not necessarily the same. Nevertheless, for material parameters these methods form a powerful tool for MOR.

In this work, an approach is presented which circumvents the topology preservation, by using univariate MOR methods on several expansion points in the parameter range. We start by observing, that in case of small geometry variations around an expansion point $\boldsymbol{\xi}_0$, the matrices $\mathbf{V}_{\xi}$ related to each $\mathbf{\Sigma}_{\xi}$ in the neighbourhood of $\boldsymbol{\xi}_0$ differ only slightly. Therefore, without introducing large error, we could use $\mathbf{V}_0$ for projection for the neighbouring subspaces corresponding to the systems $\mathbf{\Sigma}_{\xi}$, instead of $\mathbf{V}_{\xi}$. This context is illustrated by the model of Fig. 2, where the varied parameters are the length l, the width w and the depth d, in x-, y- and z-direction, respectively. The variation ranges are also indicated in the figure. We define $\boldsymbol{\xi} = (l, w, d)$. In Fig. 3a, the matrix $\mathbf{V}_0$ related to the middlepoint of the variation range, $\boldsymbol{\xi}_0 = (9, 10.5, 1.05)$, has been used for all systems $\mathbf{\Sigma}_{\xi}$ quoted in the figure. The absolute logarithmic error of the S-Parameter S_{11} compared to a solution of full FIT-calculations lies in the range of 10^{-4}, which is satisfactory in practical applications.

In order to increase the geometry variation range, we can pick several expansion points $\boldsymbol{\xi}_i, i = 1 \dots N$, and build up the matrix $[\mathbf{V}_1\ \mathbf{V}_2\ \dots \mathbf{V}_N]$. Let m_i be the number of columns of each $\mathbf{V}_i$. In most of the subspaces corresponding to $\mathbf{V}_i$, common directions appear. In order to set up $\mathbf{V}$, the directions are sorted by relative importance with the aid of the singular value decomposition (SVD) [2], i.e. $[\mathbf{V}_1\ \mathbf{V}_2\ \dots \mathbf{V}_i] = \mathbf{U}\mathbf{\Sigma}\mathbf{N}^*$, where $\mathbf{U}$ and $\mathbf{N}$ are $n \times n$ and $\sum_{i=1}^{N} m_i \times \sum_{i=1}^{N} m_i$

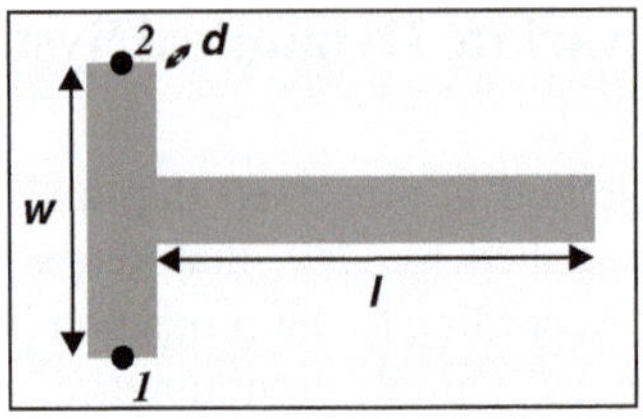

Parameter	Range
frequency s	0…5 GHz
length l	8…10 mm
width w	10…11 mm
depth d	1…1.1 mm

Fig. 2 Testmodel: Microstrip line discretized with 5,491 unknowns

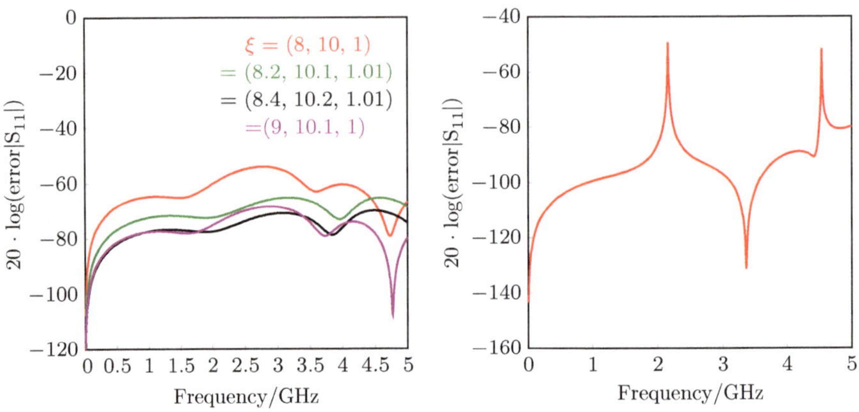

Fig. 3 Error of the S-Parameter S_{11} in a logarithmic scale for the Microstrip example of Fig. 2 (*left*) The subspace corresponding to the expansion point $\boldsymbol{\xi}_0 = (9, 10.5, 1.05)$ has been used for all systems $\boldsymbol{\Sigma}_{\xi}$ quoted here (*right*) Five expansion points have been chosen and the first ten columns of each $\mathbf{V}_1 \ldots \mathbf{V}_5$ have been taken to set up $\mathbf{V}$

unitary matrices and $\boldsymbol{\Sigma}$ is an $n \times \sum_{i=1}^{N} m_i$ matrix containing the singular values of $[\mathbf{V}_1\ \mathbf{V}_2\ \ldots \mathbf{V}_N]$. Choosing for $\mathbf{V}$ the first k columns of $\mathbf{U}$, guarantees to capture the k most important directions of $[\mathbf{V}_1\ \mathbf{V}_2\ \ldots \mathbf{V}_N]$. Apparently, besides the number of expansion points, also the number of columns of each $\mathbf{V}_i$, as well as the number of directions k kept in $\mathbf{V}$ can be chosen freely. Two examples will illustrate these aspects.

For the testmodel of Fig. 2, five expansion points, randomly spread in the parameter range, have been chosen. Each $\mathbf{V}_i, i = 1 \ldots N = 5$, has 10 columns, which is a typical value for univariate MOR, and $k = 50$, that is, all directions of the corresponding $\mathbf{U}$ have been used in order to set-up $\mathbf{V}$. The results are shown in Fig. 3b.

Many possible choices of expansion points exist. For geometry variations it makes sense to include the edge points of the parameter range and then successively take further intermediate points. Let $\mathscr{S}_0 \subseteq \mathscr{S}_1 \cdots \subseteq \mathbb{R}^3$ be the sets of points which are build up in this way. In Fig. 4 the geometrical variation range of the model is shown. $\mathscr{S}_0$ contains only point $\boldsymbol{\xi}_0$, $\mathscr{S}_1$ contains $\boldsymbol{\xi}_0$ and the three edge points (red dots) of the parameter set and $\mathscr{S}_2$ contains the points of $\mathscr{S}_1$ as well as one

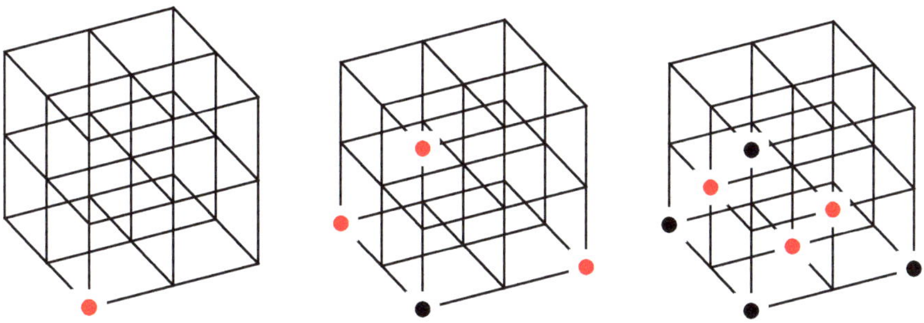

Fig. 4 Choice of expansion points, from *left* to *right*: $\mathscr{S}_0$, $\mathscr{S}_1$ and $\mathscr{S}_2$

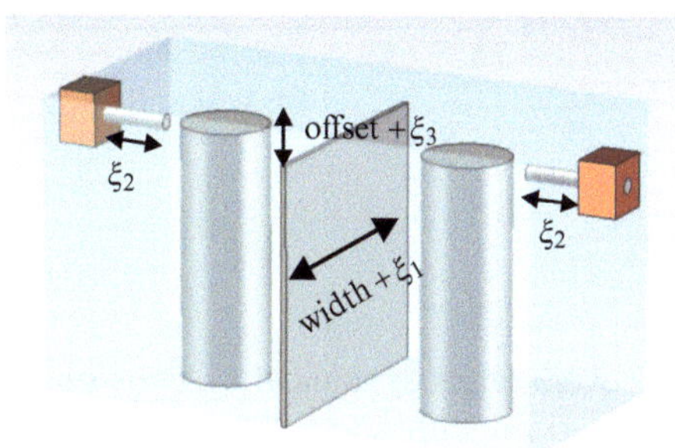

Parameter	Range
frequency s	$0.58 \ldots 0.63$ GHz
width ξ_1	$0 \ldots 6$ mm
coupling ξ_2	$0 \ldots 4$ mm
iris ξ_3	$0 \ldots 4$ mm

Fig. 5 Filter serving as a testmodel. Besides the frequency, the variable parameters are the width (ξ_1), the coupling (ξ_2) and the iris (ξ_3)

intermediate point in each direction (red dots). This choice allows an automated construction of the expansion points. Formally, the qth set is defined by the following expression [6]:

$$\mathscr{S}_q = \{(\boldsymbol{\gamma}, q - |\boldsymbol{\gamma}|, |\boldsymbol{\gamma}| \leq q)\}, \quad q \geq 0, \tag{11}$$

in which $\boldsymbol{\gamma} = (\gamma_1, \gamma_2) \geq \mathbf{0}$ are multiindices of dimension 2, [10].

This set of points has been used for the filter in Fig. 5. Figure 6 shows the logarithmic error of the S-Parameter S_{11} for different numbers of expansion points N. Obviously, here, one expansion point is not sufficient. The number m_i of columns of each $\mathbf{V}_i, i = 1 \ldots N$, is chosen as 16. Again, all columns of $\mathbf{U}_i$ have been used for $\mathbf{V}_i$, but e.g. for $N = 10$ (black curve in Fig. 6), without large accuracy loss, half of the singular vectors in $\mathbf{U}_{N=10}$ can be omitted when setting up $\mathbf{V}_{N=10}$ (80 vectors instead of 160). The average calculation time $\bar{t}_{calc}$ of each $\mathbf{V}_i$ was approximately 1 s, so that the calculation time for $\mathbf{V}$ was less than 10 s. For a parameter sweep of M parameter sets $\boldsymbol{\xi}$, usage of the common $\mathbf{V}$ would be $\frac{M}{N}$ faster than taking for each $\boldsymbol{\xi}$ its own projection matrix $\mathbf{V}$. Consider that multivariate MOR methods are recommended for large parameter sweeps ($M \gg N$).

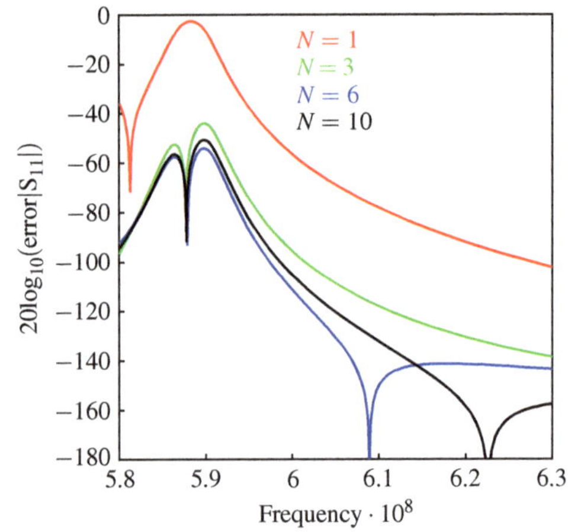

Fig. 6 Results for the Filter depicted in Fig. 5

5 Conclusion

Subject of investigation in this work were large dynamical systems obtained on the basis of the FIT which are parameterized with several variables, in particular geometrical variables. A MOR method has been proposed, which is based on the fact that for small geometrical variations, as this is the case e.g. in filter optimization, the system matrices differ only slightly, thus also the respective Krylov subspaces. The method uses the univariate MOR algorithms to calculate the projection matrix at several expansion points in the parameter domain. These matrices are put together and a SVD is applied. The resulting matrix composes the projection matrix for the multivariate problem. The method is applied to two numerical examples and the S-Parameters agree well with the comparison results of full FIT-calculations.

References

1. Poldeman, J., Willems, J.: Introduction to Mathematical Systems Theory. Springer, Berlin (1998)
2. Antoulas, A.: Approximation of Large Scale Dynamical Systems, SIAM, Advances in Design and Control (2005)
3. Wittig, T.: Zur Reduzierung der Modellordnung in elektromagnetischen Feldsimulationen, PhD Thesis, Cuvillier Verlag Göttingen (2003)
4. Weiland, T.: Eine Methode zur Lösung der Maxwellschen Gleichungen für sechskomponentige Felder auf diskreter Basis Electronics and Communication (AEÜ) **31**(3), 116–120 (1977)
5. Weiland, T.: Time domain electromagnetic field computation with finite difference methods, Int. J. Num. Modelling **9**, 295–319 (1996)
6. Codecasa, L.: Boundary condition independent compact dynamic thermal networks of packages. IEEE Trans. Compon. Packag. Tech. **28**(4) (2005)

7. Farle, O., Ordnungsreduktionsverfahren für die Finite-Elemente-Simulation parameterabhängiger passiver Mikrowellenstrukturen, PhD Thesis, Online Catalogue of the "Saarländische Universitäts- und Landesbibliothek", 2007
8. Wittig, T., Schuhmann, R., Weiland, T.: Model order reduction for large systems in computational electromagnetics. LAA 2006 **415**(2), 499–530 (2006)
9. Stavrakakis, K., Wittig, T., Ackermann, W., Weiland, T.: Linearization of parametric FIT-discretized systems for model order reduction. IEEE Trans. Magn. (IEEE Mag) **45**(3), 1380–1383 (2009)
10. Gasca, M., Sauer, T.: Polynomial interpolation in several variables. Adv. Comput. Math. **12**, 377–410 (2000)

Author Index

B. Michielsen and J.-R. Poirier (eds.), *Scientific Computing in Electrical Engineering SCEE 2010*, Mathematics in Industry 16, DOI 10.1007/978-3-642-22453-9,

Index

B. Michielsen and J.-R. Poirier (eds.), *Scientific Computing in Electrical Engineering SCEE 2010*, Mathematics in Industry 16, DOI 10.1007/978-3-642-22453-9,

MIX
Papier aus verantwortungsvollen Quellen
Paper from responsible sources
FSC® C105338

If you have any concerns about our products,
you can contact us on
ProductSafety@springernature.com

In case Publisher is established outside the EU,
the EU authorized representative is:
Springer Nature Customer Service Center GmbH
Europaplatz 3, 69115 Heidelberg, Germany

Printed by Libri Plureos GmbH
in Hamburg, Germany